HIGH
PRESSURE
TECHNOLOGY

HIGH PRESSURE TECHNOLOGY

VOLUME I

Equipment Design, Materials, and Properties

Edited by

IAN L. SPAIN

*Laboratory for High Pressure Science
and Engineering Materials Program
Department of Chemical Engineering
University of Maryland
College Park, Maryland*

JAC PAAUWE

*Naval Surface Weapons Center
White Oak Laboratory
Silver Spring, Maryland
and
Department of Chemical Engineering
University of Maryland
College Park, Maryland*

MARCEL DEKKER, INC.   New York and Basel

LIBRARY OF CONGRESS CATALOGING IN PUBLICATION DATA

Main entry under title:

High pressure technology.

   Includes bibliographies and index.
   CONTENTS: v. 1. Equipment design, materials, and
properties. --v. 2. Applications and processes.
   1. High pressure (Technology) I. Spain, Ian L.
II. Paauwe, Jac.
TP156.P75H53   660.2'8429   77-21302
ISBN 0-8247-6560-5 (v. 1)
ISBN 0-8247-6591-5 (v. 2)

MARCEL DEKKER, INC.

270 Madison Avenue, New York, New York 10016

Current Printing (last digit):

10 9 8 7 6 5 4 3 2 1

PRINTED IN THE UNITED STATES OF AMERICA

Dedicated To

Our Parents

Who Gave So Generously

For Our Education

PREFACE

The present book fills a gap in the existing literature on high pressure technology. Although there are several books which cover specific topics in high pressure technology, or summarize current research at high pressure, the present book in two volumes attempts to cover the basic technology of high pressure equipment, the effects of pressure on matter, and technological applications which use high pressure.

Our basic aim has been to write a practical book that will be of use to those using high pressure as a tool in industry or the laboratory. The material will be of particular interest to the practitioner, since he will find a summary of the state of the art within these pages. Undoubtedly he will find techniques, facts and applications with which he has not previously been familiar. Tables summarize information which until now has only been dispersed in many articles and volumes. Copious references have been supplied which guide the reader to further in-depth study.

These volumes will also be invaluable to the beginner or student, who will find a convenient summary of basic knowledge about equipment, safety aspects, etc. Also, the more experienced researcher will find much information of value. In addition he will be able to see his contributions set in a wider context. There has been a very close link between research and later applications. It is hoped that these volumes might also stimulate ideas about new techniques and applications.

Others who may find these volumes of interest are managers and administrators, since they will be able to gain a broad overview of the whole field. Teachers may find many examples of equipment design, properties and processes to use in classroom discussion and problems.

The volumes may also be of interest as text books for specialized courses at the advanced undergraduate or early graduate level. It would be particularly valuable for a course in which an attempt is made to broaden a student's horizons from a particular discipline into others. As such it could be used in Physics, Chemical, Mechanical or Industrial Engineering curricula.

This book began as a lecture to the American Carbon Society at the Session in 1972 at which Dr. H. Tracy Hall was honored for his creative inventions in the field of High Pressure Science. The lecture introducing Dr. Hall was subsequently published. It was this paper which prompted Dr. Maurits Dekker, Chairman of Marcel Dekker, Inc., to suggest a book on High Pressure Processes and Technology. The enormity of the task led to a work with two editors and several contributors. In line with our desire to write a practical book, many chapters have been written by recognized authorities in industry and government laboratories.

v

It is realized that with the enormous expansion of uses of this Technology there have been inevitable omissions. We hope that the selection covers the principal areas and introduces some of the less well known.

This is a fascinating area of Technology. These volumes will have served their purpose if they stimulate further interest in it.

Ian L. Spain

Jac Paauwe

ACKNOWLEDGMENTS

Many people are to be thanked for their part in preparing these two volumes. Firstly, the contributors must be thanked for the care taken in the preparation of chapters and subsequent correction and proof-reading.

Mrs. Jeanne Fineran must be thanked for the long hours of painstaking work which she gave to the typing of the final copy and also Mrs. Florence Goldsworthy for the equations and tables. The secretaries of each contributor cannot be listed separately, but they are thanked collectively for their contribution with first drafts.

Many colleagues read chapters and made contructive criticisms. Thanks are offered to them collectively. They are Mr. Leonard Abbot, Dr. Francis Bundy, Dr. William Carter, Dr. Jack E. Goeller, Dr. Jan Sengers, Dr. Joseph Silverman, Dr. Earl Skelton, Dr. Laird Towle, Mr. Robert H. Waser, and Mr. Wayne D. Wilson. Several reviewers asked to remain anonymous.

Finally, thanks are due to our wives, Wendy and Rie, and to Anthony, Andrew and Russell for giving up so much to make possible these volumes. Thanks are also due to Wendy for compiling and proof-reading the final version.

CONTENTS

INTRODUCTION TO VOLUME I

The present volume falls naturally into several parts. Firstly the general operation of high pressure systems is considered, including standard operating proceedures, safety, safety codes, testing proceedures, hazardous materials and processes. It is to be stressed that high pressure equipment can be used safely, provided reasonable care is taken and proceedures followed. However, the practitioner should be aware of the possible effects of an explosive decompression, know how to calculate possible damage resulting from it, and how to control it.

The second broad classification covers the technology of high pressure systems, including chapters on high pressure components, pumps and compressors, pressure vessels, ultrahigh pressure equipment, and measurements at high pressure. In all equipment the materials of construction are of prime importance and a separate chapter is devoted to this subject, while another outlines techniques for non-destructive testing of components.

The third category covers the basic properties of materials at high pressure, concentrating on those of major technological interest. Separate chapters cover the properties of fluids and solids, while mechanical properties of materials are covered in the final chapter. Phase changes are covered in the second volume.

Sometimes it is possible to extrapolate into the future using the past as a guide. It will be of interest to see whether utilization of high pressure technology pursues the same pattern following basic research as it has done till now, as outlined in the Introductory chapter. Only time will tell.

CONTRIBUTORS TO VOLUME I

JAC PAAUWE, Naval Surface Weapons Center, White Oak Laboratory,
    Silver Spring, Maryland 20910 and Laboratory for High Pressure
    Science, Department of Chemical Engineering, University of
    Maryland, College Park, Maryland 20742

IAN L. SPAIN, Laboratory for High Pressure Science and Engineering
    Materials Program, Department of Chemical Engineering, University
    of Maryland, College Park, Maryland 20742

V.C.D. DAWSON, Naval Surface Weapons Center, White Oak Laboratory,
    Silver Spring, Maryland 20910

JOHN N. FEDENIA, Professional Engineer, Rockville, Maryland; formerly
    Naval Surface Weapons Center, White Oak Laboratory, Silver
    Spring, Maryland 20910

JAMES M. EDMISTON, P.E., Manager of Engineering, American Instrument
    Co., Division of Travenol Laboratories, Industrial Products
    Division, Silver Spring, Maryland 20910

P. BOLSAITIS, Instituto Venezolano de Investigaciones Cientificas
    (IVIC), Center of Engineering and Computation, Caracas, Venezuela

EDWARD L. CRISCUOLO, Naval Surface Weapons Center, White Oak Labora-
    tory, Silver Spring, Maryland 20910

WILLIAM B. STREETT, Science Research Laboratory, United States
    Military Academy, West Point, New York 10996

H. Ll. D. PUGH, Hydrostatic Extrusion and High Pressure Engineering
    Division, National Engineering Laboratory, East Kilbride,
    Scotland

E. F. CHANDLER, Hydrostatic Extrusion and High Pressure Engineering
    Division, National Engineering Laboratory, East Kilbride,
    Scotland

CONTENTS OF VOLUME II

"Applications and Processes"

HIGH
PRESSURE
TECHNOLOGY

Chapter 1

PAST - PRESENT - PROSPECTIVE

Jac Paauwe

Naval Surface Weapons Center
White Oak Laboratory
Silver Spring, Maryland 20910
and
Laboratory for High Pressure Science
Department of Chemical Engineering
University of Maryland
College Park, Maryland 20742

Ian L. Spain

Laboratory for High Pressure Science and
Engineering Materials Program
Department of Chemical Engineering
University of Maryland
College Park, Maryland 20742

The roots of modern high pressure science and technology can be traced back over many centuries. Although many of the early records have been lost, it is obvious that from the earliest times man needed to control supplies of water; and devices such as the Archimedes screw for raising water clearly point to later developments in pumps. The Romans also had a well developed technology for conducting and distributing water and in basic principle the pump shown in Figure 1 is very similar to many in use today.

Attempts to harness steam power led to developments in high pressure technology. As early as 1125 steam was used by a monk in Rheims Germany to blow air through organ pipes. In 1663 the Marquis of Worcester was awarded a patent for the first working steam engine. A Frenchman, Denis Papin, designed and built a boat equipped with a steam engine and demonstrated his invention in Kassel Germany in the early 1700's. He did not use the steam power as a driving force, but rather the differential pressure caused by the condensation of steam.

Although substantial results were not achieved in this period, much of the work led to important developments. For instance the basis was laid for better pumps when Morland introduced leather piston cups and closures.

1

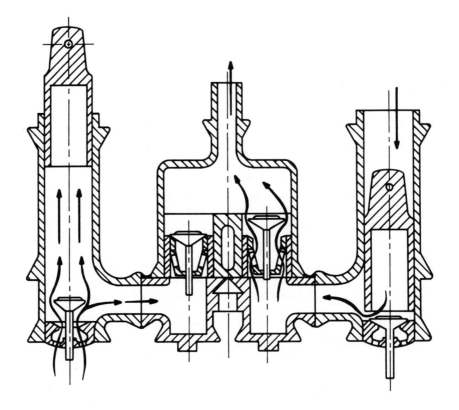

FIG. 1. An early pump -- credited to Ctesibios of Alexandria,
possibly a contemporary of Archimedes. (Figure adapted from
R. J. Forbes, Cultuurgeschiedenis van Wetenschap en Techniek (n.v.
Maanblad Success, den Haag)).

It was another hundred years however before James Watt made the
necessary improvements in the use of steam for motive power in 1769.
Firstly, he used positive steam pressure to push the piston and, secondly, he converted the up and down or the straight back and forth
movement into a rotational one. Important applications followed in a
number of areas, principally rail and sea transport and power for industrial machinery. Although pressures were relatively low in these
machines, boiler failures were frequent and led to the adoption of
safety regulations and codes.

The construction of the first autoclave is generally   attributed
to Papin in 1680. His pressure vessel for bone extraction was built
of bronze with the cover being secured by means of a saddle clamp
(Figure 2). His vessel also contained a safety valve based on the
level and weight principle, still in use today. This design also
foreshadows the pressure balance for measuring pressure.

Papin worked for some time in England where he met Robert Boyle.
Here the first scientific approaches to the field of high pressure

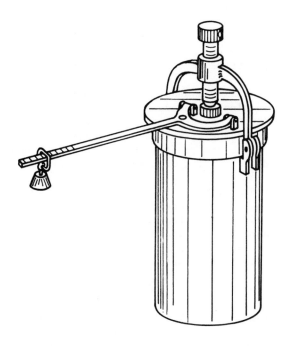

FIG. 2.   A sketch of Papin's autoclave with safety valve.
(Figure redrawn from Tongue [12]).

developed.   In 1662, Boyle propounded his law relating the pressure of
a perfect gas to its volume at constant temperature:

$$PV = Constant \qquad\qquad (1)$$

In 1802 L. J. Gay-Lussac published his experimental study of the
expansion of gases with temperature and showed that volume varied linearly with temperature at constant pressure -- a result anticipated by
J.A.C. Charles in 1787. When combined with Boyle's Law, the equation
of state of the perfect gas is obtained:

$$PV = nRT \qquad\qquad (2)$$

where n is the number of moles of gas of volume V at pressure P and
absolute temperature T and R is the Universal Gas Constant.
These early experiments with gases were conducted at pressures
not far from atmospheric, and were closely related to the problem of
defining a temperature scale. However, deviations from the ideal gas
law led to experiments at higher pressure. Such experiments related
to thermometry by H. V. Regnault, G. Magnus, and A. W. Witkowski are
described by Preston [1] in his classic text "The Theory of Heat".
Later, the subject of the equation of state of gases and liquids
became an important branch of science. Early experiments were made by
John Canton (1762-1764) on the compressibility of water.   He chal-

lenged the accepted view, propounded by the Florentine Academy, that water was incompressible. By 1826 Jacob Perkins had experimented with water up to 2,000 bars -- an enormous pressure for those days. His early experiments with water in a cannon up to ~100 bars were made in his native country, America, but his later measurements were conducted in England.

Perkin's apparatus was beautifully designed, and constructed so well that no soft packing was needed on the plunger with which pressure was produced. He measured the pressure with a development of Papin's safety valve -- the first use of a crude pressure balance. He was able to observe the change of compressibility of water with pressure and discovered the raising of the melting point of acetic acid with pressure.

Of note is the discovery in 1822 of the critical point in fluids by Cagniard de la Tour, rediscovered and quantified by Andrews in 1861. Theoretically, a great step forward was made by Van der Waals, who published his Nobel prize-winning thesis in 1873 on the continuity of the properties of the gaseous and liquid states into the coexistence region. His classical equation describing the fluid is:

$$(P+a/v^2)(v-b) = RT \tag{3}$$

where v is the molar volume and a and b constants for each particular gas. Van der Waals' work led to intensive experimental work which has been summarized in certain aspects by A. Levelt-Sengers [2]. Van der Waals' work naturally led to the idea that the properties of fluids could be universally expressed in terms of a reduced coordinate description. This approximate description of the thermodynamic properties of fluids has been of immense significance in design calculations for chemical plants.

In the years following, high pressure research was dominated by two Frenchmen, E. H. Amagat and L. Cailletet, who worked extensively on the properties of condensed fluids up to ~3,000 bars. Cailletet was probably the first scientist to seriously attempt the construction of a pressure standard, by building a mercury column in the Eifel Tower, Paris. However, difficulties with temperature control led Amagat to improve the mercury column by using a mine shaft.

Another important development which arose directly out of this research was the discovery in 1852 by J. P. Joule and W. Thomson (later Lord Kelvin) of the cooling effect produced by the isenthalpic expansion of gas below its inversion point. This fact is used in most home refrigerators (sometimes called Kelvinators).

Research into the liquefaction and solidification of gases using high pressure technology was closely related to these other developments. It is now realized that a first-order transition from the gas to liquid phase can only be accomplished by compression if the temperature is below the critical point, which for many substances is below room temperature. Probably all substances can be compressed into their solid state at room temperature, but for some substances (e.g. H_2, He, Ne) the pressures would be very high. As early as 1806, T. Northmore liquefied chlorine by compression at room temperature, while carbon dioxide was solidified by J. C. Thilorier in 1835. Oxygen was

first liquefied by Cailletet in 1877 using an expansion process starting at ~300 bars and 29°C. In 1900 J. Dewar liquefied hydrogen (normal boiling point 20.4K) and in 1908, H. Kamerlingh Onnes liquefied
helium (normal boiling point 4.2K). Finally helium was solidified by
W. H. Keesom in 1926. Helium is unique in that it can only be solidified if a pressure of more than 25 bars is applied, even at 0K.

It is doubtful whether the importance of these early discoveries
was realized at the time they were made, but this early work led to
the creation of a large industry involving liquefied natural gas,
rocket fuels, gas separation and purification processes.

By the latter half of the 19th century small autoclaves were being used in the synthetic dye industry at pressures up to ~20-30 bars,
particularly in Germany. In 1829, W. C. Roentgen published a study of
the effect of pressure on chemical reaction, and this work was quickly
followed by many others on liquid-phase reactions. In such systems
the main effect of pressure is on the reaction rate. However, in gaseous reactions the equilibrium yield can be changed significantly by
modest pressures (e.g. ~100 bars). By the end of the 19th century a
quantitative theory of gas-phase reactions had been developed and in
1903 V. N. Ipatieff started experimental work with equipment suitable
for pressures up to ~400 bars and temperatures up to 500°C.

LeChatelier attempted to synthesize ammonia from its constituents
in 1901, but was deterred by an explosion. W. Nernst and F. Jost
measured the increase in yield of this reaction at high pressure (~ 70
bars) in 1907. Two years later, F. Haber and R. Le Rossignol constructed a small plant for producing a few liters of ammonia per day
(~150-200 bars, 550°C). Badisch Anilin und Soda Fabrik started large
scale production (~7,000 tons per annum) in 1913 using plant designed
by C. Bosch. Further developments were made by G. Claude in 1921, allowing the operating pressure to be increased to ~1000 bars.

This reaction was the forerunner of many others that have since
been developed into commercial processes. Amongst them was the large
scale synthesis of methanol from carbon monoxide and hydrogen, introduced in Germany in 1923; the Fischer Tropsche Process (1923-24) and
many others. These early studies and processes have been outlined by
K. E. Weale [3].

The polymerization of ethylene is of particular interest to the
area of high pressure technology since the commercial process occurs
at relatively high pressure (~2,000 bars). W.R.D. Manning [4] and M.
W. Perrin [5] have given an interesting account of the discovery of
polyethylene by E. W. Fawcett and R. O. Gibson at the research laboratories of the Imperial Chemical Industries, England. The high pressure equipment was designed by A. Michels of the Van der Waals Laboratory. A photograph of one of the first pumps used for the production
of polyethylene is shown in Figure 3.

One of the present editors (Jac Paauwe) later joined Michels'
staff and learned basic engineering in high pressure technology from
him. He remembers the interaction between the two countries and the
way in which challenges in the production of polyethylene spurred
technological gains in the design of high pressure equipment and materials of construction. The other editor (Ian Spain) first performed
high pressure experiments in the Department of Chemical Engineering

FIG. 3. A picture of one of the first pumps used for the
production of polyethylene. Another view of this pump has been
given in Perrin [5]. (We are indebted to Dr. W.R.D. Manning for
confirming the use of this pump and supplying other relevant information. Picture obtained from the University of Amsterdam
(van der Waals Laboratory).)

and Chemical Technology, Imperial College, London, where Newitt and
collaborators performed many experiments on the effect of pressure on
chemical reactions, and contributed to the technology of polymer manufacture, described in his text [6].

Polyethylene or polythene™ was of immediate importance in the
Second World War as an insulator in radar equipment. Later developments have created an enormous demand for this and similar materials,
which will continue to grow.

The first work on polymerization reactions at high pressure was
completed by G. Tamman at the beginning of this century. He actually
began work on high pressure about 1890 and subsequently published many
important papers, which dealt with phase diagrams at elevated pressure. His book, "Kristallisieren und Smelzen" [7] summarizes much of
his work, which led to later commercial developments in the field of
synthesis and crystal growths of solids at high pressure.

In the present century uses of high pressure technology have increased dramatically. Many commonplace materials are produced or
formed in high pressure processes. Examples can be found in metals,
ceramics, polymers and single crystals. In everyday life steam turbines operating at pressures of the order of 100 bars produce domes-
tic power; hydraulic winches are used in an enormous number of appli-

cations; bottled gas at high pressure is used in a wide range of activities from scuba-diving, through acetylene welding, to medical applications. Without high pressure technology, civilization, as we
know it, could not survive.

Basic research has been the key to these developments. The relationship can be seen in Figure 4 which illustrates a. the pressures
attained by leading research-workers as a function of time on a semi-
logarithmic scale and b. some of the processes that have been introduced on a commercial scale. From this figure it is apparent that
areas opened up by research workers have led to technological exploitation about twenty-five years later.

The work of Professor Percy Bridgman towards the attainment of
higher pressures is dominant. He began his career at Harvard in 1906
and published over two hundred papers which are now conveniently published as the "Collected Works" [8]. These works, together with his
book "The Physics of High Pressure" [9] are an indispensable source of
information for the beginner or experienced practitioner in the field.

Bridgman discovered the packing based on the "principle of the
unsupported area", now known as the "Bridgman seal" (Figure 5) (see
Chapter 5). With it he was able to obtain leak-tight apparatus up to
~12 kb and later to 50 kb. Then, by enclosing a miniature piston-cylinder device inside a 30 kb apparatus, he was able to achieve pressures as high as 100 kb.

However, it was his development of opposed-anvil devices (Bridgman anvils) that led him to the highest pressures (although he claimed
400 kb, it is probable that his maximum pressure was ~150-200 kb) (see
remarks in Chapter 11 on pressure calibration in the ultra-high pressure region). His device utilized the "principle of massive support"
which has since been exploited in several multi-anvil devices. This
research has directly led to the synthesis of diamond in 1955 and its
commercial exploitation (see Chapter 8, Vol. 2). A Bibliographical
Memoir of Professor Bridgman has been prepared by E. E. Kemble and
F. Birch [10].

In contrast to this, the large body of scientific literature attributable to Professor A.M.J.F. Michels at the Van der Waals Laboratory, Holland was devoted to work of the highest accuracy in the pressure range up to ~3000 bars. The bulk of this work was directed towards studies of the properties of compressed fluids.

The methods by which Bridgman and Michels attained their results
were also very different in style. Bridgman had few collaborators,
preferring to work alone. Michels was a researcher and teacher, who,
as much as possible, let the students do their work independently
after initial guidance.

However, both laboratories depended on the skilled work of lifetime assistants. Charles Chase constructed much of Bridgman's apparatus while Leonard H. Abbot served as his experimental assistant. (Mr.
Abbot still works for the Harwood Engineering Company, Walpole, Massa-
chusetts and kindly reviewed two chapters of the present volume.) In
Amsterdam it was J. Ph. Wassenaar who fulfilled this role. In the
fields of high pressure research and technology it is the attention to
detail that is essential to success and invaluable contributions are
made by skilled craftsmen and technicians whose efforts are largely
unknown.

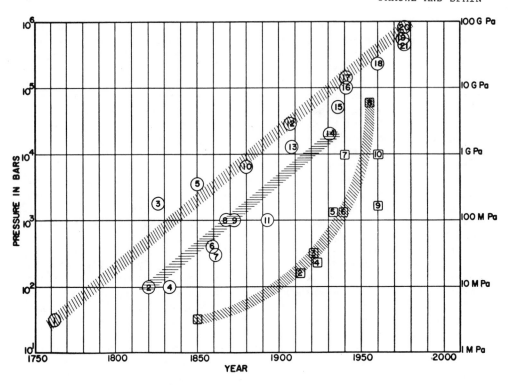

FIG. 4. A sketch showing the increase in attainable and
technologically used pressures as a function of time.

OScientific Studies

(References to studies 1-15 may be found in Chapter 1 of P. W. Bridgman "The Physics of High Pressure", 16-21 in Chapter 11 of the present
volume.)

1. Canton (1762); 2. Perkins (1819-1820); 3. Perkins (1826); 4. Par-
rott and Lentz (1833); 5. Natterer (1850); 6. Wartman (1859); 7. Andrews (1861); 8. Cailletet (1870); 9. Amagat (1870); 10. Spring
(1880); 11. Tammann (1893); 12. Eve and Adams (1907); 13. Bridgman
(1908); 14. Tammann (1931); 15. Bridgman (1936); 16. Bridgman (1940);
17. Bridgman (1941); 18. Balchan and Drickamer (1960-61), Bassett,
Takahashi and Stook (1967); 19. Piermarini and Block (1975); 20. Mao
and Bell (1976); and 21. Bundy (1976).

OTechnologies

1. Autoclave for Dye Industry (1850-1900); 2. Ammonia Synthesis,
Haber-Bosch Process (1913); 3. Ammonia Synthesis, Claude Process
(1921); 4. Methanol Synthesis, Fisher-Tropsch Process (1923); 5. Polymerization of ethylene, Gibson-Michels (1933); 6. Polyethylene Production, I.C.I.(1939); 7. Autofrettage of Gunbarrels (~1940); 8. Synthesis of Diamond (1955); 9. Isostatic Compaction (1960); and 10. Hydrostatic extrusion of metals (1962).

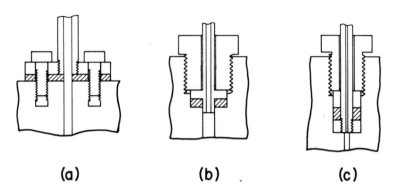

FIG. 5. Three kinds of packing for high pressure vessels:
a. primitive gasket, not suitable for high pressure; b. Amagat's
fully enclosed packing which leaks when pressure increases above
pressure supplied in the packing by the gland nut; c. Bridgman
unsupported area seal which can be used to the limits of materials
of construction. (Figure redrawn from Bridgman [9]).

The foregoing paragraphs have not been intended as an exhaustive
history of high pressure science and technology. Many important contributions have been omitted. Rather, an attempt has been made to
illustrate several of the main lines of scientific research and the
way they have led to later commercial developments. The reader interested in a complete list of references to early work can consult the
Historical Introduction in Bridgman's book "High Pressure Physics"
[8].

At the present time the large-scale applications of high pressure
technology to industry can be gauged from the chapters in Volume II of
the present treatise. Even so it must be acknowledged that many applications are not covered therein, and could themselves become the
subject of a further volume.

As applications have increased, interest in research at high
pressure also increased in industrial, government and university research laboratories. A listing of research interests in different
countries was given by Bradley in 1963 [11]. As the use of pressure
as a tool grew, the need for standards also increased. Several facilities for standards have been developed, most notably at the National
Bureau of Standards, U.S.A. and the National Physical Laboratory, Gt.
Britain.

Without doubt the use of high pressure technology in industry
will grow. Possible areas of immediate interest lie in energy storage and conversion devices. For instance, the possible use of hydrogen as a fuel will surely involve high pressure technology. New coal
to oil or gas conversion systems will almost certainly require high
pressure processes, while "in-situ" conversion processes being discussed for coal, oil-shale and other energy-bearing minerals, appear
a promising area of applications. Meanwhile the search for new oil
and mineral deposits at greater depths and in off-shore locations will

involve exciting areas of research and application. The possibilities of using water-jet cutting at high pressure are of interest.

The application of high pressure technology to metal forming will surely be considered more closely as energy-efficient and waste-saving processes become more important. Likewise the importance of new materials in critical applications (e.g. aerospace, nuclear, energy-conversion applications) will offer increased opportunities to the high pressure field, such as in hot-pressing metals and ceramics at high pressure.

Many people argue that the next few years will see a slowing down of our capabilities to achieve higher pressures, since the latest advances have utilized the strongest material known to man -- diamond. Even so, there is great current interest in the possible synthesis of super-materials using ultra-high pressure technology. Two such materials are metallic hydrogen and ammonium. The former metal would have a higher stored energy than any other known material and it is also speculated that it would be a high temperature superconductor.

In the U.S.S.R., Academician L.F. Vereschagin[*] heads a large laboratory dedicated to high pressure research. It is perhaps fitting to end this chapter with some words spoken by him at a press conference in which he was discussing the subject of metallic hydrogen. When asked whether he felt it to be far-fetched to imagine hydrogen metal in industrial quantities, his reply was succint. "In 1885," he said "a French scientist named Charles Terier predicted it would be impossible to produce industrial ammonia because the process requires pressure of 14 atmospheres." Is it possible that many fundamental studies being made today will hold the key for new industries? Only time will tell.

* now deceased

REFERENCES

1. T. Preston, "The Theory of Heat" (MacMillan and Co., Ltd., London, 1929).

2. A. Levelt Sengers, Physica 82A, 319 (1976).

3. K. E. Weale, Chemical Reactions at High Pressure (E. and F. N. Spon, Ltd., London, 1967).

4. W.R.D. Manning, "Some Experiences in the Fatigue of Metals" (Inaugural Lecture, Loughborough University of Technology, England).

5. M. W. Perrin, "The Discovery of Polyethylene", Research 6, 111 (1953).

6. D. M. Newitt, "Design of High Pressure Plant and the Properties of Fluids at High Pressure" (Clarendon Press, Oxford, 1940).

7. G. Tammann, "Kristallisieren und Smelzen" ("Crystallization and Melting").

8. P. W. Bridgman, "Collected Experimental Papers" (Harvard University Press, Cambridge, 1964).

9. P. W. Bridgman, "The Physics of High Pressure" (G. Bell and Sons, Ltd., London, 1931) (reprinted by Dover Publications, 1970).

10. E. C. Kemble and F. Birch, "Percy William Bridgman,
    1882-1961", Biographical Memoirs Vol. XLI (National
    Academy of Sciences of the United States, 1963).
11. R. S. Bradley, Chapter 1, p. 1 in "High Pressure Physics
    and Chemistry" (R. S. Bradley, ed.) (Academic Press, London
    and New York, 1963).
12. H. Tongue, "Design and Construction of High Pressure Plant"
    (Chapman and Hall, London, 1934).

Chapter 2

WORKING WITH HIGH PRESSURE

Jac Paauwe

Naval Surface Weapons Center
White Oak Laboratory
Silver Spring, Maryland 20910
and
Laboratory for High Pressure Science
Department of Chemical Engineering
University of Maryland
College Park, Maryland 20742

I.   INTRODUCTION

The rapidly increasing number of applications of high pressure and the acceptance of pressure as an essential tool in many processes, demands a basic knowledge of equipment and an understanding of the hazards involved. Although it is impossible to give a simple set of rules for designing, manufacturing, testing and operating high pressure systems, some basic and guiding principles should be followed for a successful and safe use of high pressure.

The potential or stored energy of a system increases with pressure and volume; but it is only when this energy is accidently re-

leased as kinetic energy that it becomes hazardous. In other words if
the forces and energy are confined correctly, there is no hazard. For
safe and efficient operation, for the controlled confinement of sometimes enormous energies, first and most important is the matter of design, material and manufacture. Second, and equally important is its
operation and maintenance. High pressure equipment should be designed
and manufactured by those fully familiar with high pressure techniques. Designers should know what is commercially available and
should have a thorough knowledge of materials, or should be assisted
by specialists acquainted with the effects of creep, corrosion, embrittlement, fatigue, thermal and mechanical stresses. Manufacture
and assembly should be carried out by skilled and trained personnel
with experience and a natural feel for accuracy and neatness. Reliability and cleanliness in this work is an absolute necessity.

Before going into the aspects of design, manufacture, assembly
and testing of high pressure equipment a general observation of high
pressure and its consequences is in order.

II.   FORCES AND ENERGY IN HIGH PRESSURE SYSTEMS

At what value can pressure be called "high"? For some, 15 bars
(1.5 MPa; ~220 psi) is high pressure, while for others it means pressures of 7,000 or 10,000 bars (700 - 1000 MPa; ~100,000 - 150,000 psi)
and still others work in the pressure range over 100 kbar (10 GPa;
~1,500,000 psi). But all work at high pressure demands respect, since
the hazard involved depends on both the pressure and the volume.

A "so-called" low pressure, large volume system with a working
pressure of 15 bars (1.5 MPa or ~220 psi) can be much more dangerous
than a system for 10,000 bars (1 GPa; ~150,000 psi). Of importance
is the total energy of a system, the product of pressure and volume.
A few examples will make this clear.

Consider a large volume, low pressure system containing air in a
pipe with an inside diameter of 300 mm (~12 inch) at a working pressure of 15 bars (1.5 MPa; ~220 psi). The total load or thrust load on
the closure of the pipe is:

$$L = P \times A \qquad\qquad (1)$$

where L is the thrust-load and P the pressure acting on the closure of
area A. This large force (Table 1) can be compared to the relatively
small force on a closure for 1/4 inch (~6.2 mm) tubing with an inside
diameter of 1/16 inch (~1.6 mm) produced by a pressure of 7,000 bars.
However, the force on the closure for a typical pressure vessel used
for isostatic compacting (see Figure 1) with an inside diameter of
300 mm (~12 inch) and a pressure, P, of 4,000 bars (~58,000 psi) is
truly enormous. Note that this force must be borne by the threads
holding the closure.

The quantities in (1) are given in Table 1 for three unit-systems
which are often met with in practice. Although the International
System of units (SI) may be unfamiliar to many readers it is gradually
gaining world-wide acceptance. The Newton is the Standard S.I. force
required to accelerate a 1 kg mass at the rate of one meter sec^{-2}

TABLE 1

Forces acting on closures of a low pressure, large volume system, a high pressure, small bore tube, and a large pressure vessel. Force is given in three common unit systems.

SYSTEM	INSIDE DIAMETER	PRESSURE P	AREA A	FORCE OR THRUST LOAD
Large Volume low pressure (15 bars) air system	300 mm 30 cm ∿ 12 in.	1.5 MPa 15.3 kg cm^{-2} ∿ 220 psi	.0707 m^2 707 cm^2 113 in.^{2}	106,050 N 10,817 kg 24,860 lbs.
High pressure (7,000 bars) Closure in ¼ inch tubing	1.59 mm 0.159 cm 0.0625 in.	0.7 GPa 7140 kg cm^{-2} 101,500 psi	1.985·10^{-6}m^2 .01985 cm^2 .003 in.^{2}	1,390 N 142 kg 305 lbs.
Isostatic Compaction Vessel (4,000 bars)	300 mm 30 cm ∿ 12 in.	0.4 GPa 4,080 kg cm^{-2} 58,000 psi	.0707 m^2 707 cm^2 113 in.^{2}	28,280,000 N 2,884,560 kg 6,554,000 lbs.

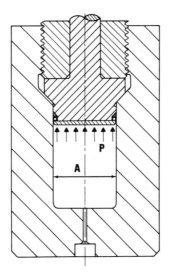

FIG. 1. Schematic of a single-ended high pressure vessel with pressure connection on the bottom. The force acting on the sealing plug is discussed in the text.

(1 N = 10^5 dynes). Since standard gravitational acceleration is 9.81 ...msec^{-2} it follows that:

$$1 \text{ kg force} \equiv 9.81...\text{N}$$

However, it should be recognized that while the kilogram-force (and lb-force) are convenient units they are not recognized units in the International System.

Equally, the stored energy in typical high pressure enclosures can be very high. The stored energy due to compression alone is:

$$\Delta E_{comp.} = \int_{V_1}^{V_2} P(V) dV \qquad (2)$$

where V_1 is the initial, and V_2 the final, volume of the medium being compressed. The more compressible the medium, the greater is the stored energy. In addition to the energy of compression is the chemical energy of reactive materials such as hydrogen gas. Most spectacular are the figures when we compare the total stored energy with T.N.T. equivalents. We will compare (i) a standard, so-called I.C.C. (*) gas cylinder; (ii) a gas-storage vessel with a volume of 10 cubic feet and (iii) a "pipe" trailer with a volume of 220 cubic feet. All cylinders are filled with hydrogen gas and in all three cases the pressure is the same, namely 14 MPa or 140 bars (~2000 psi). Their T.N.T. equivalents are:

one standard gas cylinder (1.5 ft^3; ~.04m^3) - 14 kg TNT[†]

one storage vessel (10 ft^3; ~0.28m^3)   -- 100 kg TNT

one trailer or (220 ft^3; ~6.2m^3)   --- 2130 kg TNT

Even non-flammable gases, such as nitrogen are dangerous under pressure and more so when large volumes are involved. In wind-tunnels, for example, nitrogen gas is used at a pressure of 4000 bars (400 MPa; ~60,000 psi). At this pressure, 100 ft^3 (~2.8m^3) of nitrogen represents the equivalent of about 175 lbs (79.5 kg) of TNT when suddenly released to the atmosphere.

The above figures have not been mentioned to deter anyone from working with pressure systems, but to make it clear that handling high pressure equipment demands respect. When any pressure is confined in vessels, pipes, tubing etc. and the system is designed for a certain pressure and volume we should call the system a high pressure system. Whether the pressure is 15 bars (1.5 MPa; ~220 psi) or 15 kbar (1500 MPa; ~200,000 psi) the system will be treated as a high pressure or high energy system.

(*) I.C.C. = Interstate Commerce Commission of the U. S. Department of Transportation (D.O.T.)

(†) See Chapter 3 for discussion of T.N.T. equivalents.

III.   SOME CAUSES OF ACCIDENTS IN HIGH PRESSURE SYSTEMS

High pressure equipment is used increasingly in industry and re-search laboratories. Knowledge of design factors and availability of superior materials have improved considerably. Yet, accidents take place. Why do accidents happen and what can be done about it?

With a bad design it is clear that sooner or later there will be problems; equipment will fail because of inferior design work. In-correct or faulty materials can cause serious problems and everything possible should be done to know what materials are used and whether or not they are sound and without defects, flaws or deficiencies which can lead to serious accidents.

Failures can happen because of a defect or inadequate seal. When a seal leaks (and even the best designed and manufactured seal can fail) the design should be such that the leaking fluid can escape without doing any harm. Pressure build-up over a larger area must be avoided.

Failure can happen because of mistakes made by personnel operat-ing a system or due to malfunctioning of control systems, safety-valves or pressure gauges. Failures can also happen because of poor workmanship, dirty seals or seal areas. Run-away reactions can be disastrous. Safety valves, rupture discs and special vents may not release a rising pressure fast enough, for example, with an unexpected or uncontrolled temperature increase. Finally, high pressure equip-ment will not permanently resist the effect of pressure and will fail even under normal working conditions due to fatigue, corrosion-induced failure, etc. That is why, even with the best equipment and designs using skillful operators, accidents still happen.

Many times the remark is made that systems operating with a liq-uid (oil, water) are not as dangerous as those in which the pressure medium is gaseous. As soon as a fitting fails or a leak occurs, the pressure drops immediately. In some instances this may be correct, but, depending on pressure and volume and the size or kind of leak, it is still dangerous. The stored energy, when released, can propel steel parts with high velocities and these parts can strike personnel. Someone hit by a jet-stream of oil can be seriously injured. Hydrau-lic oil can easily penetrate the skin and cause the loss of a hand or an arm. In very bad cases it can be fatal. It is easy to replace a piece of tubing, a questionable fitting or a piece of equipment; it is not so easy to replace someone's eye. Personnel testing fuel in-jection equipment for diesel engines, for example, have been severely injured, and the test pressures are "only" 70 - 300 bars (7 - 30 MPa; ~1000 - 4500 psi). In general it can be said that when the pressures become higher, the dangers are larger and when accidents happen they are more violent.

In gas systems the stored energy is much larger because of the high compressibility. When a fitting, a closure, or a vessel fails, there is a good chance that flying parts will be ejected with even supersonic velocities. Since almost all accidents do not happen with-out a cause, what can be done to limit failures to the very minimum? In the following chapters designs, materials and safety will be dis-cussed specifically. In this chapter a general review will be given

for a successful "working with high pressure."
    The basic points are:
·Design and Reliability
·Materials
·Testing and Safety
·Standard Operating Procedures (S.O.P.) and Instructions

IV.   DESIGN AND RELIABILITY

As already mentioned the design of reliable equipment should be done by competent and experienced designers. There is no substitute for experience. This is a good reason for purchasing as much standard high pressure equipment as possible and to consult with the established manufacturers first, before embarking on a design and manufacturing adventure. High pressure equipment, standard catalog items as well as specifically-designed systems, are available for almost all high-pressure applications. During the last fifty years, high pressure equipment has been developed and constructed to meet the requirements of safety and reliability as set forth by such organizations as the American Society of Mechanical Engineers (ASME), the American Institute of Chemical Engineers (AICHE), the larger chemical companies, the U.S. National Bureau of Standards, the U.S. Department of Transportation, the U.S. Department of Defense, the National Aeronautics and Space Administration (NASA), the Compressed Gas Association and more. In Europe the National Physical Laboratory and Imperial Chemical Industries (ICI) in England have been very active, whereas the Technische·Überwachungs Verein (TUV) in Germany have set rules for safety and reliability. Most, or all, of the equipment nowadays available from specialized manufacturers will comply with the requirements as mentioned above. Aspects of Safety Codes are discussed in Chapter 3.
    In the design or lay-out of high pressure systems, each part can be looked upon as a pressure vessel, whether it is a reaction vessel, storage vessel, a pump body, a valve or a fitting. Also, tubing containing pressure should be handled as pressure vessels. In Chapter 7, the containment of pressure is approached in depth, with the needed information for a safe design. A bad design sooner or later will give problems. The design has to be sound and simple. Complicated systems tend to confuse and leave too much room for mistakes.
    As an integral part of the whole design, one should consider the effects that might occur in the event of a failure of a component. All possible failures should be considered and their effects predicted. Problem areas should be isolated and chain-reactions eliminated --one failing part is one too many. A presumably minor failure, such as a leaking seal, should not lead to disastrous results. That is why every seal must be designed in such a way that whenever the seal leaks the fluid can escape without building up pressure over a larger area. In other words, provisions should be made so that a leaking fluid always has a release--or vent to atmospheric pressure. (See Figures 2(a)(b)(c), and further details are given in Chapter 5.) Also, the sealing area should always be at the smallest possible diameter, thus reducing the thrust loads to a minimum.

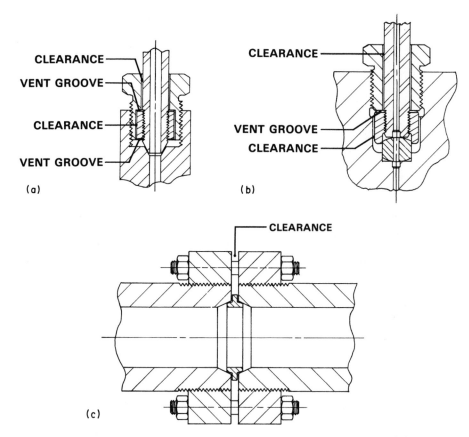

FIG. 2. Schematic showing vents for the controlled release
of leaking fluid: (a) cone-type fitting; (b) lens-ring fitting;
(c) Grayloc seal.

In designing pressure vessels, pump bodies, etc., a few basic
remarks are worth mentioning.
    ˙The bore or inside diameter of a vessel should be concentric
with the outside diameter (see Figure 3(a)).
    ˙Small vessels may have a pot-hole, or blind hole. This bore,
however, should never have sharp edges but a smooth and large radius,
(see Figure 3(b)). Also undercuts for screw threads should have a
radius.
    ˙Larger vessels should be designed with a through-bore and clo-
sures on both ends. Accurate machining and finishing of the bore is
much easier. Concentricity of inside and outside diameter can be
accomplished without effort. For long and small bores the inside diameter whould be machined first and the outside cut accordingly, referenced to the inside bore.

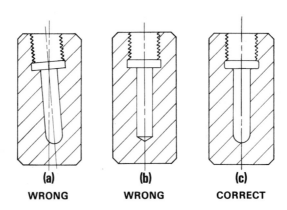

(a)   **(b)**   **(c)**
WRONG   **WRONG**   **CORRECT**

FIG. 3. Schematic showing two often-encountered mistakes in high pressure vessels. (a) Lack of concentricity and sharp thread undercut; (b) sharp corners; (c) correct design. Note that threads should not contain sharp stress-raisers.

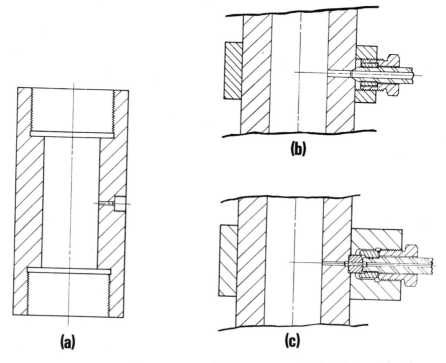

FIG. 4. Schematic of three different ways of side connection designs to a pressure vessel. (a) Pressure connection without collar; (b) cone connection with collar; (c) lens-ring connection with collar. Methods (b) and (c) are to be preferred.

⋅Side connections should be avoided as much as possible. There is usually space enough in the end-closures for pressure connections, lead-throughs, etc. These side connections can become dangerous stress-raisers [1, 2, 3]. If the side connection is unavoidable a collar-connection should be preferred, in particular for the pressure ranges of 3000 bars (300 MPa or ~45,000 psi) and higher (see Figure 4 (a), (b), (c)).

⋅The surface finish of the bore should be approximately 16-32 microinches or better. The outside finish should be approximately 63 microinches.

⋅Tapered screw-threads or so-called pipe-threads should be avoided. Tapered threads can only be used in lower pressure ranges. For pressures over 200 bars (20 MPa or ~3000 psi) standard straight threads, class 2, [4] should be used (finish ~32 microinches).

⋅For heavier and larger vessels, buttress threads or the Gasche Resilient-Thread closure should be used for end-closures. Square threads should not be used (see Figure 5 (a), (b)).

⋅If possible, the same machine tool (lathe) or lead-screw should be used for cutting the female thread of the vessel as well as the male thread of the closure-plug for that vessel.

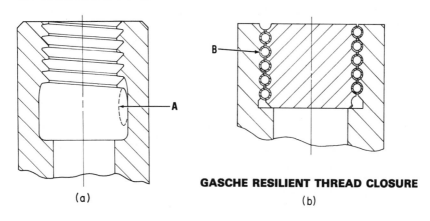

MODIFIED BUTTRESS THREAD WITH SPECIAL UNDERCUT
SEE NATIONAL FORGE PATENTS 3,664,540 AND 3,669,301
AND NBS HANDBOOK H28-PART III

A

B

(a)

GASCHE RESILIENT THREAD CLOSURE

(b)

FIG. 5. Two forms of the thread for a threaded closure.
(a) Modified buttress thread with special undercut (see National Forge patents 3,664,540 and 3,669,301 and NBS Handbook H28-Part III). The dashed line (A) refers to the elliptic form of the undercut with axial ratio ~4:1.
(b) Autoclave Engineers resilient thread closure. Part B is a spiral spring with solid core. The thread was designed by the late F. Gasche, founder of Autoclave Engineers, Inc., Erie, Pa. (Published with permission of Autoclave Engineers.)

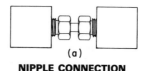

NIPPLE CONNECTION

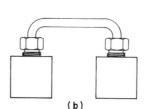

(b)

'U' TUBE CONNECTION

FIG. 6. (a) A nipple connection in which a short, straight
piece of tubing connects two fixed parts. (b) A nipple connection
which is preferable since undesirable stresses are minimized.
Method (a) is not to be recommended.

High pressure equipment -- vessels, tubing connections, pump
bodies and fittings -- should be perfectly aligned in assembly, before tightening up the connections. No extra strain should occur because of incorrectly aligned parts. Connections between heavy or
fixed parts should not be accomplished by straight pieces of tubing
(Figure 6 (a)). These connections are difficult to assemble or dis-
assemble; by tightening up, very high stresses can result in addition
to the inside pressure. Tubing connections with a possibility of expansion, or "U"-shaped connections should be chosen for ease of assembly and to avoid extra stresses. Free movement is also needed to
offset the effects of pressure and temperature, see Figure 6(b).

Reliability should be of uppermost concern, not only in design
and assembly of high-pressure systems but also in the choice or selection of personnel -- operators, technicians, as well as machin-
ists. When a machinist, for example, discovers while machining a
pressure body that the cut coil (coiled chip) breaks every turn of
the part at the same spot, it might be an indication of a crack or inclusion in the material. If there is an unusual experience or a difference from normal pattern in the fabrication, the tool operator,
machinist or technician should notice this and call it to the attention of others involved in the evaluation and decision making.

Pressure gauges or other pressure measuring devices should be
checked or calibrated frequently. Even new pressure gauges should be
checked before use. A gauge might have been damaged during transportation, or dropped. The hand or needle can be loose so that it slips
over the shaft while pressurizing the system. Once, twice, or even
more times per year, depending on the use, pressure gauges and also

secondary standards should be calibrated against a primary standard. The pressure balance is considered a primary standard; a Heise gauge or manganin cell can be used as a secondary standard provided they have been calibrated against the primary standard (see Chapter 8). In every day practical use, however, pressure-measuring devices can be checked against secondary standards. In general, whenever there is any doubt about a piece of equipment it should not be used until carefully checked and evaluated.

Parts and systems should be designed with ease of maintenance and repair in mind using units which can be taken out easily. Accessibility is of importance. Make-shift equipment or assemblies can be dangerous and should only be used in emergencies and after careful consideration.

Materials and equipment not especially suited or designed for use in high pressure systems should not be used. All parts such as valves, fittings and tubing in particular must be inspected before assembly. Some of these "fittings" might appear similar at first sight but are designed for different pressure ranges. Needless to say all parts must be clean, inside and outside, free of burs and correctly greased where necessary. Special attention should be paid to threads which are highly loaded, to avoid galling.

High pressure is an established tool and it requires special attention and attitudes in design and operation. High pressure means accuracy and precision, neatness and efficiency, cleanliness and carefulness. No slipshod work should be allowed, no "that-will-do" attitudes accepted. There is no room for mistakes. Finally, in designing large and expensive high pressure installations the engineers, scientists and designers must work together as a team. It is good practice to form an independent and neutral "ad-hoc" committee, to oversee and review the design before manufacturing starts. Also this "ad-hoc" group should review the operation and consequent standard operating procedures.

V.   MATERIALS

The choice of materials for high pressure applications depends on a number of different factors. Consideration should be given to the working conditions such as pressure and temperature range, pressure and temperature cycling (fatigue-life), corrosion, restrictions or regulations from authorities and codes. Finally, economic factors are important.

Only the best suited-materials should be used and stringent requirements are mandatory. There should never be any doubt about the identity and quality of the materials. For important parts (and in high pressure systems all parts are vital) manufacturers certificates should accompany the stock materials, forgings and component parts. Complete items, such as pumps and compressors should also have a test certificate, including capacity data.

Materials for use in high-pressure equipment subjected to tensile stresses should never be brittle. As a general rule the longitudinal elongation must be between 18 and 20%, whereas the transverse elongation should not be much under 12-15%. Brittle materials should be

avoided wherever possible. These materials can be used only in special designs and preferably under compressive loads.   Glass windows
in a system are always critical points and should be designed and
placed in such a way that no harm will be done when failures occur.
Flying pieces of material resulting from a rupture should be caught.

In Chapter 9 materials will be handled in detail whereas in Chapter 4 attention will be paid to Hazardous Materials. Non-destructive
testing techniques are considered in Chapter 10.

VI.   TESTING AND SAFETY

Since the total energy of a system is the product of pressure and
volume, the fluid volume should be reduced as much as possible when
testing a high pressure system, particularly when the test pressure is
to be higher than the working pressure. Storage vessels, pressure
containers, manifold blocks, etc. should be filled with spacers. Aluminum rods, steel balls, etc. can be used as "filler", provided that
these fill materials are solid. Liquids, including water should not
be used and are not considered "fillers". For pressurizing, a low-
compressibility fluid should be used, but even so the stored energy
can be very high.

Before testing a complete system, component testing is advisable;
the total capacity is smaller and it is much easier to detect leaks
or other problems. Testing should be done in steps with slowly increasing pressure and with a stop at different pressure levels for inspection. Leaks can be detected by carefully monitoring the pressure.
After reaching a certain pressure level and pressurization is stopped,
it is normal that the pressure slightly decreases because of temperature effects.   It should, however, be a slight drop and the pressure
should remain steady afterwards.

Component testing has another advantage. The unit to be tested
is usually small in comparison with a complete system and can easily
be put into a specially designed test chamber, a pit or an outside
test area. Remote operation can be advisable, although for leak inspection the equipment has to be within reach and easily accessible.

It is not always possible to reduce the internal volume of a
unit, such as a large vessel. When large vessels are involved, a pit
or outside test area should be recommended. The test pit can be covered with heavy material and an outside area should have a protective
earth wall on three sides while the terrain should be large enough in
case rupture takes place and parts are ejected.

In the past few years ideas about the way to conduct pressure
tests have been changing.   Instead of pressurizing a unit to twice
its working pressure, lower test pressures are acceptable and even advisable.   Depending on the working pressure, test pressures of 1.25-
1.5 times the working pressure are now common.   At high test-pres-
sures, for example when the test-pressure is twice the working pres-
sure, it is possible that the yield-strength of the material will be
reached and permanent deformation results. In general it can be said
that in pressure-tests the elastic limit should not be reached.   Almost always there are areas where stress-raisers are involved and
higher stress concentrations occur. These are the areas where the

unit might fail. However, the controlled autofrettage of vessels is
an exception to this rule (see Chapter 7 for a discussion).

Special attention should be paid to equipment with parts at different pressure levels, as in an intensifier system. One should be
sure that high-pressure fluid can not leak into the low-pressure part
of the design. This could lead to disastrous results. Safety devices
must be checked to see whether they are functioning and working according to the pre-set pressure levels. These safety items must be
inspected and checked regularly. Solenoid- or pneumatically-operated
valves should be assembled in the so-called "fail-safe" position. In
other words, when the electric power fails or the air supply is not
adequate, the safety unit or valve must be in the safe position. For
example a vent-valve should be in a normally-open position, a supply
valve in the normally-closed position so that work cannot continue until the power has been restored. In particular cases interlocks are
desirable. This means that, for example, a valve cannot be opened (or
closed) unless other valves, or parts of the systems, have been set
correctly and/or are functioning properly.

Grounding of electric wiring should be checked, and when flammable gases are involved one should be certain that the equipment has
been grounded also for so-called low-energy, static electricity. In
the case of flammables (vapor mixtures, gases) spark-proof tools such
as beryllium-copper should be used.

The complete assembly should be tested, starting with a low pressure run and, if possible, with a neutral fluid -- in gas systems, for
example, nitrogen gas. Pressure should be increased slowly (and temperature if applicable) until the whole system is functioning satisfactorily at the maximum working conditions. One guide-rule should
always be followed -- NEVER tighten a leaking fitting while the system is at pressure!

Every high pressure system has its own characteristics and not
all systems should be handled in the same way. Also, individuals will
have different ideas about test procedures and inspections. Although
experience is the best teacher it might also be very costly; one can
use the experience of others by consultation and acceptance of proven
practices.

It is good practice before designing a high-pressure system to
investigate rules and regulations which might effect the operation of
high pressure equipment in a particular area. An evaluation should be
made of what effects a failure might have on the environment (see
Chapter 3 for further discussion). Hazard areas should be marked and
dangerous operations should be located in such a way that when a failure occurs the damage will be kept to the very minimum and nobody will
be injured. Even though measures are taken to prevent accidents from
happening, they will happen. In this event the effects should be controlled and calculated. Attention should always be paid to the lia-
bility, in cases where accidents can happen and persons might be injured, both in industrial plant and research facilities.

VII.   STANDARD OPERATING PROCEDURES (SOP'S) AND INSTRUCTIONS

Standard operating procedures should be mandatory for operations
in the field of high pressure. Not only should the SOP give a clear
description of the facility and a step-by-step review of its working,
but should also denote the operators and supervisors and identify the
person in overall charge. The responsibilities should be clearly assigned and persons be appointed to the particular and different types
of work in the process. There should be no doubt about the authority
when it comes to decision-making.

A few basic rules should be considered and agreed on. Typical
examples are:

·Energy has to be contained.

·Release of energy should be controlled.

·In case of failure or run-away reactions adequate emergency
escape plans must be ready.

·The operators must know when and where to push the "panic
button" (main switch, main vent valve, etc.).

·Tests should be continually made for small leaks; they can
lead to disastrous consequences if not discovered.

·Dangerous gases must be monitored as well as neutral gases
in an enclosed area. A small nitrogen leak might replace the
oxygen in the air, for example.

·Precautions should be taken to ensure that parts are not
used beyond their fatigue life. Pressure cycles and/or hours of
service (for example diaphragms of compressors) should be monitored.

·Only approved components should be used.

·Maintenance and inspection should be planned.

·Pressure gauges and all other monitoring instruments should
be checked regularly.

·Barricades should be used where needed.

·Protective clothing must be available and used when needed.

·Work areas must be kept clear of obstructions and clean.

·Personnel should be thoroughly trained in the safe operation
of equipment and checks made to ensure that they follow SOP's and
safety procedures at all times.

No alterations should be made without the approval of the supervisor, no short cuts taken, no interlocks removed or safety-devices
altered, unless officially ordered or approved. All this should be
clearly described in the S.O.P. and when changes are made the operating procedures should be edited accordingly. A list with the names
of the operating personnel should be made with one or more alternates
for each person. Emergency situations must be described and actions
to be taken in case of accidents or failure of equipment should be
clearly indicated.

Relations with the local fire department and police force should
be regulated and in case of disaster an emergency plan should be activated automatically. When dangerous materials are involved and the
risk of explosion is imminent, if fires and the escape of harmful
vapors are a possibility, only experts and knowledgeable personnel
should be in charge of rescue operations, firefighting and the shut
down of part or whole of plant operation. It cannot be accepted that

nonexperts or persons not familiar with the area, or acquainted with the dangers of the process, take over responsibilities. (Even the relation with the press and the appearance of television crews should be regulated before a disaster strikes and special areas and locations should be assigned for this purpose!)

Personnel must be trained thoroughly and only the best mechanics should be chosen for specific high-pressure processes. Operators must be familiar with the process, the dangers involved and the end product. They must know and understand the mechanics of the system and must know what to do in emergencies. A maintenance and repair program on a regular scheme should be part of the procedures.

Maintenance and operation are two independent organizations and operations must never be resumed without the approval of the maintenance supervisor. Instrument calibration and repair must be done by those specially trained for this type of work. For the maintenance of pumps and compressors it is probably more efficient to call for outside service. Most suppliers of this type of equipment have specialized mechanics who can be hired for a fixed amount per day.

The operation crew must be capable of testing the facilities and must be knowledgeable of pressure-testing. For inspection and testing of pressure vessels specialists should be called in. The closure-threads and the material of the vessels, especially where corrosive fluids are used, should be inspected regularly. Non-destructive tests can be of great importance (see Chapter 10). Brittle failure of small and large vessels has to be prevented. Stringent regulations for materials, use and tests must be part of the procedures.

Operations and maintenance have the task of achieving reliable, safe and efficient production. Based on their functioning, experience and capabilities, new, better and, in particular, safer methods, equipment and processes will be found. The standard operating procedures should also reflect these innovations and changes and should be kept up-dated.

REFERENCES

1.   J. H. Faupel and D. B. Harris, Stress Concentration in Heavy-walled Cylindrical Pressure Vessels. Effect of Elliptic and Circular Side Holes, Industrial and Engineering Chemistry, Vol. 49, No. 12, p. 1979-1986 (December 1957).
2.   R. E. Peterson, Stress Concentration Design Factors, Wiley, New York (1953).
3.   G. Savin, Stress Concentration Around Holes, Pergamon Press, New York (1961).
4.   Screw-Thread Standards, Handbook H28, Parts 1, 2 and 3. National Bureau of Standards, U.S. Government Printing Office, Washington, D.C.

BIBLIOGRAPHY

1.   High Pressure Principles and Process Trends, a feature
     report of Chemical Engineering, McGraw-Hill Publication;
     23 September 1968, 21 October 1968, 4 November 1968,
     2 December 1968.
2.   P. W. Bridgman, The Physics of High Pressure, G. Bell &
     Sons, Ltd., London (1949).
3.   H. H. Buchter, Apparate und Armaturen der Chemieschen
     Hochdrucktechnik, Springer Verlag, Berlin/Heidelberg,
     New York (1967).
4.   E. W. Comings, E.W. High Pressure Technology, McGraw-Hill
     Book Co., Inc., New York, New York (1956).
5.   H. Tongue, The Design and Construction of High Pressure
     Chemical Plant, D. van Nostrand Co., Inc., Princeton,
     N.J. (1959).
6.   W.R.D. Manning and S. Labrow, High Pressure Engineering,
     C.R.C. Press, Cleveland (1971).
7.   D. S. Tsiklis, Handbook of Techniques in High Pressure
     Research and Engineering, ed. by A. Bobrowski, Plenum
     Press, New York (1968).

Chapter 3

SAFETY AND SAFETY CODES

V.C.D. Dawson

Naval Surface Weapons Center
White Oak Laboratory
Silver Spring, Maryland 20910

I.   INTRODUCTION

The design of a high pressure system cannot be considered complete until it includes an analysis of the safety associated with its operation. Besides the actual design of the pressure containment vessel and its associated components, this should include a standard operating procedure, a failure damage analysis, and a prescribed procedure and schedule for inspection and maintenance. It should also be axiomatic that the entire design package, including the items mentioned, is presented to management for an independent review with subsequent alteration or approval.

The standard operating procedure (SOP) provides a checklist of sequential operations that are to be followed whenever the high pressure system is to be used. The SOP should be designed as far as possible to eliminate operator error and management policy should demand rigid adherence to it after it has been carefully reviewed and approved. The development of the SOP frequently indicates to the designer possible flaws in the design and/or conditions where safety devices, such as pressure relief valves, interlocks, etc., are required. A more complete consideration of the SOP is given in Chapter 2.

The failure-damage analysis is a technique wherein various parts of the high pressure system are assumed to fail catastrophically and the damage resulting from such failures is assessed. Ideally, the analysis should include each mechanical and electrical component since the failure of very small elements of the system may lead to the subsequent rupture of additional components and eventually to the complete destruction of major parts of the system.

The method of making a preliminary failure-damage analysis, which is described below, can generally be made rather quickly and is of considerable worth to the designer in making recommendations on the site location of the high pressure system, building construction and safety precautions. It also provides an indication of where safety interlocks and redundancy of interlocks are required. If the preliminary analysis indicates that a major safety problem exists with respect to personnel and/or the building site, the designer can make a more detailed and accurate structural analysis, usually at a considerable cost in time, using the relationship between the effect of explosives and type of structure contained in reference [1].

The analysis begins by assuming instantaneous catastrophic rupture of the main pressure containment vessel. For the time being, it will be assumed that the vessel contains either an inert or an explosive gas but mention will later be made of high pressure liquids. The analysis is based upon the following steps:

A.   Inert Gas

1.   For gas pressures up to about 100 bars (10 MPa or ~1500 psi), it is assumed that the containment vessel fails catastrophically, as, for example, by fatigue, and that the gas expands explosively in an adiabatic manner to ambient pressure. The energy released in this process is calculated by assuming the gas is ideal with constant specific heats. For pressures greater than 100 bars, the gas can no longer be considered to be ideal and real-gas effects become important. The total energy released is usually reduced considerably from what would be predicted on the basis of the ideal assumption, and experimental thermodynamic data should be used where possible and available (see Chapter 12).

2.   The calculated energy of step 1 is converted to an equivalent weight of TNT based upon the assumption that TNT releases 1830 Btu/lb (4253 kJ/kg) in an oxygen-rare atmosphere.

3.   Blast-pressure contours showing the peak static over-pressure as a function of distance are constructed based upon test results with explosives, the weight of TNT from step 2, and the known scaling laws that apply to explosive effects.

4.   A preliminary estimate of damage to the surroundings can be predicted on the basis of the blast contours and the estimated pressure capability of various types of structures.

B.   Gas that Forms an Explosive Mixture with Air

1.   It is assumed that the gas expands, mixes stoichiometrically with the oxygen in the air, and then detonates. The energy released

is a function of the particular gas involved. It is usual to assume something less than a 100-percent reaction for this case since complete mixing is required for such an event.

Steps 2, 3, and 4 are the same as for Case A.

The methods indicated above are related to the peak over-pressures connected with the blast wave. Also associated with the explosion is a dynamic pressure which can accelerate fragments of the chamber and fittings to relatively high velocities by aerodynamic drag. Calculation of the effects of such shrapnel is complicated by the fact that the size and drag coefficient of the fragments, as well as the dynamic pressure characteristics of the explosion, are generally not known with any certainty. Some structures are primarily affected by drag (or wind) loading, so that both the peak dynamic pressure and the duration of the positive phase of the blast wave are important.

II. BLAST WAVE CHARACTERISTICS

The pressure-time characteristics of a surface blast-wave are shown schematically in Figure 1. As indicated, there is a static pressure, which represents the blast overpressure, and a dynamic pressure, which represents the high air or particle velocity behind the shock front. For overpressures less than about 10 psi (0.7 bars), the contours can be expressed as follows:

$$p(t) = p(1 - \frac{t}{t^*})e^{-\frac{t}{t^*}} \tag{1}$$

$$q(t) = q(1 - \frac{t}{t^*})^2 e^{-\frac{2t}{t^*}} \tag{2}$$

where p is the peak overpressure, q is the peak dynamic pressure, and t* is the duration of the positive phase of the overpressure.

If the blast wave encounters a structure, it reflects and the overpressure increases to a peak value

$$p_r = 2p + (\gamma+1)q \tag{3}$$

where γ is the specific heat ratio. For air the peak reflected overpressure can attain values of from twice the incident overpressure (weak shock) to eight times this value (strong shocks).

The actual pressure attained when a blast wave encounters a structure depends upon the peak overpressure of the incident wave and the angle between the direction of motion of the wave and the face of the building. As the wave moves forward, the reflected overpressure on the face drops rapidly to that produced by the blast wave without reflection, plus an added drag force due to the dynamic pressure. Eventually the structure is engulfed by the blast and approximately the same static pressure is exerted on the side walls and the roof. In addition, the front wall is still subjected to a dynamic pressure while the rear wall is shielded from it.

The response of a structure to blast loading depends upon both the diffraction forces and the drag forces. The relative importance of each in causing damage depends upon the structure and the blast

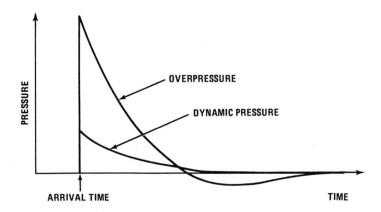

FIG. 1. Variation of overpressure and dynamic pressure
with time at a fixed location.

wave characteristics. In the damage assessment technique to be described for high-pressure gas release, only the effect of overpressure
is considered. Estimates of structural damage are based upon the experimentally observed failures that have occurred with low-yield nuclear blasts as given in references [2] and [3]. This simplifies the
analysis somewhat and provides a means of rapidly determining the damage without concern for the many variables involved in the air-drag
phenomena of the blast wave.

For most office-type and residential buildings, the extent of destruction is mainly dependent on the peak overpressure. Table 1 shows
the approximate correlation between the overpressure and the expected
physical damage to a structure.

Table 1

Relation Between Peak Overpressure and
Damage to Structures (from reference 2)

Structure Type	Damage	Overpressure (psi)	kPa
Wood frame building, residential type	moderate	2 to 3	13.8-20.7
	severe	3 to 4	20.7-27.6
Wall-bearing, masonry building apartment house type	moderate	3 to 4	20.7-27.6
	severe	5 to 6	34.5-41.4
Multi-story, wall-bearing building, monumental type	moderate	6 to 7	41.4-48.3
	severe	8 to 11	52.2-75.9
Reinforced concrete (not earthquake resistant) building, concrete walls, small window area	moderate	8 to 10	55.2-69.0
	severe	11 to 15	75.9-103.5

III.   METHODS OF CALCULATION

A.   Energy Released by Expansion of Gas

1.   Inert gas (Pressure less than 100 bars (10 MPa, ~1450 psia)

The high-pressure gas is assumed to be released and to expand
adiabatically. It is further assumed to be ideal with a constant specific heat ratio. The energy released is indicated schematically in
Figure 2 by the area a - 1 - 2 - b - a and is given by the equation

$$E = \frac{p_1 V_1}{\gamma-1} \left[1 - \frac{1}{\left(\dfrac{p_1}{p_2}\right)^{\frac{\gamma-1}{\gamma}}} \right] \tag{4}$$

Since work is done against the atmosphere (area a - c - 2 - b - a)
(see Figure 2) during the expansion, the energy released is sometimes
considered to be area c - 1 - 2 - c and is given by the equation

$$E_1 = p_2 V_1 \left(\frac{p_1}{p_2}\right)^{\frac{1}{\gamma}} \left\{ \left(\frac{p_1}{p_2}\right)^{-\frac{1}{\gamma}} \left[\frac{(\frac{p_1}{p_2})}{\gamma-1} + 1 \right] - \frac{\gamma}{\gamma-1} \right\} \tag{5}$$

The difference between E and E_1 is important only when the initial
pressure of the gas is close to the ambient pressure. As mentioned in
reference [4], the available data on the release of high-pressure gas
indicates that equation (4) more nearly represents the explosive
energy generated by the expansion. In the case of an inert gas at
pressures greater than about 100 bars, real gas effects become important and the energy released should be corrected. Experimental data
for some of the commonly used gases are available from which TNT
equivalency can be calculated.

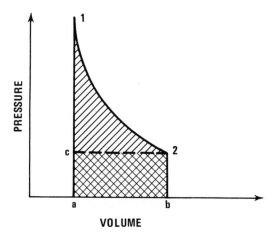

FIG. 2. Energy released during adiabatic expansion of an
ideal gas with constant specific heats.

2. Gas that Forms an Explosive Mixture with Air

In this case, it is assumed that the gas mixes with atmospheric oxygen and then detonates. The actual energy released depends upon the particular gas and the fractional explosion assumed. The method

3. SAFETY AND SAFETY CODES   35

reaction with 100-percent explosive energy release is remote. Experiments with liquid propellants (liquid H_2 and O_2 in liquid fueled rockets) have demonstrated that fractional explosions up to only about 20-percent are possible. With gases, the possibility of somewhat better mixing exists so that percentages may be larger than 20-percent. For analysis purposes it might be well to assume a 50-percent value for gases with the recognition that higher or lower values may exist for the particular conditions that prevail. In any case, the energy released is given by the equation

$$E = 0.0932 \; \frac{p_1 V_1}{T_1} \; fQ \;\; Btu = 0.1204 \; \frac{p_1 V_1}{T_1} \; fQ \;\; (kJ) \qquad (9)$$

where f is the fractional explosion assumed.

B.   Equivalent Weight of TNT

The energy released by the expansion is equivalent to a certain weight of explosive such as TNT. For purposes of damage assessment, it is assumed that TNT releases 1830 Btu/lb (4253 kJ/kg) in an oxygen-rare environment. Thus, the equivalent weight of TNT for an inert ideal-gas high-pressure system is

$$W_{TNT} = \frac{1.01(10^{-4})}{\gamma-1} \; p_1 V_1 \left[1 - \frac{1}{\left(\frac{p_1}{p_2}\right)^{\frac{\gamma-1}{\gamma}}} \right] \begin{matrix} lbs \\ (kg) \end{matrix} \qquad (10)$$
$$(2.35 \times 10^{-7})$$

and for a gas that forms an explosive mixture with air

$$W_{TNT} = \begin{matrix} 0.51(10^{-4}) \\ (2.83 \times 10^{-5}) \end{matrix} \; \frac{p_1 V_1}{T_1} \; fQ \; \begin{matrix} lbs \\ (kg) \end{matrix} \qquad (11)$$

Figure 3 gives the weight of TNT/unit volume of gas (lb TNT/ft^3, (kg/m^3) of gas) as a function of initial pressure for an inert ideal gas, with a specific heat ratio of 1.4, that expands adiabatically to atmospheric pressure.

Figure 4 provides the TNT equivalency/unit volume of gas for N_2, at high pressure, where correction has been made for real gas effects; this correction, due to high density, is seen to be large.

reaction with 100-percent explosive energy release is remote. Experiments with liquid propellants (liquid H_2 and O_2 in liquid fueled rockets) have demonstrated that fractional explosions up to only about 20-percent are possible. With gases, the possibility of somewhat better mixing exists so that percentages may be larger than 20-percent. For analysis purposes it might be well to assume a 50-percent value for gases with the recognition that higher or lower values may exist for the particular conditions that prevail. In any case, the energy released is given by the equation

$$E = 0.0932 \frac{P_1 V_1}{T_1} fQ \quad Btu = 0.1204 \frac{P_1 V_1}{T_1} fQ \ (kJ) \qquad (9)$$

where f is the fractional explosion assumed.

B.   Equivalent Weight of TNT

The energy released by the expansion is equivalent to a certain weight of explosive such as TNT. For purposes of damage assessment, it is assumed that TNT releases 1830 Btu/lb (4253 kJ/kg) in an oxygen-rare environment. Thus, the equivalent weight of TNT for an inert ideal-gas high-pressure system is

$$W_{TNT} = \frac{1.01 (10^{-4})}{\gamma - 1} P_1 V_1 \left[1 - \frac{1}{\left(\frac{P_1}{P_2}\right)^{\frac{\gamma-1}{\gamma}}} \right] \begin{matrix} lbs \\ (kg) \end{matrix} \qquad (10)$$
$$(2.35 \times 10^{-7})$$

and for a gas that forms an explosive mixture with air

$$W_{TNT} = \begin{matrix} 0.51 (10^{-4}) \\ (2.83 \times 10^{-5}) \end{matrix} \frac{P_1 V_1}{T_1} fQ \begin{matrix} lbs \\ (kg) \end{matrix} \qquad (11)$$

Figure 3 gives the weight of TNT/unit volume of gas (lb TNT/ft^3, (kg/ m^3) of gas) as a function of initial pressure for an inert ideal gas, with a specific heat ratio of 1.4, that expands adiabatically   to atmospheric pressure.

Figure 4 provides the TNT equivalency/unit volume of gas for N_2, at high pressure, where correction has been made for real gas effects; this correction, due to high density, is seen to be large.

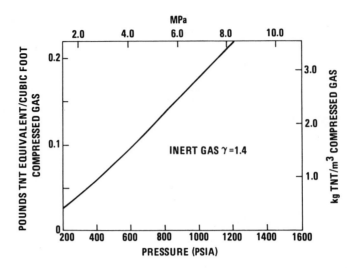

FIG. 3. Explosive equivalent of inert ideal gas.

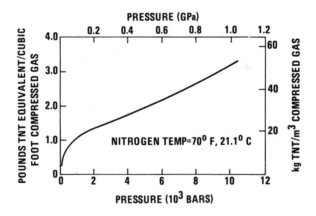

FIG. 4. Explosive equivalent of real gas (nitrogen).

C.   Peak Overpressure Blast Contours

Theoretically, in the far field a given pressure will occur at a
distance from an explosion that is proportional to the cube root of
the energy yield. Full-scale nuclear tests have shown that cube-root
scaling may be applied with confidence over a wide range of explosion
energies. According to this law, if D_1 is the distance from a refer-

ence explosion of W_1 pounds of TNT at which a certain overpressure or dynamic pressure is attained, then for any explosion of W pounds, these same pressures will occur at a distance D given by

$$D = D_1\left(\frac{W}{W_1}\right)^{1/3} \tag{12}$$

If the reference explosion is taken to be 1 kiloton (2 x 10^6 pounds of TNT) (9.07 x 10^5kg), then the data of reference [2] can be conveniently utilized. Figure 5 is a log-log plot of peak overpressure and peak dynamic pressure against distance for such an explosion, as given by reference [2]. Blast pressure contours can be calculated according to the equation

$$D = \frac{D_1}{100}\left(\frac{W}{2}\right)^{1/3} \; ; \; \begin{array}{l} \text{W in lbs} \\ \text{D in ft.} \end{array} = \frac{D_1}{97}\left(W\right)^{1/3} \; ; \; \begin{array}{l} \text{W in kg.} \\ \text{D in m.} \end{array} \tag{13}$$

Thus, if a gas explosive equivalent of 2000 pounds (908kg) of TNT occurs, the distance at which a peak overpressure of 10 psi (69 kPa) is felt can be found from equation (13) and Figure 5 as follows:

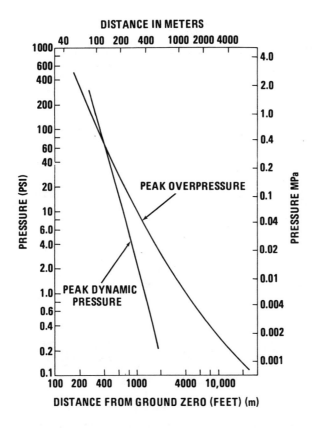

FIG. 5. Peak overpressure and peak dynamic pressure for 1-kiloton surface burst [2] (1 kiloton = 9.07 x 10^5kg).

From Figure 5, for a peak overpressure of 10 psi (69 kPa), $D_1 = 1000$ feet (304.8 m) and from equation (13)

$$D = \frac{1000}{100} \left(\frac{2000}{2}\right)^{1/3} = 100 \text{ feet } (30.5\text{m}) \qquad (14)$$

so that the 10-psi (69 kPa) peak overpressure blast contour has a radius of 100 feet (30.5 m).

It should be noted that the overpressure given in Figure 5 applies to a nuclear explosion and may underestimate the distance at which a given overpressure occurs for conventional explosives. Reference [6] contains data on tests made with hemispherical charges of TNT detonated at the surface. These data provide the curve shown in Figure 6 which, for the same example as considered above, yields a distance of 124 feet (37.8 m) at which a 10-psi (69 kPa) overpressure would be felt from an explosion of 2000 pounds (907 kg) of TNT. The differences would be in the nature of the nuclear explosion compared to that of TNT. In the latter case, the overpressure is higher but the positive duration of the pulse is shorter than in the equivalent weight nuclear explosion.

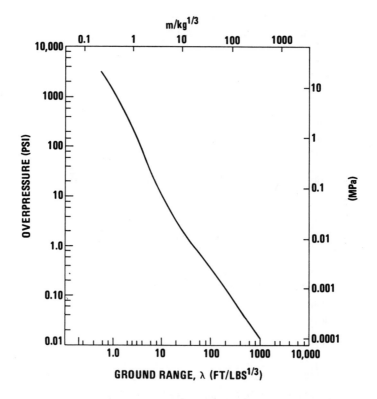

FIG. 6. Peak overpressure vs. scaled distance for TNT surface bursts (hemispherical charges) [6].

Which figure, 5 or 6, to use really depends upon the similarity of the release of the high-pressure gas to an actual explosion. Unpublished computer solutions concerned with the expansion of a spherical volume of high-pressure gas in air indicate that the energy release is more nearly related to a conventional TNT air burst, particularly for overpressures below 10 psi (69 kPa). For this reason, and because it is more conservative (i.e., predicts a larger blast radius), it is recommended that Figure 6 be used in a damage assessment analysis and that overpressure predictions be limited to 10 psi (69 kPa) and lower. This is not really a severe restriction since most structures will be destroyed by a pressure of this magnitude.

D.   Blast-Damage Assessment

As indicated in reference [2], damage to structures or objects is divided into three categories as follows:

Severe damage.   A degree of damage that precludes further use of the structure or object for its intended purpose without essentially complete reconstruction. For a structure or building, collapse is generally implied.

Moderate Damage.   A degree of damage to principal members that precludes effective use of the structure or object for its intended purpose unless major repairs are made.

Table 3

Conditions of Failure of Peak Overpressure
Sensitive Elements (from reference 2)

Structural element	Failure	Blast overpressure (psi)	(kPa)
Glass windows	shattering, occasional frame failure	0.5-1.0	3.45-6.90
Corrugated asbestos siding	shattering	1.0-2.0	6.90-13.80
Corrugated steel or aluminum paneling	connection failure followed by buckling	1.0-2.0	6.90-13.80
Wood siding panels, standard house construction	usually failure occurs at the main connections allowing a whole panel to be blown in	1.0-2.0	6.90-13.80
Concrete or cinder block wall panels 8 in. (20.37cm) or 12in.(30.48 cm) thick (not reinforced)	shattering of the wall	2.0-3.0	13.80-20.70
Brick wall panel 8 in.(20.32cm) or 12in.(30.48cm) thick (not reinforced)	shearing and flexure	7.0-8.0	48.30-55.20

Slight Damage. A degree of damage to buildings resulting in broken windows, slight damage to roofing and siding, falling down of light interior partitions, and slight cracking of curtain walls in buildings.

For certain structural elements, with short periods of vibration and small plastic deformation at failure, the conditions for failure can be expressed as a peak overpressure without consideration for the duration of the blast wave. The failure conditions (severe damage) for elements of this type are given in Table 3. Some of these elements fail in a brittle fashion so that there is little difference between the pressures for light damage and complete failure. The pressures are side-on blast overpressures for panels that face the explosion. For panels that are oriented so that there are no reflected pressures thereon, the side-on pressures must be doubled.

The accurate assessment of damage necessarily involves a knowledge of the peak overpressure and dynamic overpressure, together with the positive phase time durations. However, a reasonably good damage analysis can be made by consideration only of the peak overpressure. Reference [2] lists some values of overpressure with respect to variations in damage. Table 4 is a compilation for various structures of

Table 4

Overpressure/Damage for Various Structures
(from references 2 and 3)

Structure/Damage	Peak Overpressure psi (kPa)		
	Slight	Moderate	Severe
Windows	0.2(1.38)	–	–
Wood frame structures	–	1.7(11.7)	2.5(17.2)
Light steel frame industrial buildings, light walls	–	2.5(17.2)	–
Motor vehicles	3.1(21.3)	10.0(69.0)	16.5(113.8)
Medium steel frame industrial buildings, light walls	–	4.0(27.6)	8.0(55.2)
Reinforced concrete frame and walls, multi-story structures	–	6.0(41.4)	9.0(62.1)
Wall-bearing brick buildings	–	4.7(32.4)	6.5(44.8)
Massive wall-bearing, multi-story structures	–	9.0(62.1)	–
Steel frame office-type buildings	–	4.7(32.4)	10.0(69.0)
Reinforced concrete, blast resistant, windowless structures	–	19.0(131.1)	–
Oil storage tanks:			
Empty	–	–	1.0(6.90)
Filled	1.0(6.90)	–	11.0(75.9)

Table 5

Recommended Values of Overpressure for Damage Assessment

	Peak Overpressure psi(kPa)		
Structure/Damage	Slight	Moderate	Severe
Windows	0.2(1.38)	–	0.5(3.45)
Wood frame structures	0.5(3.45)	0.7(4.83)	1.5(10.35)
Motor vehicles	3.0(20.7)	10.0(69.0)	16.0(110.4)
Reinforced concrete frame and walls, multi-story structures	3.0(20.7)	6.0(41.4)	9.0(62.1)
Wall-bearing brick buildings	1.0(6.9)	4.0(27.6)	6.0(41.4)
Steel frame office-type buildings	2.0(13.8)	4.0(27.6)	19.0(69.0)
Corrugated steel or aluminum paneling	0.5(3.45)	0.7(4.83)	1.0(6.90)
Light steel frame industrial buildings, light walls	1.0(6.9)	2.0(13.8)	3.0(20.7)
Medium steel frame industrial buildings, light walls	2.0(13.8)	4.0(27.6)	8.0(55.2)

The values above are for a structure oriented perpendicular to the
blast wave. They should be doubled for side-on orientation.

peak overpressure to have slight, moderate, and severe damage. There
are some slight discrepancies between Tables 3 and 4. Table 5 shows
the overpressure values that are suggested for use in making a failure
damage analysis. It should be mentioned that structural damage begins
to appear whenever overpressures greater than about 0.5 psi (3.45 kPa)
are felt.

D.   Shrapnel and Subsidiary Damage

As mentioned earlier in this chapter, it is difficult to calculate the velocities attained by fragments as a result of a high-pressure release because of the many variables involved. First of all,
strain energy exists in the vessel as a result of the internal pressure applied. When the vessel fails catastrophically, the various
fragments will be given an initial velocity because of this strain energy. Generally speaking, however, only a fraction of the strain energy will be directly converted into velocity with the remaining part
represented by vibration of the particles. Secondly, once the particles are broken away, they are subjected to a gas-dynamic drag phase

which is proportional not only to the size, weight, orientation, and drag coefficient of the various particles but also to the relative velocity of the gas to the particle. As a result, the prediction of final particle velocities is extremely difficult.

Reference [7] estimates that the proportion of total energy released that appears in fragments is unlikely to exceed 20% when failure occurs, as for example, a rupture of the wall of the vessel. Values approaching 20% would probably apply when the vessel fails in a brittle manner with much lower percentages associated with a ductile failure. This is another reason why the designer should employ ductile steels wherever possible.

Higher percentages than 20% are possible when an end closure of a vessel fails. The calculation of the fragment velocity in such a case is relatively straightforward since the mass of the fragment can be estimated easily and its velocity calculated by assuming it is a projectile being launched from a gun.

As a general rule, it can be expected that chamber and structural fragments, as well as pipe fittings and miscellaneous equipment parts, will pose a damage mechanism in the form of flying shrapnel with velocities of up to several hundred feet per second ($\sim$60 m sec^{-1}). Reference [7] provides an empirical equation for the depth of penetration of a fragment into a barricading wall. In S.I. units the equation is

$$d = k \, W_f^{0.4} \, V_f^{1.5} \qquad\qquad (15)$$

where d is in meters, W_f in kg, V_f in m/sec, and k is a constant that depends upon the material being impacted and penetrated. For concrete k = 5.5×10^{-6} in S.I. units.

For side wall rupture of a cylindrical pressure vessel in a failure-damage analysis, a possible rough estimate of shrapnel effects might be made by means of the following approach:

1. Assume that 20% of the energy appears as fragment energy.
2. Calculate fragment velocities by assuming various fragment weights and equating the kinetic energy to the energy obtained in step 1.
3. Size barricade wall thicknesses for shrapnel protection by means of equation (15).

In the assessment of damage, consideration should be given to the proximity of other hazardous equipment which may cause additional destruction as a result of receiving a given overpressure and/or shrapnel damage. Failure of a high pressure vessel could, for example, cause high-voltage lines to be broken, rupture tanks containing hazardous or inflammable liquids, cause failure of adjacent high pressure vessels, etc.

F.   High Pressure Liquids

The technique outlined above for gases can also be used with liquids contained at high pressure. Generally speaking, however, the blast overpressures calculated will usually be overconservative unless the liquid undergoes a phase change and flashes. Few people realize that the energy contained in a saturated liquid of a given volume is

considerably higher than if the same volume is filled with saturated
vapor at the same pressure. An example of this is to consider a volume V (ft^3) (or m^3 for S.I. units) filled first with saturated water
in the liquid phase and then in the vapor phase at a pressure of say
500 psia (3450 kPa). For the liquid the total internal energy is

$$U_\ell = \frac{V}{v_\ell} u_\ell = \begin{array}{l} 22,720V \text{ Btu, with V in ft}^3 \\ 8.46 \times 10^5 V \text{ kJ, with V in m}^3 \end{array}$$

(16a)

(16b)

where u_ℓ = internal energy of saturated liquid in Btu/lb (kJ/kg) and
v_ℓ = specific volume of saturated liquid in ft^3/lb (m^3/kg). For the
vapor

$$U_g = \frac{V}{v_g} u_g = \begin{array}{l} 1,205 \text{ V Btu, with V in ft}^3 \\ 4.49 \times 10^4 V \text{kJ, with V in m}^3 \end{array}$$

(17a)

(17b)

where u_g = internal energy of saturated vapor in Btu/lb (kJ/kg) and
v_g = specific volume of saturated vapor in ft^3/lb (m^3/kg).
     At a pressure of 500 psi (3450 kPa):

$$u_\ell = 447.6 \text{ Btu/lb } (1.04 \times 10^3 \text{ kJ/kg})$$

$$v_\ell = 0.0197 \text{ ft}^3/\text{lb } (1.23 \times 10^{-3} \text{ m}^3/\text{kg})$$

$$u_g = 1118.6 \text{ Btu/lb } (2.60 \times 10^3 \text{ kJ/kg})$$

$$v_g = 0.9278 \text{ ft}^3/\text{lb } (57.9 \times 10^{-3} \text{ m}^3/\text{kg})$$

Hence U_ℓ = 22720 V Btu with V in ft^3 or

             = 8.46 $\times$ 10^5 V kJ with V in m^3

and U_g = 1205 V Btu with V in ft^3 or

             = 4.49 $\times$ 10^4 V kJ with V in m^3

Thus the energy in the liquid is almost twenty times greater than in
the vapor for the same total volume.
     Liquids at very high pressures pose an additional danger in that
very small leaks in the piping or containment apparatus can cause liquid jet velocities that are lethal. Jet velocity can be calculated
from the following equation:

$$V = \sqrt{\frac{2pg}{\rho}}$$

(18)

where in common units the pressure p is in psi; g, the acceleration
due to gravity is 32.2 ft/sec^2; ρ, the density in lb/ft^3 and V is in
ft/sec. In S.I. units p is in Pa, g is 9.81 msec^{-2}, ρ is expressed in
kg m^{-3} and V in msec^{-1}.

IV.   DAMAGE ASSESSMENT AND EXAMPLES

     The equations and results presented in the earlier sections of
this chapter can be used to make a preliminary estimate of the potential structural lethality of a high-pressure gas installation. It has
been the author's procedure in making a damage assessment during the
design feasibility stage to calculate the blast-pressure contours corresponding to 7, 1, 0.5, and 0.1 psi (48.3, 6.9, 3.45, 0.69 kPa), and
to assume severe damage out to the 7-psi contour, severe to moderate
damage between 7 and 1 psi (48.3, 6.9 kPa), moderate to slight damage

between 1 and 0.5 psi (6.9, 3.45 kPa), with window damage apt to occur out to the 0.1 psi (0.69 kPa) contour. An examination of Table 5 indicates that this yields a conservative estimate of the damage, i.e., more damage than would actually occur from blast effects alone if the high-pressure gas were released. A more exact analysis would, of course, use the techniques and data of reference [1]. The results of the preliminary damage analysis can be used by the designer of a system involving high-pressure gases to determine whether a more accurate analysis should be made, to establish specific safety requirements, to inaugurate a carefully formulated operational procedure, and to devise a safety interlock system which will preclude catastrophic release of the high-pressure gas. It should be remembered that the method used in this report to assess damage is only preliminary, and the actual overpressures that may occur will depend upon the confinement and construction afforded by the structure in which the gas system is located. It has also been assumed that the gas is released from a vessel at ground level.

In each of the following examples, the equivalent TNT weight and the radius of the 7-, 1-, 0.5-, and 0.1- psi contours have been calculated (48.3, 6.9, 3.45, 0.69 kPa respectively).

Example 1

Twenty cubic feet (0.566 m^3) of air are contained at a pressure of 1000 psia (6.9 MPa).

From Figure 3, the specific TNT equivalent is:

$$W_{TNT} = 0.176 \text{ lb/ft}^3 \ (0.497 \text{ kg/m}^3)$$

The total weight of TNT is therefore:

$$W_{TNT} = W_{TNT} \times \text{volume} = 0.176 \ (20) = 3.52 \text{ lb} \ (1.6 \text{ kg}).$$

From Figure 6 the values of λ_1 can be determined for the pressure values of interest. The distance at which a given overpressure will be felt is given by

$$D = \lambda (W_{TNT})^{1/3}$$

For this example the following table results:

Peak overpressure (psi)	(kPa)	$\lambda(\text{ft/lbs}^{1/3})$	$(\text{mkg}^{-1/3})$	D (ft)	D (m)
7.0	(48.3)	12	(4.76)	18	(5.5)
1.0	(6.9)	45	(17.85)	68	(20.7)
0.5	(3.45)	75	(29.74)	114	(34.7)
0.1	(0.69)	240	(95.18)	365	(111.3)

The explosive effect for this example is given by the value of D in the above table.

Example 2

Ten cubic feet ($0.283 \ m^3$) of H_2 are contained at a pressure of 500 psia (3.45 MPa) and 70°F (21.1°C).
From equation (11) and Table 1

$$W_{TNT} = 0.51 \ (10^{-4}) \ \frac{P_1 V_1}{T_1} \ fQ$$

$$= 0.51 \ (10^{-4}) \ \frac{500 \ (10)}{530} \ f \ (103200)$$

$$= 49.7 \ f \ lbs \ (22.5 \ kg)$$

In S. I. units from equation (11) and Table 1

$$W_{TNT} = 2.83 \ (10^{-5}) \ \frac{P_1 V_1}{T_1}$$

with

$$P_1 = 3.45 \times 10^6 \ Pa$$

$$V_1 = 0.283 \ m^3$$

$$T = 294.1 \ K$$

$$Q = 240 \ kJ/gmole$$

$$W_{TNT} = \frac{2.83 \ (10^{-5}) \ 3.45 \ (10^6) \ (0.283)}{294.1} \ f \ (240)$$

$$= 22.5 \ kg$$

For a fractional explosion of 50 percent, $f = 0.50$ and

$$W_{TNT} = 4.97 \ (0.5) = 2.49 \ lbs \ (1.13 \ kg)$$

The following table results:

Peak overpressure (psi)	(kPa)	D (ft)	D (m)
7.0	(48.3)	16	(4.9)
1.0	(6.9)	61	(18.6)
0.5	(3.45)	102	(31.1)
0.1	(0.69)	325	(99.1)

Example 3

Twenty cubic feet of N_2 are contained at 4000 atmospheres. From Figure 4

$$W_{TNT} = 1.54 \text{ lb/ft}^3 \ (24.71 \text{ kg/m}^3)$$

$$W_{TNT} = W_{TNT} \times \text{volume} = 1.54 \ (20) = 30.8 \text{ lb} \ (13.98 \text{ kg})$$

The following table results:

Peak overpressure		D	D
(psi)	(kPa)	(ft)	(m)
7.0	(48.3)	38	(11.6)
1.0	(6.9)	141	(43.0)
0.5	(3.45)	235	(71.6)
0.1	(0.69)	750	(228.6)

A necessary part of the high pressure system design package is a procedure and schedule for inspection and maintenance. Chapter 10 describes methods of non-destructive testing that might be considered for this phase of the design. In addition the use of periodic hydro-static pressure tests should not be overlooked. Fracture mechanics theory, for example, shows that cyclic loading leads first to crack initiation, next to crack growth at an initially slow rate until the crack reaches a critical size at which point it spreads essentially at an instantaneous rate.

The critical crack size is a function of the material properties and is related to a parameter, K_{IC}, called the critical stress intensity. It is, therefore, theoretically possible to overpressurize a cylindrical vessel hydrostatically, under controlled safety conditions, to a given value such that if failure does not occur it means that no cracks of critical size are present. If, in addition, data are available on crack-rate growth as a function of stress intensity, a certain number of safe operating cycles can be identified at lower pressure. Thus the cylinder can be cycled at lower pressure for these number of cycles and then resubmitted to the overpressure hydrostatic test again. This process can be repeated over and over again until failure occurs during the controlled hydrostatic test rather than during the actual operating cycle.

The importance of safety can be illustrated by unpublished data developed in the review of 27 minor and major accidents that occurred over a period of 26 years at a laboratory engaged in research and development that required the use of large high pressure facilities containing both inert and explosive gases.

The primary causes of the accidents were identified in certain
categories as follows:
1.   Design - 14
2.   Operating procedure - 8
3.   Inspection/maintenance - 3
4.   Lack of adequate design data - 2
The direct equipment and/or building replacement costs were estimated
at three-quarters of a million dollars over the 26 year period. Three
lives were lost and three temporary disabilities occurred. It is to
be noted that the three fatalities occurred not as a result of failures of any high pressure equipment but rather because of inadequate
operating procedures associated with high voltage equipment in two of
the cases and with a firefighting CO_2 area flooding system in the
other case.

The designer of high pressure systems should make recommendations
as to site location and building construction. The failure-damage analysis is a tremendous aid in this regard and can materially influence
where and how the system is to be located. Barricades, safety devices, redundancy in interlocks, etc., can be designed with the aid of
such an analysis. While the discussion of this analysis was largely
concerned with failure of the main vessel in the system, the designer
should not overlook each component and subassembly of the system.
Pressure-gauge shielding and venting, shielding for fittings and high
pressure piping, etc., should be automatic features of the design.
Reference [8] contains a description of a laboratory designed for
safety at high pressure and is recommended reading for personnel engaged in high pressure system design or operation.

V.   SAFETY CODES

A number of countries have established committees for the purpose
of formulating standard rules for the construction of steam boilers
and other pressure vessels.

The function of these committees is to establish rules of safety
governing the design, fabrication, and inspection during construction
of boilers and other pressure vessels, and to interpret these rules
when questions arise regarding their intent. Generally, the committees deal with the maintenance and inspection of in-service vessels
only by providing suggested rules of good practice as an aid to owners
and inspectors.

In the United States the regulatory committee, which was set up
by the American Society of Mechanical Engineers in 1911, is called the
Boiler and Pressure Vessel Committee. Considerations of materials,
design, method of fabrication, inspection and safety devices are used
in the formulation of its rules and in the establishment of maximum
design and operating pressures.

The legal standing of the codes varies from country to country.
For example, in 1968 West Germany and Italy had enforced codes with
the code of Sweden mandatory in practice while France, Japan, and the
United Kingdom did not enforce their respective codes. In the United
States, and also Canada, the regulations of the Boiler and Pressure
Vessel Committee are enforced in those states, and provinces, which

have adopted the code. Each such state and dominion is invited to ap-
point a representative to act on the Conference Committee to the Boiler and Pressure Vessel Committee. Since the members of the Conference
Committee are in active contact with the administration and enforce-
ment of the rules, the requirements for inspection in the code correspond with those in effect in the respective jurisdiction of the mem-
ber. The required qualifications for an authorized inspector under
the rules can be obtained from the administrative authority of any
state, municipality, or province which has adopted the rules.

The National Board of Boiler and Pressure Vessel Inspectors, established in 1919, consists of the chief inspector of the state and
provinces that have adopted the code. This Board serves to uniformly
administer and enforce the rules of the Boiler and Pressure Vessel
Code.

Pressure vessels which are designed, manufactured, and inspected
as regulated by the code can be so marked. Some times pressure vessels are designed according to the code but may not be manufactured
and/or inspected in strict compliance with the code. Such vessels are
referred to as code designed.

In general, the various codes are developed around conservative
design procedures. This is desirable since the vessels may be used in
areas adjacent to, or in, populated zones where failure might result
in serious injury or death to people as well as massive structural
damage. The code, for example, does not presently recognize autofrettage as a method that can be used to reduce creep or fatigue effects
and yet this may be the only means by which a designer can achieve a
given pressure level of operation. It should also be noted that, in
spite of its conservatism, coded vessels have been known to fail.

The designer of specialized high pressure vessels, intended for
advanced state of the art research, cannot sometimes meet the regulations of the code with existing materials and still meet the specifications imposed on the design. In such cases the designer should make
his best estimate of the pressure capability of the system as calculated by his particular approach and assumptions together with the
code estimate. A comparison of the two estimates frequently dictates
more detailed concern about the safety of the system.

Section III of the ASME Boiler and Pressure Vessel Code details
the design procedure to be used with nuclear vessels. Criteria are
provided for vessels not requiring analysis for cyclic operation and
those that do. Design stress-limits for the former case are the ultimate tensile strength at the working temperature divided by 3 or the
minimum yield strength divided by 1.5, whichever is lower. In the
case of cyclically loaded vessels, stress-cycle (S-N) curves are provided for various metals. These curves are corrected for mean stress
and consider only the alternating stress intensity. This latter
stress intensity is one-half of the largest of the principal stress
differences. Cumulative damage, due to operation of the vessel at
different pressures, is essentially handled on the basis of Miner's
rule which is expressed as

$$\sum \frac{n(\sigma)}{N(\sigma)} = 1 \tag{19}$$

where n is the number of cycles at a given stress level and N is the number of cycles that can be withstood at that stress (see also Chapter 5).

The codes that are enforced or recommended for use in the design of pressure vessels in various countries are aimed at providing safe and conservative systems and are based upon cumulative experience over a long period of time. They are not infallible but do represent good design procedure. They are particularly useful in providing the designer of specialized high pressure equipment with a comparison to his own estimates of the pressure capability of a given apparatus. Discrepancies are bound to occur for advanced state of the art equipment. These discrepancies should serve to redouble the designer's efforts toward a safe system adequately barricaded, operated, maintained, and inspected. Documentation should be compiled on calculations, comparisons, and reviews by higher management authorities and/or outside experts. In the event of an accident these can then be examined to try to establish what went wrong with the process and they may help to advance the state of the art in safety.

VI.   SUMMARY

In this chapter safety was examined from a general point of view. It was pointed out that the design process is not complete until all aspects of the system are examined starting from the basic design and progressing through a standard operating procedure, a failure-damage analysis, a maintenance and inspection procedure and schedule, and culminating in a final management review. Each of these items results in determining site location, safety devices, building construction, safety interlocks, etc., and is a necessary component of the design process. A brief discussion on safety codes was also included. Designers of high pressure equipment should certainly be familiar with the codes that apply in their particular country.

REFERENCES

1. Amman and Whitney, "Manual for Design of Protective Structures Used in Explosive Processing and Storage Facilities", Defense Supply Agency, AD 834465.
2. S. Glasstone, The Effects of Nuclear Weapons, USAEC, Feb. 1964.
3. G. L. Rogers, An Introduction to the Dynamics of Framed Structures, John Wiley and Sons (1959).
4. R. A. Boudreauxs, "TNT Equivalency - Gas Dynamics Comparison for Moderately Pressurized Tanks", Ninth Explosives Safety Seminar, Naval Training Center, San Diego, California, 15-17, August 1967.
5. Marks, Mechanical Engineers Handbook, Sixth Edition, McGraw-Hill Book Co., New York (1958).
6. C. N. Kingery and B. F. Pannell, "Peak Overpressures vs. Scaled Distance for TNT Surface Bursts (Hemispherical Charges)", BRL Memorandum Report No. 1518, April 1964.
7. W.R.D. Manning and S. Labrow, High Pressure Engineering, C.R.C. Press, Cleveland (1971).
8. J. C. Bowen and R. L. Jenkins, "Safety at High Pressures - Features of an Eight Cubicle Laboratory", Industrial and Engineering Chemistry, 49, 2019-2021 (Dec. 1957).

Chapter 4

HAZARDOUS MATERIALS AND PROCESSES

John N. Fedenia

Professional Engineer
Rockville, Maryland
(formerly Naval Surface Weapons Center,
White Oak Laboratory, Silver Spring, Maryland 20910)

I.   INTRODUCTION

In high pressure processes an important fundamental consideration is the hazard that may be involved. In this country alone, more than two billion tons of hazardous materials are shipped in one year. Since the amount of hazardous materials has grown so tremendously in chemical processing and in their application, the primary purposes of this chapter are to indicate the properties of hazardous materials, the types of hazards, and the safety aspects, with particular emphasis on materials and processes met with in High Pressure Technology.

Hazards in chemical processing may result from various sources. The physical properties of the chemical materials may be dangerous or reactive. The operating conditions of the chemical process may pose

critical limits or potential hazards. Unforeseen malfunctions in the
system or in the equipment may occur. Handling, transporting and
storage of the material may also be hazardous. Even normally non-haz-
ardous materials may result in dangerous conditions in an abnormally
functioning chemical process or in a particular environment.

Some of the particular problems with high pressure leading to
hazardous conditions are: the leaks that may occur (see Chapters 5
and 6); runaway reactions that develop excessive pressures (see Chapter 1 - 3 in the succeeding volume); the potential explosions that may
result in fire, blast effects, and flying objects; and inadvertent exposure to toxic substances.

Exposure to extremely small amounts of certain toxic materials,
whether in the air, in the water, or by direct contact may have serious effects on humans and in the environment. For example, in 1975,
the effects of contamination by Kepone, a chlorinated hydrocarbon pes-
ticide, was detected. It was reported that occupational exposure in
the chemical processing plant resulted in the serious illness of workers. Environmental contamination, caused by wastewater discharges
from the chemical processing plant, necessitated the closing of large
portions of the state of Virginia's historic James River to commercial
and sport fishing.

Another example of toxic substances of recent concern are nitrosamines, a group of organic compounds that are potent carcinogenic
substances. The U. S. Environmental Protection Agency reported detection of nitrosamines in air samples over Baltimore, Maryland and
Belle, West Virginia in 1975, using a new technique and analysis for
measuring minute quantities. These toxins were released to the air
from unavoidable leaks in the processing plant.

A.   High Pressure Technology in the Chemical Industry

In the historical development of the chemical industry, there has
been a continued interest in high pressure technology. The successful
application of higher pressures to accelerate a chemical reaction or
to obtain a higher equilibrium yield has in many cases provided an
economic advantage. However, the hazards involved in some instances
have been. a deterrent.

In Chemical Reactions at High Pressure by Weale [1] published in
1967, the effect of pressure on various types of reactions is reviewed
in high pressure chemistry. In early development, the mid-1880's,
autoclaves were used at pressures up to 20 or 30 bars (~290-435 psi)
in the synthetic dye industry. In 1892, the effects of 500 bars
(~7,250 psi) on a chemical reaction were first described. However, in
1901, a serious explosion deterred an attempt to provide an economical
means of using pressure in the catalytic synthesis of ammonia from
nitrogen and hydrogen (see Chapter 2, Vol. II). In later developments, pressures in the range of 100 to 1000 bars (~1,450-14,500 psi)
where the largest chemical effects are found in gas reactions, have
been applied in chemical production. Although materials are being
synthesized under extreme pressures in laboratores, most large scale
production-operating pressures are under 4,000 bars (~58,000 psi) (see
Volume II).

II.   PROPERTIES OF HAZARDOUS MATERIALS

With the tremendous expansion of new chemical products and the chemical industry, there has been a greater emphasis to identify the properties of hazardous materials [2]. There has also been more con- cern in the control of hazardous chemical processes and the safeguards and protection required in storage and handling. Hazardous properties are normally related to three general areas: Health, Fire, and Explo- sion. Hazardous properties are discussed in more detail in each of the danger areas. Some hazardous materials, of course, may involve more than one danger area.

A.   Health Hazard

Health hazards of materials or processes in this section are primarily concerned with toxicity and radioactivity.

1.   Toxicity

Toxicity is the degree that a material may be harmful to the body, or the ability of a material to cause injury to living tissue of the body, either internally or externally. A toxic substance may be in the form of a solid, liquid, gas, fumes, mist, vapor or dust. En- try into the body may be through breathing, swallowing, or skin con- tact. The action of a toxic agent may be local or it may be absorbed and produce harmful effects on other parts of the body.

Accidental and sudden exposure to excessive amounts of toxic ma- terials are of course recognizable because of the severity of the re- action. However, the effects of low levels of prolonged exposure in a toxic environment are more difficult to detect. Because of the numer- ous chemicals that have been and are being developed, there has been an increased attempt to identify toxic materials, and to establish limits of exposure and limiting concentrations. Several guides used for indicating exposure limits are the Threshold Limit Values (TLV), the Maximum Acceptable Concentrations (MAC), and the Criteria Docu- ments (CD).

Threshold Limit Values are an indication of the limit of daily exposure for a normal work day using time-weighted average concentra- tions. The Threshold Limit Values are established annually by the American Conference of Governmental Industrial Hygienists [3]. The units used for the toxic threshold limits are parts of gas or vapor per million parts of air (ppm), and milligrams of particulate per cu- bic meter of air (mgm/m^3). The unit employed for some mineral dusts is millions of particles per cubic foot of air (mppcf).

The Maximum Acceptable Concentrations are an indication of the maximum value of concentration that should never be exceeded. These values are published by the American National Standards Institute (formerly American Standards Association).

Criteria Documents are the recommended standards for occupation- al exposure published by the National Institute for Occupational Safety and Health, U. S. Department of Health, Education, and Wel- fare. The standards apply to hazardous materials and hazardous occu- pational environments.

Lethal doses of toxicity are sometimes indicated as MLD, LD 50
and LD 100. These abbreviations represent tests with animals. The
Minimal Lethal Dose (MLD) is one lethality in a group of animals, LD
50 is the lethal dose for 50 percent, and LD 100 is the lethal dose
for all of the animals in the group. Results from these tests provide
criteria to estimate lethal doses for humans.

2.   Ionizing Radiation

There has been a sizable increase of possible exposure to ionizing radiation X-rays and radioisotopes that emit nuclear radiation.
Hazards of exposure occur in facilities that handle radioactive materials, and utilize industrial radiography with isotopes, X-ray equipment (nondestructive testing (see Chapter 10) and pressure measurement
(see Chapter 8)), or other radiation-emitting devices. It is noted
that many nuclear power plants utilize a high pressure process-gas
treatment cycle. Radiation exposure levels should be maintained at
minimal levels possible since the effects of radiation may be delayed,
except in cases of extreme exposure.

It would be inappropriate to dwell at length on this topic in the
present volume. A brief description of radiation hazards concerned
with the characteristics of ionizing radiation and their biological
effects will be given. The biological effects of radiation are the
damage to living tissue or changes in individual cells. Various body
cells and organs have varying degrees of sensitivity to radiation.

In nuclear radiation, alpha particles, beta particles, gamma
rays, and neutrons present the dangers to living organisms. It is the
interaction or dissipation of this energy in the body that results in
biological harm.

Alpha particles are relatively large in mass but have little penetrating power. They originate from the nucleus of some of the elements with high atomic weight. The alpha particle which consists of
two protons and two neutrons, is the same as the nucleus of the helium
atom. Because the particles have little penetrating power, and travel
only several inches in air, they generally would not be considered a
hazard when outside the body. However, if materials such as uranium
or plutonium are inhaled or ingested, the results may be very harmful.

Beta particles are electrons emitted from the nucleus of an atom.
They are relatively small in mass, travel at very high speeds, and are
more penetrating than the alpha particle but less than gamma or X-
rays. Beta particles are emitted from radioactive elements of heavy
and light atomic weights. Radiation burns can be produced in contact
with exposed skin, and inhalation and ingestion of isotopes that emit
beta particles is extremely hazardous. Beta particle penetration into
the body may be from 1/10 to 1/2 inch (~2.54-12.7 mm).

Gamma rays are a form of electromagnetic radiation and are similar to X-rays except that they have higher energy and originate from
the nucleus of an atom. They consist of protons or small parcels of
energy, with no mass or electric charge, that travel with the speed of
light. They are extremely penetrating and are a very serious hazard
when absorbed by the body externally or internally in sufficient quantity. Protection against the penetrating characteristics of gamma

radiation is a major factor in designing safeguards from nuclear radiation [4, 5].

Neutrons have relatively high mass, a neutral electrical charge, and are basic parts of the nucleus of an atom. Although neutrons do not cause ionization directly, they can cause secondary reactions and the emission of alpha, beta, and gamma radiation.

X-rays are a form of electromagnetic radiation that originate in the region of orbiting electrons of an atom. They may also be generated in machines (see Chapter 10). Penetration of X-rays is dependent upon the energy of the photons in the beam and upon the wave length. Short wave high energy, or "hard" X-rays will penetrate more deeply than longer wave or "soft" X-rays.

The unit for expressing ionizing radiation such as gamma or X-rays is the roentgen, defined in Chapter 10. It is estimated that the total dose received from natural sources such as radioactive substances in the ground, radon in the air, and cosmic rays from outerspace, is about 10 roentgen in an average lifetime. Additional exposure is received from chest and dental X-rays, television, and occupations associated with nuclear energy. No measurable effects would be expected for a life-time dose of 250 roentgens distributed over 50 years [2].

B.   Fire Hazard

Fire-hazardous materials are combustible materials that act as a fuel when combined chemically with oxygen or air to produce thermal energy and, normally, a visible flame. The basic elements of fire and specific characteristics of flammable chemicals will be discussed.

1.   Elements of Combustion

The three basic ingredients of fire are: a fuel, oxidant, and heat.

Fuels or combustible substances may be solids, liquids or gases. The majority of these materials are chemical compounds; however, certain pure elements are flammable. Some of the more common elements that may burn are carbon, hydrogen, sulfur, magnesium, and titanium. The elements carbon and hydrogen in various combinations with oxygen form many basic combustible substances. Probably some of the most common forms in solids are wood and paper; and in liquid form are gasoline, kerosene and petroleum products. Common flammable gases with combinations of carbon and hydrogen are methane (natural gas), ethylene (anesthetic), and acetylene.

Oxygen, although it does not burn, supports combustion. It is the major source for combustion; however, chlorine and fluorine are other elements that support combustion. While there is an abundance of oxygen in the atmosphere, some flammable materials do not rely on this source for combustion. Oxygen in the molecules of the chemical compounds provide a means for supporting combustion.

Heat or thermal energy is generated after ignition and combustion begins. Different materials have varying degrees of flammability.

Some flammable materials require preheating to ignite, while others may be ignited under normal ambient conditions. Certain chemicals such as flammable gases or liquids may be quite volatile in their reaction after ignition and result in intense and rapid burning. The amount of thermal energy produced depends on the combustible material. For comparison, the heat content for several types of material expressed in British Thermal Units per pound weight and J/kg (1 BTU/lb = ~ 2326 J/kg) are: paper or wood 6,000-7,000 (~14.0 - 16.3 MJ/kg), coal (anthracite) 13,000 (~30.2 MJ/kg), fuel oil 19,000 (~44.2 MJ/kg), natural gas (methane) 23,000 (~53.5 MJ/kg), and hydrogen 61,000 (~142 MJ/kg).

2.   Flammability Potential

Certain physical properties or characteristics of chemicals express aspects of their flammability potential. Some of these properties are: flash point, fire-point, ignition temperature, autoignition temperature, specific gravity, vapor density, melting point, boiling point, flammable limits, and spontaneous ignition. Each of these properties will be discussed briefly to indicate its significance as a factor in flammability.

The flash-point is probably the primary indicator of the flammability hazard of a liquid. It is defined as the lowest temperature of the liquid that will produce an ignitable vapor at, or near, its surface. In a few instances, the flash point may also apply to solids such as camphor and napthalene that evolve flammable vapors. The knowledge of when a vapor can ignite or explode is particularly important because the liquid itself may not be flammable. In storage and transport of flammable liquids, the flash point is used as a flammability hazard indicator. The measurement of flash point is accomplished using an "open cup" or "closed cup" method. The open cup method gives slightly higher temperature values since the vapor-air mixture may be slightly less concentrated. Under actual chemical processing conditions, higher or lower ambient pressures and the oxygen available would be factors that would change the flash point.

The fire-point is the lowest temperature at which the vapor and air mixture will continue to burn in an open container when ignited. Usually the fire-point is several degrees higher than the flash-point. Both the fire-point and the flash-point are important factors when combating fires.

The autoignition temperature is the temperature of a material at which self-ignition and sustained combustion will occur in the absence of a spark, flame, or ignition source. It applies to a solid, liquid, or gas.

The specific gravity is a ratio of the weight of a substance compared to an equal volume of a standard substance. Solids and liquids are compared with water, and gases are compared with air. This characteristic is an important factor in combating fire. For example, most flammable liquids are lighter than water. A flaming liquid could easily spread a fire while carried by water.

Vapor density is the ratio of the density of vapor compared to air. The vapor density indicates where a flammable vapor will collect. For example, gasoline vapor, which is heavier than air, would settle in low areas, where it would become an ignition hazard.

The melting point in flammable materials indicates the temperature of transformation from a solid to a flammable liquid.

The boiling point is the temperature at which the vapor pressure in a flammable liquid equals the external pressure, or the formation of hazardous vapors.

Flammability limits indicate the range of vapor or gas and air mixtures that will burn when ignited. The upper and lower limits, expressed as percent concentration with air, represent the flammable or explosive range. Flammability limits may also be expressed for mixtures with other gases such as chlorine.

Spontaneous ignition occurs when materials at normal temperatures combine with oxygen in the atmosphere and start combustion due to the heat generated.

3.   Asphyxiation

Asphyxiation is a hazard or concern when there is an oxygen deficiency with operating personnel in an enclosed space. This can occur, for example, when oxygen or air is depleted or replaced by absorption or chemical reaction when there is combustion or fire. When the chemical composition of air decreases from the normal oxygen content of 21 percent by volume to 10 percent, dizziness and shortness of breath result. At about 7 percent oxygen, stupor comes into effect. At 2 to 3 percent oxygen, death occurs in a few minutes.

Respiration is also affected when the carbon dioxide composition of air is increased, even when the oxygen content is maintained. When the normal content of carbon dioxide in air is increased from the normal range of 0.02-0.04 percent by volume to 4.5-5.0 percent, breathing becomes extremely labored and almost unbearable in many cases. At about 25-30 percent carbon dioxide, gradual death may occur in a matter of hours.

C.   Explosion Hazard

An "explosion" when considered in terms of hazardous materials and processes has a broad concept. From the standpoint of an hazard, an explosion may be interpreted generally as an uncontrolled, rapid release of energy, either through chemical or physical reaction. Thus, an uncontrolled reaction due to such causes as the ignition of flammable gas-air mixtures, the ignition of combustible dust-air mixtures, decomposition of explosives, a nuclear detonation, and even the bursting of a pressure vessel, may all be included as potential explosion hazards. The rapidity of release of energy and the amount of energy in the form of heat, pressure, and radiation would be indicators of the severity of the hazard and the danger to personnel and property. Explosion characteristics vary depending upon the material and the conditions.

Deflagration and detonation are terms that indicate the propagating rates of combustion. Deflagration refers to the rate of an exothermic reaction that is less than the velocity of sound. Detonation

refers to a rate that is faster than the speed of sound. Low order
and high order detonations are a further indication of the destructive
nature and of the extremely rapid rate of increase in pressure. Detonation waves, when reflected, further increase the pressure and capability for destruction.

Detonability limits indicate the potential hazard range of flammable, combustible, or explosive type materials, and depend upon ignition and environmental conditions. Strong initiation will tend to
produce higher velocity detonations. In combustible gas or vapor-air
mixtures, both detonability limits and flammability limits may apply,
and in many cases indicate similar hazardous ranges. For vapors detonability limits are generally within the flammability limits.

Decomposition explosions may result from explosive materials or
other chemical compounds that react like explosives when they decom-
pose and release hot gases.

Nuclear explosions result from the tremendous energy produced
from the redistribution of protons and neutrons in the atomic nuclei,
in a "fission" or "fusion" process. In a fission process, the nucleus
is split into smaller parts, releasing energy and neutrons, which produce fission of more nuclei. In a fusion process, reactions result
from the fusion of hydrogen isotopes. The reaction of the fusion process, sometimes identified as a thermonuclear explosion, may result in
considerably more energy then in the fission process.

In a typical nuclear fission explosion in air, about 85 percent
of the energy is released as kinetic energy [5]. About one-half of
the total energy results in blast and shock, about 35 percent is in
thermal energy and about 15 percent is in nuclear radiation.

The thermal radiation from a nuclear explosion is large compared
to a conventional explosive. The intense heat and light rays will
cause skin burns and fires for some distance. Burn injury would be
quite high.

The biological effects of blast are discussed later in this chapter.

III.   HAZARDOUS MATERIALS

With the prolific expansion of the chemical industry, there have
been a wide variety of new chemical product applications. Compressed,
liquefied, and cryogenic gases are used extensively. In the aerospace
industry for example, advances have been made with high energy, liquid
propellants for rockets that combine highly reactive fuels and oxidizers. In other applications, new toxic pesticides, insecticides,
and herbicides that are essential in the food, pharmaceutical, and
agricultural industries, present hazards that require control in processing and in their use. In the plastics industry, some of the basic
materials have properties that are toxic, flammable, reactive, or explosive.

A.   Hazardous Materials in Industry

Although there is a great diversity of hazardous materials, they
can be categorized in terms of their dangerous effects and their

HAZARDOUS MATERIAL EFFECTS

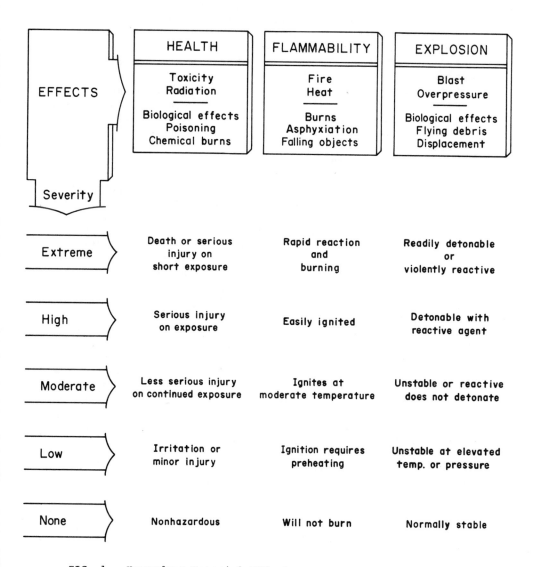

FIG. 1. Hazardous Material Effects.

potential hazard severity. Figure 1 is a simplified grouping of hazardous material effects in terms of dangerous properties related to health, flammability, and explosion. Hazard severity descriptors provide a correlation of severity and dangerous properties. The most hazardous or the "extreme" severity level indicates death or serious injury on short exposure to a health hazard. For flammable and explosive type materials, the descriptors indicate a rapid reaction, detonation or violent reaction.

In recent years, there has been an increasing effort to identify
known hazardous materials and their characteristics to provide guidelines to establish precautionary measures and occupational working
criteria. There has been considerable interest in this problem by industry, by fire protection activities, and by government agencies. A
number of publications are available on hazardous materials.

A ready reference on properties of hazardous chemical materials
in industry, including countermeasures has been prepared by Sax [2]
assisted by experts in the fields of toxicology, radiation hazards,
air pollution, food additives, and allergic disease. Authoritative
information is also presented on other important areas such as industrial air contamination control, respiratory protection, industrial
fire protection, reactor safeguards, storage and handling, and shipping regulations. In another reference, information on chemical and
process technology is presented in the form of an encyclopedia [6].

The National Fire Protection Association (NFPA) has published information on hazardous properties of chemicals in order to deal effectively with emergencies involving fires, spillages, and accidents in
transportation. In their publication which serves as a guide on hazardous materials [7], properties of flammable, toxic, explosive, and
chemically reactive materials are identified. In the "Fire Protection
Handbook" [8], several sections are devoted to hazardous materials including: flammable properties of gases, liquids, solids, and dust;
process hazards; storage and handling; and behavior of fire and explosion. The NFPA has also issued a series of volumes establishing National Fire Codes.

Two informative books on hazardous materials were prepared by
Meidl in 1970 relating to flammable [9] and explosive and toxic materials [10]. These references prepared as part of a Fire Science Series, provide a background on a variety of industrial materials.

The compilation of toxic substances is a continuing effort by the
National Institute for Occupational Safety and Health (NIOSH), U. S.
Department of Health, Education and Welfare. The Toxic Substances
List [11] is prepared as a reference for potential toxicity of chemicals in occupational areas. Also published by NIOSH are criteria documents for recommended standards on occupational exposure to specific
hazards. In 1974 about twenty criteria documents had been published.
These included some of the following material and environmental hazards: ammonia, asbestos, beryllium, carbon monoxide, inorganic mercury, ultraviolet radiation, hot environments, and noise.

Threshold Limit Values (TLV) for about 500 airborne contaminents,
including toxicity and hazard statements, is an updated document prepared by the TLV Committee of the American Conference of Government
Industrial Hygienists [3].

The effects and emergency treatment of poisoning for some of the
chemicals found in the home and in industry are summarized for use as
a ready reference on poisons [12].

Hazards of compressed, liquified, and cryogenic gases are part of
the information provided in the Handbook of Compressed Gases [13].
Information is also provided on precautions, safe handling, shipping
containers, and transportation regulations. This publication is produced by the Compressed Gas Association, a nonprofit corporation whose

members represent producers, distributors, equipment manufacturers and leading chemical industry corporations.

Guidelines for handling hazardous chemical propellants used in rockets by the aerospace industry are contained in a National Aeronautics and Space Administration (NASA) Technology Survey [14] and in a three volume report by the Joint Army, Navy, NASA, Air Force (JANNAF) Propulsion Committee [15-17]. For example, such hazardous chemicals as liquid hydrogen, and highly reactive oxidizers such as liquid fluorine, chlorine trifluoride, nitrogen tetroxide, and hydrazines are discussed [17]. Not only are the hazards identified, but specific attention is given to: safety measures, the materials and equipment for transfer and storage; storage containers, building structures for storage, systems and equipment cleaning, metal components, plastic components, emergency procedures, Department of Transportation (DOT) requirements, hazardous gas detection equipment, and education of personnel. Detailed regulations have been published by the Department of Transportation, Materials Transportation Bureau (Code of Federal Regulations 49), for the transportation of packaged, dangerous commodities.

B.   Biological Effects of Overpressure

Exposure to an explosion and its effects of blast and shock, are serious hazards that may cause injury and death. Body injuries from blast are primarily due to compression of the body, followed by decompression and shock waves through the organism. Other injuries from blast may result from displacement of the body, flying debris, and damage to the auditory canal. When comparing blast injuries from a nuclear explosion or a conventional high-explosive Glasstone [5] indicates that for a given overpressure, a nuclear source is more effective in producing direct blast injuries. The human body is more sensitive to the relatively longer pressure pulse in a nuclear explosion.

Biological effects of overpressure have been of continuing interests for industrial safety and protective design. Tentative biomedical criteria for direct blast effects are given in several papers presented at a conference on "Prevention of and Protection Against Accidental Explosion of Munitions, Fuels, and other Hazardous Materials". Estimated blast effects on young adults are summarized by White [18] from other data [19 - 21]. The data are shown graphically to indicate maximum effective overpressure for lethality for a fast-rising air blast wave for time durations from 2 to 400 milliseconds and at sea level conditions (Figure 2). Estimated values for near threshold of lethality, for 50 percent lethality (LD 50), and 100 percent lethality (LD 100) indicate that short times of exposure require considerable higher pressures for lethality. For example, the LD 50 effective pressure value for 2 milliseconds exposure is (272-397 psi; ~18.8-27.4 bars) more than five times that at 400 milliseconds (52-72 psi; ~3.59-4.97 bars). At 3 milliseconds the LD 50 pressure value is about three times higher, and at 5 milliseconds is about two times higher than at 400 milliseconds. The overpressure or effective pressure values indicated may be the incident, reflected, or incident plus the dynamic pressure.

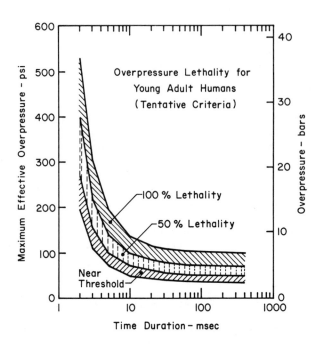

FIG. 2. Tentative criteria for lethality of young adult humans subjected to overpressures, as a function of time duration of the pulse.

Effects of overpressure on the ear when subjected to blast indicate that about 5 psi (~.34 bars) is the threshold pressure for eardrum rupture and damage to the middle ear [18-22]. The tympanic membrane or eardrum separates the external ear and the middle ear structure. An overpressure of 15 psi (~1.03 bars) will cause rupture of 50 percent of exposed eardrums. Hirsch [22] reviews the effects of overpressure on the ear and reports an accident involving a hydrogen-oxygen explosion. A 44 year old physicist was subjected to a fast rising overpressure estimated to be near 30 psi (~2.07 bars). Both eardrums were ruptured. They did not become infected and healed completely in several months. The audiograms, when compared with post rupture hearing, showed considerable recovery, but some high-tone hearing loss. In more severe cases of ear damage, there would be depressed accuity for all frequencies and the restoration of hearing would depend upon the amount of permanent damage. The proper use of ear defenders is recommended when the danger of blast injury exists.

IV.   SAFETY IN INDUSTRY

Safety encompasses a wide variety of knowledge in technology, in operational processes, and in preventative and protective procedures. In high pressure applications, particularly with hazardous materials,

the matter of safety and the prevention of accidents is a critical
factor and a continuing problem, particularly when new hazardous materials, advanced technologies and untried processes are involved. This
section will discuss the development of safety in industry and the
prevention of accidents.

A.   Occupational Safety

A coordinated emphasis on industrial safety began in the early
1900's. The National Safety Council, a national nonpolitical and nonprofit organization was founded in 1913, incorporated in 1930, and by
act of the 83rd Congress in 1953 was granted a federal charter [23].
Its main purpose was accident prevention.

The drive for safety in industry was accelerated during World War
II. During the large expansion of industrial plants, safety activities increased and encouragement was received from the federal government. Safety personnel were being trained. The philosophy of safety
and safety measures in industry became an integral consideration in
hazardous operations.

In recent years, safety and chemical hazards have become more
complex in the highly diversified chemical industry. Consider, for
example, the development of new toxic chemicals, coal-to-oil conversion processing, exotic but hazardous cryogenic rocket propellants,
and the expanded use of radioactive materials in medicine and in nuclear power plants.

Safety is important from the cost standpoint, to avoid accidents.
For example, in 1968, it has been estimated that occupational accident
costs in the United States was over 7.4 billion dollars [23].

It was also estimated that without a coordinated emphasis on
safety, the annual cost would have increased two- or three-fold.

Probably the most extensive legislation for safety in industry in
the United States is the Occupational Safety and Health Act (OSHA) of
1970 [23, 24]. The act became effective in April 1971. This act applies to most employers in industry, particularly those who are involved in a business affecting commerce. The act requires that employers provide an occupational environment that is free from recognized hazards that cause, or are likely to cause, death or serious
physical harm. The act requires compliance with safety and health
standards promulgated by the Secretary of Labor and first published in
the "Federal Register", May 29, 1971. Occupational Safety and Health
Standards, Part 1910, deals with General Industry. Subpart G, Occupational Health and Environmental Control, is concerned with air contamination, ventilation, occupational noise control, and ionizing and
non-ionizing radiation. Subpart H, Hazardous Materials, is concerned
with compressed gases, flammable and combustible liquids, explosives
and blasting agents, and storage and handling.

B.   Accidents in Industry

Although there is a continuous effort to avoid the dangers presented by hazardous materials and industrial operations, accident prevention is of increasing concern, particularly where high pressure is

TABLE 1

RELATIVE HAZARDS OF VARIOUS INDUSTRIES

INJURY RATES IN 1974*

	FREQUENCY RATE Disabling Injuries per 1,000,000 Man Hrs.	SEVERITY RATE Time Charges (Days) per 1,000,000 Man Hrs.
Aerospace	1.91	114 (60)[+]
Chemical	4.26	339 (79)
Rubber and Plastics	5.97	454 (76)
Petroleum	6.73	690 (103)
Fertilizer	7.89	682 (87)
Gas	9.08	357 (39)
Foundry	15.04	966 (64)
Food	18.77	750 (40)
Mining, Underground Coal	35.44[‡]	5,154[‡] (145)
All Industries	10.2	614 (60)

*"Accident Facts," 1975 Edition, National Safety Council, Chicago Illinois.

[+]Figures in perentheses show average days charged per case.

[‡]1972

involved. According to the National Safety Council, in 1974 the occupational deaths in the United States numbered 13,400, and occupational injuries numbered 2,300,000 [25].

A comparison of relative hazards of several industries is shown in Table 1. The frequency rate of disabling injuries in the chemical industry is 4.26 injuries per 1,000,000 man hours, or less than one half the rate of "all industry" of 10.2. The severity rate is about one half that of all industry. The average days charged per case in the chemical industry is slightly greater than that in all industry.

An assessment of accidents and their causes is important to prevent further incidents and to establish precautionary measures. The following examples of two hazardous materials, explosives and hydrogen, are used for illustration.

In his book on high explosives, Cook [26] indicates some of the fatalities that have resulted from major accidental explosions in previous years. It is pointed out that assessment of accidental explosions as well as theoretical and experimental studies have aided in establishing regulations for storing and transporting explosives.

Hydrogen has been considered extremely hazardous because of its wide limits of flammability when mixed with air [27]. Detonation occurs under confined conditions. Its low ignition energy has been of great concern in accident prevention.

In a review of 96 hydrogen mishaps [28] in an aerospace application, it was found that when hydrogen was released into the atmosphere, ignition of the hydrogen and air mixture occurred 62 percent of the time. Two predominant causes, about 40 percent of the ignitions, were due to electric shorts or sparks, and static charge. About 37 percent of the ignition causes were unknown. Further assessment indicated that operation, procedural, design, and planning factors were primary contributing factors, or about 86 percent of the total. A detailed breakdown indicated that valve leaks or malfunction, and leaking connections were some of the predominant system failures. Unsatisfactory materials and embrittlement were lesser factors. It should be kept in mind that these mishaps generally illustrate the causes of accidents with hydrogen, and apply to a particular application. When dealing with high pressure compatability of materials and hydrogen, embrittlement could be a more serious factor. This hazard is considered more explicitly in the following section.

C.   Hazards With High Pressure Working Materials

In high pressure chemical processes another hazardous area that should be considered is the compatibility of materials used, particularly for containment of pressure with the chemical product. Hydrogen and mercury, for example, are materials that may cause problems through metal failure. The use of beryllium, on the other hand, may pose health problems during its fabrication. The hazards of these working materials will be briefly reviewed. It should be noted that Chapter 7 discusses containment of high pressure more specifically.

1.   Metal Failure With Hydrogen

There have been numerous reports written and investigations are still continuing on the effects of hydrogen embrittlement on various metals. A review of over 200 reports dealing with hydrogen interaction with metal has been prepared by R. B. McLelland and C. G. Harkins [29].

Early data, about the 1940's, was directed toward hydrogen embrittlement of iron alloys. Empirical data were developed to establish design criteria for steel pressure vessels. High temperature hydrogen attack through formation of internal bubbles of hydrogen and hydrides, and low temperature hydrogen embrittlement were observed. The service life data of a number of steel alloys at various hydrogen temperatures and pressures was given later, in 1972, in a report prepated for the American Gas Association [30].

Aluminum and copper alloys are not affected by low temperature hydrogen embrittlement. They do not readily chemisorb hydrogen or form metallic hydrides.

TABLE 2

MATERIALS GROUPED BY SUSCEPTIBILITY
TO HYDROGEN-ENVIRONMENT EMBRITTLEMENT

(Source: H. Gray, Materials and Structures
Division, NASA Lewis Research Center [30])

Extremely Susceptible to Attack	High strength steels Maraging 410, 440C, 430F H-11, 4140 17-4PH, 17-7PH
Severely Attacked	Nickel and nickel alloys Nickel 200, 270 Inconel 625, 70, 718 Rene 41 Hastelloy X Waspalloy, Udimet 700
	Titanium alloys Ti-6Al-4V Ti-5Al-2. 5Sn
NOTE: The performance of pipeline steels of the X42. . .X65 series with 0.26-0.31% C is anticipated to be in this grouping.	→Low-strength steels Armco Iron, HY-100 1042, A-302, A-517 Cobalt alloys S-816, HS-188
Moderately Attacked	Metastable stainless steels 304L, 305, 310
Not Attacked	K-Monel Be-Cu Alloy 25 Pure titanium Aluminum and copper alloys Stable austenitic stainless steels 316, 347, A-286

Note that typical low-strength steels (as used in pipelines) are listed
as "severely attacked." However, it should be emphasized that the
actual conditions for hydrogen environment embrittlement should be
further examined in order to determine whether or not they exist in
typical pipeline operation.

High temperature hydrogen will not attack aluminum. However,
high temperature hydrogen will react with oxygen bubbles in copper to
form internal steam bubbles. This can be controlled by removal of
oxygen impurities.

The interest in hydrogen reaction in metal failure is continuing,
particularly because of the increased emphasis of hydrogen in rocket
propellant fuels and aerospace application. Of recent concern are
some failures experienced in high pressure storage of hydrogen gas in

welded vessels, and the phenomenon called hydrogen "environment" embrittlement. This phenomena relates to the effect of hydrogen on the "surface" of metals. The earlier data that were available were concerned primarily with "internal" hydrogen embrittlement, which was introduced during the electrolytic charging or pickling process of manufacturing steel alloys, and is commonly referred to as "classical hydrogen embrittlement".

The surface embrittlement effect is characterized as a very rapid process. Once a crack is initiated, or nucleated, at the surface, its subsequent propagation leading to failure, is very rapid. Test specimens under tensile load can fail very rapidly after hydrogen is introduced (e.g., periods less than ~1 sec.).

Furthermore, a number of studies indicate that attack is particularly severe from very pure hydrogen. Cases are known of commercial equipment failure after very brief exposure to such pure hydrogen, having successfully operated for many years under the same conditions with impure hydrogen. Oxygen as an impurity at levels of ~.06% and lower can completely inhibit hydrogen attack.

As interesting table showing the susceptibility of various materials to hydrogen-environment embrittlement has been compiled by NASA, Lewis Research Center, Cleveland, Ohio, and is reproduced as Table 2 of the present chapter. Further details may be obtained from the references cited [30].

Hydrogen embrittlement may also result under conditions of corrosion attack, such as in natural gas pipe lines. Intergranular embrittlement is caused by "atomic hydrogen", formed at the corrosion locations, that penetrates into the internal structure of the steel. Molecular hydrogen at normal operating temperatures and pressures would not have this effect.

2.   Hazards of Mercury

Mercury is widely used for low pressure measurement. Also, at high pressure, mercury is used as an oil and gas separator, for example, to provide contamination-free gas compression [31]. The properties of mercury, its negligible vapor pressure and low absorption of gases are desirable for this purpose. The toxicity of mercury, however, is a serious hazard.

The toxic properties of mercury have been known for years. When mercury vapor is inhaled or ingested, it is a cummulative poison. It may cause serious injury to the central nervous system upon exposure. In a survey made of mercury vapor content of air in a number of laboratories [32], it was found that the most effective way of reducing mercury vapor content was by proper ventilation. This reference contains a useful discussion of hazards arising from mercury.

It should be remembered that when mercury is contained at high pressure then vaporization may occur readily when it leaks. Ventilation equipment should be capable of removing such vapor from the immediate vicinity of the high pressure apparatus before it disperses into the room. As secondary measures, floors should be smooth and made of non-absorbing materials and the room temperature should be controlled, particularly to avoid high mercury vapor pressures if an accidental spill occurs.

Another characteristic of mercury is its "liquid metal embrittlement" effect on other metals. This phenomenon occurs when normally
ductile solid metals become brittle when exposed to certain liquid metal environments. The adsorption-induced embrittlement is considered
a special case of brittle fracture. It is not dependent upon penetration of the liquid or corrosion type process. Fracture behavior occurs below the critical stress value in air.

In a review of investigations of liquid metal embrittlement by
Kamdar [33], it is indicated that a liquid mercury environment will
cause embrittlement of aluminum, bismuth, copper, iron, silver, tin,
titanium, and zinc.

3.   Hazards of Beryllium

Beryllium is used both in its pure metallic form, and as an alloying agent in Be-Cu alloys. Special apparatus constructed of beryllium is transparent to X-rays, and beryllium metal powder is used as a
rocket fuel component. Beryllium-copper is widely used in a low-temperature environment and in situations where low magnetic susceptibility is a requirement.

Although beryllium has such desirable properties as high
strength, low weight and high melting point, it is extremely toxic.
Finely divided beryllium powder will burn. As a health hazard, low
concentrations of beryllium can result in fatal poisoning when inhaled
as a dust or fumes [9, 10]. Special safety measures are necessary
while machining or processing beryllium and on beryllium-copper alloys, particularly those with higher beryllium content (e.g., above
5%).

REFERENCES

1. K. E. Weale, Chemical Reactions at High Pressure, E. and
   F. N. Spon Ltd., London (1967).
2. Dangerous Properties of Industrial Materials, (N. I. Sax, ed.),
   Third Ed., Reinhold Publishing Corp., New York (1968).
3. Documentation of the Threshold Limit Values for Substances
   in Work Room Air, Third Ed., American Conference of Governmental Industrial Hygienists, 1014 Broadway, Cincinnati,
   Ohio 45202.
4. Shelter Design and Analysis, Vol. 1, Fallout Radiation
   Shielding, Defense Civil Preparedness Agency, TR-20 (Vol. 1),
   June 1976, U. S. Government Printing Office, Washington, D.C.
5. The Effects of Nuclear Weapons, (S. Glasstone, ed.), U.S.
   Government Printing Office, Washington, D.C., (1962).
6. Chemical and Process Technology Encyclopedia, (D. M. Considine,
   ed.), McGraw-Hill Book Co., New YOrk (1974).
7. Fire Protection Guide on Hazardous Materials, Fourth Ed.,
   National Fire Protection Association, Boston, Mass., (1972).
8. Fire Protection Handbook, 13TH Ed., (G. H. Tryon, ed.), National
   Fire Protection Association, Boston, Mass., (1969).

9. J. H. Meidl, Flammable Hazardous Materials, Glencoe Press, Beverly Hills, Calif., (1970).

10. J. H. Meidl, Explosive and Toxic Hazardous Materials, Glencoe Press, Beverly Hills, Calif. (1970).

11. The Toxic Substances List (H. E. Christensen and T. T. Luginbyhl, eds.), 1974 Ed., National Institute for Occupational Safety and Health, U.S. Government Printing Office, Washington, D. C.

12. V. J. Brookes and M. B. Jacobs, Poisons, D. Van Nostrand Co., Inc., New York (1958).

13. Handbook of Compressed Gases, Compressed Gas Association, Inc., Reinhold, New York (1966).

14. Handling Hazardous Materials (D. R. Cloyd and W. J. Murphy, eds.), National Aeronautics and Space Administration, Technology Survey NASA SP-5032, 1965, U. S. Government Printing Office, Washington, D. C.

15. Hazards of Chemical Rockets and Propellants Handbook, Vol. 1, General Safety Engineering Design Criteria, AD 889763, May 1972, National Technical Information Service, Springfield, Va.

16. Hazards of Chemical Rockets and Propellants Handbook, Vol. 2, Solid Rocket Propellant, Processing, Handling, Storage and Transportation, AD 870258, May 1972, National Technical Information Service, Springfield, Va.

17. Hazards of Chemical Rockets and Propellants Handbook, Vol. 3, Liquid Propellant Handling, Storage and Transportation, AD 870259, May 1972, National Technical Information Service, Springfield, Va.

18. C. S. White, Annals New York Academy Sciences, 152, Art. 1, 89 (1968).

19. D. R. Richmond, E. G. Damon, E. R. Fletcher, I. G. Bowen and C. S. White, Annals New York Academy Sciences, 152, Art. 1, 103 (1968).

20. I. G. Bowen, E. R. Fletcher, D. R. Richmond, F. G. Hirsch and C. S. White, Annals New York Academy Sciences, 152, Art. 1, 122 (1968).

21. C. S. White, I. G. Bowen and D. R. Richmond, Biological Tolerance to Air Blast and Related Biomedical Criteria, US AEC Report CEX-65.4 (1965).

22. F. G. Hirsch, Annals New York Academy Sciences, 152, Art. 1, 147 (1968).

23. Accident Prevention Manual for Industrial Operations, (F. McElroy, ed.), Seventh Ed., National Safety Council, Chicago, Ill., (1974).

24. D. Peterson, The OSHA Compliance Manual, McGraw-Hill Book Co., New York (1975).

25. Accident Facts, 1975 Ed., National Safety Council, Chicago, Ill.

26. M. A. Cook, The Science of High Explosives, Reinhold Publishing Corp., New York (1958).

27. B. Lewis and G. von Elbe, Combustion, Flames and Explosion of Gases, Academic Press, Inc., New York (1961).

28. P. M. Ordin, <u>Review of Hydrogen Accidents in NASA Operations</u>, National Aeronautics and Space Administration, NASA TMX-71585, 1975, U. S. Government Printing Office, Washington, D. C.

29. R. B. McLellan and C. G. Harkins, <u>Materials Science and Engineering</u>, <u>18</u>, 5 (1975).

30. H. Gray, "Potential Structural Material Problems in a Hydrogen Energy System", National Aeronautics and Space Administration, NASA TMX 71752 (1975), U.S. Government Printing Office, Washington, D.C.

31. J. Paauwe, <u>A New Mercury-Piston Gas Compressor for Contamination-Free Gas Compression for Pressures up to 7,000 bars (100,000 psi)</u>, 59th Annual Meeting, American Institute of Chemical Engineers (1966).

32. M. Shepard, S. Schulman, R. H. Flinn, J. W. Hough and P. A. Neal, <u>Journal of Research of the National Bureau of Standards</u>, <u>26</u>, No. 5, R.P. 1383, U. S. Government Printing Office, Washington, D. C. (1941).

33. M. H. Kamdar, <u>Progress in Materials Science</u>, Vol. 15, (B. Chalmers, J. W. Christian, T. R. Massalski, eds.), Pergamon Press, Oxford (1973).

Chapter 5

HIGH PRESSURE COMPONENTS

Jac Paauwe

Naval Surface Weapons Center
White Oak Laboratory
Silver Spring, Maryland 20910
and
Laboratory for High Pressure Science
Department of Chemical Engineering
University of Maryland
College Park, Maryland 20742

Ian L. Spain

Laboratory for High Pressure Science and
Engineering Materials Program
Department of Chemical Engineering
University of Maryland
College Park, Maryland 20742

I.   INTRODUCTION

A high pressure system generally consists of many parts. A typical system is shown in Figure 1. Rather like an electrical circuit it is important for the designer to have a detailed knowledge of all parts and then to connect them into a system in which individual parts are compatible, and in which the whole performs a desired function.

Two main parts of the typical system are the pump or compressor and the pressure vessel, dealt with respectively in Chapters 6 and 7. Another important part is the pressure measuring device, discussed in Chapter 8. In a system such as that shown in Figure 1 it is important to have the ability to measure the pressure not only in the autoclave, but also in each pressure stage, as well as in the hydraulic drive to the compressor, pump or intensifier.

There are many other important parts in the system, such as tubing, connectors, valves, seals, safety devices, electrical lead-throughs, windows, etc. These components will be the subject of this chapter.

At first sight it may appear that such components are too simple to be the subject of a separate chapter. However, it is stressed that components for high pressure service are highly specialized and not "just a branch of general plumbing". The work-crew handling a high pressure system should be specially trained in the assembly and maintenance of high pressure components. Small errors can lead to leaks, or even failure of components. It is hoped that this chapter will be of assistance to supervisors and technicians in giving basic information about such components and illustrating some common errors.

Today, several companies produce specialized equipment for high pressure systems. These parts are designed and manufactured under strict codes and standards. Several companies have been in business for over fifty years and have experienced the development and growing pains of high pressure technology. In most situations it is probably best to use items manufactured by these companies. A listing of companies manufacturing pumps, compressors and components is given in reference [1] and in the text of this chapter.

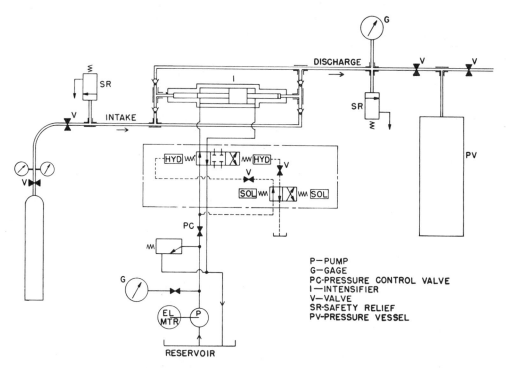

FIG. 1. A typical high pressure system showing a photograph
of pump- or compressor-system and a typical flow diagram. (Figures
courtesy of Harwood Engineering Company, Walpole, Mass.)

In the bibliography at the end of the chapter several books are listed in which high pressure components are reviewed. Valuable information can be obtained from manufacturer's catalogues and it is recommended that the practitioner obtains and studies these.

In this chapter, it is impossible to do complete, or even partial, justice to such a large subject. The authors hope that general principles will be addressed, and illustrations given of some of the main types of high pressure component. In showing equipment from one manufacturer or another, no preference is being expressed for the particular product, nor are any claims made about the equipment shown.

II.   STATIC SEALS

High pressure components are of many varieties, but basically their operation depends on the seal which is used. There are many different types of seal but a classification can be made on the basis firstly of application, secondly of principle, and thirdly of pressure range. The temperature range over which a seal can be used is largely dependent on its materials of construction, and will not be considered as a separate classification.

The main applications of seals are:

(i)   Tubing Seals

These seals are now commercially available for a wide range of tube sizes and pressure ranges. It is only in exceptional circumstances that custom designs are required. General seals for tube-fittings are treated in section II-A, B.

(ii)   Seals for Vessel Closures

Seals have been designed for a wide variety of vessels operating over wide ranges of pressure and temperature under extreme conditions, such as corrosive environments. Apart from sealing the vessel against leaks, important design criteria include the ease with which the closure and seal can be removed, and whether or not the same seal elements can be re-used for many pressure cycles and removals of the closure. Closure-seals are discussed in section II-C.

(iii) Moving Seals

The main use of these seals is for compressors, although other applications include rotary stirrers for autoclaves, specialized applications for experiments such as mechanical tensile tests at high pressure (see Chapter 14). Moving seals will be covered in the following chapter.

(iv)   Special Seals

Examples are seals for electrical leads and thermocouples (section II-D), and optical windows (section II-E). In basic design principle these seals operate in the same way as tubing and closure seals.

When reviewing different seal designs it is useful to recognize three basic seal principles:

(i)   The gasket, either confined or unconfined. This is the oldest, simplest and least expensive seal. It cannot be used to very high pressure (e.g. $\leqslant 1000$ bars (100 MPa or ~15,000 psi)).

(ii)   The self-sealing design or pressure-actuated seal. In this case the sealed pressure acts to produce a proportionally higher

pressure in the gasket, so that the seal is virtually leak-proof and
gives a better seal the higher is the pressure. The upper pressure
limit is controlled by the plastic deformation of construction materials, but uses are known to 50 kb (5 GPa or ~750,000 psi) (see Chapter
11). This design is attributed to P. W. Bridgman and many different
variations of the basic principle are known.

(iii) Designs based on combined elastic loading and plastic deformation of material. One part of the seal is deformed by another
part of higher hardness. The seal contact is along a line and leaks
only when pressure is high enough to elastically distort the seal until the force between the mating parts is reduced to a critical level.
Examples of seals using this principle are the cone-type and lens-ring
tubing seals.

Many seals in fact work by a combination of the above principles
and several examples will be discussed in the following sections.

One last general comment is in order concerning the type of fluid
to be contained. Liquids are much more readily sealed than gases, because of their higher viscosities. With the light elements helium,
hydrogen and neon, for example, small atomic or molecular size also
becomes an important factor. Thus, a seal which is satisfactory up to
10,000 bars (1 GPa or ~150,000 psi) for alcohols may not seal helium
at any pressure. Operating pressure ranges for the seals discussed in
this chapter are for normal liquids, such as oils, alcohol, etc. A
specialized literature may be consulted for sealing the light gases
[2, 3, 4].

A.   The Standard Cone-Type Seal

Most tubing seals work on the basic principle that two parts with
different angle or shape are tightened together to form a small line
or ring contact. The materials are of different hardness, allowing
one part to slightly deform. The force with which the parts are held
together is larger than the force exerted by the hydraulic fluid tend-
ing to separate the two parts.

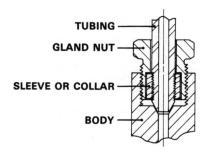

FIG. 2. Union or standard connector.

The most widely used high pressure tubing connector is the union-
or standard connection, shown in Figure 2 with a cone-type, metal-to-
metal seal. Other seal types are discussed later. As shown in Figure
2, the connection consists of the following parts:

    (i)   the tubing or male part,
    (ii)   the sleeve, ferrule or collar,
    (iii)   the gland-nut and
    (iv)   the opening, or female part.

The gland-nut screws into the body with a right-handed thread, while
the collar must be screwed onto the tubing with a left-handed thread.
If a right-handed thread is used for both parts, the seal self-loosens
when pressure is applied. Furthermore, when tightened, the collar may
turn with the gland-nut, which may result in an incorrect location of
the collar on the tubing.

    As shown in Figures 3a, b, c, there are several variants of the
basic design. Design (a) is undoubtedly the strongest since the collar is not recessed into the gland-nut. In design (b) there is a potential area of high stress concentration at the root of the recess.
Presumably design (c) is stronger because of the 45° angle on the col-

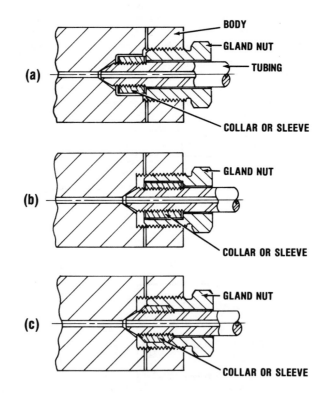

FIG. 3. Standard fittings for high pressure tubing, using
cone-type, metal-to-metal seals. (a) Collar abuts onto the end
of the gland-nut; (b) Collar is recessed into the gland-nut; (c)
Collar is recessed into the gland-nut, but the surfaces are cut at
45° to reduce stress concentration.

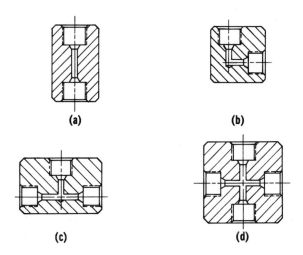

(a)   **(b)**

(c)   **(d)**

FIG. 4.   Illustration of some standard connectors.
(a) Straight coupling, (b) elbow, (c) tee, (d) cross.

lar. A potential disadvantage of design (a) is that, for a given tube
size, the bore is deeper, which may lead to weakening of the body unless extra material is allowed for.   This design is used for commercial equipment of the highest pressures (14kb; 1.4 GPa; ~200,000 psi).

Tubing connectors of various types are available, as illustrated
in Figure 4, notably the straight coupling, elbow, tee and cross. A
closure plug can be incorporated easily (Figure 5a) while Figure 5b
illustrates how an adaptor can be made for attaching smaller tubing to
a female fitting designed for larger diameter tubing. A simple union
such as that shown in Figure 5c is very useful when connecting two
pieces of tubing with limited mobility. It is made of two parts so
that the outer body can be slid over the inner after the gland-nuts
are unscrewed.   Thus, a length of tubing can be conveniently removed
or replaced without disassembly of larger parts. Bulkhead fittings
(Figure 5d) are also available in many sizes. Couplers for fittings
of different sizes are also available (Figure 5e). Swivel joints can
be obtained with which tubing can be swung through an arc while still
at high pressure (Figure 6). However, their use is generally restricted to fluids below about 2 kb (0.2 GPa or ~30,000 psi).

It is to be noted that fittings made by different companies may
use the same thread sizes but still differ in minor, but important
ways. For example, sleeves may be of different design, so that the
sleeve of company "A" should not be used with the gland-nut of company
"B" (Fig. 7a) or vice versa (Fig. 7b). The correct combinations have
been shown in Figures 3a, b, c. Even threads which are nominally the
same may not be made to the same tolerances by different companies.
Note also, that a manufacturer will guarantee his own product but cannot and will not guarantee parts from another manufacturer.

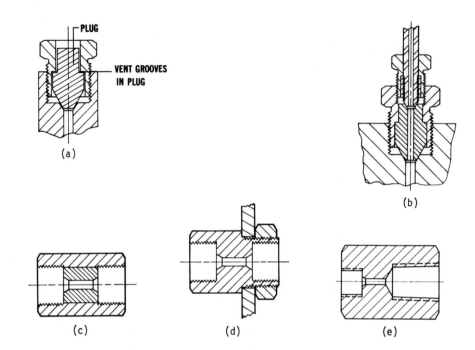

FIG. 5.   Standard connections: (a) closure, (b) adapter,
(c) straight coupling (union) in which outer piece can slide
over tubing, (d) bulk-head fitting, (e) coupler for fittings
of different sizes.

Another factor overlooked when fittings from different companies
are interchanged is that the vent-hole can become blocked. Note that
in the combinations shown in Figures 7a and b, the natural vent paths
along the tube are blocked. This may lead to serious consequences, if
not corrected. For instance, if pressure builds up behind the gland-
nut of a standard fitting for 1/4 inch (6.35 mm) tubing, the area over
which the pressure acts is increased to ~1 inch^2 (~6.45 cm^2). Without
leak, the pressure normally acts over an area of only ~0.028 inch^2
(~0.18 cm^2). At a pressure of 1,000 bars (~14,500 psi) the force to
be countered by the gland-nut is only 400 lb$_f$ (~182 kg$_f$, or 1,782 N).
If a leak develops this can increase to 14,500 lb$_f$ (~6,600 kg$_f$,
~64,600 N), which may be sufficient to damage the gland-nut, and even
blow it off. In several years experience the only explosive decompression of a high pressure (~13.5 kb, 1.35 GPa, ~200,000 psi) system
using helium as a pressure transmitting medium at the University of
Maryland occurred in a pressure gauge in which the cap retaining the
cone seal at the end of the Bourdon tube was not fitted with a vent-
hole or groove. Leaking gas enlarged the end cap fitting until it
blew off.
      The cone-seal fitting is simple and inexpensive. It is usually
adequate for most needs provided that the tubing cone is made correct-

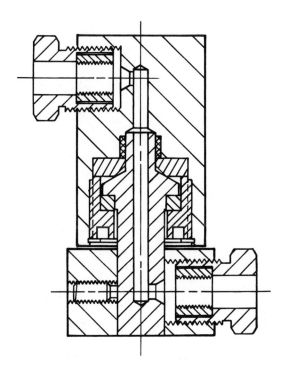

FIG. 6.   Swivel-joint assembly.

FIG. 7.   Two incorrect combinations of fittings.   In both
cases incorrect sleeves are matched with gland-nuts.   Correct
combinations are shown in Figure 3.   Note how the vent path can
become blocked.

ly and the fitting is assembled properly.   As shown in Figure 8a and b
the included cone angle on the male part must be slightly smaller than
that of the female (typically 58° and 60° respectively).   A line contact is first formed at the tip of the male part and, as the fitting
is tightened, the tube plastically deforms, enlarging the contact area.

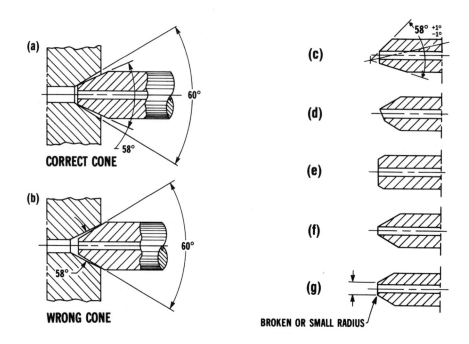

FIG. 8. Figures illustrating typical errors in cutting
tubing cones. (a) correct design, (b) angles interchanged, (c)
eccentric cone, (d) tubing not square, (e) cone too short, (f)
cone too long (insufficient edge), (g) correct cone.

TABLE 1

Torque Required to Tighten Standard High Pressure Fittings

High Pressure Connection inches (mm)	Tubing Size inches (mm) O.D.	I.D.	Torque in ft-lb$_f$ (N-m) 20,000psi (~138MPa)	30,000psi (~207MPa)	60,000psi (~414MPa)
¼" (~6.35)	¼" (6.35)	1/16 (~1.59)	5-15 (~7-20)	5-20 (~7-27)	8-20 (~11-27)
3/8" (~9.53)	3/8" (~9.53)	1/8 (~3.18)	10-30 (~14-41)	10-35 (~14-47)	15-45 (~20-61)
9/16" (~14.3)	9/16" (~14.3)	3/16" (~4.76)	25-60 (~34-81)	30-70 (~41-95)	40-100 (~54-136)
9/16" (~14.3)	9/16" (~14.3)	5/16" (~7.94)	35-75 (~47-102)	——	——

A most common fault is the overtightening of the seal. This can lead to partial, and eventually complete, closure of the inner bore of the tube. Also, sharp burrs can form and break off, thus acting as a potential hazard for other parts of the system. Correct torque for fittings of different sizes and pressure ranges is given in Table 1.

Typical errors in the preparation of male cones are shown in Figures 8c, d, e, f, while the correct shape is shown in Figure 8g.

Although the cone can be easily machined onto straight tubing with a lathe, it can be cut onto the end of bent tubing with a coning tool (Figure 9a). However, the coning tool is best used with straight tubing. The tools used for coning tubing are easy to use, and prepare a correct cone when handled in the following way:

(i)   Cut tubing to the proper length and cut the tubing square. Remove the burrs.

(ii)   Check the tool for the correct size tubing and check the cutter.

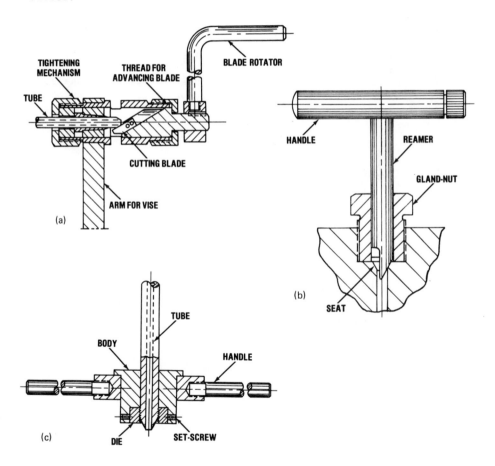

FIG. 9. Illustration of three tools used for preparing cone-type fittings. (a) coning tool, (b) reamer, (c) threading tool.

(iii)   Place the tool in a vise and adjust.

(iv)   Slide the tubing into the tool and tighten.

(v)   Lubricate, turn and feed.

(vi)   Finish (turn tool) cone without feed, to assure a smooth surface, with the correct angle (58°).

(vii)   The thread should be cut after the cone is finished. This is usually done by hand with a threading tool which holds the tubing straight while the thread is being cut (Fig. 9c) although a simple thread-cutting machine is also available. The left-hand thread should be smooth with a radius on top and bottom, or ground, of the thread. Also a sharp undercut at the end of the thread should be avoided. Throughout the cutting operation proprietary cutting fluids should be used.

For many types of tubing a stronger thread can be obtained by rolling. In this way the material may be strengthened by work-harden-ing (see Chapter 9). Thread rollers are commercially available but, before applying threads in this way, the tube manufacturer should be consulted. Some materials might be embrittled to too great an extent by this operation.

Most suppliers of coning tools or forming tools will furnish in-structions with the tool. For the cone seat a reamer is available and can be used to recondition the seat, see Figure 9b. In using the cone seal it is also important to position the sleeve or collar correctly.

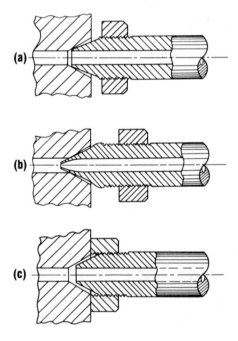

FIG. 10. Illustration of two common errors encountered in cone-type fittings. (a) Correct assembly, (b) thread too long and collar too far on tube. Note the pinch-off in the I.D. (c) Collar not sufficiently screwed onto tube; fitting seats on collar.

The sleeve with left-hand thread is screwed onto the threaded tubing. When screwed as far as possible onto the tube, between one and one and a half threads of the tubing should show from the cone end, Fig. 10a. If the thread is cut too long and the collar is screwed on too far, the tubing can mushroom (buckle) and the inside diameter can close. In addition, the gland-nut will only have a few threads of engagement which can cause a hazard (Fig. 10b). The total load on the gland-nut can be too high for the few engaged threads so that the gland-nut can be blown off. Care should therefore be taken to cut the threads to the correct length. When the sleeve is not screwed on far enough, the sleeve instead of the cone will "bottom" and there is no seal at all (Fig. 10c).

Finally, when all cutting and threading is accomplished, all fittings and tubing should be thoroughly cleaned. Cleanliness in working with high pressure apparatus cannot be over-emphasized.

B. Other Tubing Connectors and Seals

1. The Lens-Ring Seal

The principle of the lens-ring seal is shown in Figure 11. A straight coupler using the lens-ring seal is shown in Figure 12. Us-

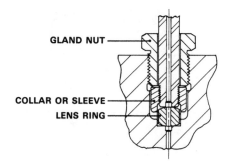

FIG. 11. Schematic of the lens-ring fitting.

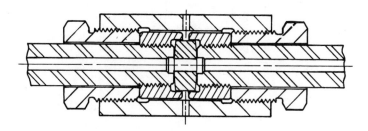

FIG. 12. A straight coupling using a lens-ring connection.

ually the lens-ring is made from a hard or heat-treated material. The seal is again made by line contact, and, when the parts are made correctly, the seal can be used many times without problems related to collapse of the inner diameter, a problem encountered with cone-type seals. After many uses, the seal area may have to be cleaned and the lens-ring polished.

The lens-ring seal is more expensive than the cone seal, but is probably a better seal for many applications. However, if the fitting is to be subjected to temperature cycling and is to seal leak-prone high pressure fluids such as helium, then a cone-fitting is to be preferred. The lens-ring seal can seal fluids up to 20 kb (2 GPa, ~300,000 psi).

2. The "Bite" Fitting

Bite fittings are popular connections for pressure applications up to 300 - 700 bars (30 MPa - 70 MPa or ~4,500 - 10,000 psi). They work on the principle that deformation of the ferrule takes place under force of the gland-nut so that the ferrule bites into the tubing, thus forming the seal and holding the fitting in place (Figure 13). These fittings need no special tools and no preparation of tubing ends, but should not be used in systems with temperature changes or vibrations.

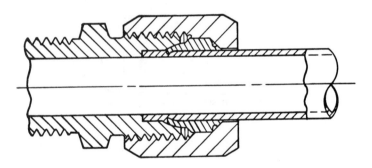

FIG. 13.   Illustration of a "bite" fitting.

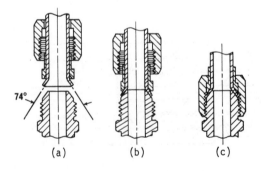

FIG. 14.   The "flare" fitting. (a) Pre-assembly, (b) in position for assembly, (c) assembled.

3.   The Flare Fitting

The simple and inexpensive tubing connection shown in Figure 14
can be used in a system with a maximum pressure of ~200 bars (20 MPa
or ~3,000 psi). The fitting however must be inspected each time it is
used, as frequent opening and closing of the fitting weakens the
flare, and eventually causes the flare ring to become separated from
the tube.

C.   Seals for Vessel Closures

1.   The O-Ring Seal

The sealing property of the O-ring is based on an initial squeeze
of the O-ring in the assembly. The depth of the groove for the O-ring
must be slightly less, for example 10%, than the diameter "W" of the
O-ring (Figure 15). This is referred to as the diametral squeeze on
the O-ring. As pressure is applied, the O-ring deforms further (Fig.
15d) assuring a leak-tight seal up to a maximum pressure (see discussion at the end of section C-2 and the following chapter).

It is difficult to give a useful pressure limit for O-ring seals,
because it depends on the design and manufacture of the parts. O-
rings are used in commercial equipment, for instance, to pressures of
13.5 kb (1.35 GPa, ~200,000 psi). However, in such designs the O-ring
serves the purpose of providing a low-pressure, or initial seal. At
higher pressure, other elements of the seal take over the main sealing
function.

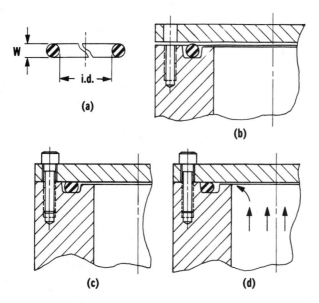

FIG. 15. A simple O-ring seal. (a) The O-ring, (b) pre-
tightening (note that O-ring diameter is greater than groove depth,
(c) tightened, (d) O-ring deforms further under action of pressure.

Most of the O-rings used today are made of natural rubber, elas-
tomers and other plastic materials. The main property of these materials is their low shear strength so that they deform readily against
the outer surfaces of their confining groove when pressure is applied.
O-rings are also made in such materials as stainless-steel, copper,
silver, gold, nylon, and teflon.

The design of an O-ring closure and the choice of the O-ring material are dictated by the maximum pressure, the temperature range,
whether high or low temperatures are to be encountered, corrosion, and
in general the compatibility of the O-ring material with the fluid.
For example, processing in the food industry has stringent regulations
and the material of the O-ring has to be accepted by the Food and Drug
Administration (F.D.A.).   Manufacturers of O-rings furnish excellent
handbooks in which information about the use, the choice of materials
and compatibility can be found. Design information of the O-ring
grooves is also given. Not all grooves are rectangular and specially
shaped O-ring grooves can be advantageous. Since the manufacture of
special grooves is more expensive, the use of these grooves should be
justified. It can be much cheaper to replace an O-ring frequently
with a new ring instead of trying to design a special seal. A damaged
O-ring should never be used and an O-ring which has been in use for
some time should be replaced by a new one.

The O-ring seal is a simple and effective design provided that it
can be confined. Up to about 300 bars (30 MPa, ~4500 psi) confinement
is not a serious problem, provided the O-ring material is selected
correctly and the cavity designed well. Also, the cavity must be machined and maintained carefully, so that there are no gaps through
which the O-ring can extrude, or scratches, burrs etc. which can nick
the O-ring. In a design such as that shown in Figure 15 the useful
pressure limit is set by the deformation of the closure which leads to
the extrusion of the O-ring between the body and closure (Figure 16b)
(see remarks in the following chapter).

Several designs which confine the O-ring have been developed, of
which a few representative examples will be discussed. Figure 17a
shows how the seal in Figure 15 can be modified and backed-up with a
simple ring made of soft metal, such as annealed 304 or 316 stainless-
steel, copper or brass. The height of the ring should initially be
slightly (e.g. 2-6%) larger than the height of the groove when closed
(H). This ring prevents extrusion of the O-ring into the gap which

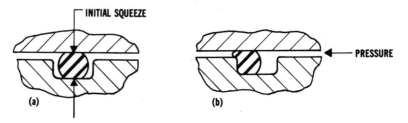

FIG. 16. Deformation of an O-ring. (a) Due to initial
tightening, (b) extrusion due to high pressure.

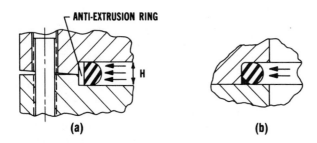

FIG. 17. Two forms of anti-extrusion rings. (a) simple
ring made of soft-metal, (b) triangular ring for corner.

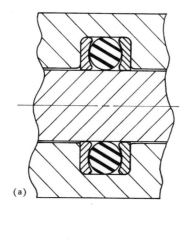

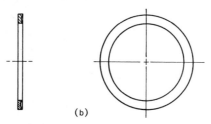

FIG. 18. Illustration of the Parbak ring.

inevitably opens between the closure and body at high pressure. An
alternative design using a back-up ring with triangular cross-section
is illustrated in Figure 17b. Both designs have been used to fluid
pressures over 10 kb (1 GPa, ~150,000 psi). A discussion of the sealing principle of this ring is given at the end of the next section.
An alternative back-up ring, the "Parbak" ring, is shown in Figure 18.
The ring is molded from hard, nitrile-rubber compound and will confine the O-ring for pressures up to approximately 650 bars (65 MPa or
~10,000 psi), depending on the hardness of the compound, or the clearance [5].

2. The Bridgman or Unsupported Area Seal [6]

Bridgman has stated that the key development which enabled him to reach higher pressures without leaks was his invention of the unsupported-area seal, now called the Bridgman seal. The principle of its operation is that pressure in the contained fluid produces an "overpressure" in the seal, which prevents leakage. Seals using this principle have been used up to 50 kb (5 GPa, ~750,000 psi).

The basic design is shown in Figure 19a. A thrust load, or force, F, is carried over by the head of the plug on a seal-ring or gasket having a smaller area, thus increasing the stresses in the seal ring higher than the working pressure in the vessel. The high stresses in the seal ring, which is solidly held in place, deform the ring towards the inside- and outside diameter, accomplishing a closure. The higher the pressure inside the vessel the larger the closure forces.

The "unsupported area" is equal to the area of the stem ($\pi d^2/4$). Thus, the total force, F_2, acting on the end-plug ($P_i \pi D^2/4$) is transmitted to an area $\pi(D^2-d^2)/4$ at the seal. Thus the pressure (assumed hydrostatic) in the seal is

$$P_{seal} = P_i \frac{D^2}{(D^2-d^2)} \qquad\qquad (1)$$

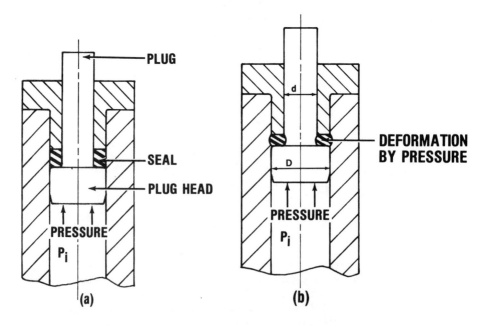

FIG. 19. The Bridgman unsupported-area seal. (a) Parts description. The plug head is sometimes referred to as the mushroom plug. (b) Illustration of the deformation of wall and plug stem that can occur due to high pressure in seal.

The load on the seal ring cannot be increased without limit. Too high a load results in the material of the seal ring being pushed into the wall of the vessel and/or into the stem of the plug. This can cause problems in disassembly and it may even break or crack the stem of the plug, see Figure 19b. This is known as "pinch-off" [6]. The two diameters D and d have to be chosen carefully and the stresses calculated [7].

Another problem encountered at higher pressure (2,000 bars (0.2 GPa, ~30,000 psi)) is the extrusion of the gasket material past the stem, outwards, or past the plug, inwards. To prevent this, anti-extrusion rings are used, made usually of a material such as bronze or copper. These extrude to a limited extent, and need to be replaced periodically. The frequency of replacement depends on operating conditions and should be determined by inspection.

Two examples of Bridgman seals are shown in Figure 20. The design in Figure 20b uses anti-extrusion rings. After assembly, initial sealing is obtained by pulling the plug up against the seal ring and spacer, held in place by the plug nut. While the plug nut needs only to be fingertight, pre-sealing should be done with a small wrench. Teflon or Nylon seal rings do not need much torque; but those made of copper, stainless-steel, silver, etc. need more. As soon as the pressure builds up, more effective sealing is obtained. It is possible that under pressure the nut on top of the vessel used for pre-tightening and obtaining the initial seal will become loose. These nut(s) should on no account be tightened under pressure. After the pressure drops back to atmospheric pressure the nut(s) usually are tight again; if not, light tightening is allowed. The nut also serves the useful purpose of an extractor for the seals. When the main retaining nut is unscrewed the seals are automatically withdrawn.

If the nut is tightened while the system is at high pressure, enormous stresses can be developed in the stem when the pressure drops, causing failure. For this reason, a lock-nut arrangement should be used, or other preventative arrangements made.

The Bridgman seal shown in Figure 20b uses, in one variant, all-metal sealing components. This enables the seal to be used to high temperatures, limited only by the strength of the construction materials at high temperature (see Chapter 9). A seal such as this can also be used to cryogenic temperatures, with the provision that construction materials in this case should not embrittle (see Chapter 9).

The unsupported area principle, in our experience, is the most reliable seal, of which the classical mushroom design (Figures 19, 20) is only one. Many other seals can be designed which incorporate this principle. For instance, the back-up ring in Figure 17b can be made from tough steel and its back surface chamfered to give an effective, unsupported area. Other variants will be met with in succeeding pages.

It is instructive at this point to discuss the sealing principle of the backed O-ring. At low pressure the diametral squeeze produces elastic stress in the ring which provides an initial seal. When the pressure is high enough to move the O-ring against the back of the groove, only part of the ring will be in contact with the back wall so that there is an "unsupported area" which produces an internal pres-

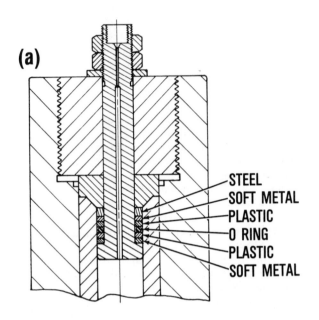

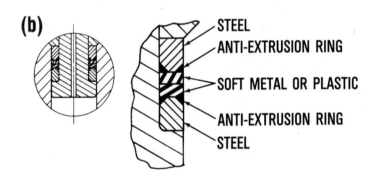

FIG. 20. (a) Assembly of typical Bridgman seal closure,
(b) alternative seal-ring assembly showing anti-extrusion rings.

sure in the ring. This seals it more effectively. However, if a
polymeric material is used for the ring, its shear modulus is very low
and the unsupported area will diminish to insufficient amounts to prevent leaks. At this point the O-ring acts as a contained packing.

If a back-up ring is used, it can be designed to have an unsupported area (Figure 17b) with the O-ring transmitting the force to it
from the high pressure fluid. Since the distortion of the back-up
ring tends to reduce its unsupported area, it must be made of a material strong enough to withstand the stresses generated in it up to the
maximum operating pressure. If a hardened steel crescent ring is used

(e.g. 4340 steel at ~55 Rc) then pressures as high as 10 kb can be contained by the seal.

3. The Wave Ring

The wave ring is a variation of the lens ring useful for closures of larger diameter. At first the lens ring was changed and made slightly hollow on the inside (Figure 21) to obtain a better self-sealing effect when the pressure inside the seal ring increases. As such, this seal utilizes the unsupported-area principle. Initial sealing is obtained by the elastic distortion of the ring against the body and closure plug. The pressure capability depends on the cross-section of the ring and the construction material. A ring with dimensions shown in Figure 21, for example, constructed from grade 4340 tool steel at ~40 Rc hardness, is effective up to ~4 kb (0.4 GPa, ~60,000 psi). Other examples of related designs are shown in Figures

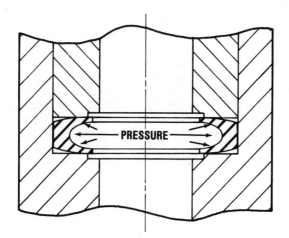

FIG. 21. Schematic of the wave-ring seal.

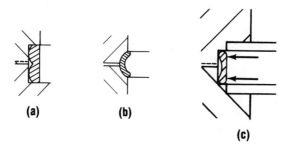

(a)   (b)   (c)

FIG. 22. Schematic of three variants of the wave-ring seal.

22a, b, c. The rings shown in Figures 21 and 22b are sometimes called
C-rings.

4.   Grayloc Seal Rings [8]

The Grayloc seal is a steel-to-steel seal-ring and is, more or
less, a modification of the wave ring (Figure 23). The seal ring has
a gasket with a flange on the inside diameter, extending on both sides
of the gasket. The flanges or lips of the "T"-shaped seal-ring are
tapered and the angle of the taper differs slightly, approximately 1°,
from the corresponding taper or angle of the body in which the seal is
used.

Initial seal is accomplished at assembly and the self-sealing effect results from the pressure build-up in the closure and inside the
seal ring. The lips are slightly deformed and a solid, steel-to-steel
surface seal is affected. It is noted that this seal again uses the
"principle of unsupported area" to give a seal at high pressure.

As with all seals, provision should be made that when the seal
leaks the fluid can escape (vent), so that pressure build-up over a
larger area will be avoided. The seal ring is made from high strength
steel, but is sufficiently ductile to withstand the slight deformation. The seal can be used for pressures up to 10 kbar (1 GPa or
~150,000 psi) and for high temperature service up to the maximum temperature for which the unit has been designed. To improve the sealing
efficiency and to prevent galling, the Grayloc seal is usually coated
with a material (such as a polymer) which can be made compatible with
the working conditions (temperature, corrosive fluids). When cleaning
the vessel with corrosive fluids a polyethylene seal-ring can be used
temporarily, but care should be taken to ensure that the ring is replaced before normal, high pressure operations are restarted [8].

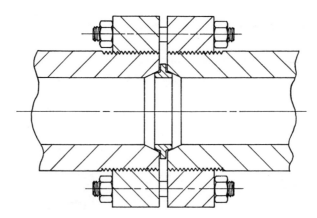

FIG. 23. Illustration of the Grayloc seal. A straight
coupler assembly is shown.

5.   Variations of Seal Designs

A large number of seal designs exist, some of them being easily
recognized as variants of designs discussed already. A few representative examples will be discussed in the following paragraphs.

a.   Gask-O-Seal [9], Figure 24(a).   Gask-O-Seal gaskets are
used for flanged closures and can be used for pressures up to about
270 bars (27 MPa or ~4,000 psi).

b.   Metal "V" Seals [9], Figure 24(b).   These   "Static Seals"
are usually made of a material with "spring" properties and the coating is usually of a soft and ductile material which adds to the sealing capabilities and guards against corrosion.

They are used in high vacuum systems and can withstand pressures
up to 1,000 bars (100 MPa or ~15,000 psi) depending on the seal diameter. They are useful for high temperatures, corrosive gases and liquids, and again use the unsupported area principle.

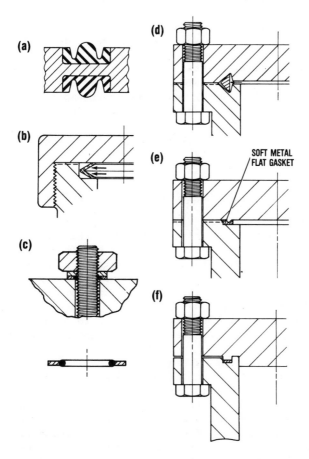

FIG. 24. Several seal designs: (a) Gask-O-Seal [9], (b) metal
V-seal [9], (c) Stat-O-Seal [9], (d) delta-ring, (e) and (f) soft-
gasket designs.

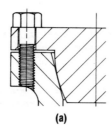 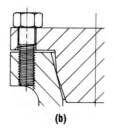

(a)   (b)

FIG. 25. Two variants of metal-to-metal closure seals for
pressure vessels. (a) curved male form, (b) straight edged form.

    c.   Stat-O-Seal [9], Figure 24(c). This one-piece seal is par-
ticularly used for sealing fasteners, although it can be efficiently
used in other closures. The rubber seal ring is bonded to, or mechanically attached to, a steel ring and confined with metal-to-metal contact.

    In Figure 24d, e, and f further design variations and modifica-
tions of seals are shown which are successfully used in static closures.

    Seals with steel-to-steel line contacts can be useful for many
applications. Two examples are illustrated in Figure 25. One uses a
curved male form (Fig. 25a), while the other employs two conical segments intersecting along a line (Fig. 25b). In both cases the vessel
can be opened and closed very easily without the necessity of replacing seals periodically. Also this design can be useful with corrosive
liquids, or in cases where it is important not to contaminate the
fluid.

D.   Electrical Lead-Throughs

    There are many different ways of obtaining a seal for electrical
leads. However, for applications involving high electric currents,
Bridgman's basic design [6] is still to be highly recommended (Figure
26). The electrical lead-through consists of a hardened metal anode
with a cone machined onto it. This is separated by a hollow, insulating cone from the seat. Initially the anode is sealed with the retaining screw, and, for gas systems, a thin layer of rubber-based ad-
hesive can be used on the mating surfaces to help with this initial
seal. At high pressure, the seal is self-tightened by virtue of the
unsupported area.

    The anode may be made of heat-treated steel (e.g. 4340 steel at
45 Rc), hardened beryllium-copper (e.g. Berylco 25 precipitation-hardened at 600°F (315°C)). Bridgman originally used pipestone (Lava-
grade A, wonderstone) for the insulating cones, but this material is
fragile, so that replacements are often used today, such as a hard
grade of Delrin™, mixtures of epoxy resin and alumina or magnesia pow-
ders (e.g. 1:1 mixture). In the latter case, the cones can be machined from rods or cast into the desired form using a mold. Delrin™

(a)   (b)

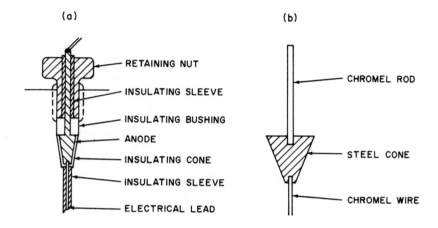

(c)

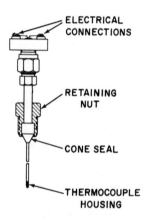

FIG. 26. Schematic of electrical lead-through and thermocouple designs. (a) Bridgman cone seal, (b) modification for
thermocouple. The case drawn is for a chromel lead. (c) Design
for inserting thermocouple into autoclave.

is also a good material to use for the insulating bushings in Figure
26.

Bridgman used a 16° included angle in his original design, but
other designs have used much larger angles (e.g. 90°). Alternatively
the included angle can be increased to 180° so that the insulating
cone becomes a flat plate.

These lead-throughs can readily be adapted to serve as thermocouple lead-throughs. For this purpose chromel/alumel, or platinum/
platinum - 10% rhodium thermocouples are recommended since the effect
of pressure on these thermocouple types has been most accurately

characterized (see Chapter 8). In all cases the thermocouple material
should be extended on both sides as far as the hardened metal anode-
cone (Fig. 26b). The temperature correction resulting from inevitable
temperature gradients in the anode-seal region is then reduced to an
insignificant amount. These leads should be brazed, if possible, with
a special low melting temperature alloy which does not appreciably
soften the hardened metal. Alternatively, soft-solder can be used.

Thermocouples can be introduced into an autoclave conveniently by
inserting the leads into a tube closed at one end (Fig. 26c). Many
designs are possible for sealing the tube into the vessel, and one
possibility is shown in this Figure. Usually the end of the tube is
sealed by welding. One advantage of this design is that the thermocouple is not subjected to pressure, so no correction is necessary.
An obvious limitation is the strength of the containing tube at elevated temperature (see Chapter 9 for comments on materials for use at
high temperature).

Many other methods of introducing thermocouple and electrical
leads into high pressure systems have been reported in the literature.
For example, varnished leads may be secured into high pressure tubing
using epoxy resin. Alternatively, thermocouple assemblies consisting
of an outer metallic sheath, ceramic insulation and multiple leads may
be impregnated with epoxy resin and fed into high pressure systems by
a number of methods [10 - 12]. A useful review is given in reference
[13] and many recent techniques have been reported in the Review of
Scientific Instruments, for such special applications as low temperature, high frequency, etc.

E.   Optical Windows

The basic principle used for sealing an optical window in a pressure vessel is shown in Figure 27a. The window is in the form of a
cylinder of optically transparent material (usually synthetic sapphire
obtainable from Linde Air Products, Synthetic Gem Division, Chicago,
Illinois). One optically flat surface (to a quarter of a wavelength)
is abutted onto a metal cylinder, also with an optically flat surface.
The two surfaces are glued together with a very thin layer of Canada
balsam. At high pressure the window and support cylinder are forced
together because of the presence of an unsupported area on the window.
This design is the basic design of Poulter [14], who used plate glass
for the window.

A window such as that shown in Figure 27a can be used to at least
10 kb with sapphire windows. If pressures up to only 0.5 kbar (50 MPa
~7,000 psi) are required, then an O-ring seal can be used as in Figure
27b. A schematic of a design used commercially up to 6 kb is shown in
Figure 27c. A retaining ring is used to hold the window and sealing
plug together and in good alignment [15].

Information about the range of wavelengths that can be transmitted through windows, and other properties, is given for a variety of
materials in reference [16], while some notes about the use of some of
them in high pressure apparatus is given in reference [17]. References [18 - 25] can be consulted for various designs of cell.

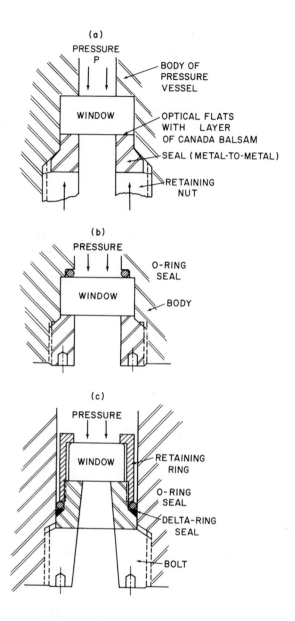

FIG. 27. Illustration of optical window designs with
different sealing arrangements. (Figures (b) and (c) courtesy
of the American Instrument Company, Silver Spring, Md.)

III.   TUBES, PIPES AND HOSES

Tubing for high pressure service is available in a wide variety
of sizes and materials. The most common and readily available tubing
is cold-drawn, seamless material, especially made for carrying high

pressure fluids. The most widely used materials are stainless 304 and
316 in a half-hard condition, although tubing can be bought from reliable sources in a host of other materials such as stainless 347,
Monel, Nickel, Inconel, Titanium, Hastelloy, Chrome-Moly, etc. Tubing
is made to rigid specifications and is tested and inspected. When
necessary, test certificates will be supplied with the material.

The capability of pressure tubing to withstand high internal
pressures is truly amazing. Commercially available tubing made from
half-hard 316 S.S., for example, with O.D. and I.D. respectively 3/16"
and 0.020" ($\sim$4.8 and 0.5 mm) has been shown to burst only above 20 kb
(2 GPa, $\sim$300,000 psi) [26, 27]. Tubing for high pressure use (e.g. >
6 kb (0.6 GPa, $\sim$90,000 psi) is usually made of single wall construction for smaller sizes, (e.g. 3/16" O.D. ($\sim$4.8 mm)) but of composite,
double wall construction for larger sizes (e.g. 3/4" O.D. ($\sim$19 mm)).
Manufacturers of tubing maintain such strict tolerances that compound
tubing can be readily made from standard sizes. After pressurization
to rated pressure the inner liner may become stuck fast to the outer
liner, showing that autofrettage has taken place. Compound tubing in
various sizes is available from several manufacturers.

The useful pressure range given by such companies for their tubing is usually rated conservatively. However, adequate consideration
should be given to fatigue and temperature cycling. For instance,
tubing connecting a reciprocating compressor should be chosen with
particular care, because the pressure may be effectively cycled many
millions of times in a short period of time. Design considerations
are given in Chapters 7 and 9.

Tubing should be free from nicks and scratches and care should be
taken that no dirt can enter when it is stored by inserting plugs or
caps. Bending must be done cold and with the correct tools. The
tools for bending should prevent nicking and damage to the outside diameter, while the radius must be ample. The radius of the bend should
be, as a very minimum, four times the outside diameter, but six times
the O.D. or more wherever possible. Bending must be done slowly and
the ends should be kept plugged or capped. Rebending must always be
avoided. See Figure 28 for a simple bending tool.

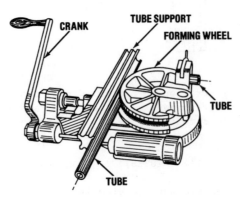

FIG. 28. Illustration of a crank tube-bender.

Tubing can be cut by hand or machine, but very small tubing should not be sawed off with a hacksaw; this can close the inside diameter with burrs. As an example 1/16" x 0.005" (~1.59 x 0.15 mm) or 1/8" x 0.020" (~3.17 x 0.5 mm) tubing should be filed with a triangle or sharp file. When a ring has been filed around the O.D. the tubing can easily be broken, thus avoiding the inside diameter from being closed or "burred".

Tubing should not be assembled under stress or tension and should be anchored where possible. This is particularly important for long pieces of tubing, so that they cannot "whip". Welding of high pressure tubing is not recommended. Annealing (after welding) is difficult to accomplish and inter-granular corrosion can occur with internal cracks resulting, especially when the tubing is used up to its maximum working pressure. A welded assembly may be less expensive initially but cannot be disassembled as easily when parts have to be replaced. Standard connections and fittings are definitely preferred.

Hose is supplied in all kinds of materials with any kind of endfitting or flange and is made for a wide variety of uses. One advantage is its flexibility with a minimum of metal fatigue, so that it can handle pulsating and surge pressures very well. In extreme cases of pulsating pressures it is recommended that the working pressure of the hose is reduced to 50% of its maximum working pressure.

High pressure flexible metal hose can be used for pressures up to about 800 bars (80 MPa or ~12,000 psi) and has excellent resistance against shock and temperature differentials. When hose is used for higher temperatures, the maximum working pressure must be reduced accordingly. Manufacturer's catalogues should be consulted for details of operating conditions. A list of manufacturers is given in reference [1].

Pipe is rated by the so-called "schedule", which defines the wall thickness [28]. The larger the schedule number, the heavier is the wall thickness. The outside diameter remains the same and consequently the inside diameter decreases with the larger schedule numbers (Table 2).

Although some suppliers have schedule 5 and 10 pipe, the most commonly used schedules are given in Table 2. When specifying pipe, two dimensions are needed; the nominal size and the schedule. Pipe connections can be made by welding (preferable minimum schedule 80) or threaded connections. Pipe should be used in straight lines and not be bent; fittings such as elbows etc. should be used.

Pipe sizes and pressure ratings are given for the most common sizes in Table 2.

IV. VALVES

The valve is an essential piece of equipment in high pressure technology and the quality of a system, be it a chemical plant or a rocket, can be determined by how effectively the valves operate. In large chemical plants, valves may represent 8 - 10% of new plant capital expenditures, and 10% of the maintenance budget. Recently, the development and manufacture of valves has grown to be a separate technology.

Table 2 Pressure Ratings for Standard Pipe Sizes

Nominal Pipe Size in.	Outside Diameter of Pipe -in.	Number of Threads Per Inch	Length of Effective Threads -in.	Schedule 40 (Standard)		Schedule 80 (Extra heavy)		Schedule 160		Double Extra Heavy	
				Pipe ID-in.	Burst Press-psi	Pipe ID-in.	Burst Press-psi	Pipe ID-in.	Burst Press-psi	Pipe ID-in.	Burst Press-psi
1/8	0.405	27	0.26	—	—	—	—	—	—	—	—
1/4	0.540	18	0.40	.364	16,000	.302	22,000	—	—	—	—
3/8	0.675	18	0.41	.493	13,500	.423	19,000	—	—	—	—
1/2	0.840	14	0.53	.622	13,200	.546	17,500	.466	21,000	.252	35,000
3/4	1.050	14	0.55	.824	11,000	.742	15,000	.614	21,000	.434	30,000
1	1.315	11½	0.68	1.049	10,000	.957	13,600	.815	19,000	.599	27,000
1¼	1.660	11½	0.71	1.380	8,400	1.278	11,500	1.160	15,000	.896	23,000
1½	1.900	11½	0.72	1.610	7,600	1.500	10,500	1.338	14,800	1.100	21,000
2	2.375	11½	0.76	2.067	6,500	1.939	9,100	1.689	14,500	1.503	19,000
2½	2.875	8	1.14	2.469	7,000	2.323	9,600	2.125	13,000	1.771	18,000
3	3.500	8	1.20	3.068	6,100	2.900	8,500	2.624	12,500	—	—

Working pressures for various schedule pipes are obtained by dividing burst pressure by the safety factor. Data from ref. 28.

Valves are available in micro-size with a port-opening of 1.5 mm
(1/16") or smaller and 12 mm (1/2") tall or with a port of ten feet
(~3 m) in diameter for water dams. Other variables are the type of
fluid, temperature range, pressure range, control function, flow-rate,
type of action (e.g. hand-operated, pneumatically or solenoid operated). Valves serve the following functions:
   (i)   to start and stop flow,
   (ii)   to regulate or control flow,
   (iii)   to prevent back flow (check valves),
   (iv)   to regulate pressure (regulators)
   (v)   to relieve excessive pressure (safety relief and safety
valves).
   In order to understand the various types and to recognize the
different designs, a short description follows on the most commonly
used valve designs.

A.   Plug Valve

   The plug valve is a simple, straight-forward design (see Figure
29a). A 90° turn of the valve handle opens the valve for 100% flow,
which gives this design a fast open-close capability. When fully open
the valve offers a relatively smooth or unrestricted passage to flow.
By turning the valve handle less than 90°, the flow can be restricted
or regulated.
   The weak points of this design are problems of torque and leak.
Sealing the tapered plug in the body is difficult, especially after
the valve has been in use for sometime. In order to obtain a good
leak-tight seal, the tapered plug should be pushed down, and this pro-

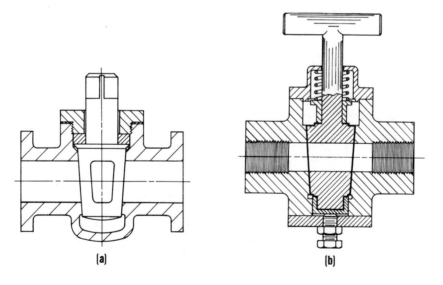

(a)   (b)

FIG. 29. Two illustrations of the plug valve. (a) simple
design (valve is closed), (b) higher pressure design with spring
assembly to control force driving plug into seal (valve is open).

duces a tendency for the plug to stick in the valve body with the possibility of galling and an increase of torque. This galling problem
can be minimized by the correct choice of materials and the hardness
of the different parts of the valve. Also, special lubricants can be
used to avoid galling, but this is not always possible and sometimes
the fluid flow through the valve may cause the lubricant to disappear.
It is now becoming a standard practice to line plugs and valve bodies
with teflon™ or other suitable materials. Some designs use a spring
loaded plug thus providing a constant load downward on the plug, assuring a clearance-free fit of plug in the valve body and seal; see
Figure 29b.

In addition to the straight-through port in the plug or the two-
way valve, multiple port arrangements are possible, with an angle of
90°, for example, from the normal flow axis. With two inlet ports,
different fluids can be mixed in regulated proportions.

The pressure rating for this valve is about 400 bars (40 MPa or
~6,000 psi) but some plug valves have been made for use at a pressure
of 700 bars (70 MPa or ~10,000 psi).

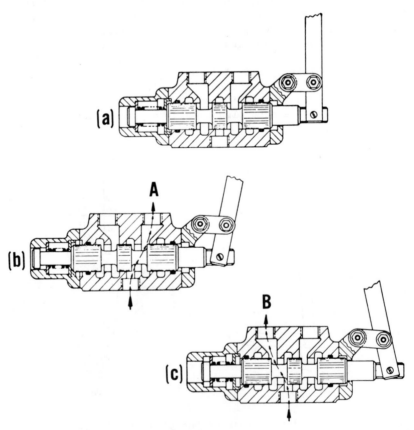

FIG. 30. Illustration of a hydraulic directional valve in three
positions. (a) closed, (b) open-direction A, (c) open-direction B.

The plug valve can be developed into a valuable and efficient
piece of equipment such as the Hydraulic Directional Control Valve
(see Figure 30). In this case the valve has a cylindrical plug whose
axis is usually horizontal. Such a valve can be an integral part of a
pumping system. The valve can also be used as a volume control valve
and pressure control valve.

B.   Ball Valve

This valve, shown in Figure 31, is available in cast iron, steel,
stainless-steel, bronze, etc. but is also made in nylon, teflon or
tetrafluoroethylene plastics and has an application to ~700 bars (70
MPa ~10,000 psi) and a temperature range from cryogenic to about
1,000°F (~540°C).
The design is a natural development of the plug valve, but each
design can vary slightly, particularly in the tightness of the seal
rings. The initial seal is achieved by a light mechanical force and
the fluid pressure increases the sealing capabilities. The valve in
general does not need lubrication and can handle slurries and abra-
sives. It has, fully opened, unrestrictive flow and can be used for
throttling. It can be used in a large number of applications, including vacuum systems. Disassembly is usually simple so that repairs can
be performed easily. The ball valve has a tendency to open when exposed to vibration, due to the low torque and the use of low-friction
seals. The limitation of its use is the strength and durability of
the sealing materials.

C.   Gate Valve

Another valve with fully open, unrestricted flow, is the gate
valve (Figure 32), but this may be its only advantage. The sealing
parts, wedge or disc, are sensitive to flow, and dirt or slurry tend
to build up in the seat cavities. Throttling is not desirable, leav-
ing this valve for opening and closing only.

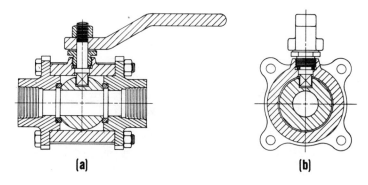

(a)   (b)

FIG. 31.   Diagram of the ball valve.   (a) side view (valve
open), (b) end view.

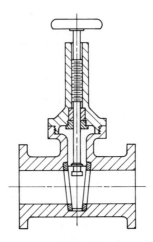

FIG. 32. Illustration of a gate-valve.

        The valve is particularly useful for large flow or very large
port openings, and sizes range up to 8 - 10 ft. (~3 m) for dams. The
pressure range is consequently very low, although the smaller sizes
can be used from ~10 bars (1 MPa or ~150 psi) to ~150 bars (15 MPa or
~2,250 psi). The temperature range can extend up to ~1800°F (~980°C)
depending on the materials used for the sealing parts.
        Another design is the gate valve with a sliding disk, shown in
Figure 33. The seal-disk is operated by a lever and the sliding or
rotating motion cuts or shears away any obstructions, resulting in a
cleaning effect on the seal area. The valve is well suited for handling slurries. Shut-off, or closure, is obtained by the force of the
down-stream pressure, which pushes the disk against the seat. The
sealing parts can be made of metal (metal-to-metal) or plastics, such
as Nylon or Teflon. Operation of this variant is limited to pressures
of ~150 bars (15 MPa or ~2,250 psi).

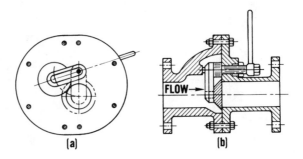

FIG. 33. A valve with sliding disk. (a) end-view showing
disk positions, (b) side-view (valve closed).

D.   Globe Valve

The globe valve is available in a large number of designs and is used basically for an "on-off" application. Since the valve produces a resistance to flow, it can also be used as a throttling device (Figure 34a), but it should only be used for clean liquids. The flow through the valve is turbulent causing wear and erosion of the seat areas. Frequent repair and replacement of seal and seat parts is necessary.

Several designs have been made to improve the flow characteristics, shown in Figure 34b. This design also uses a diaphragm for the stem seal. The split-body design, shown in Figure 35 is particularly suitable for easy repair and maintenance. The pressure rating is up to about 150 bars (15 MPa or 2500 psi) and the valve can be used for temperatures up to 1000°F (~538°C), depending on the seal materials.

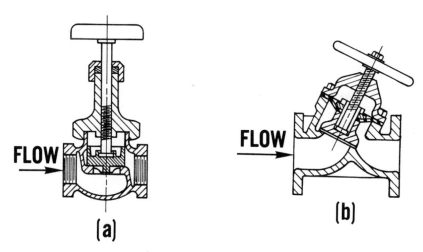

FIG. 34.   Two variants of the globe-valve.   (a) simple design, (b) design with diaphragm seal.

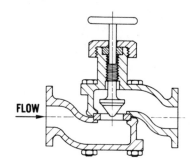

FIG. 35.   Globe valve with split body (valve open).

E.   Diaphragm Valve

This valve has no packing and therefore no leak, and is very
suitable for slurries. It is essentially an on-off device without
throttling capability. It is a simple valve consisting of only four
parts - body - diaphragm - head or bonnet and the moving device for
the diaphragm. Two designs are shown in Figures 36 and 37.

The short-coming of this valve is the diaphragm. Materials which
are strong and inert to the corosive effects of chemicals are not

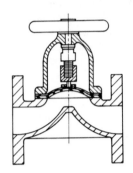

FIG. 36. Schematic of a simple diaphragm valve (valve open).

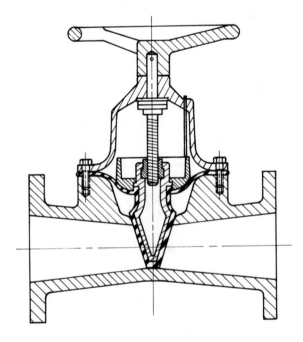

FIG. 37. Diaphragm valve with better flow characteristic
than design shown in Figure 36 (valve closed).

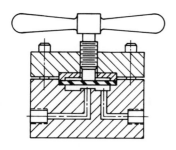 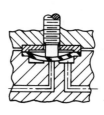

FIG. 38. Simple diaphragm valve with steel diaphragm for
operation to 1000 bars (100 MPa, ~15,000 psi).

flexible, while flexible materials needed for the diaphragm are not
strong enough and not always chemically compatible. The pressure
range of these valves is normally up to about 700 bars (70 MPa or
10,000 psi). Figure 38 shows a valve with a steel diaphragm which has
been used for pressures over 1000 bars (100 MPa or ~15,000 psi). Attempts have been made to manufacture a valve with a steel diaphragm
but only in very small sizes and for a rather low pressure range.

F.   Butterfly Valve

The Butterfly valve is a low-cost, low-weight valve which is very
simple in design. The positive points of this valve are the straight-

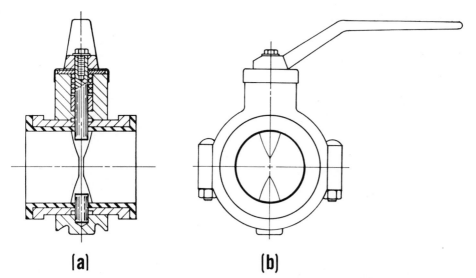

FIG. 39.   Illustration of the Butterfly valve.   (a) side view,
(b) end-view.

through, almost unrestricted flow and the self-cleaning properties, as
shown in Figure 39. A disк (vane, flapper or blade) rotates in the
circular valve body and can be compared with the flapper in stoves.
There are two basic designs. First the valve with the circular disk,
where the closure is made when the disk is rotated almost perpendicular with the valve body and the flow direction. Second, the ellipti-
cal disk, where the disk, in closed position, rests against a seat under an angle of about 10-15°. The latter design is more expensive but
has a better closing and leak-tight capability. The valve can be used
for throttling and the pressure range is up to about 140 bars (14 MPa
or ~2000 psi), while the temperature range is up to 2000°F, (~1090°C)
depending on the seat materials.

G. Pinch-Valves, Squeeze- or Clamp-Valves

A different type of valve is the pinch-valve, which can be used
for on-off control and for metering (Figure 40a, b). The valve is,
more or less, a further development of the diaphragm valve, which has
a seat, whereas the pinch-valve does not. A flexible tube is squeezed
or pinched until it completely closes. The tube is usually made from
natural or synthetic rubber compound. A partially-closed valve gives
excellent flow-control without the turbulence met usually with other
valves. The valve has a large number of applications, and functions
well with slurries and fluids with solid parts or powder, chemicals
and corrosive liquids. The design should be such that the rubber or
flexible tube can be easily replaced.

The sizes range from 1/4 inch (~6 mm) to 10 - 12 inch (~250 - 300
mm) port or more; the maximum pressure range is approximately 35 bars
(3.5 MPa or ~500 psi) and the temperature range, dependent of the materials of construction, from about 35°C to 180°C.

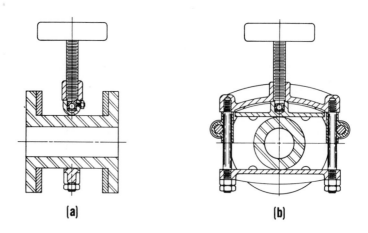

(a)   (b)

FIG. 40. Pinch- or clamp-valve. (a) side view, (b) end
view (valve open).

H.   Check Valves

The check valve works automatically, preventing back-flow or reversal of flow. There are a number of designs which depend on pressure, temperature and the fluid to be handled. The "checking element"
can be a piston or poppet, a flat disk, or multiple disks, a ball or a

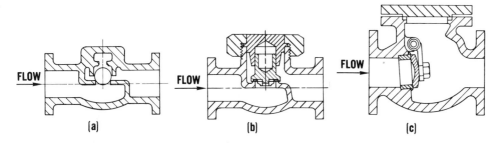

FIG. 41. (a) A ball check-valve, (b) a poppet-type check-
valve, (c) a disk check-valve (valve closed).

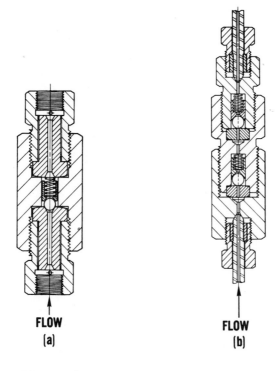

FIG. 42. Illustrations of ball check-valves. (a) Single
ball design for use to 4000 bars (~60,000 psi), (Pressure Products
Industries (PPI), Hatboro, Penn.); (b) a doubleball check-valve for
use to 2800 bars (~40,000 psi), (American Instrument Co. (Aminco),
Silver Spring, Md.). (Figures courtesy of PPI and Aminco.)

swinging disk. The valves all work with the same principle; the down-
stream flow lifts the valve checking element. When the flow stops the
valve closes. This closing of the valve can be actuated by gravity,
by a spring, or both. The differential pressure in the direction of
reverse flow increases the load on the "checking element", ball, piston, poppet or disk, and secures the closure.

     The ball- (Figure 41a) and piston- or poppet-type valves (Figure
41b) are mostly used in pumps and compressors, while the disk-valve
(Figure 41c) is mainly used to prevent back-flow. The multiple disk-
valve is used in compressors and will be discussed in the next chapter. Ball check-valves can be used for almost any application.

     Check-valves can be obtained for operation from ~20 bars up to
~14,000 bars (2 MPa - 1.4 GPa or from ~300 - 200,000 psi). The port
sizes can vary from 12 - 14 inches (~300 - 350 mm) down to 1/16 - 1/8
inch (~1.6 - 3.2 mm) for the high pressure ranges.

     Figure 42a illustrates a ball check-valve for liquid or gaseous
service in the pressure range up to 4000 bars (400 MPa or ~60,000 psi)
and Figure 42b illustrates a doubleball, in-line check-valve using
lens-rings for 3000 bars (300 MPa or ~45,000 psi). This is usually
used for difficult-to-pump liquids, such as water, volatile hydrocarbons, low- or high-viscosity liquids, or liquids containing sediment.
Figures 43a and b illustrate poppet-type check-valves for liquid or

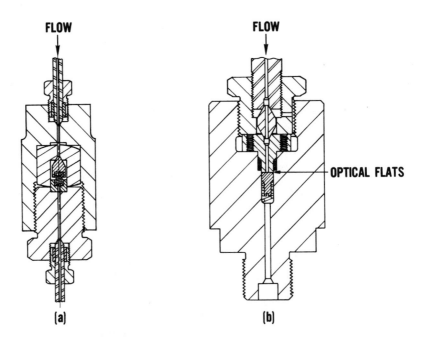

FIG. 43. Two check-valve designs for high pressure. (a) A
poppet check-valve for ~7,000 bars (100,000 psi) (American Instrument Co., Silver Spring, Md.); (b) optical-flat check valve for use
to ~14,000 bars (~200,000 psi) (Harwood Engineering Co., Walpole,
Mass.). (Figures courtesy of Aminco and Harwood Eng. Co.)

gas service (a) up to 7,000 bars (700 MPa or ~100,000 psi) and (b) 14,000 bars (1400 MPa or ~200,000 psi). For more information on check valves, see the following chapter.

I.   Relief and Safety Valves

The relief valve is an automatic device to relieve pressure or flow above certain set levels. Basically, the valve operates by opening when the pressure reaches a critical level and either returns the fluid to the reservoir, or other vessel, or allows it to blow-off.

The valve opens when the force exerted by the fluid on the checking element exceeds the opposing force exerted by the spring. As with all valves and check-valves, there are numerous designs. Figure 44a shows a spring-loaded relief valve for low pressures and large flows, whereas Figure 44b is a relief valve for higher pressures and small flow (~3500 bars (.35 GPa or ~50,000 psi)) maximum.

A simple alternative is a rupture-disk assembly (Figure 44c). The disk is selected to rupture at a pressure usually about 10% higher than the desired maximum operative pressure, and must be replaced after rupture. Care should be taken to ensure that the escaping fluid vents safely. One advantage of the rupture-disk assembly over the spring-loaded relief valve is that leaks can be prevented, whereas they invariably occur with the latter as pressure approaches the release pressure.

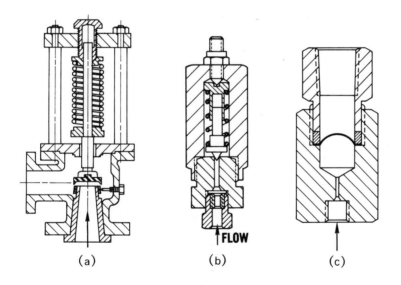

(a)   (b)   (c)

FIG. 44. Three designs of safety-relief valves. (a) Large-flow relatively low pressure design (~300 bars, 30 MPa or ~4,500 psi); (b) higher pressure design (3,500 bars, .35 GPa or ~50,000 psi); (c) rupture-disk assembly shown with a low pressure disk (~1,500 bars, 150 MPa, ~22,500 psi). (Figures (b) and (c) courtesy American Instrument Company, Silver Spring, Md.)

J.   Regulator Valves

The function of the regulator valve, or regulator, is to control
the outlet pressure at a predetermined level when the input pressure
is varying. The most common application is with a storage-gas cylinder.

The simplest design, the single-stage safety regulator, is shown
in Figure 45. This design is used wherever the outlet pressure is not
critical. A small change of the required outlet pressure will still
occur because of the decreasing supply pressure (e.g. from the storage
vessel), so that a small correction has to be made periodically to
bring the selected pressure back to the desired level. Two pressure
gauges are usually furnished with the unit; one gauge shows the supply
pressure, whereas the second gauge indicates the outlet pressure.
Clockwise rotation of the handle on top of the regulator opens the
regulator against a spring and gives the desired pressure.

When a more closely controlled pressure constant is desired, a
two-stage regulator is used. This unit contains two regulators in
series, a first stage and a second stage. The first stage receives

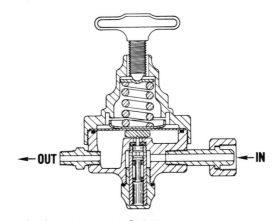

FIG. 45.   A single-stage regulator.

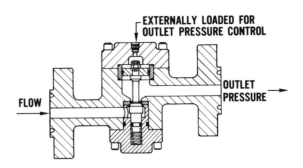

FIG. 46.   A dome-loaded regulator valve.

the supply pressure and is pre-set to deliver a reduced high pressure
intake to the second stage, the so-called intermediate pressure. The
final delivery or working pressure is set in the same way as with the
one-stage regulator by rotation of the handle on top. Clockwise rotation increases the delivery pressure up to the desired level. This
level will be maintained until the supply pressure reaches approximately the same level as the adjusted delivery pressure.

     Another design is the Dome-loaded regulator, in which adjustable
air pressure on top of the dome controls the output or delivery pressure, as shown in Figure 46. This downstream or delivery pressure can
be controlled accurately even with changes in the supply pressure.
Another feature of these units is that they can be operated remotely.

K.   Needle-Valve

     The needle-valve is generally used for applications requiring the
highest pressures and where its low-flow capacity is not a problem.
It can be used for the usual on-off function and also as a throttling
valve. The basic operation is the movement of the needle or plug in
the valve-seat. The pressure range in which needle-valves are used is
large and valves are available for pressures from a few bars up to
pressures of ~14 kbar (1.4 GPa or ~200,000 psi). The basic design is

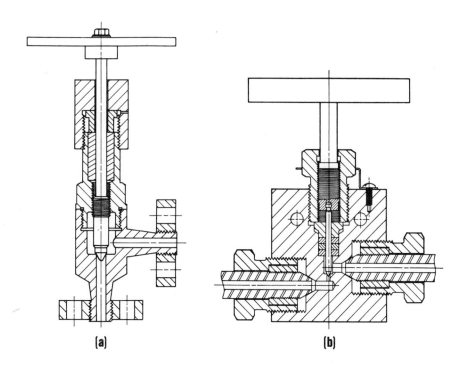

(a)   (b)

     FIG. 47. Two basic designs of needle-valves. (a) Simple de-
sign with screw thread located below the packing; (b) improved design with screw thread above the packing. (Figure (b) courtesy of
American Instrument Company, Silver Spring, Md.)

shown in Figure 47a for a valve for low-pressure service but with a
major disadvantage -- the screw thread for moving the stem or needle
up and down is under the packing. Figure 47b shows a needle valve
with the packing and screw thread in the correct location above the
packing.

The stem design has always been important. The most simple nee-
dle- or stem-design is the solid stem. This design, when made correctly, gives the most "feel" for opening and closing, but a disadvantage is that the seat can get damaged easily as the needle rotates
while closing, and may grind and gall. It should be possible to open
and close a needle-valve with one hand, but with two hands controlled
opening is possible.

Most valves are now equipped with a two-piece stem. The advantage is that the bottom part slides up and down in the packing without
rotating. The tip of the valve does not rotate on the seat, thus
avoiding the grinding and galling of the stem and seat parts. A disadvantage of most of these designs is that the "feel" for good valve-
handling decreases and, after some use of the valve, clearance and associated back-lash appears between the two stem parts.

In the following, a few stem designs are illustrated, (alphabeti-
cally). Figure 48a shows the Aminco two-piece stem. It is a robust
and simple design with a tungsten-carbide thrust disk between bottom
and top part of the stem. Figure 48b shows the Autoclave Engineers
design where the bottom part extends through a hole in the top part of
the stem. Clearance or play between the two parts can be corrected by
tightening the nuts. Care should be taken to ensure that the nuts are
not tightened too much so that the two-piece stem acts as a one-piece
stem, thus defeating the initial purpose. Figure 48c shows a design

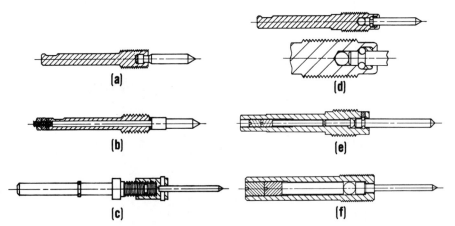

FIG. 48. Several designs of valve stem. (a) American Instru-
ment Company, Silver Spring, Md.; (b) Autoclave Engineers, Erie, Penn.;
(c) High Pressure Equipment Co. (HIP), Erie, Penn.; (d) Pressure
Products Industries (PPI), Hatboro, Penn.; (e) stem design by Webster
(Patent #3,049,332); (f) American Instrument Co., Silver Spring, Md.
(Figures courtesy of above.)

of the High Pressure Equipment Co. (HIP) where the bottom part of the
stem is attached to the body with an internal screw thread. The top
part of the stem cannot move vertically, and, by rotating, moves the
bottom part up or down. Figure 48d shows the Pressure Products Industries (PPI) design with an improved bearing arrangement and is similar
to another design of HIP. The bottom part of the upper unit of the
stem is rolled around a thrust-ring, after assembly of the bottom
stem.

There are several other, but basically similar, stem designs by
the same companies mentioned above and by others. Most designs can be
seen as a combination of these designs with slight and minor changes.
The following two stem designs will demonstrate this. Figure 48e illustrates a design by Webster (patent 3,049,332). A weak point in
this design is that the bearing balls can get out of the bearing race
and will cause problems. Figure 48f is an Aminco design with slight
changes from the Webster design. Aminco's stem design, although simpler and more robust than the Webster stem, has never been used for
production.

An important feature of the stem design is the shape of the stem
tip. Several such shapes are shown in Figure 49. Figure 49a is the
standard "V" shape for on- and off-control; b and c are shaped for
regulating and metering the flow of fluid; and d is the design for
larger port openings and is usually used in combination with a soft
seat.

Commonly, needle-valves are very stiff to open or close at their
maximum operating pressure. This makes it very difficult to know when
the valve is closed, so that overtightening can result. One solution
is to choose a valve with a higher pressure rating than the maximum
operating pressure. However, since such a valve would generally have
a smaller port than one designed for a lower pressure range, this is
not usually an effective solution.

One solution is to operate the stem with a lever. The valve
shown in Figure 50 can be used up to ~14 kb (1.4 GPa, ~200,000 psi).
A torque-control in the handle ensures that the valve cannot be over-
tightened. Alternatively, pneumatically- or hydraulically-operated
valves have been developed, such as the low-pressure valve shown in
Figure 51.

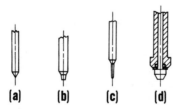

(a)   **(b)**   **(c)**   **(d)**

FIG. 49. Four different designs for valve stem seats. (a)
Simple Vee; (b) regulating stem; (c) micro-metering stem; (d) soft-
seat stem.

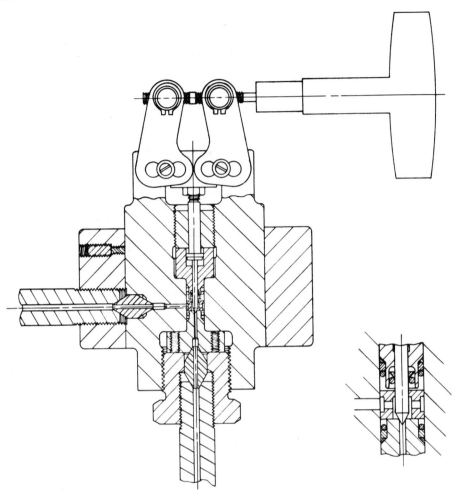

FIG. 50. A toggle valve for use to ~14,000 bars (~200,000 psi).
The insert shows details of the needle stem, seat and seals. Note
the lever movement. (Figure redrawn by permission of Harwood Engineering Company, Walpole, Mass.)

The "balanced" valve is another solution to the problem of heavy
torque. A design is shown in Figure 52 illustrating the way in which
pressure acts on both sides of the closure stem, so that the only
force to be overcome is the friction force of the seal. Another advantage is that it can be arranged to have constant volume as the stem
is advanced or retracted. When a standard needle valve is closed in a
high pressure system, the pressure can increase because the valve stem
acts as a piston. This can be avoided with a constant volume valve.
(One specialized application of such valves is in systems for measuring the equation of state of fluids precisely.)

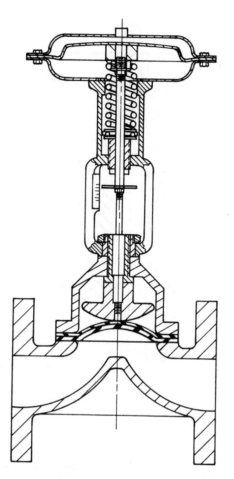

FIG. 51. A hydraulically or pneumatically operated valve
(~100 bars ~1500 psi).

A disadvantage of the valve is that two packings are needed, thus
increasing the possibility of leakage. However, this is a minor draw-
back compared to the above advantages, and the fact that the valve can
be designed with large port-openings.

Another possible way of reducing the torque is to use a drive mechanism to a non-rotating stem through two threads, one of which is
right-handed, the other left-handed. If the threads are of nearly
equal pitch, one rotation of the valve handle advances the stem by
only a fraction of the pitch. Very fine adjustments in fluid flow can
also be made with this design.

At high pressure, one basic problem which increases the friction
between the male and female thread parts is the fact that the leading
threads are subjected to a much higher stress than those behind. If
the male thread is made with a slightly altered pitch, this stress

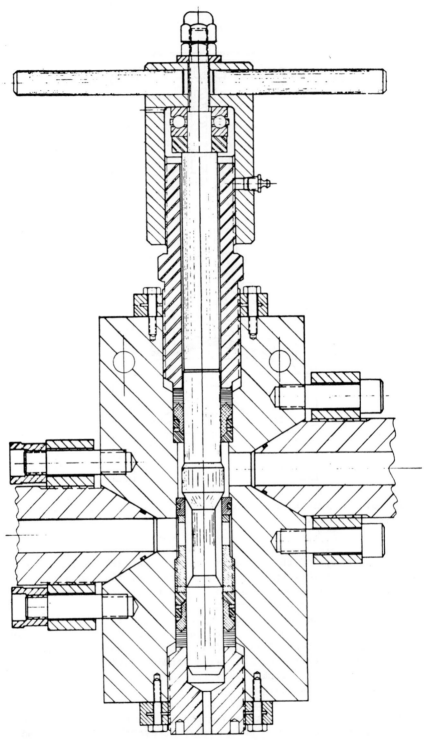

FIG. 52.   A balanced-force valve for use to 2,800 bars
(~40,000 psi).   (American Instrument Company, Silver Spring, Md.)

distribution can be made more uniform, and the friction forces are reduced. One of the present authors (Jac Paauwe) constructed several
valves for pressures up to 10 kbar (1 GPa, ~145,000 psi) using this
principle. They were easy to open and close, but too costly for mass-
production. Many pressure vessel designs use this principle to make
thread stresses in closures more uniform (see Chapters 2 and 7).

When a needle-valve is closed with too much torque the seat may
be damaged. The next time the valve has to be closed more torque is
needed for a leak-tight closure. This also makes it more difficult to
open so that when the valve is opened a pressure surge will almost
certainly result, which is undesirable. If the valve is closed too
tightly high stresses are also built up in the valve materials, increasing the possibility of damage and leakage. Handling a needle-
valve the correct way requires practice and exercise. The tendency to
overtighten them should be resisted, if possible with a torque-limiter
(Figure 50).

A valve which cannot be opened or closed smoothly is a danger. A
sudden increase in pressure can damage a component such as a pressure
gauge, and in some instances can cause an explosion due to self-ignition of lubricant and compressing media.

When the packing of a needle-valve leaks, the gland nut must be
tightened. This should only be done with the valve in open position
and without pressure in the system. This is because the stem is also
pushed down when the gland nut is tightened, so that the stem will be
forced into the seat if the valve is closed.

When tightening the packing gland nut with one hand, the other
hand should rotate the valve stem up and down, to assure that the
valve will still function after tightening the packing. If this is
not done, the packing can be tightened around the stem too heavily and
the stem cannot be moved. The heavy torque needed to rotate the stem
may even break the stem or loosen the bottom part from the top part
resulting in clearance between the stem parts.

When possible, the valve should be assembled in such a way that
the working system at pressure is always connected to the port under
the stem of the valve, and not on the packing side. This ensures the
least chance of leaks occurring from the working system if the valve
is closed.

L.   Leaking Valves

The loss of fluid is a direct loss of money, either through material replacement costs, or extra cost of compression. When gases are
used, the leak can cause an accumulation which may reach a danger level before it has been detected. Gases such as oxygen, hydrogen, carbon monoxide have to be monitored, but even a nitrogen leak can cause
accidents when the leakage goes unnoticed and in a closed space. The
loss of pressure in a liquid system is usually detected more easily
than in a gas system. However, if the fluid is corrosive, it can
build up and cause extensive damage in parts of the system that are
not specifically designed for corrosives.

It is difficult to define what is an unacceptable leak. It usually depends on the medium which is compressed. However, even if a

leakage is not particularly troublesome, efforts should be made to ensure that it is not caused by a malfunction that can later cause a
dangerous condition, such as a slowly loosening part.

In general all parts of the system should be monitored for leaks,
either continuously or periodically, depending on the system, its eco-
nomics or potential hazards.

VI.   FLUID FLOW IN HIGH PRESSURE SYSTEMS

In any high pressure system, the tubing, valves and other components provide a resistance to the flow of fluid through them. In this
section a brief description of some of the main concepts and engineering equations for flow will be discussed.

In many systems the greatest resistance to flow is attributable
to tubing. For tubing of circular cross-section the equations describing fluid flow are well developed and mathematical treatments can
be found in a number of texts, ref. [28] for instance. Considering
the practical case of a long straight tube which carries a fluid at
approximately constant density, ρ, and constant flow speed, V, the
pressure drop arises from friction losses which convert mechanical energy into heat. This pressure loss, $\Delta P = P_1 - P_2$, (where P_1 and P_2 are
the pressures at each end of the tube) is represented by

$$P_1 - P_2 = \rho \, H_f \qquad\qquad (2)$$

where H_f is a measure of the specific (per unit mass) energy loss that
occurs along the tube due to friction.

Equations giving H_f for specific situations have been developed.
There are two main regions of flow that have to be considered -- laminar (smooth) and turbulent flow. At low flow rates, to be defined later, the flow in long circular tubes is characterized by a smooth
change of the velocity from zero at the wall of the tube to a maximum
at the center (Figure 53). The velocity is a function of radial dis-

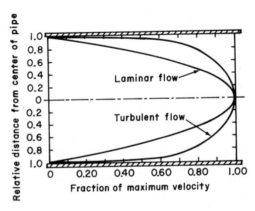

FIG. 53.   The velocity profile characterizing fluid flow along
circular tubes for laminar (smooth) and fully-developed turbulent flow.

placement (r). At higher flow velocities, this smooth flow cannot
persist and the velocity change with radial displacement V(r) occurs
sharply near the walls, and is flatter near the center (Figure 53).
In turbulent flow of this kind there are local, transitory eddy currents, so that the velocity distribution shown in Figure 53 is for
averaged velocity.

The transition between the two types of flow can be correlated
with a dimensionless number, called the Reynolds number (Re)

$$Re = \frac{D\bar{V}\rho}{\mu} = \frac{D\bar{V}}{\nu} = \frac{Dg}{\mu} \tag{3}$$

Quantities in this equation are defined in Table 3 with units given
for two principal systems. It is noted that the Reynolds number for a
fluid is independent of the unit system used. Laminar flow always occurs for Reynolds numbers up to 2,100, and may persist for values up
to 4,000. Turbulent flow, occurring for higher Reynolds numbers, is
often encountered in practical situations, but laminar flow may be encountered in high pressure systems where the tube diameter is small
and the viscosity high (eqn. 3).

In engineering practice it is customary to define the friction -
loss parameter, H_f, using the equation known as the Fanning equation:

$$H_f = \frac{2fL\bar{V}^2}{Dg_c} \tag{4}$$

where symbols are defined in Table 3.

The friction factor, f, has been evaluated from a wide range of
tests and shown to be a function of the Reynolds number and the roughness of the tube (Figure 54). In the region of laminar flow, irrespective of the roughness

$$f = \frac{16\mu}{D\bar{V}\rho} = \frac{16}{(Re)} \tag{5}$$

so that

$$P_1-P_2 = \frac{32L\bar{V}\mu}{g_c D^2} \qquad (Re) \lesssim 2100-4000 \tag{6}$$

This equation is known as the Hagen-Poisseuille equation.

In the region of turbulent flow, the friction factor f depends on
the roughness of the tube. If the tube is polished until further pol-
ishing does not change the flow characteristics, the tube is said to
be hydraulically smooth and the friction factor in the region of tur-
bulant flow is approximately ($\pm$5%) given by the equation

$$f = 0.00140 + \frac{0.125}{(Re)^{0.32}} \quad (Re) \gtrsim 4,000 \tag{7}$$

The roughness of a pipe can be defined as the ratio of the height
of the roughness irregularities (k) (Figure 54) compared to the tube
diameter (D). Typical values for k/D are indicated in Figure 54.

The above equations can be useful for estimating pressure losses
in circular tubes. The power needed to maintain the flow is readily
calculated from the equation

$$\text{Power} = \text{Mass flow} \times H_f \tag{8}$$

Table 3. Symbols and Conversion Factors for Quantities Relevent to Flow Equations Used.

Symbol	Quantity	S.I. Unit	Engineering Unit	Conversion Factors
L	Tube length	m	ft	$1m = 3.281$ ft.
D	Tube diameter	m	ft	1 ft $= 0.3048$m
ρ	Density	$kg\ m^{-3}$	lb_m/ft^3	$1\ lb_m/ft^3 = 16.02kg\ m^{-3}$
$\bar{V}$	Mean Velocity	$m\ sec^{-1}$	ft/sec.	$1m/sec = 3.281ft/sec$
$G = \bar{V}\rho$	mass velocity of flow	$kg\ m^{-2}sec^{-1}$	lb_m/ft^2-sec	$1\ lb_m/ft^2\text{-}sec = 4.883kg\ m^{-2}sec^{-1}$
P	Pressure	$Nm^{-2} = Pa$	lb_f/ft^2	$1\ lb_f/ft^2 = 47.88Pa$
g_c	Conversion factor	1 (Dimensionless)	$ft\text{-}lb_m/lb_f\text{-}sec^2$	
$\mu^{\dagger}$	Viscosity(dynamic or absolute)	$kg\ m^{-1}\ sec^{-1}$	lb_m/ft-sec.	$1\ lb_m/ft\ sec = 1.488kgm^{-1}sec^{-1}$
$\nu^{\dagger}$	Kinematic viscosity*	$m^2 sec^{-1}$	ft^2/sec	$1\ ft^2/sec = 0.0929m^2 sec^{-1}$
H	Friction energy loss per unit mass	$J\ kg^{-1}$	$ft\text{-}lb_f/lb_m$	$\dfrac{1ft\text{-}lb_f}{lb_m} = 2.989\ \dfrac{J}{kg}$
(Re)	Reynolds number	Dimensionless		
k/D	Smoothness ratio	Dimensionless		
f	Friction factor	Dimensionless		

†Absolute viscosity, μ, is defined by the equation:

$$\text{shear-stress } (\tau) = \mu \frac{dv}{dx} \text{ (velocity gradient)}$$

$$\text{Dimensions of } \mu = \frac{\text{force-time}}{(\text{distance})^2}$$

In S.I. units $1Nsec\ m^{-2} \equiv 1kg\ m^{-1}\ sec^{-1}$

In Engineering units, the factor g_c must be employed:

$$\tau = \frac{\mu}{g_c}\frac{dv}{dx}$$

Many compilations use the c.g.s. unit, the Poise   1 Poise $\equiv$ 1 gm cm^{-1} sec^{-1} $\equiv$ 1 dyne sec cm^{-2}
                                                               $\equiv$ 0.1 kg m^{-1} sec^{-1} $\equiv$ 100 centipoise

\# Kinematic viscosity is defined by $\nu = \mu/\rho$ Commonly the stoke is used in compilations of data
                                                1 stoke $\equiv$ 1 cm^2 sec^{-1} = 10^{-4} m^2 sec^{-1} = 100 centistokes

* The kinematic viscosity is sometimes given in SSU (Seconds Saybolt Universal) or SSF (Seconds Saybolt Furot). These units relate directly to convenient methods of measuring viscosity and conversion factors can be found in Journal Inst. Petroleum Tech., 22, 21 (1936), and Fluid Power, Design News Annual, Aug. 1975. Representative data comparing centipoises and SSU is listed below.

Centipoises	1	4	10	20	30	50	70	90
Saybolt Universal								
SSU	31	38	60	100	160	260	370	480
Centipoises	120	160	200	240	300	340	400	440
SSU	580	790	1000	1200	1475	1630	1950	2160
Centipoises	500	600	800	1000	1200	1500	1700	2000
SSU	2480	2900	3880	4600	5620	7000	8000	9400
Centipoises	2200	2500	3000	4000	5000	6000	7000	8000
SSU	10,300	11,600	14,500	18,500	23,500	28,000	32,500	37,000
Centipoises	10,000	20,000	50,000	70,000	100,000	125,000	150,000	200,000
SSU	46,500	92,500	231,000	323,500	462,000	578,000	694,000	925,000

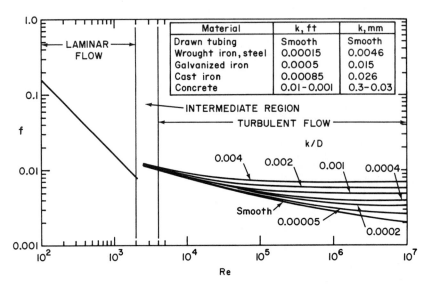

FIG. 54. The Fanning friction factor as a function of Reynolds
number for flow through circular tubes. In the region of turbulent
flow, the curves are drawn for different values of the relative
roughness factor. (Figure adapted from L. F. Moody [29].)

In this equation an appropriate unit conversion factor is required (e.g. 1 hp = 550 ft-lb$_f$) if engineering units are used, whereas
in SI, the result if given directly in watts, as shown below. The
equation neglects inefficiency of the pumping unit.

     Data for the density and viscosity are required to evaluate the
pressure drop and power requirement, discussed in Chapter 12 of this
volume. A brief example will be given, worked in both the unit systems used in Table 3.

Example:

     A 1/4" ID tube (D = 0.02083 ft = 0.00635 m) of length 50 ft (L =
50 ft = 15.24 m) is to pass 1 U.S. gallon of oil per minute of density
1.1gm/cm^3 (ρ = 1.1 x 10^{3}kg m^{-3} = 68.66lb$_m$/ft^3) when its kinematic viscosity is 400 centistokes. What is the pressure loss along the tube
and the power required to maintain the flow if the effective roughness
parameter (k) is 0.00015 ft ($\sim$0.0046 mm)? Working this problem in
both unit systems,

$$\nu = 4 \times 10^{-4} \, m^2 sec^{-1} = 4.306 \times 10^{-3} ft^2/sec.$$

$$\mu = 4.4 \times 10^{-1} \, kg \, m^{-1} sec^{-1} = 2.957 \times 10^{-1} \, lb_m/ft\text{-}sec.$$

     Since 1 U.S. gallon = 3.785 x 10^{-3}m^3 = 0.1337 ft^3, the volume
flow rate is 0.6308 x 10^{-4} m^3/sec = 2.228 x 10^{-3} ft^3/sec, so that the
mass flow rate per unit area (G) is:

$$G = \frac{\text{Volume flow rate x Density}}{\text{Tube Area}}$$

$$= 2191 \text{ kg m}^{-2}\text{sec}^{-1} = 448.9 \text{ lb}_m\text{ft}^{-2}\text{-sec.}$$

Thus, the Reynolds number describing the flow condition is:

$$(Re) = \frac{DG}{\mu} = 31.62 \text{ (both unit systems)}$$

This implies that the fluid flow is laminar in character, so that the friction factor, f, can be obtained from eqn. 5

$$f = \frac{16}{(Re)} = 0.5060.$$

Thus $P_1-P_2 = \dfrac{32L\bar{V}\mu}{g_c D^2}$

$$= 10.6 \text{ MPa} = 2.215 \times 10^5 \text{ lb}_f/\text{ft}^2 = 1538 \text{ psia}$$

From equations (4) and (5) the power consumed in maintaining this flow is:

$$\text{Power} = f\text{x}\frac{2L\bar{V}^2}{Dg_c} \text{ X Mass flow rate}$$

$$= \frac{16\mu}{D\bar{V}\rho} \text{ x } \frac{2L\bar{V}^2}{Dg_c} \text{ x } \frac{\pi D^2 G}{4}$$

$$= \frac{8\pi LG^2\nu}{g_c\rho}$$

$$= 670 \text{ Watts} = 493 \text{ ft-lb}_f/\text{sec} = 0.90\text{HP}$$

This exercise shows that low Reynolds numbers can be encountered in typical hydraulic equipment, so that laminar flow holds rather than turbulent flow. The considerable pressure drop is of interest as well as the power needed to maintain the flow.

The above equations only hold if the tube is sufficiently long for the fluid to establish its final velocity distribution. For turbulent flow this is approximately 40 - 50 diameters, while for laminar flow the transition region is of length L_t, given by:

$$L_t = 0.05 \text{ (Re)D} \qquad\qquad (9)$$

Any restriction in the tube, such as that encountered at a fitting or valve will interrupt the pattern of flow and create an additional pressure drop. Certain pieces of equipment are provided with design information about the resistance to flow, illustrated below with the case of a valve.

The flow coefficient of the valve, or other restrictions in the hydraulic or pneumatic system, C_v, is defined as the volume of water, measured in gallons per minute at 60°F (15.5°C) which flows through a

valve with a pressure drop of 1 psi across it. This definition gives
a convenient way for measuring the C_v of any restriction.

For any hydraulic fluid the C_v can be found from the following
formula:

$$C_v = Q/\sqrt{\Delta P/\rho} \qquad (10)$$

where Q is the flow in gallons per minute, ΔP, the pressure drop a-
cross the valve in psi and ρ the specific gravity of the fluid.

For pneumatic valves the following formula is recommended:

$$C_v = \frac{Q}{22.67\sqrt{\dfrac{\rho T}{(P_1-P_2)k}}} \qquad (11)$$

where Q is the flow in standard cubic feet per minute, ρ is the specific gravity of the fluid with reference to air at S.T.P., T is the
absolute temperature in Rankine ($°F+460$), P is the absolute pressure
($psig+14.7$), P_2 is the outlet pressure in psi ($P_2 > 0.53P_1$) and k is a
multiplying factor depending on the pressure drop across the valve.
It is defined as follows:

$$k = \begin{cases} P_2 \text{ (absolute)} & \text{if } P_2 \text{ is } 0\text{-}10\% \text{ of } P_1 \\ (P_1+P_2)/2 & \text{if } P_2 \text{ is } 10\text{-}25\% \text{ of } P_1 \\ P_1 & \text{if } P_2 \text{ is } >25\% \text{ of } P_1. \end{cases}$$

Further information for the computation of C_v for a valve can be
found in references [28 - 30]. However important the valve is in re-
stricting flow in a system, it should be recognized that a total system, including all components, must be checked for flow restrictions,
pressure or flow losses, and not just the valve only. Also, it should
be noted that when the flow is stopped, the pressure will build up to
its maximum value; in other words, pressure drop occurs only when the
fluid is in motion.

Undersize valves give loss in power (pressure) and loss of flow
(speed). An oversize valve, although satisfying pressure and flow requirements, is economically a waste. However, an undersize valve is
not always undersirable. When, for instance, a slow build-up of pressure is desirable during flow and full pressure is only required at no
or almost no flow, an undersize valve may even be preferable. When
power is demanded during the flow of the fluid an undersize valve is
not acceptable.

It is usual practice to designate the pipe or tubing size based
on its flow capacity, then to choose the valve. The inside diameter
of the pipe or the tubing indicates approximately the port size of the
valve. The flow capacity of the valve, however, should still be
checked. Different valves may have a rather large variation in flow
capacity, even with the same port size. Hence the C_v of the valve can
be a useful way to decide which is the best valve for the required application.

References [28 - 30] contain information in the form of charts, equations and nomograms for calculating quantities related to flow in hydraulic and pneumatic systems. The reader is referred to these articles as well as to others listed in the Bibliography for further information.

VI.   SUMMARY

After the assembly of a high pressure system has been completed and the components of the system have been tested separately, the entire system must be pressure-tested. While increasing the pressure step by step, leaks may occur. The pressure should be dropped first before the leaking connection or fitting is tightened. A leaking seal, connection or fitting should <u>never</u> be tightened while the system is under pressure.

It is common to find that some fittings or connections are rather loose after the first pressure test, sometimes even without the appearance of leakage. This is caused by the "setting" of the material and the seal areas. Overtightening should never take place. When too much torque is needed to make a leak-tight seal, then this is an indication of a problem. Overstressing of gland nuts, etc. will put too much stress in the material, particularly in the threaded areas. The material can be stressed over the elastic limit; deformation then takes place, dimensions change and failure may follow. The phenomena of "cold working", which increases the strength of the material as it plastically flows should not be counted on in normal standard fittings. An exception is the controlled "autofrettage" of a high pressure vessel which is a valuable tool (see Chapter 7), but only when the principle is applied correctly.

The correct wrenches and tools should be used and, wherever it is critical, torque wrenches to tighten fittings. A general guide for torque data for standard high pressure "fitting" has been given in Table 1. The importance of clean work has been emphasized for all high pressure fitting- and assembly-work. Finally a remark should be made for systems handling compressed air, hydrogen, oxygen or other flammable fluids. These systems should be very carefully cleaned and completely freed from oil and grease. When oil or grease is present inside a system, self-ignition (dieseling) may occur. When a valve is opened, more or less abruptly, a flow of fluid, for example air, can rush into the high pressure system and the heat of compression can ignite oil, grease or other combustible matter. Besides expert cleaning, supply valves should be opened slowly, allowing the pressure to build up gradually.

For more information on testing and safety, Chapters 2 and 3 should be consulted.

BIBLIOGRAPHY

A. General

1. P. W. Bridgman, "The Physics of High Pressure", (G. Bell
    and Sons, London, 1949).
2. H. H. Buchter, "Apparate and Armaturen der Chemieschen
    Hochdrucktechnik", Springer Verlag, Berlin/Heidelberg,
    New York (1967).
3. E. W. Comings, "High Pressure Technology", (McGraw-Hill
    Book Company, Inc., New York, 1956).
4. W.R.D. Manning and S. Labrow, "High Pressure Engineering",
    (C.R.C. Press, Cleveland, Ohio, 1971).
5. D. M. Newitt, "High Pressure Plant and Fluids at High
    Pressure", (Oxford University Press, 1940).
6. R. H. Perry and C. H. Chilton, "Chemical Engineers Handbook",
    (McGraw-Hill Book Company, New York, 5th edition, 1973).
7. H. A. Rothbart, editor-in-chief, "Mechanical Design Systems
    Handbook", (McGraw-Hill Book Company, 1964).
8. H. Tongue, "The Design and Construction of High Pressure
    Chemical Plant", (Chapman and Hall, Ltd., London, 1959).
9. D. S. Tsiklis, "Handbook of Techniques in High Pressure
    Research and Engineering", (A. Bobrowsky, ed.) (Plenum Press,
    New York, 1968).

B. Specific Subject Areas

1. F. E. Anderson, "Pressure Relieving Devices", Chemical
    Engineering, Vol. 83, No. 11, May 24, 1976.
2. C. H. Artus, "All About Chemical Hose", Chemical Engineering,
    Vol. 83, No. 9, April 26, 1976.
3. "A Valve is a Valve is a Valve", Petroleum Management, The
    Engineer, General Section, July 1963.
4. D. W. Bean, "How to Select High Performance Valves",
    Chemical Engineering, February 5, 1973.
5. C. S. Beard, "Control Valves", (F. D. Marton, ed.),
    Instruments Publishing Company, Pittsburgh, Penna.
6. A. Brodgesell, "Valve Selection", Chemical Engineering/
    Deskbook, October 11, 1971.
7. J. Ciancia, Hernandez, E. Steyman, "Valves in the Chemical
    Process Industries, a CE Report", Chemical Engineering,
    August 30, 1965.
8. W. Coopey, "Holding High Pressure Joints", Petroleum Refiner,
    Vol. 35, No. 5, May 1956.
9. S. Crocker, "Piping Handbook", (McGraw-Hill Book Company,
    New York).
10. L. Dodge, "Fluid Throttling Devices", Product Engineering,
    March 30, 1964.
11. L. Driskell, "Coping with High Pressure Letdown", Chemical
    Engineering, Volume 83, No. 23, October 25, 1976.
12. H. G. Federlein, "How to Analyze Steady-State Flow in a
    Hydraulic System", Hydraulics and Pneumatics, June 1976.

13. A. Fleming, "Ball Valves Versus other Throttling Devices", Instruments of Control Systems, Vol. 40, June 1967.
14. D. F. Frederick, "Pressure Vessel Closures", Machine Design Notebook.
15. D. F. Frederick, "Resilient Vessel Closures and Connections for High Pressure Temperature Service", Technical reprint 2751, Autoclave Engineers, Inc., Erie, Pa.
16. "Fundamentals of Valves", Petroleum Management, The Engineer, General Section, July 1963.
17. V. Ganapathy, "Rupture Disks for Gases and Liquids", Chemical Engineering, Volume 83, No. 23, October 25, 1976.
18. M. Glockman and A. H. Henn, "Valves", Chemical Engineering Deskbook, April 14, 1969.
19. T. Goldoftas, ed., "Leakage and Leakage Control", Hydraulics and Pneumatics, Volume 30, No. 2, February 1977.
20. W. S. Goree, B. McDowell and G. A. Scott, "Seals for High Pressure, Low Temperature Systems", Rev. Sci. Instr. 36, 99 (1965).
21. C. S. Hedges, "Industrial Fluid Power", Volumes 1 and 2, Womack Machine Supply Co., Dallas, Texas.
22. "Hose Handbook", 3rd Ed., Rubber Manufacturers Association, Washington, D.C., 20006.
23. ISA Handbook of Control Valves, 2nd edition, (J. W. Hutchison, ed.), Instrument Society of America, 400 Stanwix Street, Pittsburgh, Penna.
24. R. Kern, "Pressure-Relief Valves for Plants", Chemical Engineering, February 28, 1977, Volume 84, No. 5, p. 187-194.
25. O. P. Lovett, Jr., "Control Valves", Chemical Engineering/ Deskbook, October 11, 1971.
26. J. L. Lyons and C. L. Ashland, "Lyons Encyclopaedia of Valves", van Nostrand, Reinhold, 1970.
27. Machine Design, Reference Issue, "Seals", 9th edition, (Penton Publishing Company, Cleveland, Ohio) (1969).
28. W. C. Mack, "Selecting Steel Tubing for High-Temperature Service", Chemical Engineering, June 7, 1976, p. 145-150.
29. B. A. Niemeier, "Seals to Minimize Leakage at High Pressure", Trans. A.S.M.E., p. 369, April 1953.
30. P. M. Paige, "How to Estimate the Pressure Drop of Flashing Mixtures", Chemical Engineering, August 14, 1967.
31. A. Pikulick, "Selecting and Specifying Valves for New Plants", Chemical Engineering, Vol. 83, 1976.
32. L. M. Polentz, "Calculating Fluid Flows Through Orifices, Hydraulics and Pneumatics", February 1965.
33. "Recommended Practices for Bulk Loading and Unloading Flammable Liquid Chemicals to and from Tank Trucks", Bulletin TC-8, Manufacturing Chemists Association, Washington, D.C. 20009.
34. R. A. Rothman, "Gaskets and Packings", Chemical Engineering/ Deskbook, February 26, 1973.
35. "Valves", Petroleum Refiner, October 1961.
36. "Valves", Special Report, Power, June 1961.

REFERENCES

1. Hydraulics and Pneumatics, January 1977, Vol. 30, #1, pp.
    161-229, 21st Designers Guide to Fluid Power Products,
    (Penton/IPC, Cleveland, Ohio).
2. D. Langer and D. M. Warschauer, Rev. Sci. Instr. $\underline{32}$, 32
    (1961).
3. R. B. Jacobs, Phys. Rev. $\underline{54}$, 325 (1938).
4. H. S. Yoder, Trans. Am. Geophys. Union $\underline{31}$, 827 (1950).
5. Parker Seal Company, Back-up Rings (Parbak).
6. P. W. Bridgman, "The Physics of High Pressure", (Bell and
    Sons, London, 1949).
7. D. M. Warschauer and W. Paul, Rev. Sci. Instr. $\underline{28}$, 62 (1957).
8. Grayloc Catalogue, Gray Tool Company, Houston, Texas.
9. Parker Seal Company, Gask-O-Seal, Metal "V" Seal and
    Stat-O-Seal.
10. R. H. Cornish and A. L. Ruoff, Rev. Sci. Instr. $\underline{32}$, 639 (1961).
11. P. L. Heydemann, Rev. Sci. Instr. $\underline{38}$, 558 (1967).
12. P. L. Heydemann, Rev. Sci. Instr. $\underline{41}$, 1896 (1970).
13. J. L. Downs and R. T. Payne, Rev. Sci. Instr. $\underline{40}$, 1278 (1969).
14. T. C. Poulter, Phys. Rev. $\underline{35}$, 297 (1930).
15. "Superpressure Catalogue", American Instrument Company,
    Silver Spring, Maryland.
16. Optical Materials for Infrared Instrumentation, U. S. Depart-
    ment of Commerce State-of-the-Art Report, PB181087. (This
    report includes physical and mechanical data on a wide range
    or materials for use in the infrared, visible and ultraviolet
    regions.)
17. W. Paul, W. M. DeMeiss and J. M. Besson, Rev. Sci. Instr.
    $\underline{39}$, 928 (1968).
18. E. Fishman and H. G. Drickamer, Anal. Chem. $\underline{28}$, 805 (1956).
19. D. F. Williamson, I. A. Nichols and B. Schurin, Rev. Sci.
    Instr. $\underline{31}$, 528 (1960).
20. R. W. Parsons and H. G. Drickamer, Rev. Sci. Instr. $\underline{46}$, 464
    (1956).
21. H. W. Schamp and W. G. Maisch, Rev. Sci. Instr. $\underline{32}$, 414 (1961).
22. A. S. Balchan and H. G. Drickamer, Rev. Sci. Instr. $\underline{31}$, 511
    (1960).
23. S. J. Gill and W. D. Rummel, Rev. Sci. Instr. $\underline{32}$, 752 (1961).
24. D. S. Hughes and W. W. Robertson, Rev. Sci. Instr. $\underline{46}$, 557
    (1956).
25. A. W. Lawson and G. E. Smith, Rev. Sci. Instr. $\underline{30}$, 989 (1959).
26. W. Paul and D. M. Warschauer, Rev. Sci. Instr. $\underline{27}$, 418 (1956).
27. D. M. Warschauer and W. Paul, Rev. Sci. Instr. $\underline{29}$, 675 (1958).
28. Fluid Power Handbook and Directory 1976-1977, by Hydraulics and
    Pneumatics, Cleveland, Ohio, published by Industrial Publishing
    Company (division of Pittway Corp.).
29. L. F. Moody, Trans. ASME $\underline{66}$, 671 (1944).
30. Fluid Power, Design News Annual, Cahners Publishing Company,
    Inc., Boston, Mass.

Chapter 6

PUMPS AND COMPRESSORS

James M. Edmiston, P.E.

Manager of Engineering
American Instrument Co.
Division of Travenol Laboratories
Industrial Products Division
Silver Spring, Maryland 20910

Jac Paauwe

Naval Surface Weapons Center
White Oak Laboratory
Silver Spring, Maryland 20910
and
Laboratory for High Pressure Science
Department of Chemical Engineering
University of Maryland
College Park, Maryland 20742

Ian L. Spain

Laboratory for High Pressure Science and
Engineering Materials Program
Department of Chemical Engineering
University of Maryland
College Park, Maryland 20742

I.   INTRODUCTION

In Chapter 1 it was pointed out how increasingly important high pressure technology is becoming in our everyday lives. Since the pump or compressor is the "heart" of all high pressure equipment, these machines are among the most important ever invented.

Even a cursory glance around us shows the importance of these units. Water distributed to our homes has been compressed to positive pressure at the reservoir, and may have been previously pumped from deep wells. Simple hydraulic devices power components in cranes, winches, automobiles, trucks, aircraft, elevators, etc. Basic energy sources such as oil have to be pumped from the earth, then pumps are used in transportation, processing and distribution of oil and its many byproducts.

It is more difficult to see the importance of pumps and compressors used for processing many materials that have become a part of our
daily life. For example, consider their use in the petrochemical industry to create synthetic fibers which are then used in the textile
industry to fabricate our clothing. Also, many metallic objects have
been pressed into shape or modified in form using hydraulic equipment
powered by pumps. The spark plugs used in most automobiles are iso-
statically pressed. Even the margarine that we eat depends on the operation of a high pressure reactor served by a compressor. The list
of applications is without end, and is constantly expanding.

The   range of pumps and compressors is very large even if the examination is limited to the categories of size, pressure range, use,
and type of fluid to be compressed. Whole volumes have been devoted
to this subject. Many of these are referenced elsewhere in this book.

In this chapter, pumps and compressors will be classified; some
basic principles will be discussed; each significant configuration
will be commented upon; and then, some of the practical problems involved in selection and maintenance will also be given. Types of
pumps and compressors that have lent themselves to development of
pressure above 1,000 bars (100 MPa or 15,000 psi) will receive greater
emphasis.

A Bibliography at the end of the chapter will guide the reader to
a larger literature on the subject. In showing equipment from one
manufacturer, we are not expressing a preference, nor are any claims
made for the equipment. Performance figures quoted in the text for
such equipment are taken from manufacturer's catalogs. A list of pump
and compressor manufacturers can be found in references 1 and 2.

II.   GENERAL CLASSIFICATIONS AND CRITERIA

The words "pump" and "compressor" are often used interchangeably.
However, in this chapter the following distinction will be adhered to:

> Pumps are units for pressurizing or circulating
> liquids; these are fluids that are relatively
> incompressible.
> Compressors are units for compressing gases or
> gaseous fluids; i.e., fluids with relatively
> high compressibilities.

Although the distinction between gas and liquid cannot be drawn
clearly for substances above their critical temperatures (see Chapter
12 of this volume and Chapters 4 and 5 of Volume II), the above distinction is generally useful.

A second basic classification concerns the type of motion that is
employed - either rotary or reciprocating (Figure 1). All reciprocating units are also positive displacement types; that is, the fluid
is trapped in discrete volumes, then compressed or forced from the
unit. In reciprocating units, the fluid is forced in and out through
one-way check valves.

Rotary units may be of the positive displacement type, such as
gear, vane, lobe pumps. Alternatively, they may be kinetic or dynamic
units in which kinetic energy is converted into potential energy of
compression. Examples of dynamic units are the centrifugal compres-

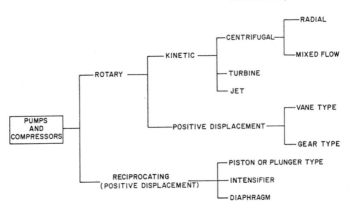

FIG. 1. Classification of pumps and compressors based on type of motion.

sor, and the turbine pump. Jet pumps are used for very low pressure applications and will not be discussed further [3].

Two performance criteria of prime concern to the designer are the pressure range and volume capacity. The maximum pressure range is normally defined as the pressure at which internal leakage reaches undesirable levels, or at which internal parts are damaged by stress or temperature. In reciprocating units, a major consideration is the fatigue life of the moving components such as check valves, crossheads, bearings, pistons, and the dynamic seals related to them, which limit long-term operation.

The capacity is usually expressed as a volume of processed fluid at inlet conditions per unit time. However, many manufacturers specify the volume in terms of the fluid at ambient conditions, even though the pressure and temperature may be very different at the suction point of the unit. If specified in terms of inlet conditions, the unit of measure is usually prefixed by the letter I; e.g., ICFM = inlet cubic feet per minute. SCFM refers to standard cubic feet per minute. Standard conditions can differ, but in the U.S.A., for example, conditions of 1 atmosphere pressure, T = 70°F are commonly used (1.013 bars, 0.1013 MPa, T $\sim$21°C). The S.I. unit of capacity is $m^3 sec^{-1}$, although other units of time are acceptable (e.g., minutes, hours). Since the volume of compressible substances depends so sensitively on pressure, capacity can more conveniently be described on the basis of weight per unit time (e.g., kg/sec; tons/hour) or on a molar basis (e.g., moles sec^{-1}). (In the S.I., one mole is that quantity of substance containing Avogadro's Number of molecules, i.e., $\sim$6.023x10^{23} molecules, so that one mole of H_2O weighs $\sim$0.018 kg.)

Small high pressure pumps were developed about 200 years ago and consisted of a lever-operated plunger compressing fluids in a cylinder, with inlet and outlet check valves. Simple pumps such as this are still widely used and an example is shown in Figure 2 for use up to 4,000 bars (.4 GPa or 60,000 psi).

FIG. 2.   A simple hand pump for use to ~4,000 bars (400 MPa or ~60,000 psi). (Photo courtesy of American Instrument Co., Silver Spring, Md.)

The industrial use of compressors for large-scale operations can be dated from 1895 when Linde constructed a plant for the continuous liquefaction of air.   His plant operated at about 200 bars, using a two-stage compressor with horizontally-opposed cylinders. Soon after-wards (1913), Haber and Bosch began production of ammonia in a pilot plant (see Chapter 1), again operating at about 200 bars (20 MPa ~3,000 psi).

The following years saw the development of much larger compres-sors for the chemical industry, operating at pressures up to about 1,000 bars (100 MPa or 15,000 psi); but more often in the region from about 200 - 300 bars (20 - 30 MPa or 3,000 - 4,500 psi). These ma-chines were often constructed with horizontal opposed cylinders run-ning at slow speeds (about 2 Hz or 120 RPM) [4].

However, for compression of large volumes of gases up to about 300 - 400 bars (30 - 40 MPa or 4,500 - 6,000 psi), rotary kinetic units are now widely used. Production costs are reduced considerably as the size of the plant is increased, so that these compressors are now the work horses of the chemical industry. A photograph of a large axial compressor is shown in Figure 3. The particular unit shown is a multi-stage unit for compressing air for a nitric acid plant. It

FIG. 3. Photograph of a large axial compressor.
(Photo courtesy of Delaval Turbine Inc., Trenton, N.J.)

handles 100,000 - 150,000 ICFM ($\sim 47 - 71$ Im^{3}sec^{-1}) from approximately atmospheric pressure to 70 psia (4.8 bar, 0.48 MPa), at $\sim 8,000$ HP (~ 6 MW). Larger units of this type can go to $\sim 500,000$ ICFM (~ 240 m^3/sec) at powers up to $\sim 50,000$ HP (~ 40 MW). The end wheel on the unit shown is for power recovery.

Industrially, centrifugal compressors are used for compressing fluids up to pressures of ~ 200 bars (20 MPa, $\sim 3,000$ psia) while axial compressors are used for lower pressure applications (e.g. up to ~ 20 bars (2 MPa, 300 psia). These axial compressors are also used as the first stage of a train, feeding fluid to a centrifugal compressor at a reduced volume, higher pressure.

It is difficult to exactly define the pressure-volume region where centrifugal compressors are to be preferred to reciprocating units. Certainly at pressures above ~ 200 bars (20 MPa, 3,000 psi) the reciprocating unit is unchallenged, although research is underway to increase the operating pressures of centrifugal units [5]. Below this pressure, rotary units are preferred for larger volume applications on economic grounds. Although centrifugal units have preferable maintenance requirements, a disadvantage is their limited range of operating speeds. Below certain flow levels an instability, known as surge, becomes apparent [3]. However, these units are increasingly preferred for volume flow rates above $\sim 1,000$ ICFM (~ 30 m^3/min).

FIG. 4. Photograph of a two-stage, four crank, secondary compressor for ethylene. (Photograph courtesy of Burckhardt Engineering Works Ltd, Basel, Switzerland.)

One example of the use of reciprocating units for large volume compression of gaseous fluids is found in the production of polyethylene, where pressures in the range of 2,000 - 3,500 bars (200 - 350 MPa or 30,000 - 50,000 psi) are commonly used (see Chapter 3, Vol. II). In the early days of manufacture, before and during World War II, mercury piston compressors were used, driven by oil. Today, large hypercompressors using crank- or hydraulically-driven pistons are used, a photograph of a typical unit being shown in Figure 4. Such units can typically compress 10^5 kg/hour (~100 tons/hour) of ethylene between the suction pressure of ~200 - 300 bars (20 - 30 MPa, 3,000 - 4,500 psia) and discharge pressure.

Other higher pressure applications of reciprocating units include water jet cutting, hydrostatic forming, isostatic pressing. For the highest pressures, intensifiers must be used, which are available for pressures up to about 14,000 bars (1.4 GPa or 200,000 psi). Such units are used for laboratory experiments or for autofrettage of cyl-

FIG. 5. Photograph of an intensifier unit for compressing
gaseous fluids (e.g., He,Ne...) up to 14 kb (~200,000 psi). Two
intensifier stages are shown, the one in the foreground with
larger piston diameter operating to ~4 kb (~60,000 psi). (Photo
courtesy Harwood Engineering Co., Walpole, Mass.)

inders (see Chapter 7). A photograph of a unit for compressing gases
such as helium up to this pressure is shown in Figure 5.

Positive displacement, rotary pumps are widely used for low pressure fluid power applications, such as in hydraulic winches. Metal
cutting machines utilize somewhat higher pressures. Still higher
pressures, up to about 200 bars (20 MPa or 3,000 psi) are used in
metal-forming and plastic-molding equipment.

Closely associated with the pressure range and capacity is the
power rating of the pump or compressor. Since World War II the size
of units has grown considerably, so that both reciprocating and rotary
machines are available with power ratings of 12 MW and higher (>15,000
HP).

Each type of pump has an acceptable range of operating speeds. Reciprocating compressors are relatively slow, normally operating at speeds of about 300 RPM (5 Hz) up to ~700 RPM (~11 Hz). Rotary displacement pumps normally operate at about 3,000 RPM (60 Hz), but some miniature vane pumps operate at speeds as high as 30,000 RPM (600 Hz). Centrifugal pumps and compressors also operate over a wide range of speeds from about 1,000 - 30,000 RPM (about 17 - 600 Hz). The speed of the pump influences its dimensions and weight, and the selection of the coupling mechanism between the drive and the compressor. Some problems become more acute as speed increases, particularly sealing. Increased fluid velocity through valves in reciprocating compressors will increase performance losses and accelerate wear. It is significant to note that piston speeds have not increased substantially over the last twenty years; speeds of 4m/s (about 13 ft/sec) are seldom exceeded. Without component/material development, speeds cannot rise.

There are a number of design criteria which must be considered when selecting a pump or compressor. The type of unit required is very often evident once pressure range, capacity and type of fluid to be compressed are specified. Many selection factors relate to the fluid so that characteristics of the fluid must be known, such as:

   (i)   Viscosity and density (liquid or gas)
  (ii)   Corrosive nature (chemical attack)
 (iii)   Compatibility with metals, non-metals, lubricants and seals of the unit
  (iv)   Flash and fire points (explosion or fire hazards)
   (v)   Entrainment of particles that could cause pump wear
  (vi)   Lubricity and heat capacity of fluid (self-lubrication of pump)

Another consideration is the duty of the machine. Specific consideration should be given to:

   (i)   Continuous, interrupted, or intermittant operations. For example, considerations for a small laboratory intensifier would be completely different from those for an industrial unit operating 24 hours per day.
  (ii)   Operations at variable capacity or pressure. The capacity of some pumps can only be changed with speed, while in others the stroke can be adjusted. A recirculating loop could be a compromise to permit use of a simpler pump assembly.
 (iii)   Importance of reliability, service and "down" time, ease of maintenance and repair.
  (iv)   Temperature factors may include the need to pump hot or cold fluids, or temperature rises associated with adiabatic compression of fluid or heat generated in seals. Temperature changes can be particularly serious because of differential thermal expansion, leading to increased leak rates in seals, etc. Temperature will also significantly effect the viscosity of the fluid being pumped.
   (v)   The size and weight of the pump may be of importance, for instance in aerospace applicatons. Some pumps for hydraulic applications have power-to-weight ratios as

low   as   1/10 hp/lb, (about   165 W/kg)   while 1/3 hp/lb
(about 550 W/kg) is more common, and values as high as
4 hp/lb (6.6 kW/kg) may be encountered in very high
speed units for critical applications.

(vi)   Operating conditions can include the possibility of use
in outdoor locations where temperature and humidity may
have to be considered, or in applications where shock
and vibrations are critical, e.g., compressors for applications where pressure drops can be sudden.

(vii)   Where personnel have to spend time in close proximity
to the pump, noise levels or noise control must be considered. Noise levels over 85 dBA at 3 ft (~1 m) are
usually unacceptable, and lower levels are to be preferred.

All of the above considerations should be taken into account when
deciding on the pump or compressor to be used.

The American Petroleum Institute [6] has many specifications
available which suggest other information that may facilitate proper
selection of a unit, but most manufacturers will supply a data sheet,
a catalog and expertise necessary to arrive at the best match of unit
to requirements. The manufacturer can only meet the customer's needs
if they are adequately specified, and this should be done in written
form.

The following sections cover detailed descriptions of pumps and
compressors.

III.   RECIPROCATING PUMPS AND COMPRESSORS

A.   General Principles

The general configuration and principle of a reciprocating pump
is illustrated in Figure 6. A piston is driven in a reciprocating motion against the fluid in the cylinder. On the back stroke, fluid is
forced in through the inlet check valve by a differential pressure,
then it is compressed on the forward stroke and forced out through the
outlet check valve.

The drive for the reciprocating piston can be of several types.
For large-scale compressors, a crank-drive is most widely used; however, a hydraulic drive is sometimes selected. Various drive mechanisms are employed with pumps such as (but not exclusively), wobble-
plates (Fig. 7(a)), radial piston drive (Fig. 7(b)), bent-axis drive
(Fig. 7(c)). Finally, a hydraulic drive is commonly coupled to an intensifier (pressure multiplier) for operation above 350 bars (35 MPa
or 5,000 psi) (Fig. 7(d)), the last is discussed in Section D.   In
Figure 7(d), a balanced, or double-ended unit is shown. The driving
fluid is passed through a directional valve to one side of the pump,
compressing fluid in the opposing high pressure head. When the stroke
is complete, the valve diverts the driving fluid to the opposite side,
so that the suction and discharge operations of the high pressure
heads are also reversed.

Double-ended units are often staged, so that one side feeds fluid
to the other side. Thus, the discharge pressure of the lower pressure
stage becomes the suction pressure of the higher pressure stage. Such

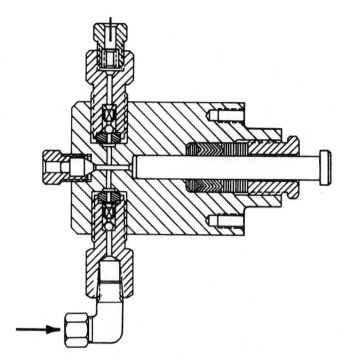

FIG. 6. Cross-section of the head of a reciprocating pump.
A Vee-packing seals the plunger. Ball check-valve elements seal
on lens rings. (Figure courtesy American Instrument Company,
Silver Spring, Md.)

units can also be built with several stages, as shown in Figure 8.
The unit shown has four stages, but as many as six have been used in
commercial equipment [4].

Multistaging is of particular use for compressors, since it is
difficult to achieve higher ratios than 10-20 per stage without substantial loss of efficiency. However, multistage units generally consist of a train of single or double stage units, rather than the integral design of Figure 8. It is also desirable to employ multiple
stages with interstage cooling, as discussed in Section B-4. In opposed units, both hydraulic and crank-drive are regularly used. Multistage units are not commonly encountered in pumps for highly incompressible fluids. For pumps, pressure ratios of over 1,000 are easily
achieved without excessive heating or loss of efficiency. Sometimes
the inlet of a high pressure pump will be boosted to 10-100 bars (1 -
10 MPa, ~145 - 1450 psi) to increase the flow rate. The booster pump
is usually a separate unit.

One of the advantages of the opposed unit is that forces on the
drive unit can be reduced by using the pressure acting on the pistons
on the suction side to partly counteract the forces from the fluid being compressed. By careful choice of piston areas and stage pres-

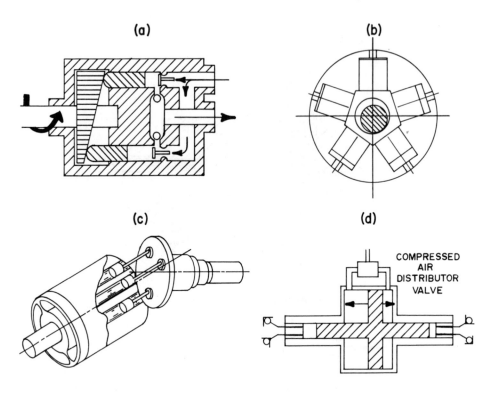

FIG. 7.   Several types of drive used for reciprocating pumps:
(a) wobble-plate; (b) eccentric drive; (c) bent-axis; (d) hydraulic
(compressed air).   (Figures (a), (b), (c) adapted from ref. 7.)

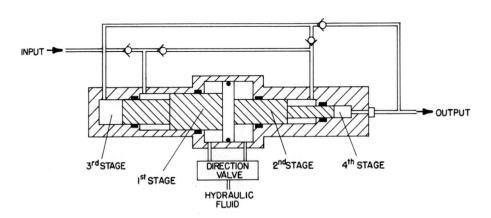

FIG. 8.   Sketch of a hydraulically-driven, multi-stage, opposed
unit with four stages.   Units with up to six stages are used industrially.

sures, this effect can be maximized. However, many units employ several stages in an unopposed configuration to minimize space. In this
case, the total force is applied to the crank in one direction.

B.   Thermodynamcis of the Compression of Fluids and Power Calculations

In this section, a brief introduction to the quantitative evaluation of compressor performance will be given, based on simple ideas of
the thermodynamic properties of fluids. The interested reader can
find further details in refs. 4, 8-11.

The work of compression in a cyclic change imposed on a fluid can
be found directly from the relationship

$$\text{Work} = - \oint P dV \qquad (1)$$

By international agreement the sign of the work is today taken to
be positive if done on the fluid, P is the pressure, V the volume, and
the integral sign denotes a value obtained by summing over the cycle
(Figure 9). For a fixed mass of fluid, the internal energy (U) after
a cycle is the same as that at the beginning (initial and final states
the same), thus, the work done on a fluid for a thermodynamically reversible cycle can be obtained either from a P-V or T-S diagram:

$$\oint dU = \oint T dS - \oint P dV = 0 \qquad (2a)$$
$$\therefore \ -\oint P dV = -\oint T dS = \oint dW \qquad (2b)$$

where T is the absolute temperature (Kelvin or Rankine) and S is the
entropy.

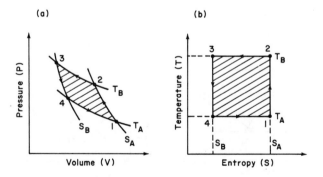

FIG. 9.   (a) Pressure-volume plot for a Carnot-cycle operating
between two temperatures T_A, T_B. The adiabatic changes (1→2) and
(3→4) correspond to entropies S_A, S_B respectively. The work per
cycle is indicated by the dashed area within the loop.
(b) The T-S diagram for the same cycle. Again, the work per cycle
is indicated by the dashed area within the loop.

When applying the above equations to estimate the work needed to compress a fluid, it should be borne in mind that the value given by (1) and (2) is the thermodynamically <u>minimum</u> amount of work to achieve the desired cycle, where the fluid is undergoing <u>reversible</u> changes. Any irreversibilities due to friction, or non-equilibrium conditions in the fluid, will result in a <u>greater</u> expenditure of work (in a cycle <u>producing</u> power from the expansion of fluid, then the work given by (1) and (2) will be the <u>maximum</u> work obtained). A further discussion of this point is given in Section B.3.

Equations (1) and (2) will now be used to estimate reciprocating compressor performance.

Consider the simple reciprocating compressor cycle shown in Figure 10(a). At point 1, the piston is at the bottom of its stroke and has received fluid at pressure P_1. During the compression stroke (1→2), the fluid is usually compressed rapidly, so that little heat flows from the fluid to the cylinder walls. Compression is thus quasi-adiabatic. When pressure P_2 is reached (1→2), the outlet check valve is forced open and fluid is expelled from the compressor until the piston reaches the top of its stroke (2→3). During the expansion process, the fluid expands quasi-adiabatically (3→4) until pressure P_1 is reached, when the inlet check valve is forced open by the fluid in the reservoir, allowing fluid to enter (4→1).

At the top of its stroke (3) the remaining volume between the piston and valves is called the clearance volume ($V_c = V_3$). The swept volume (V_S) is equal to $V_1 - V_3$, while the intake volume (V_I) is equal to $V_1 - V_4$. The ratio V_C/V_S is known as the clearance-ratio (C), and the ratio V_I/V_S is known as the volumetric efficiency ($\varepsilon_{vol.}$).

During the cycle, the mass of fluid in the cylinder is not constant, (see suction and discharge strokes (4→1, 2→3)), and Figure 10(b) illustrates the number of moles of fluid enclosed in the cylinder during the cycle. The P-V diagram (Fig. 10(a)) thus represents the instantaneous pressure-volume relationship for the fluid within

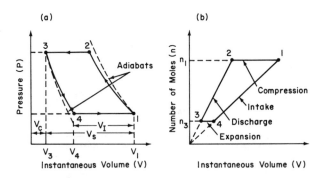

(a)   (b)

FIG. 10. (a) Idealized pressure-instantaneous volume relationship for a reciprocating unit operating between pressures P_1 and P_2. The compression and expansion processes are compared to ideal adiabatic ($\Delta Q = 0$) processes. (b) The number of moles in the cylinder of the compressor as a function of instantaneous volume.

the cylinder. This P-V diagram is known as the "indicator diagram", since the volume is often represented proportionally by an indicator attached to the piston.

Actual indicator diagrams differ from that shown in Figure 10(a) slightly, since the valves do not open instantaneously, and since there is a certain amount of valve-bounce. However, indicator dia- grams can still be used to estimate the work per cycle even though the quantity of fluid is not fixed, as shown from the following equations:

$$W_{cycle} = W_{1\to2} + W_{2\to3} + W_{3\to4} + W_{4\to1} \tag{3a}$$

$$= -\int_1^2 PdV - P_2(V_3-V_2) -\int_3^4 PdV - P_1(V_1-V_4) \tag{3b}$$

$$= - \oint PdV_{indicator} \tag{3c}$$

In deriving (3b) from (3a), it is assumed that suction and dis- charge strokes are at constant pressure. Equation (3c) follows (3b) by comparing representative areas for each of the terms in (3b) on the indicator diagram. If the compressor is operated at Ω cycles per sec- ond, the power (P) consumed in the compression cycle, neglecting fric- tional corrections, is

$$P = \Omega \, W_{cycle} \tag{4}$$

In evaluating the power rating for a simple compressor, many unit systems will be encountered. Examples of calculations for power are given in Section B.5.

1.   Work of Compression for Ideal Gases

To proceed further, an equation of state for the fluid, $P(V,T)$ is required. Extensive data for a number of fluids is obtainable (see Chapter 12). We will briefly consider the performance of a compressor operating with a perfect gas, for which the equation of state is

$$PV = nRT \tag{5}$$

During a reversible adiabatic compression or expansion, the pressure- volume relationship is

$$PV^\gamma = Constant \tag{6}$$

where γ is numerically equal to the ratio of the principal specific heats (C_p/C_v) for the perfect gas. For an ideal monatomic gas, $C_p/C_v = 1.67$; for diatomic gases, 1.40; and for more molecularly complex gases, γ decreases to values lower than 1.4. During the compression ($1\to2$), the temperature-rise from (5) and (6) is

$$\frac{T_2}{T_1} = \left(\frac{P_2}{P_1}\right)^{\frac{\gamma-1}{\gamma}} \tag{7}$$

Figure 9 illustrates two adiabatic (isentropic) processes in the Carnot cycle, while two such processes in a normal compression cycle are illustrated in Figure 10(a).

Using equations (1) and (6) the performance of the compressor can be readily obtained by integrating over a cycle, assuming paths $2 \rightarrow 3$ and $4 \rightarrow 1$ (discharge and suction) occur at constant pressure:

$$\varepsilon_{vol} = \frac{V_I}{V_S} = 1 - C\left[\left(\frac{P_2}{P_1}\right)^{1/\gamma} - 1\right] \tag{8}$$

$$W_{cycle} = \frac{\gamma}{\gamma - 1} P_1 V_S \left[1 - C\left[\left(\frac{P_2}{P_1}\right)^{1/\gamma} - 1\right]\right]\left[\left(\frac{P_2}{P_1}\right)^{\frac{\gamma-1}{\gamma}} - 1\right] \tag{9}$$

$$\text{i.e. } W_{cycle} = \frac{\gamma}{\gamma - 1} P_1 V_I \left[\left(\frac{P_2}{P_1}\right)^{\frac{\gamma-1}{\gamma}} - 1\right] \tag{9a}$$

The quantity of gas passing through the compressor, rated at the intake pressure, is obtained from:

$$\begin{array}{ll} \text{Volume per unit} \\ \text{time at } P_1 \end{array} = \Omega V_S \varepsilon_{vol} = \Omega V_I \tag{10}$$

$$= \frac{dn}{dt} v_1$$

where dn/dt is the number of moles of gas compressed per unit time, and v_1 is the molar volume of the intake gas.

It is evident from equations (8) and (9) that V_I and ε_{vol} approach zero for a critical value of pressure, which is the maximum attainable pressure of the compressor:

$$\left(\frac{P_2}{P_1}\right)_{max} = \left(\frac{C+1}{C}\right)^\gamma \tag{11}$$

Thus, the clearance ratio is the most important parameter in determining the maximum operating pressure, other than limitations due to effective sealing, mechanical strength of the piston, cylinder, etc.

The maximum pressure condition is indicated in Figure 11. From this figure it can also be seen that, as P_2 increases, the area representing work per cycle first increases, then decreases. During normal operations, the inlet and outlet pressures will vary. It is important that the compressor be operated below the maximum power condition (Figure 12). If operated above it, an increase in outlet pressure results in a decrease in power consumption so that unstable conditions can result. From Figure 12 it is clear that the condition for W to be at its maximum value depends on both the clearance volume and the properties of the fluid being compressed.

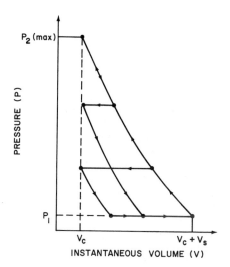

FIG. 11.   Indicator plot showing that as the outlet pressure, P_2, increases, the discharge volume decreases, until it reaches zero at P_2 (max).

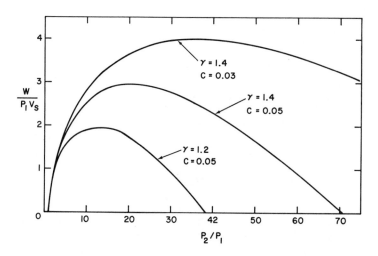

FIG. 12.   A plot of the work ratio ($\bar{W}/P_1 V_s$) performed by the compressor per cycle as a function of the pressure ratio P_2/P_1. Values of the polytropic exponent, γ, and clearance ratio, C, are indicated for each curve.

2.   Compression of Real Fluids

If a real fluid is being compressed, then the equation (5) must be replaced. For a non-ideal gas the compressibility factor, Z, which acts as a correction factor to the ideal gas equation, may be used:

$$PV = nZRT \qquad\qquad (12)$$

Values for Z in reduced coordinates are available for a large number of substances (see Chapter 12). Formulae for computing the power required in a compressor using equation (12) to describe the fluid are discussed in references 4, 10 and 11.

A useful approximation for the work of adiabatic compression can be obtained by multiplying the work of adiabatic compression for the perfect gas (equation 9) by $\sqrt{Z_1 Z_2}$, where Z_1 and Z_2 are the compressibility factors for the gas at the suction and discharge conditions respectively

$$W = \sqrt{Z_1 Z_2} \cdot W_{ideal} \qquad\qquad (13)$$
$$\text{adiabatic}$$

During a reversible adiabatic compression of non-ideal gas, the pressure - volume relationship can still be approximated with an expression such as (6); however, the exponent γ is no longer equal to C_p/C_v. (Values can be found in several tabulations. Typically, values of $\gamma \sim 1.2 - 1.3$ are taken for gaseous fluids).

For liquids (in this case, compressors are usually called pumps), the above equations are not applicable. The liquid-state is characterized by a much lower compressibility than the gaseous state, so that a relatively higher proportion of the forward stroke $(1\rightarrow3)$ is taken up by discharge $(2\rightarrow3)$ (Figure 10). For a perfectly incompressible fluid, the compression $(1\rightarrow2)$ and expansion $(3\rightarrow4)$ lines become vertical. For actual fluids, the lines will be in between those for the incompressible fluid and the gaseous fluid. Reference should be made to the equation of state.

For the incompressible fluid the "work of compression" cannot be stored in the fluid, but is related to the work done by the piston in forcing it from the compressor, where this work is dissipated as friction (heat). In actual cases, the work done by the compressor is spent on both stored energy and friction energy.

3.   Polytropic Processes and Irreversibility

In an actual compressor, other factors should also be taken into account when estimating the performance, other than deviations from ideal gas behavior. For example, during the compression $(1\rightarrow2)$ and expansion $(3\rightarrow4)$ strokes, some heat will in fact be passed between the walls of the cylinder and the fluid. A general process in which pressure and temperature changes are accompanied by heat-exchange is called a polytropic process. An equation of the form of equation (6) can still be used approximately for a general polytropic change, with the adiabatic exponent (γ) in (6) replaced with an effective polytropic exponent, γ'.

$$pv^{\gamma'} = \text{Constant} \tag{6a}$$

where γ' is between 1.2 and 1.3 for many typical situations.

On the P-V indicator diagram, the expansion and compression paths will differ from the adiabatic paths, as indicated in Figure 10(a). Similarly, the path on the T-S diagram will not be one at constant entropy.

Another factor which has to be taken into consideration is the irreversible nature of the cycle. Friction is a major source of irreversibility, converting useful work of compression into waste heat. Even if the compression stroke were adiabatic ($\Delta Q = 0$), the process would not be isentropic ($\Delta S = 0$), since the heat of friction is counted as irreversible heat generation ($\Delta S = \int dQ/T$ only for a reversible process). Since irreversible heat is generated, $\Delta S \geq 0$ for an adiabatic process, where the zero value only applies to a reversible change.

It can be seen that practical compressors can only be described approximately using thermodynamic considerations. However, the approximate equations can be used to estimate required compressor performance and to decide if one process is potentially more economical than another. A further example follows on the use of multistage compression with interstage cooling.

4. Multistage Compression and Interstage Cooling

In practice, the pressure-ratio across a single stage reciprocating compressor is usually limited to ~10 without an unacceptable loss of volumetric efficiency. Using successive stages of compression the final temperature would be very high without interstage cooling being used (unless the cylinder jacket cooling was extremely efficient). For instance, if the effective polytropic exponent $\gamma' = 1.3$, and initial temperature is 20°C, then the outlet temperature, calculated from equation (7) would be 225°C for a pressure ratio of 10. If two successive stages were used (pressure ratio = 100), the outlet temperature would be 517°C.

By using interstage cooling, another advantage is evidently the decrease in power consumed for the process (Figure 13). Using equation (6) for the adiabatic compression and expansion, it is straightforward to show that the work of compressing a fluid in a multistage system with m stages to a pressure ratio P_2/P_1 is minimized if each of the m stages has the same pressure ratio. In this case,

$$\text{Work/cycle.} = mP_1V_1 \cdot \frac{\gamma}{\gamma-1}\left[\left(\frac{P_2}{P_1}\right)^{\frac{\gamma-1}{m\gamma}} - 1\right] \tag{14}$$

It is interesting to note that if the number of stages is increased indefinitely ($m \to \infty$), the work per cycle gradually decreases to a limiting value

$$W_{min} = P_1V_1 \ln P_2/P_1 \tag{15}$$

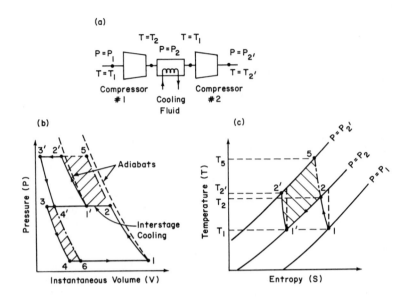

FIG. 13. (a) Schematic of a two-stage compressor with inter-stage cooling. (b) P-V diagram for the two compressors. Ideal adiabats are indicated by dashed lines. The path 1→2→5 would be followed by a single-stage compressor, or double-stage compressor without interstage cooling. The shaded areas represent the work saved during compression and that lost during expansion as a result of interstage cooling. (c) The corresponding entropy-temperature diagram for compression only. Adiabats are represented by vertical, dashed lines. The work saved during compression is represented by the shaded area.

This is exactly the work of compression for an isothermal (T = constant) compression. This equation then represents the thermodynamically ideal value for the work of compression of a perfect gas for any process. More generally, the isothermal compression process uses the least energy for any fluid, since energy is wasted in heating fluid between two temperatures. In practice, the requirements of minimizing running costs through power saving must be balanced against the higher initial costs of multistage systems and inherently higher maintenance costs.

5.   Calculation of Pump and Compressor Power [9-12]

There are several different applications of pumps and compressors in which the first calculation required is the power rating. As with many engineering calculations a major difficulty is with the different unit systems employed. A few typical examples will be given. In all cases discussed it will be recalled that the SI unit of power is the watt (1 watt (W) $\equiv$ 1 Js^{-1} $\equiv$ 1 Nms^{-1} $\equiv$ 1 kgm^{2}s^{-3} [13]) while the horsepower is commonly used in America and Britain (1 HP = 550 ft-lb$_f$sec^{-1} ~746W). In this latter system it is necessary to use the force con-version:

$$1 \text{ lb}_f \equiv 32.174 \text{ lb}_m\text{-ft sec}^{-2}$$

a. Compressor for Perfect Gas. In this case the compressor is working to store compression energy in the gas, and an equation such as (14) may be used with (4).

Example 1:

A single stage compressor has a swept volume of 0.1 ft^3 (2.83 x 10^{-3} m^3) and is compressing a quasi-perfect gas from 14.5 psia (1 bar, 0.1 MPa) to 110 psia (0.7584 MPa; 7.584 bar). The effective poly-tropic exponent is 1.25, the speed of the compressor is 300 r.p.m. (5 Hz), the clearance ratio is 0.05 and the power efficiency of the compressor is 0.85.

Working the problem firstly in SI units, the intake volume, V_I is, from (8)

$$V_I = V_S \left[1 - C\{ \left(\frac{P_2}{P_1} \right)^{1/\gamma} - 1 \} \right] = 0.797 \ V_S = 2.257 \times 10^{-3} m^3$$

From (9a) the ideal work per cycle is:

$$W = \frac{1.25}{0.25} \times 10^5 \frac{N}{m^2} \ 2.257 \times 10^{-3} m^3 \times [0.4997]$$

$$= 5639 \text{ N-m} = 563.9 \text{ Joules}$$

Since the power efficiency is 85%, the actual power is

$$\text{Power} = \frac{\Omega \ W_{ideal}}{\text{Efficiency}} = \frac{5 \text{ sec}^{-1} \times 563.9J}{0.85}$$

$$= 3.32 \text{kW}$$

If the problem were worked with American Engineering units, the intake volume could be found immediately from the fact that the volu-metric efficiency is given above as 0.797, i.e., $V_I = \varepsilon_{vol} V_S = 0.0797 ft^3$. The work per cycle is:

$$W_{cycle} = \frac{1.25}{0.25} \times 14.5 \ \frac{lb_f}{in^2} \times 0.0797 \ ft^3 \times \frac{144 in^2}{ft^2} \times [0.4997]$$

$$= 415.8 \text{ ft-lb}_f$$

$$\text{Power} = \frac{5 \text{ sec}^{-1} \times 415.8 ft-lb_f}{0.85} \times \frac{1HP}{550 ft-lb_f sec^{-1}}$$

$$= 4.45 \text{HP} \ (3.32 \text{ kW})$$

The two values are in agreement. However, it is noted that the Inter-national System is to be preferred because of its simplicity.

Example 2:

It is desired to compress the same quantity of gas to the same pressure, using a two-stage compressor with interstage cooling.

For this case it would have to be assumed that the product of intake volume and compressor speed is the same as in the previous example. Thus, in SI

$$W_{ideal/cycle} = 2 \times 10^5 \times 2.257 \times 10^{-3} \times \frac{1.25}{0.25} \times [.2246]$$

$$= 506.9 \text{ Joules}$$

$$Power = \frac{5 \times 506.9}{0.85} W = 2.98 kW = 4.00 \text{ HP}$$

Thus, the power consumed is reduced by about 10%. If American units are to be used, a summary formula can be given for calculating the horsepower.

$$Power(HP) = \frac{mP_1(psia)V_I(ft^3)\Omega(rpm)}{229.2 \times Power\ Efficiency} \frac{\gamma}{\gamma-1} \left[\left(\frac{P_2}{P_1}\right)^{\frac{\gamma-1}{m\gamma}} - 1 \right] \qquad (16)$$

b. Pumps. As discussed in the text, an equation of state is needed for the calculation of the compression energy of a liquid. However, for modest pressures commonly met in engineering applications it can be assumed that the stored, or compression, energy of the liquid is small in relation to other energy terms. Typically, pump energy is expended in:

      (i) Performing work elsewhere (e.g., hydraulic fluid operates winch, crane, etc.).

     (ii) Fluid pumped to holding tank at higher level.

    (iii) Energy dissipated as loss in tubing, etc.

As far as the pump is concerned, it acts to provide a certain flow-rate across a pressure rise. Thus, the power is:

$$Power = \frac{Rate\ of\ Flow \times Pressure\ Rise}{Power\ Efficiency} \qquad (17)$$

The remaining problem is to choose a consistent set of units and conversion factors.

Example 3:

An hydraulic pump takes $10^{-3} m^3 sec^{-1}$ of incompressible fluid across a pressure rise of 50 bars (e.g., from 1 to 51 bars). Calculate the power if the pump has a power efficiency of 0.85. From (17):

$$Power = \frac{10^{-3} m^3 sec^{-1} \times 5.0 \times 10^6 Nm^{-2}}{0.85} = 5.88 \frac{kN}{sec} = 5.88 kW.$$

The simplicity of this unit system is again apparent.

Example 4:

The same problem will be worked in American units. The volume is now 2.12 ft^3/minute across a pressure rise of 725 psi.

$$Power = \frac{2.12ft^3}{min} \times \frac{1}{60} \frac{min}{sec} \times \frac{725}{in^2} \frac{lb_f}{} \times \frac{144in^2}{ft^2} \times \frac{1HP}{550} \frac{sec.}{ft\text{-}lb_f} \times \frac{1}{0.85}$$

$$= 7.88HP(5.88kW)$$

A summary formula would be

$$Power (HP) = \frac{Flow\ Rate\ (ft^3/min)\ x\ Pressure\ Rise\ (psia)}{229.2\ x\ Power\ Efficiency} \qquad (18)$$

Often in such cases the flow rate is expressed in gallons/minute (1 gallon (U.S.) = 0.133 ft^3). In this case the conversion factor would be

$$Power\ (HP) = \frac{Flow\ Rate(galls(U.S.)/min)\ x\ Pressure\ rise(psi)}{30.6\ x\ Power\ Efficiency} \qquad (19)$$

Very often the pressure rise is expressed as a head (h) of fluid. The head is the height of a column of fluid in standard gravity (32.17 ft/sec^2 or 9.81 $msec^{-2}$) whose pressure differential is the same as that across the pump.

$$Pressure\ difference = h\rho g \equiv \Delta P \qquad (20)$$

$$\rho = density\ of\ fluid$$

$$g = acceleration\ due\ to\ gravity$$

There are two common confusions about the use of this concept. In normal use, the head and the density in this formula relate to the fluid being compressed. However, some manufacturers use the equivalent head of water, with ρ = 1000 kg/m^3, 1 gm/cm^3, 62.5 lb_m/ft^3 or 8.34 lb_m/gallon (U.S.). Secondly the concept is sometimes used for compressible substances. Normal practice is to assume that the density remains constant at its value at standard conditions. Quite clearly the customer should make sure that he understands the convention being used by the manufacturer when he specifies h.

In essence, the calculation of power simply converts head to the equivalent pressure drop and substitutes it in the above equations. Unfortunately a plethora of units and conversion factors can arise. Again, the International System is simple.

Example 5:

A fluid of specific gravity (relative to water) of 2.00 is raised to a head of 500 ft (152.4 m) at the flow rate of 0.5 ft^3/min. Calculate the power of the pump if its efficiency is 0.85.
Using S.I. units

$$\Delta P = 152.4m\ x\ 9.81\ msec^{-2}x2,000kg\ m^{-3}$$

$$= 2.99x10^6Nm^{-2}(Pascals)$$

$$= 29.9\ bars.$$

Since the flow rate is 2.361×10^{-4} $m^3 sec^{-1}$ the power is 0.831kW, using (17).

If the problem were worked in American units,

$$\Delta P = 500 ft \times 32.17 \frac{ft}{sec^2} \times 2 \times 62.5 \frac{lb_m}{ft^3} \times \frac{1\ lb_f}{32.174 lb_m ft\ sec^{-2}} \times \frac{ft^2}{144 in^2}$$

$$= 434\ psi$$

Power $= \dfrac{0.5 \times 434}{229.2 \times 0.85}$ HP

$$= 1.11\ HP (0.831 kW)$$

In practice, many other units are encountered, such as litres per minute, millions of standard cubic feet per day, for flow rate. However, calculation of power can readily be made using the above formulae and conversion factors where necessary.

When sizing the drive for the pump or compressor, it is desirable to use a generous estimate for the power required and to use an oversized drive shaft. High starting loads may be encountered and sometimes it is convenient to exceed normal operating conditions for short periods of time.

C. Crank-Operated Pumps and Compressors

In the following pages, a discussion will be given of several commercial designs to illustrate many of the salient features of these units and their range of application. A brief commentary will then be given on considerations of materials, design and operating problems.

A horizontal, crank-operated pump is shown in Figure 14. The unit shown can be supplied with different head assemblies to cover the pressure range from ~20 bars (2 MPa, ~300 psi) to 700 bars (70 MPa, ~10,000 psi. A pump rated at 80 HP (~60 kW) will deliver ~10 gallons per minute (~3.8 $\times 10^{-2} m^3$/minute) of water.

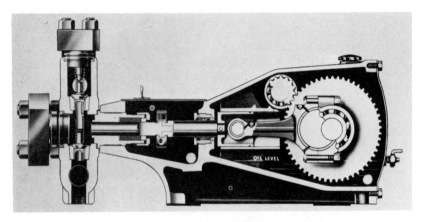

FIG. 14. Photograph of a cross section of a typical horizontal, reciprocating pressure with crank drive. (Photograph courtesy of Bean Triplex Pumps, FMC Corporation, Jonesboro, Ark.)

The pump shown uses poppet-type valves, the inlet valve being located inside the high pressure head. A separate oil seal is used, so
that the fluid is compressed free of oil. The main shaft seal is of
the compression type and the retaining flange can be tightened periodically to compensate for seal wear.

Three designs of pump head are shown in Figure 15(a), (b) and
(c). They are fundamentally the same design with variations for different applications and pressure ranges. The basic design, Figure 15
(a), is used to ~700 bars (70 MPa, ~10,000 psi). The plunger and
liner design is metal-to-metal, so that the O-ring near the bottom of
the liner bushing is used only as a secondary seal at temperatures
near ambient. The cylinder block is typically of a carbon steel forging. The check valves are of the single-ball type. The liner can be
removed, and is compressed against the static liner-to-wall seal packing by the retaining flange and bolts (bottom of the figure). The
pump is for use with oil, water and oil, light hydrocarbons, etc. It
can be used to temperatures limited by the choice and availability of
seal materials.

A version of the pump shown in Figure 15(b) can be used for pumping cryogenic liquids to 400 bars (40 MPa, ~6,000 psi). Typical liquids are nitrogen, hydrogen, ethylene and carbon dioxide. The check
valves are disc-type with tapered insert seats, (the inlet valve is on
the left-hand side) while the shaft seal consists of many teflon rings
with stainless steel contracting rings. The body and valve ports are
of austenitic stainless steel and the body is jacketed, to reduce heat
influx.

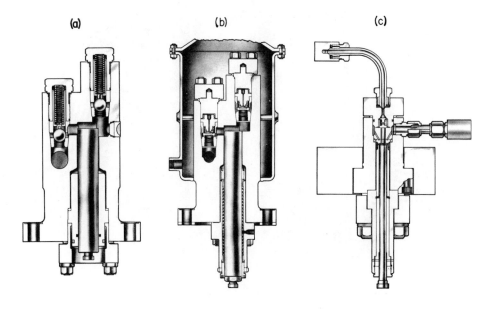

(a)   **(b)**   **(c)**

FIG. 15. Three pumps based on same basic design. (a) standard
unit; (b) cryogenic unit; (c) high pressure unit. (Figures courtesy
Kobe Inc., Huntington Park, Calif.)

These designs can be adapted to higher pressure (2 kbar, 200 MPa, ~30,000 psi) as shown in Figure 15(c). Fluid inlet is from the right-hand side through off-axis ports into the inlet check valve, which is of the disc-type, mounted immediately above the piston. Fluid outlet is through the axial port to the disc-type outlet check valve. All sections of the stainless steel valve head, subjected to stress, have

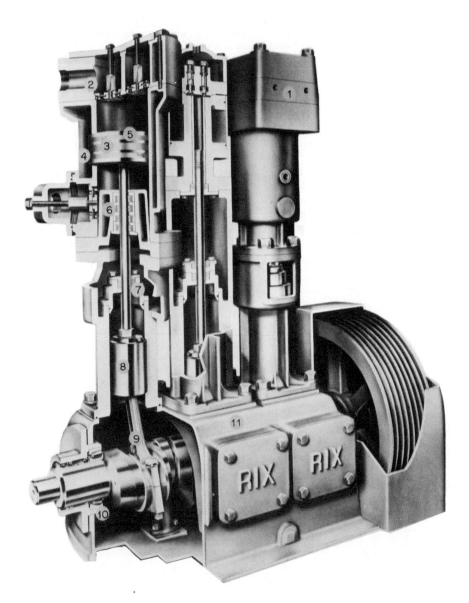

FIG. 16. A vertical, three-stage compressor for air to 400 bars (40 MPa, ~6,000 psi) or light gases to 200 bars (20 MPa, ~3,000 psi) at 1.7 m^3/min (~60 scfm). (Figure courtesy Rix Industries, Emeryville, Calif.)

circular cross-section and are free of intersecting holes. The plunger and liner are retained with a flange for easy removal and interchangeability.

     A three-stage, vertical compressor is shown in Figure 16. The
first stage, shown to the left, is double-acting. Fluid above it is
compressed on the up-stroke; fluid below it on the down-stroke. This
effectively doubles the volume compressed. Valves are of the disc-
type and are equipped with adjusters, so that on start-up the valves
are inoperative. Thus, the driving motor can warm up the compressor
without the additional load of compression. Piston rings seal the
large piston, while four shaft seal elements, each consisting of two
sealing rings (see Section G) are employed for the piston-rod seal. A
separate oil seal allows oil-free compression.

     This unit is representative of multi-stage compressors which pro-
vide compressed gases for industry in the range up to ~400 bars (40
MPa, ~6,000 psi). The particular unit shown is for use with air to
~350 bars (35 MPa, ~5,000 psi), or light gases such as helium to ~200
bars (20 MPa, ~3,000 psi). Intake flow-rate is 60 SCFM (~1.7 m^3/min)
of gas, such as air, while developing 60 HP (~45 kW).

     A photograph of a much larger unit is shown in Figure 17. It is
a ten cylinder compressor in an ammonia synthesis plant, developing
~10,000 HP (~7.46 MW). The unit compresses air in four stages to over
350 psia (~24 bars, 2.4 MPa) and synthesis gas in five stages to over
9,000 psi (~600 bars, 60 MPa). The final stage of compression uses an
interesting mechanical design - the "bathtub" design - which is unique
to Worthington's compressors.

     FIG. 17. Photograph of a large compressor in an ammonia
synthesis plant. Surge-tanks can be seen mounted above the
compressor stages (see Section III-I). (Photograph courtesy
Worthington Compressor, Inc., Buffalo, New York.)

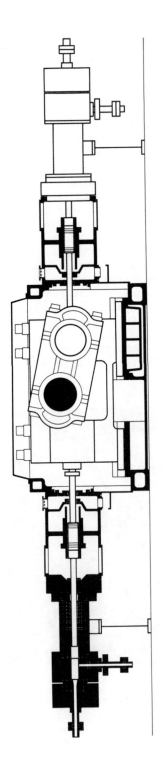

FIG. 18. Cross-section of a horizontal, opposed, hyper compressor with crank-drive for compressing ethylene. (Figure courtesy Burckhardt Engineering Works, Ltd., Basel, Switzerland.)

At the upper end of the pressure scale, large units are used for
compressing ethylene for polyethylene production. Although practical
details are often proprietary, their basic principles of operation are
similar to those already discussed.

A cross-section of such a unit is shown schematically in Figure
18 for use to 3,500 bars (.35 GPa, ~50,000 psi). It is a double-ended
unit for balancing purposes, but each end operates to the same discharge pressure. These units are designed for suction pressure of
~200 - 300 bars (20 - 30 MPa, ~3,000 - 4,500 psi) and deliver up to
100 tonnes/hr at 3,000 bars (300 MPa, ~45,000 psi). A photograph of a
typical unit is shown in Figure 19.

Up to this point we have discussed only simple pumps whose capacities are varied directly with operating speed. For greater flexibility metering pumps are sometimes used, whose capacity can be varied by
adjusting the stroke.

FIG. 19. Photograph of a horizontal, opposed hyper-compressor
for polyethylene production. (Figure courtesy Burckhardt Engineering
works Ltd., Basel, Switzerland.)

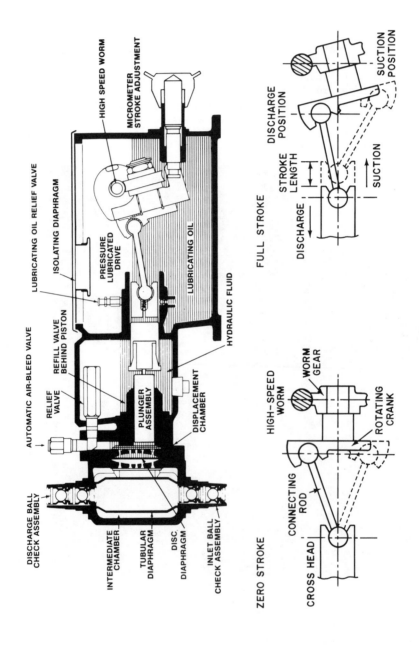

FIG. 20. Cross-section of a proportioning pump, with insets showing the principle of operation. (Figure courtesy Milton Roy Company, Ivyland, Pa.)

The design in Figure 20 uses a polar crank drive. The micrometer adjustment control changes the angle of the drive, and consequently adjusts the stroke. Zero stroke can be achieved with the rotating crank in a vertical position, increasing stroke with a larger angle from the vertical.

The unit shown uses a diaphragm head, and the upper pressure limit is then 250 bars (25 MPa, ~3,500 psi). Through the use of a secondary tubular diaphragm, corrosive liquids can be isolated from the high pressure head, enabling standard materials of construction to be used. However, this reduces the maximum pressure to ~70 bars (7 MPa, ~1,000 psi). If a standard plunger piston head is used without a diaphragm, pressures of over 500 bars are possible (50 MPa, ~7,500 psi).

An alternative way of varying the stroke is afforded by the design in Figure 21. The piston is actuated by motion of the vertical crank which has a fixed travel at its lower end. The adjustment at the top of the unit moves the position of the top of the crank up or down. Since the amplitude of motion of the crank varies along its length, this adjustment varies the stroke of the piston plunger.

A cam-yoke follower is located on each side of the crank arm. The purpose of the spring is to permit the distance between the ball and bearing centers to vary, while keeping a heavy pressure on the crank arm during its return stroke.

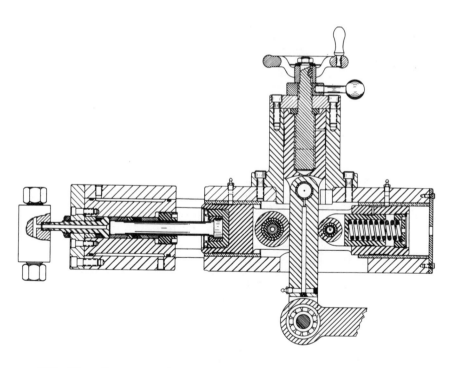

FIG. 21. Cross-section of a proportioning pump for use to 4 kbar (.4 GPa, ~60,000 psi). (Figure courtesy American Instrument Company, Silver Spring, Md.)

The maximum stroke is arranged to coincide with a vertical position of the crank arm. Thus, irrespective of stroke, the plunger always returns to the same clearance position, maintaining the clearance
volume constant. This compares with the earlier design shown in Figure 20, where the clearance volume varies with stroke. The limiting
pressure for the design in Figure 21 is ~4 kbar (0.4 GPa, ~60,000
psi).

D.   Fluid-Operated Units (Intensifiers)

The principle of the single-acting intensifier is shown in Figure
22, while double-acting units have been introduced in Figure 7(d).
Hydraulic fluid is used to drive a piston with area A, connected to a
piston with reduced area, a, compressing fluid in the high-pressure
head. Apart from friction forces, the force-balance equation gives
the fluid pressure (P) in terms of the hydraulic (drive) pressure (P')
as:

$$P = \frac{A}{a} P' \qquad\qquad\qquad (21)$$

Intensifiers generally operate with area (intensification) ratios
between 10-1000 and are available with maximum pressures up to 14 kb.
Long strokes can be used, resulting in very high compression ratios
and high efficiency.

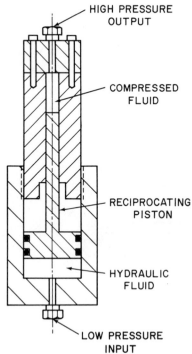

FIG. 22.   Principle of the hydraulic intensifier.

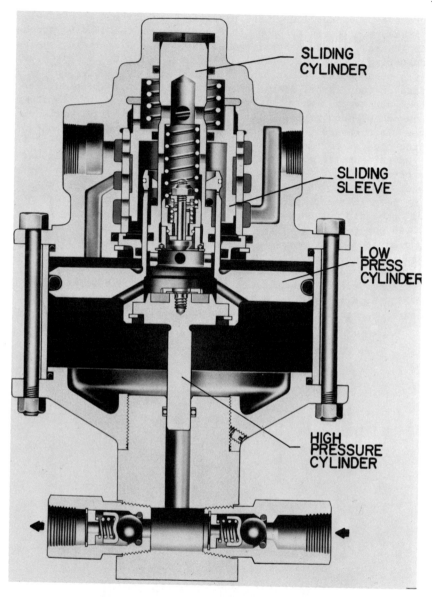

FIG. 23.   Cross-section of an air-driven, single stage plunger pump. (Figure courtesy S.C. Hydraulic Eng. Corp., Los Angeles, Ca.)

A schematic of a single-acting, air-driven pump is shown in Figure 23. The plunger piston operates with ball-type check valves. Air drive is controlled by the slide valve. When the piston reaches its lowest position, the sleeve is repositioned so that compressed air is fed to the bottom of the piston, forcing it upward, while air above the piston is vented via the side port. Conversely, when the piston reaches its top position, the slide valve is positioned so that air is

fed to the upper surface of the piston. The position is controlled by
air pressure. It can be seen that a flange on the sliding sleeve extends inwards to the central cylinder valve, and is sealed against it
by an O-ring. When the central, or sliding cylinder is raised, air is
fed via the ports (seen near the top of the cylinder) to the upper
surface of the flange, forcing it downwards. This also drives the
piston downwards. When it reaches near the end of its travel, a
flange on the main piston extension strikes the sliding cylinder and
forces it downwards. The vent ports are then below the flange on the
sliding sleeve, forcing it upwards and reversing flow.

Units such as the one shown are available in many sizes and produce pressure up to ~5,000 bars (.5 GPa, ~70,000 psi). An external
view of a similar unit is shown in Figure 24. The directional valve
can be seen in the foreground, with the high pressure cylinder mounted
above the air cylinder in this case. Shop (drive) air at 6 bar (0.6
MPa, ~90 psi) enters from the left through the dust and moisture filter, pressure control and oil drip-feed unit, to lubricate the low
pressure piston. Provision must be made in these units for venting
the spent air suitably.

A schematic of a double-acting, air-driven unit is shown in Figure 25. Both this unit and the one shown in Figure 24 use the high
pressure head shown in Figure 6. By controlling the pneumatic pres-

FIG. 24. Photograph of an air-driven, single stage plunger
pump. (Figure courtesy American Instrument Company, Silver Spring,
Md.)

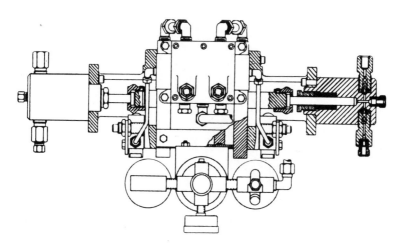

FIG. 25. Partial cross-section of a double-stage, air-driven pump. (Figure courtesy American Instrument Company, Silver Spring, Md.)

sure with the regulator valve, the pump will operate until a pressure is reached at which the forces across the pistons are balanced, and the pump stops. When pressure falls, the pump automatically restarts. With the double-stage unit, a maximum pressure of 7 kb (.7 GPa, ~100,000 psi) is possible. With air pressure at 6.2 bars (0.62 MPa, ~90 psia), these pumps operate at 130 strokes/minute free-flow. They are small volume, pumping only 7 gallons/hour ($\sim 2.5 \times 10^{-2} m^3$/hour) under free-flow conditions.

For research purposes and for the autofrettage of cylinders, hydraulically driven pumps and compressors, capable of reaching 14 kbar (1.4 GPa, ~200,000 psi), are available [14]. An intensifier for use to this pressure is illustrated in Figure 26. A photograph of the unit has been given earlier (Figure 5). The lower piston uses compression seals and is provided with an arrangement for forcing the piston down as well as up. The high-pressure head is of double-wall construction and is attached to the lower hydraulic cylinder with retaining ring and bolts. The piston shaft is of very hard steel or tungsten carbide, while the piston seal is of the Bridgman mushroom seal type. The piston shaft is sealed hydraulically with a compression seal with Vee packings. Bridgman unsupported area seals are again used in the top plug. The unsupported area is derived from the backing ring which sits on a 45° shoulder, and only contacts initially over about half the surface area. This seal ring is made of a steel such as AISI 4340 at ~50 Rc, so that only a small amount of deformation occurs at the highest pressure (see Chapter 5 for further discussion).

In units of this sort, the piston is advanced slowly. For production, or processing application where it is a disadvantage to have the supply of compressed fluid interrupted for the lengthy periods of time during which the piston is retracted and the unit recharged, a

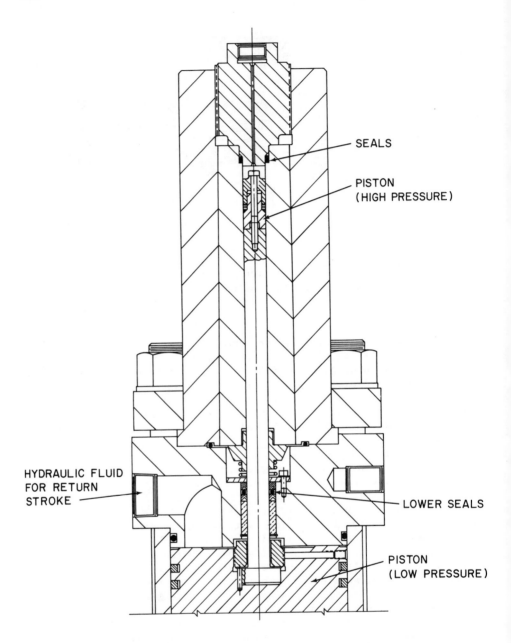

FIG. 26. Cross-section of an intensifier for use to 14 kbar.
(Figure courtesy Harwood Engineering Company, Walpole, Mass.)

double-acting unit can be used (Figure 27). Both of the above two
units can be supplied with pistons of different diameters applicable
to different pressure ranges.

FIG. 27. Photograph of a horizontal, hydraulically driven
intensifier. The unit shown is for 50,000 psi (~3,500 bars,
.35 GPa), developing 100 HP (~75 kW). With different piston
sizes the pressure range from 0.5-14 kbar can be covered. (Figure
courtesy Harwood Engineering Company, Walpole, Mass.)

Finally, the hypercompressor used for the production of polyethylene (see Figure 18) can be driven hydraulically. Such a unit is
shown in Figure 28 [4, 15].

E.   Diaphragm Units

Another form of reciprocating compressor is the diaphragm compressor. The basic principle is illustrated in Figure 29. Hydraulic
oil is fed to one side of a flexible diaphragm through a distribution
plate. The diaphragm is free to flex in a lens-shaped cavity (part 3)
made by the top of the distribution plate and the lower side of the
head. As the diaphragm is flexed, fluid is forced in through the inlet (part 4) and discharged through the outlet (part 5) check valves.
The particular importance of this compressor is that it is non-contam-
inating and will seal toxic fluid from the local environment. This
feature is used in critical applications such as gas purification
cycles in the nuclear power industry, and for charging bottles of compressed gas. Another advantage is that pressure ratios of 20:1 can be
handled easily without undue temperature rises because the large area
of the head acts as a heat sink.

The unit shown in Figure 29 is driven by a crank. Some oil from
the sump is fed during each backstroke by an eccentrically driven piston pump (part 7) into the cylinder above the reciprocating drive-pis-
ton (part 1). This ensures that the supply of oil is sufficient for
each stroke, since some may be forced to return to the sump via the
relief valve (part 6).

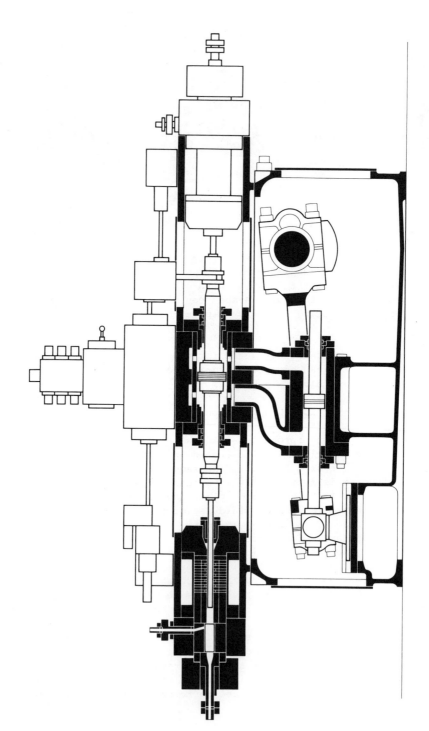

FIG. 28. Cross-section of a horizontal, opposed hypercompressor using hydraulic drive for ethylene compression to 3,500 bars (.35 GPa, ~50,000 psi). (Figure courtesy Burckhardt Engineering Works, Ltd., Basel, Switzerland.)

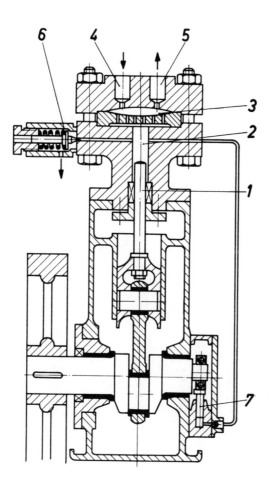

FIG. 29. Simplified cross-section of a crank-driven diaphragm compressor. The diaphragm lies flat against the oil distribution plate. Descriptive numbers are explained in the text.

In operation, the diaphragm compresses the fluid above it during the discharge stroke. The diaphragm has approximately balanced forces on it until it reaches the head, which forms the top surface of the lens-shaped cavity. At this point, oil pressure can build up rapidly below the piston, so that the relief valve serves the necessary purpose of preventing the diaphragm from being over-pressurized, leading to rupture of the diaphragm in the valve ports.

A diagram showing more complete details of such compressors is given in Figure 30. A cylinder liner of hydraulic tubing is used inside the cylinder of heat-treated steel. Diaphragms can be made of mild steel, stainless steel, monel or other materials, depending on the application. Units of this type can handle gases up to 2 kbar (0.2 GPa, ~30,000 psi) at volume rates up to ~1,000 SCFM (~28m^3/minute) with power input of up to 150 HP (~112 kw).

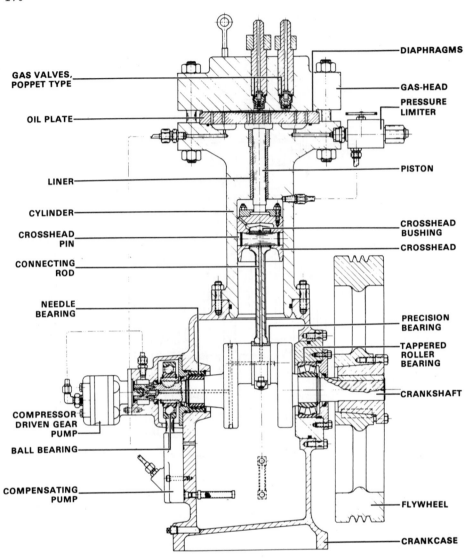

GAS VALVES,
POPPET TYPE

OIL PLATE

LINER

CYLINDER

CROSSHEAD
PIN

CONNECTING
ROD

NEEDLE
BEARING

COMPRESSOR
DRIVEN GEAR
PUMP

BALL BEARING

COMPENSATING
PUMP

DIAPHRAGMS

GAS-HEAD

PRESSURE
LIMITER

PISTON

CROSSHEAD
BUSHING

CROSSHEAD

PRECISION
BEARING

TAPPERED
ROLLER
BEARING

CRANKSHAFT

FLYWHEEL

CRANKCASE

FIG. 30. Cross-section of an industrial crank-driven,
diaphragm compressor, for use to 1 kbar (100 MPa, ~15,000 psi).
(Figure courtesy American Instrument Company, Silver Spring, Md.)

A double-ended air-driven unit is shown in Figure 31, using the
same body as the plunger pump depicted in Figure 25. The two sides
may be run in parallel or series, depending on the needs of the application. The unit shown in Figure 31 has heads for use to 700 bars
(70 MPa, ~10,000 psi) operating in parallel, while the unit in Figure
32 has one head for use to 1.4 kbar (140 MPa, ~20,000 psi), and operates in series. Note the simple hand-driven crank in Figure 32.

FIG. 31. Photograph of a horizontal, opposed, air-driven
diaphragm compressor for use to 700 bars (70 MPa, ~10,000 psi).
(Figure courtesy American Instrument Company, Silver Spring, Md.)

FIG. 32. Photograph of a hand-operated diaphragm pump.
The right hand unit is for use to 700 bars (70 MPa, ~10,000 psi)
delivering fluid to the left hand stage for use to 1.4 kbar (140 MPa,
~20,000 psi). (Figure courtesy American Instrument Company,
Silver Spring, Md.)

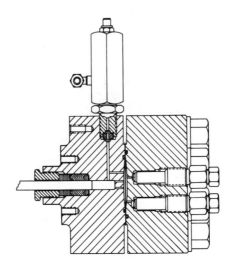

FIG. 33.   Cross-section of the higher pressure head of the
diaphragm compressor shown in Figure 32.   (Figure courtesy
American Instrument Company, Silver Spring, Md.)

Details of the high pressure head of the unit shown in Figure   32
are shown in Figure   33.   In this unit, the oil distribution plate is
integral with the oil chamber. The standard Vee compression seal   for
the   plunger   piston   can be seen. The relief valve is secured to the
head with a lens-ring seal (see   previous   chapter),   while   the   top
flange   is   secured with a series of bolts. O-ring seals are provided
as secondary seals for the diaphragm.

One manufacturer of diaphragm compressors, Pressure   Products   Industries, uses their patented Bootstrap   Closure.   At pressures of 667
bars (66.7 MPa   or   10,000 psi) and higher, sealing of the compressor
head   assemblies   becomes   difficult   and   high torque of the bolts is
needed to accomplish a good, tight seal of the   diaphragm.   The   bootstrap closure solves this problem (see Figure   34).   A force is generated in   the   bootstrap   cavity (between   1   and 3) by hydraulic fluid
pressure, which is greater   than   the   loads generated during the compression cycle. This assures a leak-tight seal assembly.

<u>F.   The Mercury Piston Compressor [16]</u>

The mercury piston gas compressor uses a piston of liquid mercury
to separate   hydraulic oil   from the gas to be compressed. Thus, the
difficult   problem of directly compressing gas to high pressure can be
circumvented. The mercury piston compressor   cannot be used as an intensifier, since the pressure   across   the   fluid/liquid mercury interface must be continuous. Historically, such units were of   great   importance   in   equation of state studies for high pressure and in early
equipment for the manufacture of polyethylene [4].

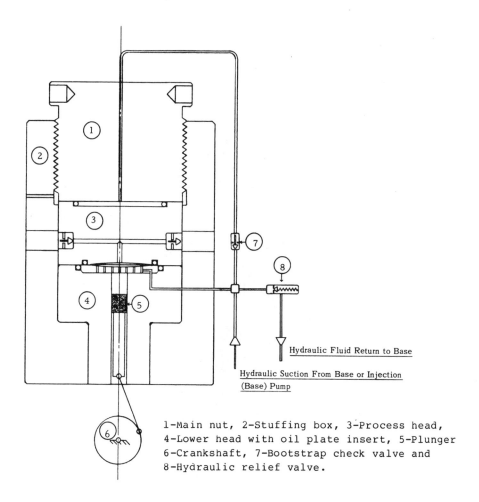

Hydraulic Fluid Return to Base

Hydraulic Suction From Base or Injection
(Base) Pump

1-Main nut, 2-Stuffing box, 3-Process head,
4-Lower head with oil plate insert, 5-Plunger
6-Crankshaft, 7-Bootstrap check valve and
8-Hydraulic relief valve.

FIG. 34.   Bootstrap closure (U.S. Patent No. 3,052,188).
(Figure courtesy of Pressure Products Industries, Hatboro, Pa.)

The basic principle is illustrated in Figure 35. In early designs, the mercury was allowed to come into contact with the walls of
the containing vessel. Since mercury can embrittle steel (see Chapters 3 and 9) a new design was developed by one of the present authors [17], shown in Figure 36. It was developed for use on hydrogen
gas research to 7,000 bars (700 MPa, ~100,000 psi) and can be used to
10,000 bars (1 GPa, ~145,000 psi) provided the temperature is kept
above about 18°C. (Mercury freezes at 11,550 bars (~167,000 psi) at
20°C (see Chapter 8)). In this new design, only oil comes into contact with the body of the vessel. Seals are simple O-ring and triangle, back-up rings (see previous chapter). The top closure has a
long, tube-type extension which dips into the mercury held in a container with a closed bottom. Oil can freely flow around and under the

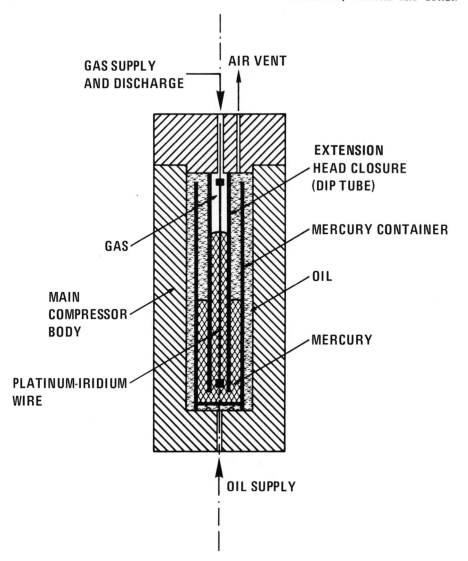

FIG. 35.   Schematic of a mercury piston compressor.

mercury container so that all parts in the vessel are at the same
pressure.   Thus, there is no pressure differential across the mercury
container.

The mercury level is monitored using the resistance of a plati-
num-iridium wire; warning signals can be activated when critical levels are reached. A vent hole is provided at the top for initially
displacing any unwanted air in the system.

Mercury used in these units must be clean (at least triple dis-
tilled); cleaning methods are discussed by Comings [18].   Some consid-

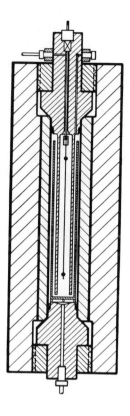

FIG. 36. Detailed cross-section of a mercury piston compressor for use to ~7 kbar (0.7 GPa, ~100,000 psi) [11].

erations of the safe use of mercury are dealt with in Chapter 4 of
this volume.

G.   Dynamic Seals [19, 20]

A principle consideration with dynamic seals is the type of shaft
motion. Rotating shafts have to be sealed in rotary compressors, and
for mechanical stirring devices in autoclaves. There are many types
of seals for reciprocating compressors, of which three are shown in
Figure 37, which depicts a double-acting, oil-free unit.
> (i)   The main piston seal is of the piston-ring type which
>   must seal a pressure differential acting in either di-
>   rection. At both ends of the piston, a wear-ring is
>   shown which does not serve a sealing, but only an
>   alignment function [21].
> (ii) The shaft seal must confine the compressed fluid and,
>   in the case shown, is of the annular cup type (see
>   later discussion) [22, 23, 24]. Seals such as this are
>   frequently internally cooled for temperature control on
>   oil-free units.

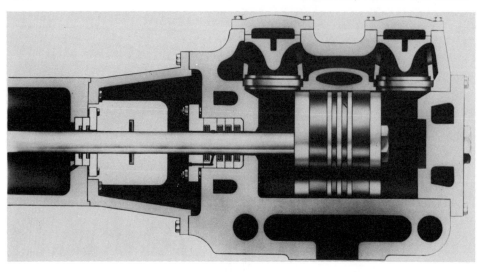

FIG. 37. Cross-section of a double-acting, oil-free compressor head showing piston sealing rings, oil-free shaft-seal
and oil shaft seal. (Figure courtesy of France Products Div.,
Garlock, Inc., Newtown, Pa.)

(iii) Oil seal rings control oil leakage from the crankcase
       along the reciprocating shaft. In the oil-free unit,
       shown in Figure 37, the design incorporates an extended
       distance piece which moves the cylinder more than a
       strokes' length from the crankcase and a slinger on the
       rod between the two to insure freedom from contamina-
       tion of the fluid by oil.

In some compressors and pumps the piston is in the form of a
plunger, which is sealed by a shaft seal, rather than a piston seal
    Some of the main considerations in the choice of a seal are:
        (i)   type of shaft motion,
        (ii)   fluid to be sealed - the corrosive nature of the fluid
               is of particular importance, since this limits the ma-
               terials that can be used in the seal. If a fluid of
               low-lubricity, such as water, is used, then particular
               care must be used in selecting sealing materials that
               self-lubricate, such as graphite.
        (iii) Temperature range - high temperatures may be encount-
               ered due to process conditions or to friction effects
               in the seal. The use of leather or rubber is limited
               to temperatures below about 100°C (220°F), while some
               synthetic elastomers may be used to $\sim$300-350°C (572-
               662°F) [26, 27]. Metal seals must be used above this
               temperature range. Heat sinks and hydraulic cooling
               may be employed to maintain the seal temperature within
               reasonable limits. Consideration should also be given

to whether temperature changes produce sufficient change in clearances to open or tighten the seal.

   (iv)   shaft speed - either rotational or linear shaft speed can determine the types of seal which may or may not be used. The speed will also influence other considerations such as temperature, wear, etc.

   (v)   wear - can be caused by lack of lubrication, incompatible sealing materials or from particles which enter the seal area from the compressed fluid, or other parts of the compressor. The finish of the rotating or reciprocating shaft is of prime importance. Surface furnishes of at least 32 μinch (0.8μm) are usually specified.

   (vi)   shock and vibration - effects of shock and vibration can be particularly severe with brittle packing materials such as ceramics.

   (vii) power requirement - a seal can be made increasingly leak-tight at the expense of friction. Eventually the power required to overcome this friction loss will become excessive. The power loss will manifest itself as heat generated in the seal, which may lead to other undesirable consequences.

   (viii) expected life of seal - all of the above considerations will affect the life of the seal. The importance of lifetime depends on the type of service, the maintenance schedule and other related factors.

   With the above factors, and the inevitable one of cost, no generalizations can be made about the choice of seal. Some examples will be given of the main types of seal for both rotating and reciprocating motion.

1.   Compression Type Packings - Stuffing Boxes [23, 27, 28]

   The compression packing works on the principle that the seal is pre-tightened to produce a sealing effect between the moving and fixed parts and, as wear occurs, is tightened again. There are many different designs for the sealing parts, but Vee-seals are very often used, and two such designs are indicated in Figure 38(a) and (b).

   The piston seal shown in Figure 38(a) is a double-acting design in which the initial compression of the seal is obtained by tightening the retaining rings onto the packing. It is important that the opposed Vee-rings be arranged as shown, rather than reversed. A number of similar designs are given in reference [23, 27, 28].

   The shaft seal shown in Figure 38(b) is shown with spring loading so that the retaining ring should be tightened onto the body, as shown, whereas in Figure 38(a), there is clearance between the two. Shaft seals of this general kind are often referred to as "Stuffing Boxes".

   One of the advantages of this type of seal is that the packing can be tightened to different levels, enabling leak rates to be adjusted to within desired levels. However, the adjustment must also be checked periodically since the packing stress inevitably decreases as

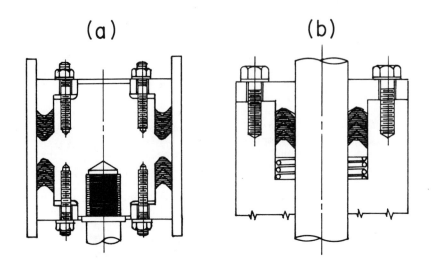

FIG. 38. Two types of compression packing using Vee-rings.
(a) A double-acting piston seal. (b) A unidirectional shaft-seal
with spring-loaded seating elements. (Figure adapted from ref. 27.)

the seal materials are worn. The spring design shown in Figure 38(b)
is less susceptible to the latter disadvantage.
     The degree of tightening is important in determining friction.
Overheating can result if the packing is squeezed too severely. Friction may be reduced for a shaft-seal by inserting an oil ring (lantern
ring) connected via ducts to a positive pressure oil supply. Starting
friction will be higher than dynamic friction, so that this should be
allowed for in the starting specifications of the driving motor, particularly if power absorbed by the shaft is an appreciable fraction of
the total.
     The number of Vee rings to be used depends on the pressure [27]:

TABLE 1

	Pressure Range	
Number of Packings	psi	bars
3	0 - 500	0 - 35
4	500 - 5,000	35 - 350
5	5,000 - 10,000	350 - 700
6	>10,000	>700

Data adapted from ref. 27.

Vee seals can be used in many applications. Leather packings are often used, and are applicable in situations where cylinder materials are abrasive, shafts slightly eccentric, or where there is excessive clearance between plunger and cylinder. Many other materials are used for Vee rings, including polymers and metals. This seal is not to be recommended for high-speed compressors.

Examples of applications of Vee rings in actual equipment are shown in Figures, 25, 26, and 33 earlier in the chapter.

2.   Piston Rings [21]

The basic principle of the piston ring is illustrated in Figure 39. The basic ring is rectangular in cross section, and has a gap which permits it to be assembled over the end of the piston. When installed, it is under compression against the cylinder. When pressure is applied, the force against the cylinder wall is increased and the ring is also forced against the groove side, resulting in good sealing.

For the ring to be maintained under compression, there must be a gap between the ends. Many designs are used with filletted ends which overlap, and reduce the leakage opening (see Figure 39(b). Leakage may be reduced further by using two or more rings arranged so that the gaps cannot line up. Although the ring is simple in conception, the manufacturer has to design the form in such a way that tension between ring and wall is uniform around the periphery of the ring.

There are many designs of piston rings including single and multiple units as well as multiple segment rings. When used in high pressure pistons, designs are selected to distribute the pressure drop across multiple rings on the piston.

The piston may accommodate other rings with a different purpose than sealing. Wear-rings serve the purpose of maintaining piston alignment and preventing contact between piston body and cylinder wall. Oil control rings are sometimes used on single-acting, trunk-type pistons where the trunk side of the piston is exposed to the crankcase.

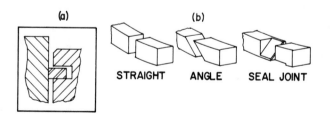

FIG. 39. (a) Principle of piston ring. Spring force in the ring seals it against the cylinder wall while differential pressure seals it axially. (b) Three examples of piston ring joints. (Figure adapted from ref. 21.)

3.   Shaft or Mechanical Seals [21, 22, 23, 24]

The shaft or mechanical seal operates in a similar way to piston
rings, but is held in tension against the shaft. As with the piston
seal, the fluid pressure provides axial force which seals the rings
against the side of the retaining groove, and may provide additional
radial force.

Several representative designs for shaft seals are illustrated in
Figure 40. If the rings are paired, they are pinned together. In design (a), radial and tangential slots are matched to reduce leakage,
while in (b), tangential slots are matched so that leakage paths are
separated by 60°. The ring, (c), is used as a "pressure breaker" to
reduce the load on the leading ring by allowed controlled leakage,
while design (d) acts as a heat barrier or "fire check", by allowing
heat to pass from the compression chamber to the water-cooled piston
rod. The above are only representative of the many different designs
and applications.

The spring in these designs around the circumference holds the
rings together and onto the shaft, even after considerable wear has
occurred. The rings are free to move radially in the ring grooves so
that the seal adjusts to the rod even if there is slight lateral motion of the rod.

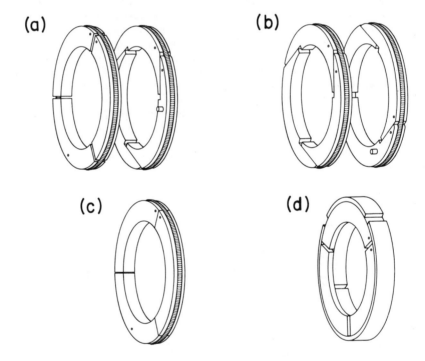

FIG. 40. Several types of packing rings used for sealing
shafts. (Figure courtesy France Products Div., Garlock, Inc.,
Newtown, Pa.)

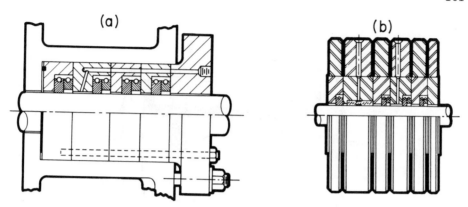

FIG. 41.   Examples of two types of shaft seal. (a) Seal
for 6,000 psi (~400 bars, 40 MPa) with passages for lubrication
and venting.   (b) Seal for ~45,000 psi (~3,000 bars, 300 MPa) with
pre-stressed cups and plunger-lubricated bushing guide. (Figure
courtesy France Products Div., Garlock, Incs., Newtown, Pa.)

Two designs utilizing rings for shaft seals are given in Figure
41(a) and (b). The design in Figure 41(a) is for use to ~400 bars (40
MPa or ~6,000 psi). Several pairs of rings are used, and the cups and
flanges are internally drilled for oil lubrication and venting. Note
how the cups and flanges are held together by tie rods which also
serve to align internal drilled passages for lubrication and venting.

The design shown in Figure 41(b) is for high pressure application
~3,000 bars (300 MPa or 45,000 psi). The cups are prestressed by the
outer rings, and the design includes a plunger guide bushing and case
cooling. Shaft seal rings such as this are used in compressors for
polyethylene production.

For compressors, both piston and shaft seal rings are most often
manufactured of material with self-lubricating properties, e.g.,
chromium-plated cast iron, tetrafluorethylene (TFE) resins filled with
wear-resistant, heat-conducting additives such as glass, graphite,
bronze, molybdenum disulphide, etc. [20]. If oil lubrication is adequate, pearlite grey iron, with spring properties introduced by ham-
mering, is used.

In addition to the designs shown, mechanical seals have been illustrated in Figures 14, 16, 18 37.

4.   Pressure-Activated Seals [16, 20, 23, 27, 29]

This type of seal works on the principle that the fluid pressure
acts on the seal to increase the contact force between sealing element
and moving part. Thus, the seal becomes more effective as the pressure increases. The upper pressure limit is set by the deformation of
the seal to the point where the unsupported area vanishes.

Friction effects in the seal can be important because of the
large sealing forces, and also because only one sealing element is

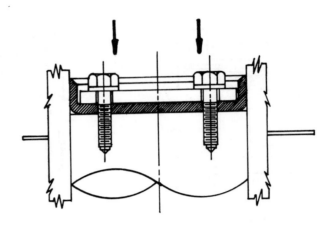

FIG. 42. Schematic of a cup-packing. (Figure adapted from ref. 27.)

normally used. However, the contact area is usually high so that wear is more of a problem than high surface temperatures.

The two basic types of pressure-activated seal are the lip-type and the compression-deflection type.

Lip-type seals include piston cups, flange packing U-cups, and Vee rings. A representative design is shown in Figure 42. Lip seals are often used for oil seals on shafts, where the initial spring force is responsible for the sealing effect and such seals only operate in the pressure range up to a few bars.

The compression-deflection type is exemplified by the O-ring. Under the action of pressure, the ring deforms against the back surface of the retaining groove. However, with a typical elastomer material such as Buna-N, the unsupported area of an O-ring is usually inadequate above ~35 bars (3.5 MPa or 500 psi) and extrusion between piston and cylinder can be serious by ~100 bars (10 MPa or 1,500 psi). (See Figure 43.) The use of a back-up ring prevents extrusion, but unless this ring has an unsupported area (in the assembled condition), the seal is not pressure-activated. With all ring shaft seals, whether rotating or reciprocating, a fine finish is required on the shaft - usually 32 μinch (0.8 μm).

Other materials can be used to increase the pressure range of the O-ring, and the cross-section can take many initial shapes. (See reference 16.)

The classic Bridgman mushroom packing design (Figure 44) can be used to seal even leak-prone gases such as helium to pressures in excess of 20 kb (2 GPa or ~290,000 psi). Anti-extrusion rings are usually made of bronze, while disc-rings are made of Neoprene® (or preferably Adiprene®) for a low-pressure seal, with rings of teflon and cooper on either side. Many other combinations of materials have been used. The principle of the unsupported area has already been discussed for an analogous static packing in the previous chapter. This type of seal can only be used with very slow piston speeds so that its use is limited to intensifiers. (An example of a unit using such a piston has been given in Figure 26.)

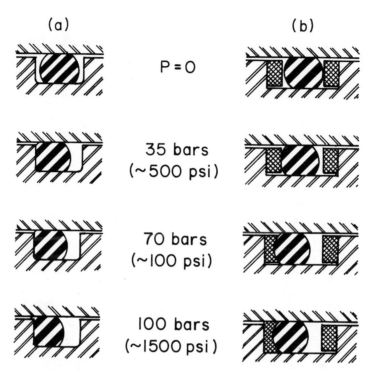

FIG. 43.   The deformation of an O-ring made of typical elastomeric material at different pressures.   (a) Un-backed. (b) Backed with stiffer elastomeric material.   (Figure courtesy E. F. Houghton Co., Lynchburg, Va., ref. 23).

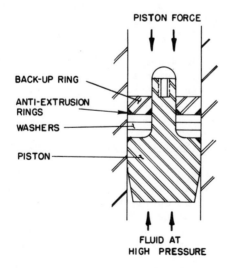

FIG. 44.   Schematic of a Bridgman piston seal using the unsupported area principle [16].

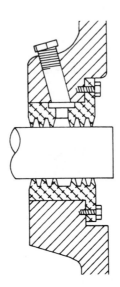

FIG. 45. Schematic of a labyrinth rotary shaft seal used in a
low pressure centrifugal compressor. (Figure courtesy Dresser
Industries, Inc., Roots Blower Operation, Connersville, Ind.)

5.   The Labyrinth Seal [19, 20]

The labyrinth seal is based on the principle that controlled
leakage of fluid can be obtained for a given pressure drop along a
passage of small area if the path is sufficiently long.   Essentially,
the seals shown in Figure 38 employ the labyrinth principle, as well
as multiple piston rings and mechanical seals for high-pressure service, discussed earlier.

The labyrinth seal is widely used for sealing rotating shafts in
turbine compressors. The design shown in Figure 45 is used for a low
pressure 0.6 bars (60 kPa or 10 psi) compressor. However, the vent
spacing can be used for buffer gas application if necessary, which decreases the leak rate.

6.   The Face-Type Shaft Seal [19, 20, 21, 24, 30, 31]

The face seal is the preferred type for rotating shafts where low
leak rates are critical. It consists of a stationary sealing ring,
usually of carbon, which is spring-loaded to seat against a rotating,
flat, precision-finished plate or collar attached to the shaft (Figure
46). A secondary seal is made between the stationary ring and the
body. This seal may be an O-ring, piston ring, machined lip ring, or
bellows seal. The face-seal ring can move axially, relative to the
seal body and secondary seal.

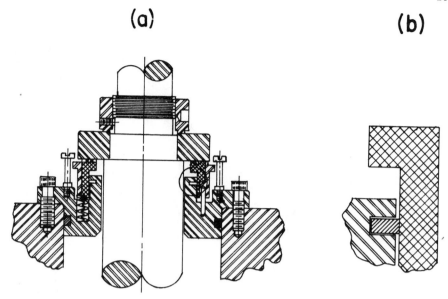

(a)   **(b)**

FIG. 46. Schematic of a gas face seal. (Figure courtesy of Koppers Co., Inc., Baltimore, Md., ref. 21).

The advantages of this seal include positive sealing, small rubbing friction between shaft or shaft sleeve and seal parts, small length, long service life and automatic adjustment within design limitations to compensate for shaft-whip, axial end-play, vibration and wear of sealing faces. The surface areas can be designed so that the hydraulic forces are balanced across the seal. The parts are precision-machined and must be handled with care, particularly to avoid damage or contamination to critical faces.

Bellows or diaphragm secondary seals completely eliminate leaks except past the face seal. If made of metal, the unit can be used to high or low temperature (e.g., $-400 \rightarrow 1,200°F$; $-240 \rightarrow 650°C$). With bellows seals pressure differentials up to ~70 bars (7 MPa or 1,000 psi) can be used, and the seal can be axially balanced within certain limits.

Face seals can become quite complicated when used in multi-stage centrifugal compressors. The unit shown in Figure 47 can withstand differential pressures up to ~100 bars (~10 MPa or 1,500 psi).

In practice, shaft seals may include several types of sealing element and principle in addition to the face seal. An example is shown in Figure 47, used by Delaval. These seals operate flooded on the oil side at nominally ~3 bars (0.3 MPa or 45 psi) above the gas pressure. The oil flow acts to cool the seal, while its pressure hydraulically balances the floating seal element. A thin oil film then lubricates the seal face. The differential pressure across the seal gives low leakage rates with pressures up to ~120 bars (12 MPa or 1,800 psi), with leak-prone gases such as light hydrocarbons.

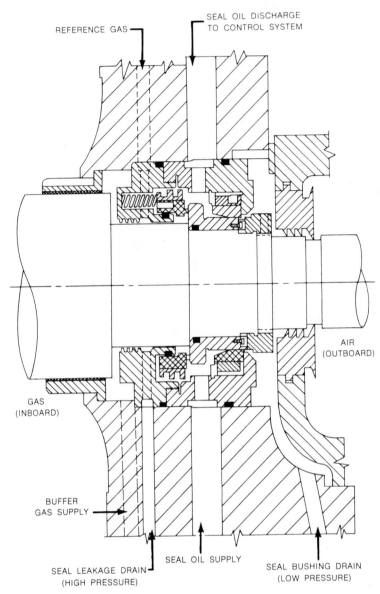

FIG. 47. Schematic of a shaft seal with oil compensation.
(Figure courtesy Delaval Turbine Inc., Trenton, N.J.)

H.   Compressor Valves [4, 32, 33]

     The function of the compressor valve is to permit one-way flow of
fluid into the compressor chamber (inlet check valve) and out of it
(outlet check valve). Valves are critical components of compressors
since they can control the overall efficiency and cause extensive damage if breakage occurs.

The checking elements are subjected to harsh conditions. Each valve must open and close once for every revolution of the crankshaft. With a crankshaft rotating at 700 rpm (~11.6 Hz) for 8 hours per day, 250 days per year, the valves will have opened and closed 336,000 times in one day, 84,000,000 times in a year. A typical valve movement (valve lift) of 0.035" (~0.9 mm) represents the displacement through which the checking element is accelerated and decelerated. Travel in one year would be ~1/3 mile (~1/2 km). On closing, the impact forces between element and seal can be high.

The valve should have several desirable qualities, some of which are mutually incompatible:

   (i)   for fast response the element should be light, with small lift, no "bounce", and have small clearance space,

  (ii)   when closed the valve should be leak-tight,

 (iii)   the checking element should be rugged to withstand the repeated shocks to which it is subjected,

  (iv)   the pressure drop across the valve should be low,

   (v)   the valve should be easy to service and maintain and be able to accept replacement parts without elaborate "fitting" or "adjustment" procedures,

  (vi)   the valve should run smoothly and quietly even at high speed,

 (vii)   finally, it should be inexpensive.

The selection of materials is important for the above reasons, and also because the compressed fluid and foreign entrainments can erode, corrode and abrade the valve. The three critical components are the moving, or checking element, the seat, and the spring.

In most valves the spring fails in fatigue, often coupled with corrosion. Corrosion resistance can be increased by plating with a resistant metal. High discharge temperatures cause earlier failures, so that monitoring the head temperature may be recommended.

The spring force is critical, since too high a force produces inefficient opening whilst too weak a force delays closing and may lead to valve bounce. Normally, when the spring is fully loaded, the spring force is ~0.005-0.25 times the force exerted by the fluid on the throughput channel area. For high speed compressors this ratio may rise to 0.6 and 0.8. The spring stress at maximum compression should be below the fatigue limit and the natural spring frequency well removed from the compressor frequency or multiples of this frequency. Whereas an outlet check valve may operate for some time with a broken spring (albeit inefficiently) a broken spring on a suction valve can be much more serious, since broken parts can fall directly into the cylinder.

There are basically two types of valve mechanism:

   (i)   automatic - opened by fluid pressure and closed by a spring,

  (ii)   mechanical - opened by linkage from the crankshaft and closed by a spring.

Almost all modern reciprocating compressors use the automatic type because of their simplicity.

Small, high pressure pumps and compressors often use ball-type check valves. Examples have been illustrated in Figures 6, 15(a), 20, 23. The main disadvantage of these valves is the high mass of the moving element, which delays valve closing. However, the elements are strong and can be used if necessary to very high pressures (e.g., 14 kb), (~200,000 psi). Another advantage is that parts cannot fall into the cylinder past the balls. Valves employing multiple balls mounted in a plate have been constructed for critical applications, such as for ammonia synthesis compressors. A discussion of ball check valves has been given in the previous chapter.

Poppet type check valves are also used for small, higher pressure units, and examples have been given in Figures 14, 30, 37. In the poppet valve shown in Figure 48(a), for discharge, fluid passes through the orifice in the bottom, and exits through radial holes near the base, passes the spring and leaves through the top. Similarly, for the suction valve, Figure 48(b), fluid enters the core, exits via inclined holes near the base and passes through the space between the valve poppet and housing at the bottom.

For larger high-pressure compressors, the checking element is usually in the form of a disk or plate with holes in it. Cross-sections of a design using a single disk are shown in Figure 49, while an exploded view of a similar design is shown in Figure 50. In the simpler design shown in Figure 49, a helical spring forces the valve disc against the valve seat. The seat has several holes drilled in it which are covered when the disc is seated. Fluid can pass around the disc when it is unseated, exiting through the retainer.

More than one disc can be used in a valve as shown in the exploded views given in Figures 51(a) and (b). In these designs, the discs are guided by protrusions from the seat. The springs are small compression units, rather than one large spring as in the section (Figure 49). Each spring is seated in a well in the seat and, as in Figure 51(a), activates two discs. By comparison, the springs in Figure 51(b) activate only one disc (3 springs to each disc).

The design shown in Figure 52 is of the type patented by Hoerbiger near the end of the 19th Century, and is referred to as a Hoerbiger valve. The disc has several slots cut into it at different values of the radii. When seated, the uncut segments of the plate close off the inlet holes or grooves. In this design the plate is fixed at the center, so that lift of the plate is achieved by distortion of the plate. Initial spring force is provided by the small helical springs on the plate. When the plate strikes the damper plates, further slowing down of the plate occurs. The inclusion of a damper plate and secondary spring mechanisms reduces the wear and increases service life.

An exploded view of a similar design is shown in Figure 53(a), in which the operation is very similar to that in Figure 52 except that two sets of springs are used, separated by a damper plate. The springs can be in the form of fingers, which extend underneath the spring plates. The valve plate is fixed at the center and the design of the plate, which enables it to flex, can be seen. The thickness of the lift washers control the valve lift.

(a)   (b)

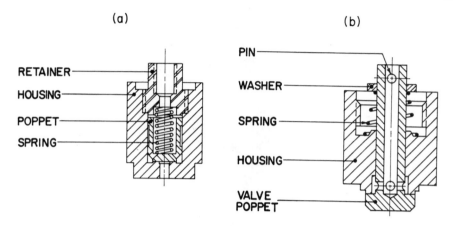

FIG. 48.   Cross section of a typical poppet-type check valve.
(a) Discharge valve.   (b) Inlet valve.

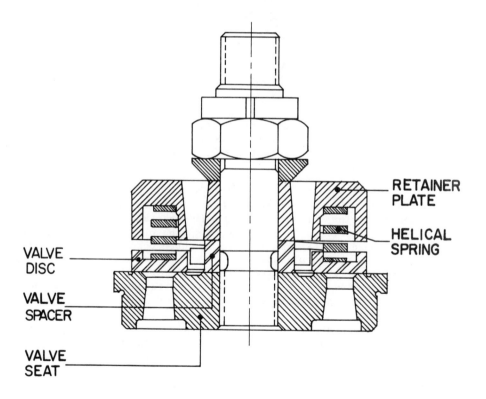

FIG. 49.   Cross section of a single disc valve with helical
spring. (Figure courtesy of American Instrument Company, Silver
Spring, Md.)

FIG. 50. Exploded view of a single disc valve similar to the
one illustrated in Figure 49. (Figure courtesy of Hoerbiger
Corporation of America, Roslyn, N.Y., ref. 33.)

     Typical curves for the spring force versus displacement for these
valves are shown in Figures 53(b) and (c). In Figure 53(b), the action of a second spring set acting on the damper plate can be seen as a
decrease in the slope. For a design with spring plates the curves are
smoothed, as shown in Figure 53(c) for three possible designs with
one, two, or three spring plates.
     Other important types of valves are strip valves, channel valves,
reed valves, etc. The reader is directed to reference 9 for further
discussion of these types which are generally not used for high-pressure applications.
     Important parameters in the design of valves are the valve lift
area and the fluid speed. The valve lift area is the minimum area between the valve plate and its seat when it is fully open. This is usually the smallest flow area in the machine so that the flow speed (V)
is largest there. From the continuity equation

$$V = \frac{A}{Na} C \qquad\qquad\qquad (22)$$

(a)   (b)

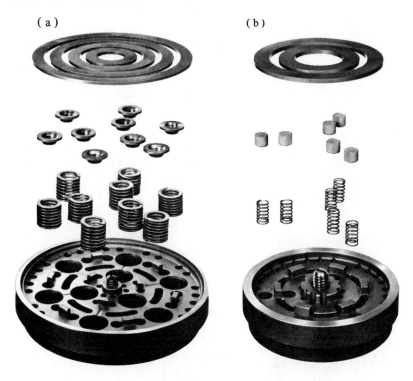

FIG. 51.   Two multiple disc valves, shown in exploded view
without the cover.   (Figure courtesy France Products Division,
Garlock, Inc., Newtown, Pa., ref. 32).

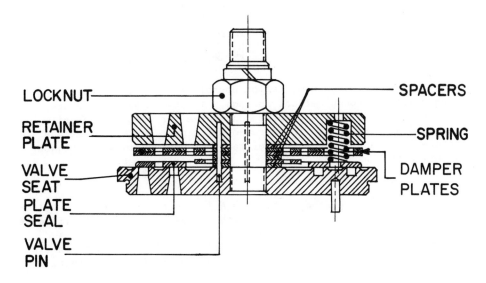

FIG. 52.   Cross section of a Hoerbiger type valve.   (Figure
courtesy of Hoerbiger Corp. of America, Roslyn, N.Y., ref. 33).

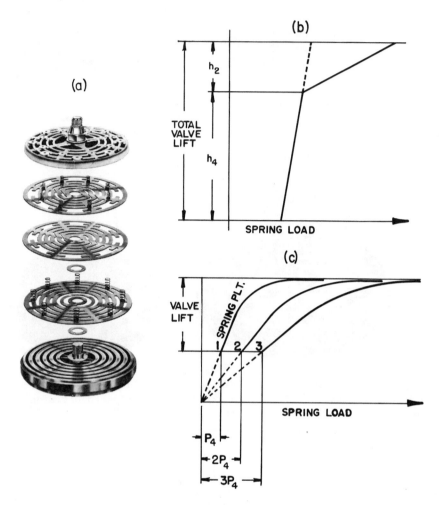

FIG. 53. (a) Exploded view of Hoerbiger valve with three
plates coupled through helical springs. The two upper plates are
damper plates. (b) Force-displacement curve for design with coil
spring and damper plate. (c) Force-displacement curve for designs
with one, two, and three spring plates (not coil springs). (Figure
courtesy Hoerbiger Corp. of America, Roslyn, N.Y., ref. 33).

where V is the mean piston speed and A is its area; Na is the valve-
lift area of all N discharge or suction valves per cylinder (if more
than one is used).

In practical units, if the valve-lift area is expressed in in^2,
the stroke in inches, and the swept volume, D, in ft^3/min., then,

$$V = \frac{144D}{Na} \text{ ft/min} \qquad (23a)$$

In S.I. units, the displacement would be expressed in m^3/sec, the valve lift area in m^2 and

$$V = \frac{D}{Na} \text{ m/sec} \qquad (23b)$$

Some manufacturers use valve lift velocities of 6,000 ft/min (~30 m/sec) while others prefer lower values (2,000 ft/min or ~10 m/sec). In high-speed compressors the speed may rise to 12,000 ft/min (~60 m/sec).

A high pressure (e.g., 700 bars, 70 MPa, ~10,000 psi) high speed (e.g., 1,000 r.p.m., ~17 Hz) compressor with disc valves would typically employ valve lift ~0.035 in (~0.9 mm). A lower speed, lower pressure unit (e.g., ~7 bars, 0.7 MPa, 100 psia, at 300 r.p.m., or 5 Hz) would use higher values of valve lift, such as 0.18" (~4.6 mm).

I.   Some Considerations of Design and Maintenance for Reciprocating Compressors

The reciprocating action of the compressor is at the same time its motive force and its inherent weakness. Cyclic stresses are produced on moving parts and in the cylinder head. As a consequence, parts must be designed for the maximum stress level, taking into account fatigue, so that they are much heavier than they would be if the stresses remained at their mean level. The fatigue life of the valves and other critical parts becomes an important factor. Furthermore, the flow of fluid through the unit is interrupted, and this may produce undesirable effects on the supply unit and in the pipes and vessels into which fluid is being delivered.

The valves are the most critical components, since their failure is most frequently the cause of compressor breakdown. A compressor must be shut down immediately a valve breaks, otherwise the consequences can be very serious. For instance, if broken parts fall onto the piston, the cylinder liner and piston can be badly damaged.

Another serious problem can arise if the inlet and discharge valves are confused. For instance, if a suction valve is placed by mistake into the outlet port, fluid can be sucked into the head, but not delivered. Pressures far above the rated level of the unit may then result in a major breakdown, with disastrous consequences. Some specifications insist that seat areas for inlet and delivery valves differ, so that interchange is not possible.

The fatigue factor becomes the principle concern in the design of the heads of compressors and pumps. Design criteria will be discussed in the next chapter. Maximum permissible stress levels are reduced well below those for units with static pressure. Furthermore, the allowable stress level is not increased by using compound cylinder construction with segments in tension if fatigue factors are predominant. However, the allowable stress level can be raised if the inner liner is in compression.

The notch-sensitivity of the materials of construction becomes a most important property. Low carbon steels have lower notch sensitivities than alloy steels; however, the latter materials may be preferred for applications to the highest pressure because they allow higher stress levels to be reached. In this case, the design should avoid stress concentration regions (stress raisers).

Failure generally starts at these stress-raisers, such as port openings, or changes in cross-section (e.g., at sealing areas). These parts must be carefully considered by the designer. Generous radii must be used and all sharp edges smoothed. Defects in the material can also be responsible for initiating failure. Forged steel is generally recommended, with large area reduction to ensure uniform mechanical properties.

Standard design criteria and materials for construction of pressure vessels are given in ASME Boiler and Pressure Vessel Code, Section VIII "Unfired Pressure Vessels" while test procedures are given in ASME Power Test Code PTC-9 (Reciprocating units) and PTC-10 (Centrifugal units). Acceptable materials are listed with chemical, mechanical and physical properties and processing information (e.g., heat treatment conditions). For parts subjected to static (non-cyclic) loading, the allowable stresses are restricted to approximately one quarter of the minimum yield strength. For parts subjected to cyclic loading, information on the allowable stress level as a function of the number of cycles is given (see discussion in Chapters 7 and 9). Other bodies perform this service in countries outside the U.S.A. However, even with these data and restrictions, designs can be improved by careful consideration of the critical components. Unfortunately, cost requirements are too often limiting factors in the design of equipment.

A design in which care has been taken to avoid stress raisers has been discussed earlier (see Figure 15). All sections of the valve cylinder subject to stress are circular and free of intersecting holes. The cost of manufacture of designs using several cylinders is increased, but, by breaking the design into several cylindrical sections, notch effects are minimized and effects can be calculated. For applications to higher pressure (e.g., $\gtrsim$ 2 kb (0.2 GPa, ~30,000 psi)), the sealing areas must also be split into several rings with interference fits, and an example is discussed in a previous section (see Figure 41(b).

In some cases, a liner of tungsten carbide is used for the cylinder. This material has a very high modulus of elasticity (~60 kbar, 6 GPa, ~900,000 psi), high compressive strength (~30 kbar; 3 GPa, ~450,000 psi) but is weak in tension. Accordingly, the liner must be in compression (see following chapter). With this type of design, fatigue life is increased both for the liner and the outer steel segments supporting it. Tungsten carbide has good wear properties and is extensively used in equipment for ultra-high pressures (see Chapter 11). Pistons for hypercompressors in the polyethylene industry are often constructed of tungsten carbide. Since the material is brittle and fails at "low", cyclic bending stress levels, care must be taken to ensure that these stress components are minimized. Furthermore, if failure occurs, this material shatters, so that precautions must be taken to stop flying parts.

The diaphragm of a diaphragm compressor is a part which is critically dependent on fatigue. The shape of the lens-ring cavity can be designed to reduce stress in the edge regions of the cavity. Calcula-

tion of the fatigue life begins with an estimate of the maximum stress
in it. Usually, data on fatigue life published in the ASME Code does
not extend to the number of cycles encountered in normal life ($\sim 10^{8}$
cycles) so that an equation for extrapolating the data is also supplied in the ASME Codes. (Typical data is given in the following
chapter.) The calculated life of the diaphragm is usually conservative, since it assumes that the unit continually operates at its maximum pressure. However, the life may be considerably reduced by extraneous factors such as hard particles in the oil or compressed
fluid, which can even lead to a puncture in the diaphragm.

Other critical components in the head of the compressor are the
fasteners used to clamp the head to the body, as in the diaphragm compressor head shown in Figure 29. The tension in the bolts must produce sufficient clamping force that the parts do not separate under
the load provided by the maximum pressure of the unit. In addition,
they may be required to provide additional clamping force for sealing
elements. The initial torque on the bolts must allow for the considerable friction forces which act between male and female parts of the
fastener. Data for estimating tightening torque is given in the ASME
Code, and a number of references [9, 12, 34, 35].

These threads are subject to cyclic loading, so that fatigue factors should also be considered. In addition, leakage of corrosive
fluids may be a serious problem, so that fasteners should be periodically checked, both for corrosion and adequate torque.

Pressure Products Industries reduces the bolt loading on the gas
head in some of their head assemblies by containing their head assembly in a pressure vessel and pressurizing both sides of it. The outer
head of the vessel must contain the load with one large threaded connection instead of multiple bolts, but this thread is not subjected to
cyclic stresses at the compressor frequency (see Figure 34).

In operation, a number of problems can be avoided by routine
checks. Problems frequently occur because the oil used in hydraulic
units is dirty. Adequate provision should be made for filtering, with
frequent checks made of the filters and the oil. If the oil becomes
discolored, it is often a sign that seals are wearing badly, or parts
such as valves, bearings are overheating.

The temperature of the compressed fluid or hydraulic oil can
often be a valuable guide to potential problems. It is useful to provide equipment with temperature monitoring devices in critical areas
such as valve-heads, oil reservoirs, etc. In some cases, a safety
device linked to an excessive temperature detector can forestall serious problems.

Extraneous noises from the unit can often be used to pinpoint
problems. For instance, if the net positive suction head (NPSH) is
insufficient to feed the unit, loud rattling or clanking may result.
In hydraulic units, compressed air is sometimes used to pressurize oil
reservoirs to forestall this problem. Particular care should be taken
in this case to ensure that reservoirs are adequately filled, otherwise air may enter the unit. Detonation may even result, as the oil-
air mixture heats and ignites.

Most manufacturers have made detailed calculations of the performance and mechanical design factors of their pumps and compressors.

In addition, they will provide maintenance information, such as charts describing noises and their possible causes, which are very helpful in diagnosing problems and checking specifications. The user should obtain as much of this information as possible.

One possible problem with reciprocating compressors is that the flow of fluid is interrupted. Thus, vessels, pipes and fittings can also be subjected to cyclic pressures. Sometimes in liquid systems, this is referred to as "hydraulic hammer". There are several publications which review ways of reducing this effect [6, 36-40].

A surge-tank is most frequently used to reduce pressure oscillations in liquid systems and to smooth flow-oscillations in gas systems. In essence, this consists of a large bottle connected very near to the compressor or pump, fitted with a shaped orifice. A quantity of fluid introduced by the pump produces a reduced pressure pulse at the outlet because of the large capacity of the bottle, and also because its shape produces some filtering action. Some heat dissipation also occurs.

Such systems have been analyzed by analogy with electrical systems. The pressure of the incoming fluid pulse is treated as a voltage, the volume of the tank as a capacitance, and dissipation as a resistance. Such designs are particularly effective in filtering the higher frequency components of the pulse. This essentially smoothes the pulse, which at the inlet has an "impulsive" waveform. Heat dissipation (resistance) reduces the overall amplitude of the pulse. Specific values for the size of the bottle, and recommendations for its shape can be found in the references quoted. A unit equipped with surge tanks has been shown in Figure 17.

Some reduction in gas flow pulsations can also be obtained with tuned resonators, analogous to resonators in car exhaust systems. However, again they are mainly effective in reducing higher frequency components, rather than reducing the amplitude of the fundamental frequency (the reciprocator frequency). Tests indicate that the flow rate from compressors can sometimes be increased by as much as 30% by "tuning" the delivery lines. The effectiveness of filters and surge tanks can be approximately predicted, but maximization of their efficiency is best obtained from practical tests.

In certain applications it is important to deliver compressed fluid from hydraulic intensifiers without the periodic fluctuations in pressure which result when the unit reaches the end of its discharge stroke. W. Newhall [41] has described a method of "Pipeless Pumping" and an application to the feed of catalyst to reactors is given in Chapter 3, Vol. II.

IV.   ROTARY PUMPS AND COMPRESSORS

A.   Positive Displacement Units [3, 7, 42]

There are many different designs of rotary, positive displacement pumps, but they share the characteristic that a rotor is carried on a driving shaft. The rotor can be in the form of a gear, vane, lobe, or screw.

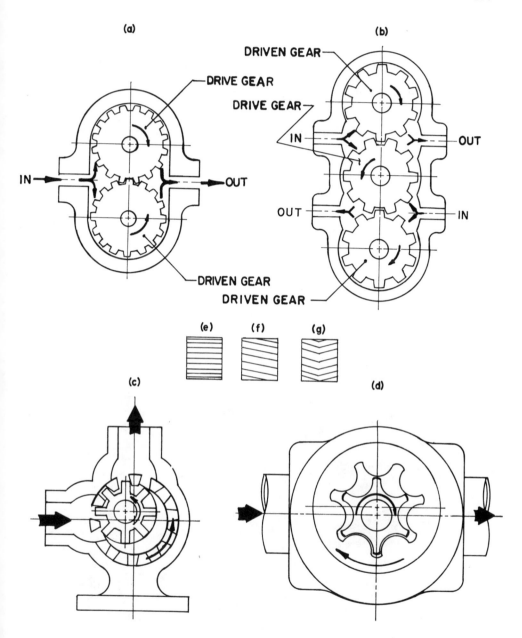

FIG. 54. Several designs of gear pumps. (a) External gear pump. (b) Three-gear pump. (c) Internal gear-and-crescent pump. (d) Gear-within-gear pump. Internal and external gears have only one tooth difference. (e) (f) (g) Some designs for gear teeth used in gear pumps. (Figures redrawn from ref. 7.)

Several designs of the gear pump are illustrated in Figure 54. The external gear pump (Figure 54(a)) consists of two gears rotating in a casing. One gear is externally driven, while the other is driven by meshing its teeth with the first. Fluid from the suction port is trapped between the gears and forced out through the exit port. Since one gear is driven by the other, they are forced together, so that back-leaks are prevented. When operated at high pressure the gears are highly stressed, so that they must be of high quality gear steel. Gears can be cut in several ways, as illustrated in Figure 54(e)(f)(g).

The output of the gear pump can be doubled by the use of three gears (Figure 54(b)). Alternatively, this design can be used as a two-stage pump, but this results in unbalanced forces across the gears. Since there is a relatively short path between the inlet and outlet paths across the central gear, the leakage is more severe than with the double gear pump, and can only be used to ~70-100 bars (~7-10 MPa or ~1,000-1,500 psi) compared with ~200-270 bars (20-27 MPa or 3,000-4,000 psi) for the double gear unit.

The principle of a gear-within-a-gear pump is shown in Figure 54 (c) and (d). The unit shown in (c) is sometimes known as a crescent and gear pump. Power is applied to the rotor gear. As the gears unmesh, liquid is drawn into the pump, liquid flows around the crescent

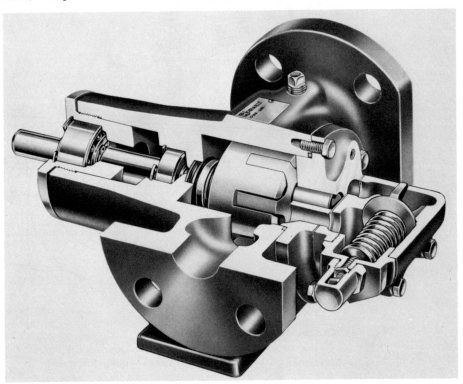

FIG. 55. Cutaway photograph of a crescent and gear pump. (Photograph courtesy of Viking Pump Division, Houdaille Industries, Cedar Falls, Iowa.

and is forced out at the discharge point as the teeth   mesh   into   the
gear. The crescent can be fixed or can rotate. A cut-away photograph
of a crescent gear pump for use to ~18 bars (1.8 MPa or   250   psi)   is
shown   in   Figure 55.   The spring in the foreground is a safety relief
valve and the shaft seal can be   seen   immediately   behind   the   rotor
gear, in the form of a mechanical seal.

An alternative design shown in Figure   54(d),   uses   an   internal
gear with only one tooth difference between the two gears. A separat-
ing element is not required.

A screw   pump   is illustrated in Figure 56. It is an axial flow
pump   whose basic principle was invented by Archimedes. Although   the
gears rely on edge contact, or in some   cases   do   not contact at all,

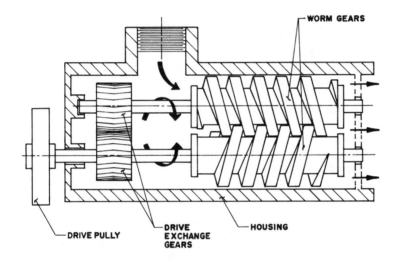

FIG. 56.   Schematic of a screw pump.

FIG. 57.   Cutaway photograph of a balanced, Sier-Bath screw pump.
(Photograph courtesy of worthington (Canada), Ltd., Brantford,
Ontario.)

sealing can be made more effective by employing long path lengths. The gears are subject to high-end loading, and in order to overcome this, balanced, or double-ended units are used.

The balanced unit shown in Figure 57 is for service to ~241 bars (24.1 MPa or 3,500 psi) with viscous fluids with flow rates up to 10,000 gallons per minute (~38m^3/minute). It can handle non-lubricating liquids or semi-solids over an enormous range of viscosities from 32 to 200,000,000 SSU (~1 - 43 x 10^6 centipoises). Since the screws do not touch, wear is reduced, except at the drive gears and the seals which have no axial thrust on them since the unit is balanced.

Although most rotary, positive displacement units are used as pumps, screw units are widely used for compressors also. For instance, one major manufacturer offers compressors with discharge of ~10 bars (1 MPa or 145 psi), at rates up to ~120 ft^3/min. (~3.4 m^3/min.). Oil is used to lubricate and seal the screws in most of these units; however, oil-free compression can be obtained in units where water serves these functions.

Another type of rotary unit employs lobes for the rotating parts. An example of the action of one design is shown in Figure 58. In (1), the upper impeller or lobe has started its suction cycle and is completing the discharge cycle while the lower impeller is transferring liquid from the suction to the discharge side. In (2), the upper impeller has completed its suction cycle and the lower impeller is starting to discharge. (3) and (4) represent similar stages to (1) and (2), except that the functions of upper and lower impellers are reversed.

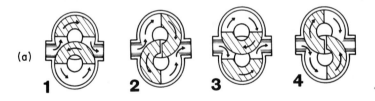

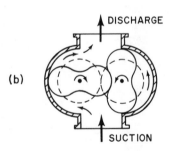

FIG. 58. (a) Action of a lobe type rotary pump, showing successive positions of the lobes. (Figure courtesy of Tuthill Pump Company, Chicago, Ill.) (b) Sketch of a rotary lobe blower as manufactured by the Roots Blower Operation of Dresser Industries, Inc. (Figure courtesy of Dresser Industries, Inc., Roots Blower Operation, Connersville, Ind.)

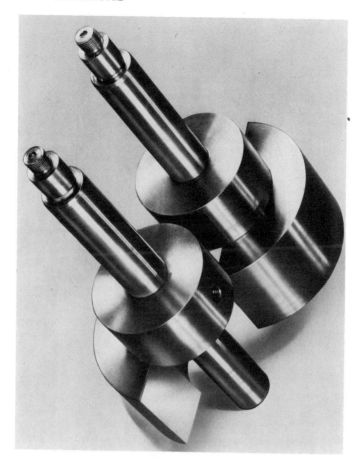

FIG. 59.   Photograph of two impellers used in a lobe-type
process pump.   (Photograph courtesy of Tuthill Pump Company,
Chicago, Ill.)

A photograph of two impellers used in a process pump is illustrated in Figure 59.   Since the lobes are never in contact, wear problems
are eliminated, except in the shaft seals.   These pumps can run dry
without damage and provide heavy duty pumping for many industrial   applications up to ~30 bars (3 MPa or ~450 psi), including corrosive and
viscous liquid applications.

A sketch of a rotary lobe blower is given in Figure 58(b).   Two
figure eight impellers are mounted on parallel shafts which rotate in
opposite directions within a cylindrical housing.   As each lobe passes
the inlet, it traps a definite volume of air or gas and carries it
around the case to the discharge.

The principle of the vane pump is shown in Figure 60.   In the
sliding vane pump, the rotor is mounted off the axis of the casing
cylinder.   As the vanes rotate, centrifugal or spring force keeps them
against the walls and they trap fluid between segments, compress it

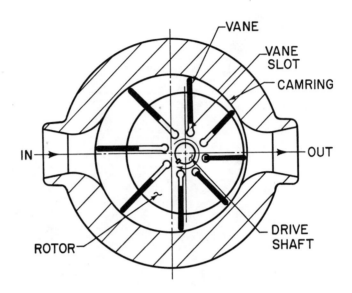

FIG. 60.   The principle of the rotary, sliding-vane pump.

and eject it at high pressure at the outlet. Generally, the pump operates more efficiently below ~70 bars (7 MPa or ~1,000 psi), and volumetric efficiencies can approach 95%. By adjusting the position of
the rotor, the output of the pump can be varied in a finely controlled
manner. Another advantage of this type of pump is that vane wear is
uniform, and does not affect its performance, since the vanes simply
adjust to the wear.

      One disadvantage is that the rotor shaft is subjected to a high
axial force, as in the simple gear pump (Figure 54(a). Balanced units
have been built using elliptically-shaped housings, but these cannot
incorporate the simple volume control mechanisms of the unbalanced
pumps. More complicated designs incorporating both pressure-balancing
and volume control have been built. In some cases, the position of
the shaft can be adjusted by the outlet pressure, so that the pump
acts as a constant pressure pump with varying load. Other vane pumps
have been constructed with flexible vanes. Such units are generally
limited to pressure ranges below ~7 bars (.7 MPa or ~100 psi).

      A version of the vane pump is depicted in Figure 61 in which the
vanes are in the form of rollers. Centrifugal action holds the rollers against the casing, and compression of fluid occurs in the same
way as in the sliding vane pump. The particular unit shown can be
used to ~20-27 bars (2-2.7 MPa or ~300-400 psi), but leakage or "slip"
occurs at the upper pressure range. Units such as this can be primed
relatively easily and deliver high output for their size.

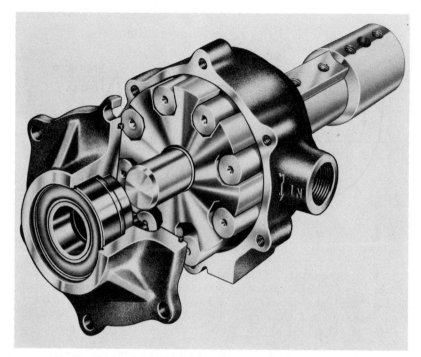

FIG. 61. Cutaway photograph of a partially disassembled
roller-vane pump. (Figure courtesy of Hypro Division, Lear Siegler,
Inc., Saint Paul, Minn.)

Rotary, positive displacement pumps generally share the inherent
capability of giving relatively trouble-free performance, with pulse-
free output. Check valves are not needed. They can handle very viscous liquids and are self-priming. Large volumes can be handled with
high efficiency and quietness. Axial screw pumps are particularly
quiet, being used in submarines, for instance. The common fault is
that leakage (called slip) occurs between vanes and housing, both at
the rotor tips and along the edges.

These pumps are generally limited to pressures   below ~300 bars
(30 MPa, 4,500 psi) but in this range they are used in hundreds of applications -- hydraulic supply for presses, lubricators, heavy equipment, cutting equipment; for pumping fluids in refrigerators; oil
burners; for volume pumping of brine, tar, etc.

B.   Kinetic or Dynamic Units [3, 4, 43-45]

The basic principle of the centrifugal compressor representing a
kinetic unit is shown in Figure 62. Fluid in the rotating impeller is
given rotary motion by the impeller. Centrifugal forces then produce
radial motion, bringing more fluid in along the axis by vacuum action.
The centrifugal forces create positive pressure in the volute housing
whose shape is designed to maximize the conversion of kinetic energy
into compressional energy. The tapered shape of the exit port is also
designed to maximize this effect.

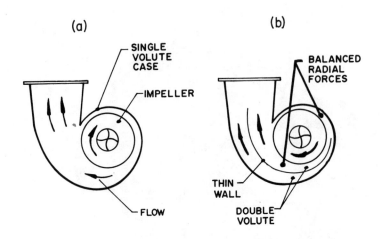

FIG. 62.   Schematic of the volute type centrifugal pump or
compressor. (a) Single volute in which radial forces acting on
the impeller are unbalanced. (b) Double volute in which radial
forces on impeller are balanced.

The above centrifugal unit is often called a volute pump. A
double-volute pump (Figure 62(b)) refers to a unit in which the volute
is split into two channels. An advantage of this design is the reduced off-axis force that results if the unit is run below its normal
capacity. This force is designed to be zero at the normal operating
condition in both single and double volute designs. Typically, the
impeller is surrounded by a set of fixed vanes, called a diffuser
(Figure 63(a) and (b)), whose purpose is to convert kinetic energy into compressional energy.

In a single-stage unit, the impeller experiences a net axial
force, which rises with capacity. This force must be balanced by the
shaft bearing. A double-suction pump results in a balanced design.
Other sub-classifications of these pumps are related to the location
of the suction nozzle: end-or axial suction, side-suction, bottom-
suction, or top-suction.

The design of the blade is of prime concern. A brief description
will be given of the equations relating the pressure differential, or
head, across the blade, since it illustrates the main features. Also,
brief consideration will be given to thermodynamic factors.

1.   Impeller Design

There are several blades on an impeller with channels between
them, forming a channel for the passage of fluid from the axial region
to the periphery. Figure 64 illustrates one blade with fluid velocities at inlet and outlet; u represents the azimuthal speed which is
the product of the angular rotation speed, ω, and the radius, r. ($u_o = \omega r_0$; $u_1 = \omega r_1$), while w represents the component of velocity parallel

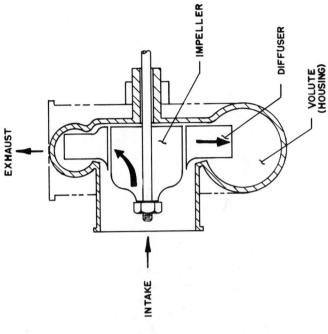

(b)

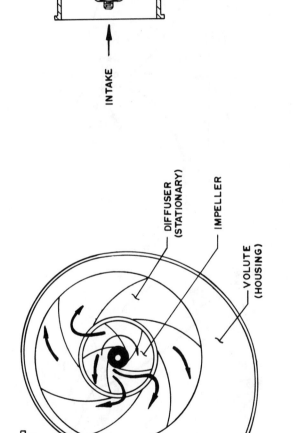

(a)

FIG. 63.   Schematic of centrifugal compressor with diffuser.   (a) View along axis.
(b) Side view.

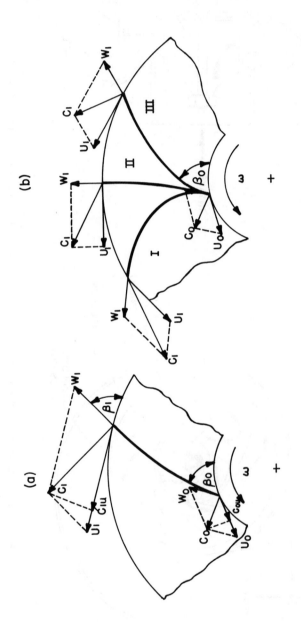

FIG. 64. Section of impeller blades showing velocity triangles. (a) Backward – curved blade showing definitions of β, c_u. (b) I – forward – curved blade; II – radial vane; III – backward–curved blade. In each case the input angle, β_o, is the same.

to the blade at the periphery. From the vector triangle, c is the re-
solved fluid velocity. If c_u is the component of this velocity parallel to the azimuthal speed, u, then the equation for the ideal pressure across the blade is:

$$P_1 - P_2 = \frac{\rho}{g} \left[(u_1 c_{1u} - u_o c_{ou}) + (w_o^2 - w_1^2) \right] \qquad (24a)$$

$$= \frac{\rho}{g} \left[(c_1^2 - c_o^2) + (u_1^2 - u_o^2) + (w_o^2 - w_1^2) \right] \qquad (24b)$$

where ρ is the density of the fluid, g the gravitational acceleration.
Expression (24b) is obtained from (24a) using trigonometric formulae
for the velocity triangle shown in Figure 64. The second term in
(24b) can be regarded as the static head developed by centrifugal
forces, which is corrected by the third term, which is related to the
change in the channel width. The first term represents a kinetic energy term which is converted to compressional energy (pressure) in the
volute. The ratio of static head to total head is known as the degree
of reaction.
     The above expressions are ideal, and neglect friction between the
fluid and impeller, or mechanical losses. However, they are useful in
understanding the influence of the impeller. A dimensionless parameter which can be used to discuss the performance is the head coeffi-
cent [4]:

$$\psi = \Delta P \cdot \frac{2g}{\rho u^2} = H \cdot \frac{2g}{u^2} \qquad (25)$$

where H is the "head" of the fluid being compressed.
     The input blade angle, β_o, is usually maximized at ~30°. How-
ever, the outlet blade angle, β_1, is chosen differently for various
applications (Figure 64(b)). If a forward curved vane is used ($\beta_1 >$
90°), the head coefficient, ψ, increases, but the static head and degree of reaction decrease. The conversion of kinetic energy to compressional energy in the volute housing or diffuser is inefficient, so
that this blade design is rarely encountered, except in large volume,
low pressure (circulation) units.
     If a backward-curved blade is used ($\beta_1 <$ 90°), the static head
and efficiency is maximized, and this design is preferred for industrial compressors. If β_o ~30°, then β_1 is usually in the range 35-
50°. An increase in efficiency results if the channels in the impeller are made with approximately constant cross section (i.e., $w_1 \sim w_0$
in equation (24)).
     If β_1 = 0, the maximum static head is developed and blades con-
structed in this way are stronger, because bending stresses are reduced. Thus, extremely high tip speeds are possible, and this type of
blade is used in critical applications such as aircraft turbo-jet engines.
     Examples of three blades, or impellers, used for compressors are
illustrated in Figure 65.
     An important parameter in centrifugal pump design is the specific
speed

$$\omega_s = \frac{\omega Q}{(gH)^{0.75}} \qquad (26)$$

where Q is the volume flow through the compressor.

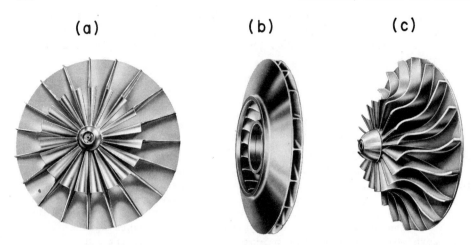

FIG. 65. Three types of impeller. (a) Open, radial-bladed impeller. (b) Closed, backward-bladed impeller. (c) Open, backward-bladed impeller. (Figure courtesy of Dresser Industries, Inc., Roots Blower Operation, Connersville, Ind.)

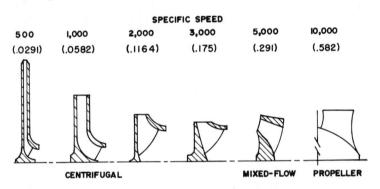

FIG. 66. Sketch showing the type of blade used for different specific speeds. Figures given are in American units. Equivalent dimensionless units are given in brackets. (Fig. adapted from ref. 3.)

This is a dimensionless number, like the Reynolds number defined in the previous chapter. Unfortunately, American practice is to neglect the gravitational acceleration in (26) and to use the equation with the following inconsistent units.

$$\omega_s^{Amer.} = \frac{\omega \, Q}{H^{0.75}} \tag{27}$$

We will refer to ω in (26) as the "true" specific speed (ω_s^{true}).

This parameter is important in determining the efficiency of the compressor, the shape of the blade and the type of fluid motion across the blade (centrifugal, axial, or mixed). A summary of blade shape and flow for different specific speeds is given in Figure 66.   Low

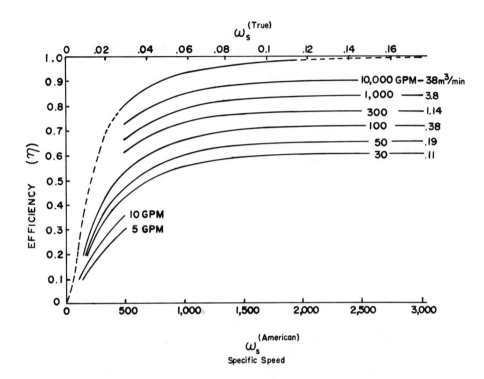

FIG. 67.   The efficiency of centrifugal pumps as a function of specific speed.   (Figure adapted from ref. 3.)

values of specific speed correspond to high-pressure heads and low flow rates. In the specific speed range ~2,000-3,500 (American) (~.12-.2, ω^{true}) the short centrifugal blade type is often referred to as the Francis screw. Axial and mixed flow types are used for the highest specific speeds where flow rate is maximized and pressure head is small.

The efficiency (η) of a centrifugal pump or compressor is defined as

$$\eta = \frac{\text{Actual head developed}}{\text{Theoretical head}} = \frac{H_{act}}{H_{theor.}} \tag{28}$$

From measurements of the efficiency for a large number of units, curves of $\eta(\omega_s)$ have been built up. Experience shows that larger units tend to be more efficient than smaller units so that flow rate has to be taken as a secondary factor. The curve shown in Figure 67 is an updated version (ref. 3) of an earlier curve proposed by Wislicenus [46]. Above a specific speed of 0.1 (true) (~1,700 American scale) very little gain in efficiency is realized. Below this, the efficiency falls increasingly rapidly.

Some units perform better than the curve suggests, but the curve serves the useful purpose of acting as a rough guide against which a particular manufacturer's product can be judged.

2.   Thermodynamic Considerations [8]

Looking   at only the thermodynamic aspects of the process of converting shaft work into work of compression, the turbo-compressor   can
be considered to operate in a steady-flow condition.   The first law of
thermodynamics for this case is expressed in the form

$$\Delta W = \Delta H - \Delta Q + \Delta (KE) + \Delta (PE) \qquad\qquad (29)$$

This work relates to changes that occur in the system   over   some
length   of   time   (e.g.,   time taken to pass a unit quantity of fluid)
where ΔW is the actual work performed on the fluid (i.e., not   including work consumed as friction in the bearings, etc.). ΔH, the enthalpy change, ΔQ, the heat influx, $\Delta (KE)$ and $\Delta (PE)$, the changes   in   the
kinetic and potential energy of the fluid respectively.   For most tur-
bo-compressors the last two terms can be neglected, and this   approximation will be used in the following.

Considering a single turbine stage operating between pressures P_2
and P_1, and temperatures T_2 and T_1, and assuming that the process is
ideally adiabatic, then equation (29) becomes

$$\Delta W' = h_2 - h_1 \qquad\qquad (30)$$

where h_2 and h_1 are the molar enthalpies of the fluid at the final and
initial   conditions,   and $\Delta W'$ is the work of compression per mole.   In
order to calculate the efficiency of the compressor and the work, some
assumptions   have to be made about the actual path that is taken.   For
the reciprocating compressor, the polytropic path defined by   equation
6(a) was chosen.   However, in turbine engines it is more useful to define a polytropic path through an effective specific heat (C):

$$T \left(\frac{dS}{dT}\right)_{\text{polytropic}} = C \qquad\qquad (31)$$

The actual value of C will depend on the path taken.   If the path
is   a reversible, adiabatic, then $\Delta S = 0$, i.e., $C \equiv C_s = 0$; if pressure
is constant, $C \equiv C_p$; in a general polytropic change, such as that   followed by   the gas   through the turbine, it will be assumed that C is a
constant.   Hence:

$$q = C(T_2 - T_1) \qquad\qquad (32a)$$

$$\Delta W' = h_2 - h_1 - C(T_2 - T_1) \qquad\qquad (32b)$$

In defining the efficiency, η, of the process,   it   is   usual   to
compare the actual work required to compress the gas between two pressures P_1 and P_2, with the work required to achieve the same   differential   pressure in an ideal compressor.   The actual work, $W_{\text{act.}}$, is usually defined as the shaft work applied to the fluid, excluding   friction   in bearings, etc.   The ideal compression is often defined as the
reversible adiabatic path (Figure 68(a)), but if heat flow occurs (usually   heat   removed)   then   the ideal, reversible, polytropic path is
often used.

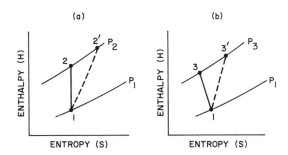

FIG. 68. Enthalpy - entropy diagram for a centrifugal compressor operating between two pressures P_1, P_2. (a) 1-2 represents
a reversible adiabatic path, 1-2' an irreversible adiabatic. (b)
1-3 represents a reversible polytropic, 1-3' an irreversible polytropic path.

Irreversibility in the turbine compressor largely comes about because the fluid flow generates friction, both internally to the fluid,
and by contact with walls and blades of the compressor. Thus, heat is
generated over and above what would be generated if the compression
occurred reversibly and results in an additional temperature rise.
Thus, the compressed gas has a higher enthalpy than it would have if
compressed reversibly, and the work of compression is raised accordingly (equation (32(b)).

The Adiabatic Efficiency (η_a) for each stage can be written, with
reference to Figure 68(a), where primed symbols refer to outlet conditions for irreversible processes, and unprimed symbols to outlet conditions for reversible processes.

$$\eta_a = \frac{\text{Work Performed in Ideal Adiabatic Compression}}{\text{Work Performed in Actual Adiabatic Compression}} \qquad (33a)$$

$$= \frac{h_2 - h_1}{h_2' - h_1} < 1 \qquad (33b)$$

For the polytropic compression (Figure 68(b)), the polytropic efficiency η_p is often defined as (Figure 68(b)):

$$\eta_p = \frac{\text{Work Performed in Ideal Adiabatic Compression}}{\text{Work Performed in Actual Polytropic Compression}} \qquad (34a)$$

$$= \frac{h_1 - h_2}{(h_3' - h_1) - C(T_3' - T_1)} \qquad (34b)$$

To evaluate the efficiencies and work of compression, knowledge
of the state of the fluid before and after compression is required, as
well as the thermodynamic properties of the fluid, including the polytropic specific heat, C.

The pressure ratio across a stage of a typical centrifugal compressor is usually ~2 for a gas such as air. Thus, many stages have
to be used to achieve high pressure ratios. From (14), it is clear

that work of compression can be reduced if heat is extracted, similar
to the case of the reciprocating compressor. Diagramatically, this is
illustrated in Figures 68(a) and (b) by the divergence of lines of
constant pressure for increased temperature (i.e., increased S or H on
the Mollier diagram). In some cases, cooling is effected at each
stage or is sometimes accomplished between multistage units, each of
which may have 5 - 10 stages.

Very often the factor which controls the number of stages between
cooling is the temperature rise. With normal gases such as air, hydrocarbons, etc., the discharge temperature is usually limited to
~450°F (232°C). An exception to this rule is oxygen, for which temperatures are limited to ~250°F (121°C). For air, the temperature
rise is usually such that cooling is effected after each stage, whereas with propylene, as many as eight stages may be used without cooling.

3.   Examples of Some Typical Kinetic Units

In the introduction, a photograph of a large axial turbine compressor was shown (Figure 3), handling large flows of gas without a
large pressure differential. These units operate at high specific
speeds (Figure 66) and can handle large quantities of gas. They may
serve the useful purpose of reducing the volume of gas delivered to a
higher pressure unit.

Small pumps are sometimes constructed with axial impellers, and
are usually called turbine pumps. A sketch of a section of a multi-
stage unit is shown in Figure 69. Such pumps are typically used for
deep water wells. Heads up to ~300m (~1,000 ft.), corresponding to
pressure heads of ~30 bars (3 MPa or ~450 psi), are typical.

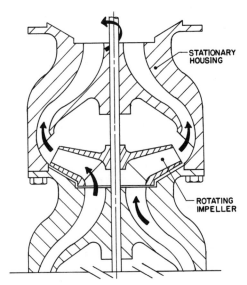

FIG. 69.   Cross section of a stage of a turbine pump used
for deep wells.

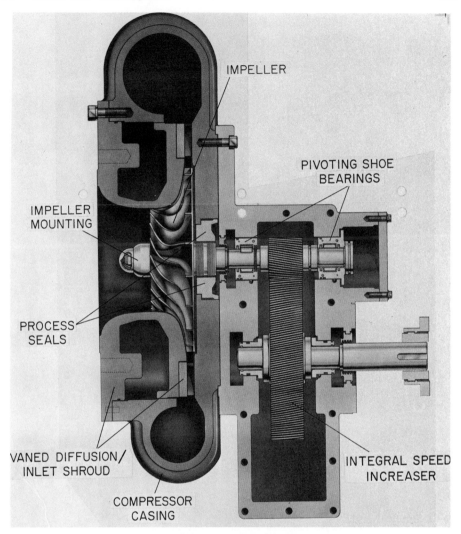

FIG. 70.   Cross section of a single stage centrifugal compressor. (Figure courtesy of Turbonetics, Inc., Latham, N.Y.)

A cut-away section of a single-stage compressor is shown in Figure 70. This design operates at high speed and produces an unusually large pressure ratio (up to 3.2) when operated with inlet air or gas. More typically, values of pressure ratio are ~2. The unit may work to pressures of ~35 bars (3.5 MPa or ~500 psi). The impeller blade is constructed of 17-4 PH stainless steel and the unit can be used to handle corrosive gases, steam, etc.

An example of a multi-stage unit is shown in Figure 71. The pump is for the water feed of a high-pressure boiler unit which delivers 3,000 gpm (~12m^3/min) at 180 bars (18 MPa, 2,600 psi), developing 20,000 HP (~15,000 kW). Some units go into the supercritical region 250 bars (25 MPa or 3,600 psi). Larger units develop 40 to 50,000 HP (30-40 MW).

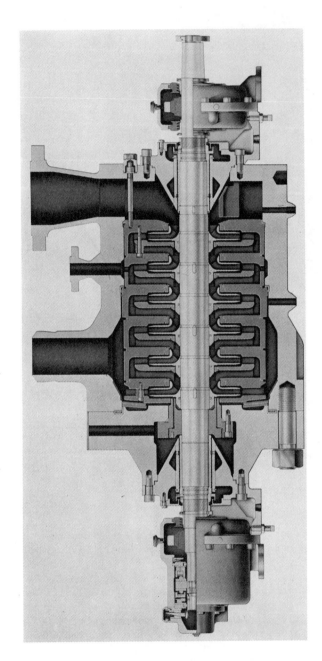

FIG. 71. Cross section of a six-stage, boiler feed, centrifugal pump.
(Figure courtesy of Delaval Turbine Inc., Trenton, N.J.)

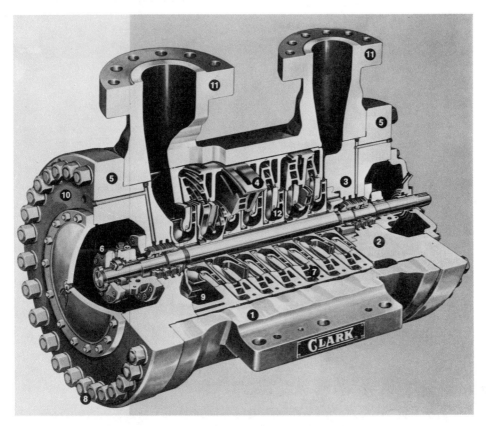

FIG. 72.   Cut-away picture of a centrifugal compressor.
1. Case, 2. Pressure-vessel type head, 3. Oil-film type seal,
4. Diffuser with saw-tooth diaphragm, 5. Cover, 6. Thrust bearing,
7. Diffuser and guide vanes showing horizontal-split construction,
8. Retaining studs, 9. Extra-heavy guide vanes and diaphragms to
handle high pressure, 10. Cover, 11. Connections, 12. Blade with
labyrinth seal. (Figure courtesy Dresser Clark Division, Dresser
Industries, Inc., Olean, N.Y.)

The pump shown has its axis horizontal and is referred to as a
horizontal pump. Another feature of the design is the double-casing
construction. The working parts are enclosed in an inner casing, surrounded by an outer casing. The space between the casings is at the
discharge pressure. This type of assembly can give much greater ease
of assembly and maintenance and is pressure-balanced.
    A cutaway diagram of a centrifugal compressor with several stages
is shown in Figure 72. Up to ten stages can be accommodated per case.
The unit is horizontally split for maximum accessibility and ease of
maintenance. Through bolts tie the inner case components together.
Diaphragms are the elements which separate stages. In the case shown

FIG. 73.   Cutaway photograph of a pitot pump.   (Figure courtesy
of Kobe, Inc., Huntington Park, Calif.)

they employ saw-tooth sealing forms and are integral with the diffuser.   The unit illustrated is representative of many employed in a wide
variety of industries, including chemicals, petrochemicals, etc.

       This unit is to be compared with the reciprocating design shown
in Figure 17.   Centrifugal compressors are now used from capacities as
low as 1,000 ICFM to the current technological limit of $\sim$500,000 ICFM
($\sim$30 - 15,000 m^3/min).   Although power efficiency may be lower, their
relative ease of maintenance is making them the preferred industrial
unit over wider operating ranges.

       The small pressure head developed by most dynamic rotary pumps
and compressors is obviously a disadvantage.   The pitot-head pump develops a much higher pressure ratio (Figure 73).   The pump shown can
be used to final pressures of 83 bars (8.3 MPa or 1,200 psi), and can
take fluids such as water to this pressure with only a small net positive suction head (NPSH).   (The NPSH is the head of fluid required at
the inlet to fill the pump including friction losses to the point
where the rotating part adds more energy.)   The pump operates by rotating the casing assembly.   Entering fluid is accelerated in the
rotor by centrifugal force, and, as in the centrifugal compressor, it
develops a static head.   When fluid enters the stationary pitot head,

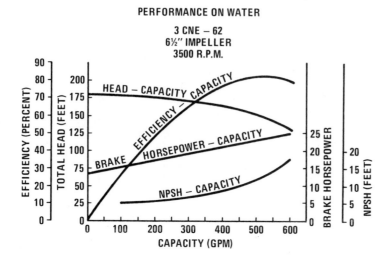

FIG. 74. Performance curve for a centrifugal pump. (Figure courtesy Worthington Pump Corp., USA, East Orange, N.J.)

velocity of the fluid is converted into a further pressure rise. The power efficiency of larger units (e.g. $\gtrsim$ 100 gallons/min ($\sim$.4 m^3/min)) is higher than 60% at design flow, but may be lower for smaller units.

Among the advantages claimed by the manufacturer for the pump compared with the centrifugal unit, is the fact that the rotating seal is at the inlet pressure, and also the pump will not seize if the fluid supply dries up. This can be particularly serious with centrifugal pumps.

The centrifugal pump should never be started without being primed and without a positive supply of fluid. Furthermore, the performance of the pump or compressor will be affected by the fluid supply. Manufacturers of the pump will supply with each pump the required Net Positive Suction Head (NPSH) which is the fluid supply energy at the datum of the pump. This figure, in feet or meters of fluid, varies with design and size of the pump. Figure 74 is a typical performance curve of a Worthington centrifugal pump for water service. The impeller has a diameter of 6-1/2" ($\sim$165 mm) and the maximum speed of the unit is 3,500 rpm (58.3 Hz).

V.   UNCONVENTIONAL COMPRESSION SYSTEMS

A.   Introduction

In addition to conventional means of compressing fluids, it is worthwhile to mention a few other ways for compressing gases particularly.

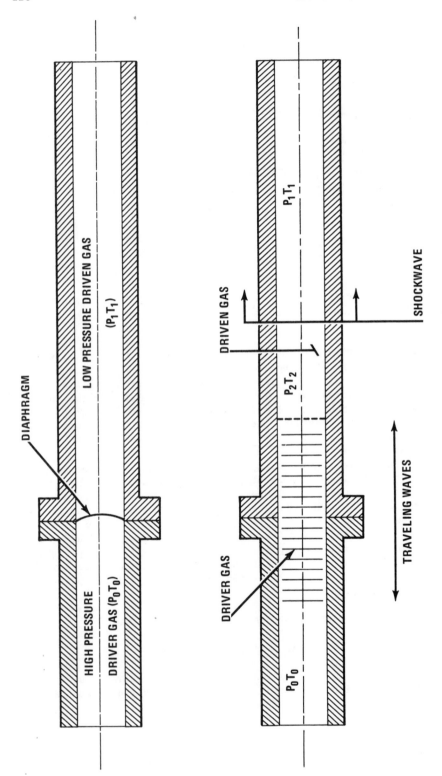

FIG. 75. Principle of the shock tube.

Conventional   apparatus is available for pressures up to ~14 kbar
(1.4 GPa, ~200,000 psi)   and temperatures of   1,500°C   can be obtained
in furnaces within vessels.   In Chapter 11, a commercial apparatus for
compressing fluids to 50 kb is described   and   specialized   techniques
for research purposes are also mentioned.

Specialized techniques are also available for   compressing   materials to sometimes very high pressures and temperatures for short
periods of time.   Methods of shock compression are described in Chapter 14, Vol. II together with some applications.   In this section
brief mention of some other techniques will now be made.

B.   The Shock Tube

Figure 75 is a schematic of a shock tube, which makes use of two
separate gases, the driver gas and test or driven gas.   The high pressure driver gas, stored in a reservoir or pressure vessel, is separated from the shock tube (and test gas) by means of a fast opening
valve or a diaphragm.   When the valve opens or the diaphragm bursts, a
shock   wave   is produced and travels through the test gas in the shock
tube, causing it to heat and pressurize the test or driven gas.

This   system   will raise the temperature drastically and temperatures up to 15,000-20,000 K can be reached.   The system is relatively
simple, and produces   the high temperature in a very short time.   Very
high pressure increases are usually not required.

C.   The Ballistic Piston Compressor [47]

This system can reach pressures of 10,000 bars (1 GPa or 145,000
psi) and temperatures of over 10,000 K.

A schematic diagram is given in Figure 76.   The unit consists of
a gas reservoir for the driver gas, a piston with piston release de-

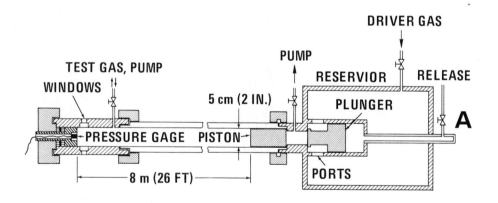

FIG. 76.   Schematic diagram of a ballistic piston compressor.

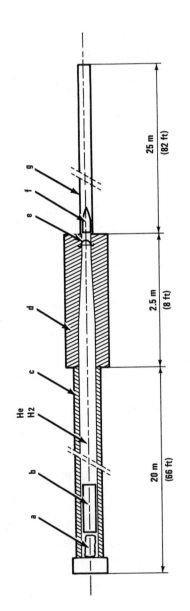

FIG. 77. Diagram of the two-stage light gas launcher. (a) Explosive charge, (b) Piston, (c) Pump tube, (d) High pressure section, (e) Diaphragm, (f) Model, (g) Barrel or launch tube.

vice and a high-pressure section. The high-pressure section is connected to the reservoir by a cylinder. The total assembly is mounted
on a special frame with provisions for recoil. When valve A is
opened, the driver gas pushes the plunger away from the seat, enabling
the driver gas to release the piston. The piston is then propelled
forward with high velocity and compresses the test gas.

D.   The Two-Stage Light Gas Launcher [48, 49]

This design is used in Ballistic Research to accelerate a model
through a barrel for free flight at high velocity (from 5,000 - 7,000
m/sec or 16,000 - 23,100 ft/sec and higher). The system is somewhat
similar to the Ballistic Piston Compressor described in Section C.
The main difference is that the gas is not compressed against a
closed-end cylinder but is pushed through a barrel. Consequently, the
pressures and temperatures are not as high as with the previous system.

The gas acts as a propellant force behind the model which is located at the barrel breech. This propelling gas is usually helium or
hydrogen. A diagram of the system is shown in Figure 77. There are
three basic sections: the pump tube (c), the high-pressure section
(d), which also serves as a "stop" for the piston, and the barrel (g).

The piston is usually made of polyethylene with or without brass
inserts, depending on the required piston weight.

An explosive charge (a) is placed at one side of the piston,
whereas at the other side of the piston a light gas is loaded. The
diaphragm (e) is located at the end of the high-pressure section just
in front of the model. When the explosive charge is set off, the piston is moved forward with high velocity (up to 700 m/sec or 2,300 ft/
sec), towards the barrel side of the system, compressing the gas. At
the burst pressure of the diaphragm, the gas accelerates the model
through the barrel. The very fast compression of the gas increases
pressure and temperature in a short time and for a short period only.
A typical compression and temperature curve of the gas is shown in
Figure 78.

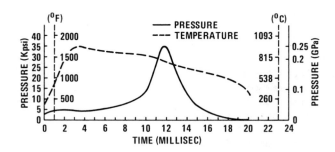

FIG.   78. Typical pressure-temperature curves.

E.   Pulsed Magnetic Fields and High Energy Lasers

Two exciting possibilities for generating very high pressure pulses for short periods of time are offered by pulsed magnetic fields and high energy lasers.

If a magnetic field is generated by a current-carrying coil, the stresses on the coil are the same as those produced by an internal pressure. A comparison of magnetic fields and equivalent pressures is given in Table 2.

High magnetic fields may be used to compress highly ionized gases of interest in studies of controlled thermonuclear fusion. Alternatively, an imploding magnetic field can be used to compress a metallic liner containing a sample [51-53]. The magnetic field-generated pulse has the advantage over the explosively generated pulse (see Chapter 14, Vol. II) in that it is slower and more nearly isentropic. Pressures well over 1 Mbar have been generated by magnetic implosion.

Very high-powered lasers can be focussed onto the surface of solids to generate extremely high pressure pulses, with very high temperatures. The highest temperatures and pressures available are produced by this means. Studies have included equation of state measurements [54] and high pressure metallurgy [55].

TABLE 2

Equivalent pressures extended by Magnetic Fields. Note the squared dependence, $P \alpha B^2$.

Magnetic Field	(Teslas)	1	10	100	500	1000
	(kg)	10	100	1000	5,000	10,000
Magnetic Pressure	(kbar)	.004	0.4	40	1,000	4,000
	(GPa)	4×10^{-4}	0.04	4	100	400

VI.   FINAL REMARKS

In the previous pages a large number of different types of pump and compressor have been discussed briefly. It should be apparent to the reader that this is an enormous field of technology and he is cautioned that it has only been possible to cover a few important topics.

The potential user of a unit for compression or pumping of fluids is faced with a number of important decisions which can be conveniently subdivided into: choice of the best unit; location and installation; operation and maintenance. Although many relevent points have been made earlier, a few remarks will be made in summary about these decisions.

In selecting a unit, common sense and experience are still the most valuable attributes. Thus, the knowledge and experience of the manufacturer should be used as much as possible. The manufacturer can only meet the suppliers needs if he is informed of all relevent facts. Most suppliers use a data sheet which is designed to cover all information needed for a correct quotation. These sheets should be filled out carefully, and possible additional factors brought to the attention of the manufacturer.

Details of the location of the pump should be thoroughly discussed with the manufacturer. Points of interest might include floor loading, vibrations transmitted to the building structure, noise, extreme conditions (climate, dust, corrosives, etc.), availability of adequate electrical supplies (including variation of line tension), etc. An obvious consideration, which is often overlooked, is the dimension of the entrance to the location in consideration, which must be sufficient for the equipment. Possible hazards to operating and other personnel should be considered, as well as the suitability of the site for later maintenance operations.

The manufacturer should again be consulted for maintenance information, including a list of regular checks, replacement schedule for critical components (e.g. diaphragms), and possible diagnostic measurements, such as temperature of critical parts, bearings, heads. It is good practice to record the actual running time of the pump or compressor, and an automatic time recorder can be installed easily.

Standardization of pumps and compressors in facilities is important. A smaller stock of parts and easier maintenance can be achieved because of interchangeability. Most pump manufacturers have an outside maintenance service available. However, trained mechanics should always be available for normal maintenance and repairs.

It is stressed that the manufacturer's recommendations should be strictly adhered to when replacing defective parts. If "short-term" solutions are used, the unit cannot be guaranteed by the manufacturer, and the consequences can be disastrous. The reader is directed to reference 56 for a discussion of a situation in which manufacturer's recommendations were ignored.

Safety must always be a prime consideration. Although discussed in earlier chapters, this factor is again brought to the readers's attention. References 57, 58 contain several informative articles on this subject including safety considerations for compressors.

Finally, a few remarks will be made of future trends. It is clear that many pumps and compressors have been improved by use of new materials. An example is the use of new polymeric materials for moving seals. However, many systems could be improved by using existing materials, while further improvements will become possible as newer steels, ceramics, polymers, etc. will become available.

Closely allied to the choice of material is the design of the compressor. It is clear from an examination of presently available designs that considerable improvements could be made by careful design. In many cases the manufacturer is deterred by increased cost of manufacture, but increased use of automation in manufacture could reduce this factor.

It is of interest to chart the spectacular increase of kinetic, rotary units over the past 20 years. It appears that there is some reluctance on the manufacturer's part to increase the size of units beyond present limits, but renewed interest in raising the pressure limit.

There is still scope for methods of compression based on new principles. Will it be possible to magnetically compress gases with a small concentration of ionic species, for example? (The reverse process is hydro-dynamic generation of electrical power.) It will certainly be interesting to look back in 10 or 20 years time to see the developments of the future.....

REFERENCES

1. Hydraulics and Pneumatics, January 1977, 30: 161-229, "21st Designers Guide to Fluid Power Products", Penton IPC, Cleveland, Ohio.

2. Thomas Register of American Manufacturers and Thomas Register File, 67th Edition, Thomas Publishing Company, New York (1977).

3. "Pump Handbook", (I. J. Karassik, W. C. Krutzsch, W. H. Fraser, J. P. Messina, eds.) McGraw-Hill Book Co., New York (1976).

4. W.R.D. Manning and S. Labrow, "High Pressure Engineering", C.R.C. Press, Cleveland (1971).

5. R. A. Strub and C. Matile, "Centrifugal Compressors for Ethylene Service", Hydrocarbon Processing 54: 67 (June, 1975).

6. American Petroleum Institute Publications (Washington, D.C.).
    (a) #822-61700; Standard 617, Centrifugal Compressors for General Refinery Services (3rd Edition, 1973).
    (b) #822-61800; Standard 618, Reciprocating Compressors for General Refinery Service (2nd Edition, 1974).
    (c) #822-61900; Standard 619, Rotary-Type, Positive Displacement Compressors for General Refinery Services.

7. Machine Design, "Fluid Power Edition:, 22, Sept. 1974.

8. K. E. Bett, J. S. Rowlinson and G. Saville, "Thermodynamics for Chemical Engineers", The MIT Press, Cambridge, Mass. (1975).

9. V. Chlumsky, "Reciprocating and Rotary Compressors", E. and F. N. Spon, Ltd., London (1965).

10. "Compressibility Charts and Their Application to Problems Involving Pressure - Volume - Energy Relations for Real Gases"; Worthington Research Bulletin P-7637, originally published in July 1949 (Worthington Corp., Harrison, N.J.).

11. "Compressed Air and Gas Data", (Charles W. Gibbs, ed.) Ingersoll Rand Corp., New York (1969).

12. "Standard Handbook of Engineering Calculations", (T. G. Hicks, editor-in-chief) McGraw-Hill, New York (1972); also "Marks Standard Handbook for Mechanical Engineers", (T. Baumeister, editor-in-chief) McGraw-Hill Book Co., New York (1967).

13. "The International System of Units - Physical Constants and Conversion Factors", E. A. Mechtly, NASA Special Publication SP-7012; also "Quantities, Units and Symbols" (Royal Society of

London, 2nd Ed., 1975); "Metric Practice Guide", American
Society for Testing and Materials, ASTM Standard E380-74 (1974).

14. D. H. Newhall, "Hydraulically Driven Pumps", Industrial and
Engineering Chemistry __49__: 1949 (1957).

15. Further details of hypercompressors for polyethylene service
are given in Chapter 3, Volume II.

16. P. W. Bridgman, "The Physics of High Pressure", G. Bell and
Sons, Ltd., London (1940).

17. "A New Mercury Piston Gas Compressor", presented at the 59th
Annual Meeting of the American Institute of Chemical Engineers,
December 6, 1966. Reprint 310, American Instrument Company,
Inc., Silver Spring, Md. 20910.

18. E. W. Comings, "High Pressure Technology", McGraw-Hill Book
Co., New York (1956).

19. "Mechanical Design and System Handbook", (H. A. Rothbart,
editor-in-chief) McGraw-Hill Book, Co., New York (1964).

20. Machine Design, "Seals", 4th Edition, Reference Issue,
The Penton Publishing Co. (1969).

21. "Engineers' Handbook of Piston Rings, Sealing Rings, Mechanical
Shaft Seals", 7th Edition, Koppers Co., Inc., Baltimore, Md.,
(1959).

22. "Mechanical Packing and Piston Rings", catalogue 731,
France Products Div., Garlock.

23. "A Handbook on Hydraulic and Pneumatic Packings",
E. F. Houghton and Co., Lynchburg, Va.

24. "Shaft Seals Catalogue", Gits Brothers Mfg. Co., Bedford
Park, Ill.

25. "Oil Seals", catalogue No. 76, Crane Packing Co., Morton
Grove, Ill.

26. "Machine Design, Plastic/Elastomers", Reference Issue,
The Penton Publishing Co., (1971).

27. "Leather Packings Handbook", National Industrial Leather
Association (1956).

28. "Mechanical Packing Catalogue", No. 60R, Crane Packing Co.,
Morton Grove, Ill.

29. "Seal Catalogue", Parker Packing Co., Culver City, Calif.,
and Cleveland, Ohio.

30. "End Face Shaft Seals", Catalogue No. 70, Crane Packing Co.,
Morton Grove, Ill.

31. "Spring-Loaded Teflon or Graphite-Teflon Seal Design Manual",
Bal-Seal Engineering Co., LaHabra, Calif.

32. "Compressor Valves and Components", Compressor Products
Division, Garlock Inc., Newtown, Pa.

33. "Valve Theory and Valve Design", Hoerbiger Corporation of
America, Roslyn, Long Island, New York.

34. "Machine Design, Fastening and Joining Issue", Sept. 1969.

35. "Select Fasteners for Full Torque", T. Green, Research and
Development, Nov. 1966, p. 38.

36. J. P. Elson, W. Soedel, "A Review of Discharge and Suction Line
Oscillation Research", Proc. of the Compressor Technology
Conference, July 25-27, 1972, Purdue Univ. Press, p. 311 (1972).

37. "Equipment Design Handbook for Refineries and Chemical Plant",
     (F. L. Evans, Jr., ed.) Vol. I., p. 24, Gulf Publ. Co.,
     Houston, Texas (1974).

38. "Design of Piping Systems", prepared by The M. W. Kellogg
     Company, p. 277-283, Revised 2nd edition, John Wiley and
     Sons, New York (1956).

39. V. L. Streeter and E. B. Wylie, "Hydraulic Transients",
     McGraw-Hill Book Co., New York (1967).

40. L. Bergeron, "Waterhammer in Hydraulics and Wave-Surges in
     Electricity", John Wiley and Sons, Inc., New York (1961).

41. W. C. Newhall, "Pipless Pumping", Proc. National Conference
     on Fluid Power, Oct. 13-15, 1970, Vol. XXIV, p. 329.

42. L. F. Moody and T. Zowski, "Hydraulic Machinery", Section 26
     of "Handbook of Applied Hydraulics", (Davis and Sorenson, eds.)
     McGraw-Hill Book Co., 3rd edition (1969).

43. A. J. Stepanoff, "Centrifugal and Axial Flow Pumps", John
     Wiley and Sons, Inc., 2nd edition, New York (1967).

44. I. J. Karrasik and R. Carter, "Centrifugal Pumps: Selection,
     Operation and Maintenance", McGraw-Hill Book Co., New York
     (1960).

45. A. Kovats, "Design and Performance of Centrifugal and Axial
     Flow Pumps and Compressors", The Macmillan Company, New York,
     (1964).

46. G. F. Wislenicus, "Fluid Mechanics of Turbomachinery", Dover
     Publications, Inc., New York (1965).

47. G. T. Lalos and G. L. Hammond, "The Ballistic Compression and
     High Temperature Properties of Dense Gases", Experimental
     Thermodynamics, Vol. II (B. leNeindre and B. Vodar, eds.),
     Butterworths and Co., London (1975), p. 1193-1218.

48. A. E. Seigel, "The Theory of High Speed Guns", Agardograph 91,
     May 1965, NASA, Washington, D.C. 20546. N.A.T.O. Advisory Group
     for Aerospace Research and Development.

49. A. Sawaoka, T. Soma, S. Saito, "Very High Pressure Production
     by Two-Stage Light Gas Gun and Recovery of Shock-Compressed
     Specimens", Proceedings of the 4th International Conference
     on High Pressure, Kyoto, 1974. (Published by the Physico-
     Chemical Society of Japan, 1974.) [This recent report
     contains several references to earlier work.]

50. J. G. Linhart, J. Appl Phys. $\underline{32}$: 500 (1961).

51. F. Bitter, Scientific American $\underline{213}$: 65 (1965).

52. L. V. Altshuler, Sov. Phys. Uspeki $\underline{8}$: 52 (1965).

53. R. S. Hawke, D. E. Duerre, J. G. Huebel, R. N. Keeler,
     "Isentropic Compression of Fused Quartz and Liquid Hydrogen
     to Several Megabars", Phys. Earth and Planetary Interiors $\underline{6}$:
     44 (1972).

54. E. Teller, "Laser Interaction and Related Plasma Phenomena",
     Vol. 3, p. 3; (H. J. Schwarz and H. Hora, eds.) Plenum Press,
     New York (1974).

55. J. F. Ready, "Effects of High Pressure Laser Radiation",
     Academic Press, New York (1971).

56. "The Flixborough Disaster", Report of the Court of Inquiry,
     Her Majesty's Stationery Office, London (1974).

57. Chemical Engineering Progress, Vol. <u>68</u>, 1972. This edition
     contains several papers on the subject of "Safety in
     Polyethylene Plant Operations".
58. "Safety in Air and Ammonia Plants", articles reprinted from
     Chemical Engineering Progress (American Institute of Chemical
     Engineers).

BIBLIOGRAPHY

H. Addison, "A Treatise on Applied Hydraulics", John Wiley
and Sons, Inc., New York, 1954.

R. M. Arrowsmith, "Special Chemical and Power Engineering
Pumping Systems", Chemical and Process Engineering,
July 1970, Vol. 51, No. 7.

J. R. Birk and J. H. Peacock, "Pump Requirements for the
Chemical Process Industries", Chemical Engineering,
February 18, 1974.

E. J. Byrne, "Measuring and Controlling", Chemical
Engineering Deskbook, April 14, 1969.

A. A. Clark and J. F. Castle, "Pumping of Lead", Chemical
and Process Engineering, July 1970, Vol. 51, No. 7.

A. J. Clemens, "Pump Breakdowns", Chemical and Process
Engineering, July 1970, Vol. 51, No. 7.

G. J. DeSantis, "How to Select a Centrifugal Pump",
Chemical Engineering, November 22, 1976.

J. H. Doolin, "Select Pumps to Cut Energy Cost", Chemical
Engineering, January 17, 1977, p. 137-139.

C. S. Hedges, "Industrial Fluid Power", Vol. 1, Womack
Machine Supply Co., Dallas, Texas, 1967.

W. M. Kauffmann, "Safe Operation of Oxygen Compressors",
Chemical Engineering, October 20, 1969.

G. D. Kelsey and D. H. Butler, "Pumping of Coal Slurries",
Chemical and Process Engineering, July 1970, Vol. 51, No. 7.

R. I. Lewis, J. E. Williams, and E. H. Fisher, "Fluid
Dynamic Design Problems of Mixed Flow Pumps and Fans",
Chemical and Process Engineering, July 1970, Vol. 51, No. 7.

J. M. McKelvey, M. Urs, F. Haupt, "How Gear Pumps and
Screw Pumps Perform in Polymer Processing Applications",
Chemical Engineering, September 27, 1976.

E. Margus, "Plastic Centrifugal Pumps for Corrosive Service", Chemical Engineering, February 28, 1977, Vol. 84, No. 5, p. 213-215.

R. F. Neerken, "Pump Selection for the Chemical Process Industries", Chemical Engineering, February 18, 1974.

I. S. Pearsall and G. Scobie, "Pumps for Low Suction Pressures", Chemical and Process Engineering, July 1970, Vol. 51, No. 7.

G.F.E. Polley, "Development of Helical Gear Pumps", Chemical and Process Engineering, July 1970, Vol. 51, No. 7.

Roper Pump Company, "How To Solve Pumping Problems", Bulletin 65-14, Commerce, Georgia.

M. Rost and E. T. Visisk, "Pumps", Chemical Engineering Deskbook, April 14, 1969.

W. Spannhake, "Centrifugal Pumps, Turbines and Propellers. Basic Theories and Characteristics", The Technology Press, MIT, Cambridge, Mass., 1934.

M. H. White, "Surge Control for Centrifugal Compressors", Chemical Engineering, December 25, 1972.

M. H. Wohl, "Properties of Liquids", Chemical Engineering Deskbook, April 14, 1969.

Worthington Corporation, Standard Pump Division, "Rotary and Centrifugal Pump Theory and Design", Brochure 2000-B7, East Orange, New Jersey, 07017.

Chapter 7

HIGH PRESSURE CONTAINMENT IN CYLINDRICAL VESSELS

V.C.D. Dawson

Naval Surface Weapons Center
White Oak Laboratory
Silver Spring, Maryland 20910

I.   CYLINDRICAL VESSELS

The most widely used form of pressure vessel is a thick-walled cylinder. It is relatively easy to construct and design criteria are relatively well known. This chapter is concerned with the design of such chambers as predicted by the theories of elasticity and plasticity and with the pressure limitations of such vessels. Also included in this chapter is a discussion of methods that can be used to increase the pressure capability above that of conventional cylindrical

vessels together with creep and fatigue analyses, and considerations
of threaded end closures.

A.   General Stress-Strain Equations

In the derivations to follow it is assumed that the vessel has
uniform longitudinal stress distribution. The inside radius of the
cylinder is a and the external radius is b.

The elastic equations for the cylinder can be derived by considering the forces acting on an element of the wall as shown in Figure
1. The element is shown without the longitudinal stress, σ_z, for
clarity.

If the forces on the element are summed in the direction of the
bisector of the angle $d\phi$, the following equation of equilibrium results:

$$\sigma_r r d\phi + \sigma_t dr d\phi - \left(\sigma_r + \frac{\partial \sigma_r}{\partial r} dr\right)(r + dr)d\phi = 0$$

If higher order terms are neglected this becomes

$$\sigma_t - \sigma_r - r\frac{d\sigma_r}{dr} = 0 \tag{1}$$

If u denotes the displacement of the cylindrical surface of radius r,
then the displacement for a surface of radius r + dr is $u + (\partial u/\partial r)dr$.
Thus the element undergoes a total elongation in the radial direction
of $(\partial u/\partial r)dr$ and the unit strain in this direction is

$$\varepsilon_r = \frac{\partial u}{\partial r} \tag{2}$$

The unit strain of the same element in the tangential direction is
equal to the unit elongation of the corresponding radius divided by
the radius, i.e.,

$$\varepsilon_t = \frac{u}{r} \tag{3}$$

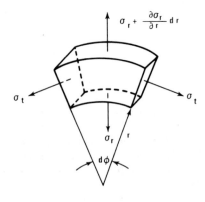

FIG. 1.   Stress element.

When the tube is stressed elastically the general radial displacement is known to be

$$u = Ar + \frac{B}{r} \tag{4}$$

where A and B are constants. Hooke's law gives the relationship between stress and strain as

$$E\varepsilon_r = E\frac{\partial u}{\partial r} = E(A - \frac{B}{r^2}) = \sigma_r - \mu(\sigma_t + \sigma_z) \tag{5}$$

$$E\varepsilon_t = E\frac{u}{r} = E(A + \frac{B}{r^2}) = \sigma_t - \mu(\sigma_r + \sigma_z) \tag{6}$$

$$E\varepsilon_z = \sigma_z - \mu(\sigma_r + \sigma_t) \tag{7}.$$

These equations can be put in the form:

$$(1+\mu)(1-2\mu)\frac{\sigma_r}{E} = A - (1-2\mu)\frac{B}{r^2} + \mu\varepsilon_z \tag{5'}$$

$$(1+\mu)(1-2\mu)\frac{\sigma_t}{E} = A + (1-2\mu)\frac{B}{r^2} + \mu\varepsilon_z \tag{6'}$$

$$(1+\mu)(1-2\mu)\frac{\sigma_z}{E} = 2\mu A + (1-\mu)\varepsilon_z \tag{7'}$$

The constants A and B are determined from the boundary conditions. In the most general case pressure can be exerted on the cylinder both internally and externally. Thus, the boundary conditions are $\sigma_r = -p_i$ at $r = a$, $\sigma_r = -p_o$ at $r = b$ where p_i and p_o are the internal and external pressures, and therefore from (5')

$$A = -\mu\varepsilon_z + \frac{(1+\mu)(1-2\mu)(p_i - \omega^2 p_o)}{E(\omega^2 - 1)}$$

$$B = b^2 \frac{(1+\mu)(p_i - p_o)}{E(\omega^2 - 1)}$$

where $\omega = b/a$ and μ is Poisson's ratio. Thus

$$\sigma_r = \frac{p_i - \omega^2 p_o}{\omega^2 - 1} - \frac{(p_i - p_o)}{\omega^2 - 1}\frac{b^2}{r^2} \tag{8}$$

$$\sigma_t = \frac{p_i - \omega^2 p_o}{\omega^2 - 1} + \frac{(p_i - p_o)}{\omega^2 - 1}\frac{b^2}{r^2} \tag{9}$$

$$\sigma_z = E\varepsilon_z + \frac{2\mu(p_i - \omega^2 p_o)}{\omega^2 - 1} \tag{10}$$

B.   Elastic Operation of a Cylinder Subjected to
Internal Pressure Only

For this case, $P_o = 0$, $P_i = p$ and the stress equations become

$$\sigma_r = \frac{p}{\omega^2 - 1}(1 - \frac{b^2}{r^2}) \qquad (8')$$

$$\sigma_t = \frac{p}{\omega^2 - 1}(1 + \frac{b^2}{r^2}) \qquad (9')$$

$$\sigma_z = E\varepsilon_z + \frac{2\mu p}{\omega^2 - 1} \qquad (10')$$

No assumption has thus far been made as to the longitudinal stress other than that it is uniform. If at this point, it is assumed that the vessel is sealed at each end such that a uniform longitudinal stress σ_z is developed wherein

$$p\pi a^2 = \sigma_z \pi (b^2 - a^2) \qquad (11)$$

then the cylinder is referred to as a closed-end cylinder. Under this condition

$$\sigma_z = \frac{p}{\omega^2 - 1} = \frac{\sigma_r + \sigma_t}{2} \qquad (12)$$

Examination of equations (8'), (9'), and (12) indicates that the stresses have their greatest absolute values at the bore. Thus for $r = a$

$$\sigma_r = -p \qquad (13)$$

$$\sigma_t = p\left[\frac{\omega^2 + 1}{\omega^2 - 1}\right] \qquad (14)$$

$$\sigma_z = \frac{p}{\omega^2 - 1} = \frac{\sigma_t + \sigma_r}{2} \qquad (12)$$

Yielding of the bore will occur when the combined stress reaches the yield point of the material. From the Distortion Energy Theory of Yielding where

$$2Y_o^2 = (\sigma_t - \sigma_r)^2 + (\sigma_r - \sigma_z)^2 + (\sigma_z - \sigma_t)^2 \qquad (15)$$

and Y_o is the isotropic yield strength, the bore will yield when the pressure reaches a value

$$p = \frac{Y_o(\omega^2 - 1)}{\sqrt{3}\,\omega^2} \qquad (16)$$

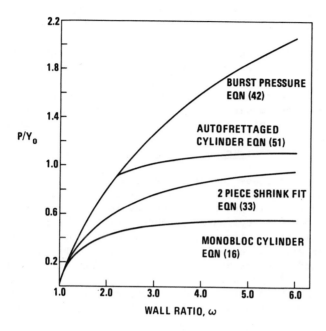

FIG. 2.   Summary plot of the limiting pressure of cylinders.

The tangential strain is given by the equation

$$\varepsilon_t = \frac{1}{E}[\sigma_t - \mu(\sigma_r + \sigma_z)]$$

and substituting the relationship given in (12) for σ_z

$$\varepsilon_t = \frac{1}{E}\left[(1 - \frac{\mu}{2})\sigma_t - \frac{3}{2}\mu\sigma_r\right] \tag{17}$$

At the bore, $r = a$,

$$\varepsilon_t = \frac{u}{a} = \frac{p}{E}\left[(1 - \frac{\mu}{2})\left[\frac{\omega^2 + 1}{\omega^2 - 1}\right] + \frac{3}{2}\mu\right] \tag{18}$$

At the outside surface of the cylinder, $r = b$

$$\sigma_r = 0$$

$$\sigma_t = \frac{2p}{\omega^2 - 1}$$

$$\varepsilon_t = \frac{u}{b} = \frac{p}{E}\frac{(2 - \mu)}{(\omega^2 - 1)} \tag{19}$$

Equation (16) demonstrates that there is a limit to the maximum pressure that can be held in a cylindrical vessel if it is desired to limit the combined stress at the bore to the yield point (Y_o) of the material.   The limiting pressure can be found by letting $\omega \to \infty$ in which case

$$\frac{p}{Y_o} = \frac{1}{\sqrt{3}} = 0.577.$$

For a nominal yield strength of 140,000 psi ($\sim$1 GPa) therefore, the maximum pressure that can be contained elastically is about 80,000 psi ($\sim$57 GPa).

In Figure 2 equation (16) is plotted as a function of the wall ratio ω. From the figure it is apparent that increasing the wall ratio beyond about 3 does not provide any great increase in pressure capability. In other words a cylinder with $\omega = 3$ will withstand a pressure equal to about 90% of the pressure that a cylinder of the same material, but having a wall ratio of 10, will withstand.

C.   General Remarks on Methods for Strengthening Cylinders

In order to increase the pressure capability of a cylinder above that imposed by equation (16), two manufacturing processes are used, namely: a. Shrink fitting two or more cylinders together; b. Autofrettaging a monobloc cylinder. Figure 3a shows the stress distribution in the wall of a cylinder of wall ratio 2 when internal pressure is applied, as given by equations (8') and (9'). The maximum absolute values of σ_t and σ_r occur at the bore but they drop off rapidly through the wall of the tube.

On the other hand, from equations (8) and (9), if external pressure only is applied to a thick walled tube, then $p_i = 0$, and

$$\sigma_r = - \frac{p_o}{\omega^2 - 1} \left(\omega^2 - \frac{b^2}{r^2} \right) \tag{20}$$

$$\sigma_t = - \frac{p_o}{\omega^2 - 1} \left(\frac{b}{r^2} + \omega^2 \right) \tag{21}$$

Figure 3b is a plot of these equations for a cylinder of wall ratio 2. As can be seen, the tangential stress at the bore is compressive. Thus, it is possible to construct two cylinders such that the O.D. of the smaller one is slightly larger than the I.D. of the larger one. These can be fitted together by heating the larger one (called the jacket), cooling the inner one (called the liner) so that diametral clearance is provided. Then the jacket can be positioned over the liner. (An alternative is to machine the outer surface of the liner and the inner surface of the jacket with a conical surface with included angle approximately 10°, but for the purposes of computation a uniform radius will be assumed.) When both cylinders are again at room temperature, the initial interference between the two cylinders will result in a pressure (called the interference pressure) at the interface. For the liner this is equivalent to an external pressure, so that, if internal pressure is now applied, the tangential stress begins not at zero, but at some negative compressive stress due to the interference pressure. Thus, the pressure overcomes the compressive stress before the tangential stress becomes tensile. With this technique of shrink fit it is possible to increase the elastic pressure capability of a cylinder to a value greater than that given by equation (16).

The other technique employed in increasing pressure capability is autofrettaging a monobloc cylinder. In this process, the cylinder is pressurized during manufacture to a pressure equal to the desired op-

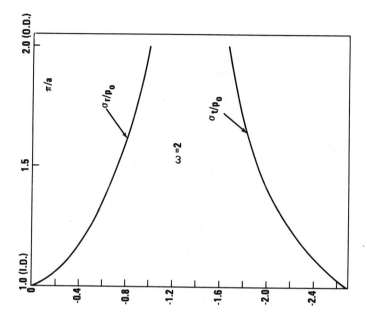

FIG. 3b. Stress distribution in cylinder of wall ratio 2 subjected to external
pressure.

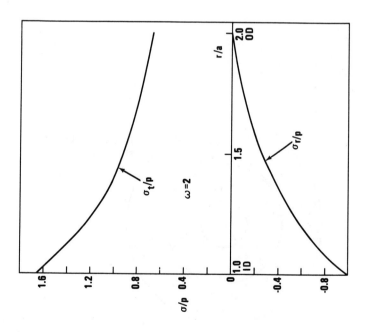

FIG. 3a. Stress distribution in cylinder of wall ratio 2 subjected to internal
pressure.

erating pressure when that pressure is greater than the limiting elastic cylinder pressure as given by equation (16). As a result of the
overpressurization the cylinder undergoes non-uniform plastic flow
which begins at the bore and progresses through the wall as the pressure is increased. Because of the non-uniform plastic flow, a residual stress pattern is developed when the pressure is released such
that the bore is in compression and the cylinder can now be operated
elastically up to the autofrettage pressure. The bore, in this process, is permanently enlarged by the overpressurization.

The applicable equations for each of the manufacturing techniques
mentioned above are developed in the following.

D.   Compound Cylinders with Diametral Interference

When a compound cylinder is constructed by shrink-fitting two
monobloc cylinders, the process occurs under an open-ended case, i.e.,
for each cylinder $\sigma_z = 0$. After assembling and cooling, a contact or
interference pressure, p_c, is produced between the cylinders. Its
magnitude may be found from the condition that the increase in the inner radius of the outer cylinder plus the decrease in the outer radius
of the inner cylinder caused by p_c must be equal to the initial radial
interference, δ.

For the liner the interference pressure is externally applied and
for the jacket it is internally applied. Figure 4 indicates the geometry involved.

For the liner at $r = c$, from equations (20) and (21) with $r = c$
and $p_o = p_c$, and where $\omega_\ell = c/a$

$$\sigma_r = -p_c$$

$$\sigma_t = -p_c \left[\frac{\omega_\ell^2 + 1}{\omega_\ell^2 - 1} \right]$$

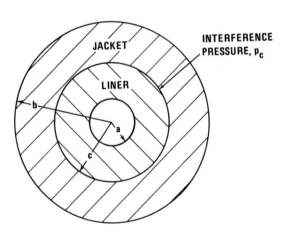

FIG. 4.   Compound cylinder construction.

The radial deflection of the liner due to the interference pressure p_c is

$$u_r = \frac{-cp_c}{E} \left[\frac{\omega_\ell^2+1}{\omega_\ell^2-1} - \mu \right] \qquad (22)$$

For the jacket with $\omega_j = b/c$ as a result of the interference pressure, at $r = c$,

$$\sigma_t = p_c \; \frac{\omega_j^2+1}{\omega_j^2-1}$$

$$\sigma_r = - p_c$$

and

$$\frac{cp_c}{E}\left[\frac{\omega_\ell^2+1}{\omega_\ell^2-1} - \mu\right] + \frac{cp_c}{E}\left[\frac{\omega_j^2+1}{\omega_j^2-1} + \mu\right] = \delta \qquad (23)$$

The absolute value of the shrinkage of the liner plus the absolute value of the expansion of the jacket must be equal to the radial interference that originally existed between the liner and jacket. Thus, if $\omega = b/a \equiv \omega_j \omega_\ell$ as before, then:

$$\frac{cp_c}{E}\left[\frac{\omega_\ell^2+1}{\omega_\ell^2-1} - \mu\right] + \frac{p_c}{E}\left[\frac{\omega_j^2+1}{\omega_j^2-1} + \mu\right] = \delta$$

or

$$p_c - \frac{\delta E}{2c} \frac{(\omega_j^2-1)(\omega_\ell^2-1)}{(\omega^2-1)} \qquad (24)$$

Equation (24) gives the contact pressure developed by an initial radial interference δ. This pressure puts the liner into compression and the jacket into tension. When an internal pressure is applied at the bore of the liner, after the liner and jacket have been assembled, the actual stress distribution that results is the superposition of the shrink fit stresses and the monobloc stress that would result in the assembled tube if there were no interference.

The equations above were developed for a two piece shrink fit construction with both tubes made from the same material, i.e., E and μ the same. It is, of course, possible to use different materials and /or to use more than two cylinders with interference existing between each cylindrical layer. Generally speaking a tube assembly of more than three cylinders becomes uneconomical and the designer is better off to use an autofrettage construction which is described later in this chapter.

For a given wall ratio, ω, of the assembled tubes, there is an optimum wall ratio for the liner, ω_ℓ, and jacket, ω_j, for maximum internal pressure capability. Another way of looking at such an optimum design is that it provides maximum performance with minimum material. The optimum design conditions in this case are obtained by considering the stress distribution in the compound cylinder that results from the applied internal pressure as well as the contact or shrink pressure.

The maximum shear stress in the liner occurs at the inside radius, a, and is given by the equation

$$
\tau_{\ell_{max}} = \frac{\sigma_t - \sigma_r}{2} = \frac{1}{2}\left[p\,\frac{\omega^2+1}{\omega^2-1} - \frac{2p_c\omega_\ell^2}{\omega_\ell^2-1} + p\right]
$$

$$
= p\,\frac{\omega^2}{\omega^2-1} - \frac{p_c\omega_\ell^2}{\omega_\ell^2-1}
$$

$$
= p\left[\frac{b^2}{b^2-a^2}\right] - \frac{p_c c^2}{c^2-a^2}
\tag{25}
$$

For the jacket the maximum shear stress occurs at the inside radius, c, and equals

$$
\tau_{j_{max}} = \frac{1}{2}\left[p\,\frac{\omega_j^2+1}{\omega^2-1} + p_c\,\frac{\omega_j^2+1}{\omega_j^2-1} + p\,\frac{\omega_j^2-1}{\omega^2-1} + p_c\right]
$$

$$
= p\,\frac{\omega_j^2}{\omega^2-1} + p_c\,\frac{\omega_j^2}{\omega_j^2-1}
$$

$$
= p\,\frac{b^2 a^2}{(b^2-a^2)c^2} + p_c\,\frac{b^2}{b^2-c^2}
\tag{26}
$$

For optimum design the maximum shear stress during pressure application should be the same for both the liner and jacket so that equating equations (25) and (26)

$$
p\left[\frac{b^2}{b^2-a^2}\right] - \frac{p_c c^2}{c^2-a^2} = \frac{pb^2 a^2}{c^2(b^2-a^2)} + \frac{p_c b^2}{b^2-c^2}
\tag{27}
$$

Thus

$$
p_c\left[\frac{b^2}{b^2-c^2} + \frac{c^2}{c^2-a^2}\right] = \frac{pb^2}{b^2-a^2}\left[1 - \frac{a^2}{c^2}\right]
$$

and

$$
p_c = \frac{pb^2}{b^2-a^2}\left[\frac{1 - \dfrac{a^2}{c^2}}{\dfrac{b^2}{b^2-c^2} + \dfrac{c^2}{c^2-a^2}}\right]
\tag{28}
$$

This value of contact pressure, p_c, can now be substituted back into either equation (25) or (26) to obtain the maximum shear stress in the liner and jacket. The result, after some algebraic manipulation, is

$$\tau_{max} = p\left[\frac{b^2c^2}{2b^2c^2 - (c^4+a^2b^2)}\right]$$

and since for maximum pressure capability the maximum shear stress should equal the elastic limit in shear, which by the Distortion-Energy Theory is $Y_o/\sqrt{3}$,

$$\tau_{max} = \frac{Y_o}{\sqrt{3}} = p\left[\frac{b^2c^2}{2b^2c^2 - (c^4+a^2b^2)}\right] \tag{29}$$

Thus the maximum internal pressure that a compound cylinder, having a liner and jacket made of the same material and yield strength, can withstand is

$$P_{max} = \frac{Y_o}{\sqrt{3}}\left[\frac{2b^2c^2 - (c^4+a^2b^2)}{b^2c^2}\right] \tag{30}$$

Differentiating (30) with respect to c and equating the result to zero gives the optimum value of c for maximum pressure. The result of this operation is

$$c = \sqrt{ab} \tag{31}$$

or

$$\omega_\ell = \omega_j \text{ and } \omega_\ell\omega_j = \omega \tag{32}$$

For optimum design the ratios of the outside and inside radii for both liner and jacket are the same. Substitution of (31) into (30) gives the maximum optimum pressure capability in terms of the inner and outer radii of the compound cylinder. The result is

$$\begin{matrix} P_{max} \\ optimum \end{matrix} = \frac{2Y_o}{\sqrt{3}}\left[\frac{b-a}{b}\right] = \frac{2Y_o}{\sqrt{3}}\left[\frac{\omega-1}{\omega}\right] \tag{33}$$

The optimum contact pressure is found from equations (24) and (31) to be

$$p_c^{opt} = \frac{\delta E}{2\sqrt{ab}}\left[\frac{b-a}{b+a}\right] = \frac{\delta E}{2a\sqrt{\omega}}\frac{\omega-1}{\omega+1} \tag{34}$$

If (31) and (33) are substituted into (28) then

$$p_c^{opt} = \frac{Y_o}{\sqrt{3b}}\frac{(b-a)^2}{b+a} = \frac{Y_o}{\sqrt{3\omega}}\frac{(\omega-1)^2}{(\omega+1)} \tag{35}$$

The optimum interference can now be found by equating (34) and (35) to obtain

$$\delta^{opt} = \frac{2aY_o}{E}\left[\frac{\omega-1}{\sqrt{3\omega}}\right] \tag{36}$$

The optimum interference given by (36) may not be practical because the assembly depends upon heating the jacket to a high enough temperature so that the bore expands sufficiently to slide over the liner.

If such a temperature is too high then the metallurgical properties of the jacket may change and the design would be compromised. The upper temperature limit would be the tempering temperature of the material.

Example I

Suppose that it is desired to contain a pressure of 80,000 psi (0.55 GPa) with a steel having a yield strength of 140,000 psi (0.97 GPa). Calculate the minimum wall ratio that can be used with (a) a monobloc construction and (b) a two-piece, shrink-fit construction.

(a) For the monobloc cylinder from (16)

$$p = \frac{Y_o}{\sqrt{3}} \frac{\omega^2-1}{\omega^2}$$

Hence

$$\frac{\omega^2-1}{\omega^2} = \frac{\sqrt{3}p}{Y_o} = \frac{\sqrt{3}}{140,000} \frac{80,000}{140,000} = 0.990$$

and

$$\omega = 9.847$$

(b) For the two piece shrink fit cylinder from (33)

$$p_{max \atop opt.} = \frac{2Y_o}{\sqrt{3}} \frac{\omega-1}{\omega}$$

Hence

$$\frac{\omega-1}{\omega} = \frac{\sqrt{3}p}{2Y_o} = \frac{\sqrt{3}(80,000)}{2(140,000)} = 0.495$$

and

$$\omega = 1.980$$

Thus the use of a shrink fit construction would cause a drastic reduction in the overall dimensions of the cylinder. In Figure 2 a comparison between the pressure containment capability of a monobloc and two piece shrink fit cylinder as a function of wall ratio as determined by equations (16) and (33) is shown. It is to be noted from these equations that in the limit as $\omega \rightarrow \infty$, the shrink-fit construction provides twice the pressure capability of a monobloc for the same overall dimensions.

E.   The Autofrettage Process

As indicated earlier the other manufacturing technique that can be used to increase the elastic operating pressure of a cylinder is called autofrettaging. In this process the monobloc cylinder is intentionally overpressurized beyond the elastic breakdown pressure, as given by (16), during fabrication so that non-uniform plastic flow occurs beginning at the bore and extending radially outward to a radius that depends upon the applied pressure. When the pressure is re-

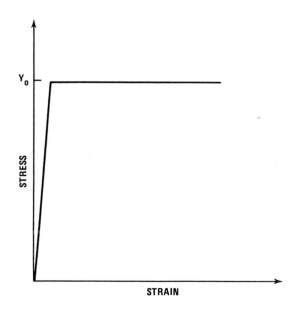

FIG. 5.   Perfectly elastic-plastic stress-strain curve.

leased the   residual stresses left as a result of the non-uniform expansion permit elastic operation up to the pressure that   was   applied
during autofrettage. The theory of autofrettage is based upon using a
material having a perfectly elastic-plastic stress-strain curve as indicated in Figure 5. In addition the assumption is usually made that
autofrettage pressurization occurs under closed-end conditions with   a
uniform longitudinal stress which can be written as

$$\sigma_z = \frac{\sigma_t + \sigma_r}{2} \tag{34}$$

in   both   the plastic and elastic portions of the cylinder. The autofrettage pressurization process can be continued up to the point where
the   entire   wall   of the cylinder becomes plastic which, in the basic
theory, is assumed to be the burst pressure of the cylinder. In actual fact, because of work hardening of the material, the burst pressure
is normally somewhat higher than the theoretical limit   and   empirical
relations   have   been   proposed   to   predict   the   burst pressure [1].
     Suppose that the applied pressure at the bore has been   increased
to a value greater than that given by equation (16) so that the cylinder is partly plastic and partly elastic as indicated in Figure 6. In
the   following derivation primed symbols refer to the elastic zone and
unprimed to the plastic zone.
     The assumptions made in the theory are:
     1.   The directions of the principal strains coincide at
           all times with those of the principal stresses.
     2.   The volume change at any section is due to elastic
           strains only.

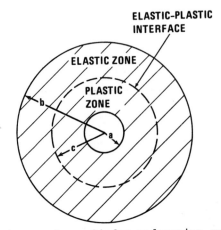

FIG. 6.   Nomenclature for cylinder undergoing autofrettage.

3.   The longitudinal strain is uniform throughout the
wall cross-section and the longitudinal stress equals
$1/2(\sigma_t + \sigma_r)$ in both the plastic and elastic zones.

The condition of plasticity as given by the Distortion-Energy
Theory is given by equation (15):

$$2Y_o^2 = (\sigma_t - \sigma_r)^2 + (\sigma_r - \sigma_z)^2 + (\sigma_z - \sigma_t)^2$$

If $\sigma_z = 1/2(\sigma_t + \sigma_r)$ is substituted into this equation, then

$$\sigma_t - \sigma_r = \pm \frac{2Y_o}{\sqrt{3}} = \text{constant} \qquad (35)$$

The condition of equilibrium of the element as given by   equation   (1)
still holds so that

$$\sigma_t - \sigma_r = r \frac{d\sigma_r}{dr} = \pm \frac{2Y_o}{\sqrt{3}} \qquad (36)$$

Integration of (36) gives

$$\sigma_r = \pm \frac{2Y_o}{\sqrt{3}} (\ln r) + C_1 \qquad (37a)$$

and this substituted in (35) yields

$$\sigma_t = \pm \frac{2Y_o}{\sqrt{3}} (\ln r + 1) + C_1 \qquad (38a)$$

Since   the   radial stress must decrease absolutely through the plastic
zone, the positive sign on the first term of (37a) is the only one ad-
missible.   Thus in the plastic zone

$$\sigma_r = C_1 + \frac{2Y_o}{\sqrt{3}} \ln r \qquad (37b)$$

$$\sigma_t = C_1 + \frac{2Y_o}{\sqrt{3}} (\ln r + 1) \qquad (38b)$$

In the elastic part of the tube it is assumed that the stresses have the same form that they did when the tube was completely elastic, i.e.,

$$\sigma'_r = C_2 - \frac{C_3}{r^2} \tag{39a}$$

$$\sigma'_t = C_2 + \frac{C_3}{r^2} \tag{40a}$$

C_1, C_2, and C_3 are constants which can be evaluated from the boundary conditions at the plastic-elastic interface where the radial and tangential stresses must be continuous. These boundary conditions are the following:

$$\text{at } r = c \text{ , } \sigma_r = \sigma'_r \text{ , } \sigma_t = \sigma'_t$$

$$\text{at } r = b \text{ , } \sigma_r = o$$

Substitution of these values into the equations yields

$$C_2 = \frac{Y_o c^2}{\sqrt{3} b^2}$$

$$C_3 = \frac{Y_o c^2}{\sqrt{3}}$$

$$C_1 = \frac{Y_o}{\sqrt{3}} \left[\frac{c^2}{b^2} - 1 \right] - \frac{2Y_o}{\sqrt{3}} \ln c$$

In addition, since $\sigma_r = -p$ at $r = a$, substitution into (37b) gives

$$C_1 = -p - \frac{2Y_o}{\sqrt{3}} \ln a$$

so that

$$-p - \frac{2Y_o}{\sqrt{3}} \ln a = \frac{Y_o}{\sqrt{3}} \left[\frac{c^2}{b^2} - 1 \right] - \frac{2Y_o}{\sqrt{3}} \ln c$$

or

$$p = \frac{2Y_o}{\sqrt{3}} \left[\frac{\omega^2 - n^2}{2\omega^2} + \ln n \right] \tag{41}$$

where $\omega = b/a$ and $n = c/a$.

Equation (41) represents the pressure required to cause plastic flow to a value $n = c/a$.

The radial and tangential stress distributions in the plastic zone at any radius within the zone are from the value of C_1 and (37b) and (38b).

$$\sigma_r = -p + \frac{2Y_o}{\sqrt{3}} \ln w \tag{37}$$

$$\sigma_t = -p + \frac{2Y_o}{\sqrt{3}} (\ln w + 1) \tag{38}$$

where $w = r/a$.

Similarly with the values of C_2 and C_3 and equations (39a) and (40a), the stresses in the elastic zone at any radius within the zone while the cylinder is under pressure are

$$\sigma_r' = \frac{-Y_o n^2}{\sqrt{3}\omega^2} \left[\frac{\omega^2}{m^2} - 1 \right] \tag{39}$$

$$\sigma_t' = \frac{Y_o n^2}{\sqrt{3}\omega^2} \left[\frac{\omega^2}{m^2} + 1 \right] \tag{40}$$

where $m = r/a$.

The following facts should be noted with respect to equation (41):

a.   When the bore is just at the point of yielding so that $n = c/a = a/a = 1$,

$$p = \frac{Y_o}{\sqrt{3}} \frac{\omega^2 - 1}{\omega^2}$$

which is equation (16) and represents the maximum pressure capability of the cylinder if it is to remain elastic.

b.   The pressure can be increased beyond the elastic breakdown pressure. As it increases, the plastic zone increases and the plastic-elastic interface moves through the wall until $n = c/a = b/a = \omega$ when the entire wall becomes plastic. Theoretically the applied pressure which causes the entire wall to become plastic is the burst pressure of the cylinder and is given by

$$P_b = \frac{2Y_o}{\sqrt{3}} \left[\frac{\omega^2 - \omega^2}{2\omega^2} + \ln \omega \right]$$

$$= \frac{2Y_o \ln \omega}{\sqrt{3}} \tag{42}$$

Figure 2 provides a comparison of equations (16) and (42). The theoretical burst pressure for a ductile cylinder is seen from this figure to be considerably higher than the initial elastic breakdown pressure.

Figure 7 is a plot of the radial and tangential stresses in a cylinder of wall ratio $\omega = 3$ and a plastic-elastic interface ratio $n = 1.6$ during pressure application. These curves are drawn by means of equations (37), (38), (39), (40) and (41). The ratio of the autofrettage pressure to the elastic breakdown pressure, i.e., equation (41) divided by equation (16) for the above condition is

$$\frac{P_{autofrettage}}{P_{yield}} = 1.86$$

The longitudinal stress in the cylinder is $\sigma_z = 1/2(\sigma_t + \sigma_r)$ and can be calculated directly from equations (37), (38), (39) and (40) or can be drawn as the mean of σ_r and σ_t in both the plastic and elastic zones in Figure 7.

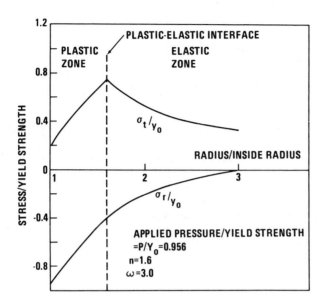

FIG. 7.   Stress distribution at applied autofrettage pressure.

When the autofrettage pressure is released the cylinder is left with residual stresses as a result of the non-uniform yielding that occurred during pressurization. It is assumed that a linear stress-strain relation exists as the autofrettage pressure is decreased so that the residual stresses are given by

$$\sigma_r^* = \sigma_r - \sigma_r'' \tag{43}$$

$$\sigma_t^* = \sigma_t - \sigma_t'' \tag{44}$$

In these equations the values with an asterisk refer to residual stresses, the unprimed values refer to the plastic or elastic stresses existing during the autofrettage process, and the double-primed values are the elastic equivalent stresses caused by the applied pressure; i.e., the stresses that would exist under the applied pressure if the material did not have a yield strength.

The double-primed stress values are given by equations (8') and (9') while equations (37), (38'), (39') and (40) represent the auto-frettage stresses in the plastic and elastic zones. Thus the substitution of these values into (43) and (44) leads to the following:

a. In the zone that was plastic during the autofrettage pressurization, the residual stress distribution is

$$\sigma_{r_p}^* = -p + \frac{2Y_o}{\sqrt{3}} \ln w + \frac{p}{(\omega^2-1)} \left[\frac{\omega^2}{w^2} - 1 \right] \tag{45}$$

$$\sigma_{t_p}^* = -p + \frac{2Y_o}{\sqrt{3}} (\ln w + 1) - \frac{p}{\omega^2-1} \left[\frac{\omega^2}{w^2} + 1 \right] \tag{46}$$

b.   In the zone   that was elastic during the autofrettage pressurization, the residual stress distribution is

$$\sigma_{r_e}^* = \frac{-Y_o}{\sqrt{3}} \left[\frac{n^2}{m^2} - \frac{n^2}{\omega^2} \right] + \frac{p}{(\omega^2 - 1)} \left[\frac{\omega^2}{m^2} - 1 \right] \qquad (47)$$

$$\sigma_{t_e}^* = \frac{Y_o}{\sqrt{3}} \left[\frac{n^2}{m^2} + \frac{n^2}{\omega^2} \right] - \frac{p}{(\omega^2 - 1)} \left[\frac{\omega^2}{m^2} + 1 \right] \qquad (48)$$

Figure 8 is a plot of these equations for the conditions used to
draw Figure 7.   It shows the residual stress distribution that results
from the autofrettage pressurization after pressure release.
The residual stresses left in an autofrettaged vessel after pressure release could be of sufficient magnitude as to cause the yield
point to be reached in compression.   It is assumed that the yield in
compression is the same in absolute magnitude as the yield in tension.
The highest residual stress values which would cause yielding occur at
the bore of the cylinder.   The distortion energy stress at the bore
due to the residual stresses is

$$S^* = \frac{\sqrt{3}}{2} (\sigma_t^* - \sigma_r^*)$$

evaluated at r = a.   Substitution of equations (45) and (46) with w =
1 into this equation gives

$$S^* = Y_o - \frac{\sqrt{3} p \omega^2}{\omega^2 - 1} \qquad (49)$$

Suppose that a cylinder is pressurized so that the entire wall becomes
plastic and then when pressure is released the bore is left at the
yield point in compression.   From equation (42) the applied pressure
is

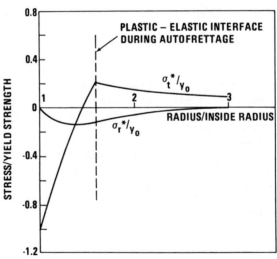

FIG. 8.   Residual stress distribution after release of
autofrettage pressure.

$$p = \frac{2}{\sqrt{3}} Y_o \ln \omega$$

and from equation (49) $S^* = -Y_o$ so that

$$-Y_o = Y_o - \frac{2\omega^2}{\omega^2-1} Y_o \ln \omega$$

or

$$\ln \omega = \frac{\omega^2-1}{\omega^2}$$

The solution to this is $\omega = 2.22$. Thus if a cylinder has a wall ratio less than 2.22 and it is pressurized so that the entire wall is plastic the combination of residual stresses left will not be sufficient to leave the wall at the compressive yield point. If the wall ratio is 2.22 under the same assumptions, the residual wall stress is just at the compressive yield point. If the wall ratio is greater than 2.22 pressurizing the cylinder so that the entire wall is plastic will leave the cylinder with a compressively reyielded core when pressure is released. The size of the reyielded core is a function of the wall ratio of the cylinder. The theory for reverse yielding of fully auto-frettaged cylinders of large wall ratio is given in reference [2].

In most pressure vessel operations it is desirable to leave the cylinder in a completely elastic state after autofrettage rather than with a reverse-yielded inner core. In order to do this the autofret-tage pressure must be limited to a value less than $2Y_o/\sqrt{3} \ln \omega$ when $\omega > 2.22$. This means that the plastic-elastic interface n is less than ω. The relationship between n and ω, so as to leave the bore at the compressive yield can be found by letting $S^* = -Y_o$ in equation (49) and substituting equation (41) for p so that

$$- Y_o = Y_o - \frac{\omega^2}{\omega^2-1} 2Y_o \left[\frac{\omega^2-n^2}{2\omega^2} + \ln n \right]$$

or

$$1 = \frac{\omega^2}{\omega^2-1} \left[\frac{\omega^2-n^2}{2\omega^2} + \ln n \right] \tag{50}$$

A plot of equation (50) is shown in Figure 9. As indicated, the theo-retical value of n can be equal to ω if $\omega \le 2.22$ but when $\omega > 2.22$ the value of n, and therefore the autofrettage pressure, is limited and governed by equation (50) if the cylinder is to be completely elastic after pressure release. When $\omega > 2.22$ the maximum autofrettage pres-sure, if reverse yielding is to be avoided after pressure release, is found from equations (41) and (50), i.e.,

$$\frac{\omega^2-1}{\omega^2} = \frac{\omega^2-n^2}{2\omega^2} + \ln n$$

and substituting this into equation (41) gives

$$p = \frac{2Y_o}{\sqrt{3}} \left[\frac{\omega^2-1}{\omega^2} \right] \tag{51}$$

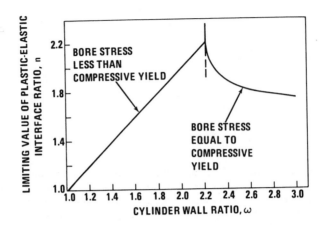

FIG. 9. Limiting value of n to leave combined residual
stress of bore equal to or less than compressive limit.

It is to be noted that the maximum autofrettage pressure, for subsequent elastic operation when $\omega > 2.22$, is just twice the initial elastic breakdown pressure of the cylinder as given by equation (16).
Thus for elastic operation after pressure release the maximum autofrettage pressure is

$$\omega \leq 2.22 \qquad p = \frac{2Y_o}{\sqrt{3}} \ln \omega \qquad\qquad (42)$$

$$\omega \geq 2.22 \qquad p = \frac{2Y_o}{\sqrt{3}} \left[\frac{\omega^2 - 1}{\omega^2} \right] \qquad\qquad (51)$$

The relations developed thus far are summarized in Figure 2. It is
seen that the autofrettage process represents the limit for the shrink
fit process if essentially an infinite number of shells were shrunk
together.

F.   Radial Expansion During Autofrettage

The calculation of the expansion of a cylinder when it is being
autofrettaged is based upon the following assumptions:
1.   The volume change at any section is due to elastic
strains only.
2.   Longitudinal strain is uniform throughout the cross
section.
From assumption 1 the following equation results

$$(\varepsilon_t + \varepsilon_r + \varepsilon_z)_{total} = (\varepsilon_t' + \varepsilon_r' + \varepsilon_z')_{elastic} + \underbrace{(\varepsilon_t + \varepsilon_r + \varepsilon_z)_{plastic}}_{= 0}$$

$$= \frac{1 - 2\mu}{E} (\sigma_t' + \sigma_r' + \sigma_z')$$

From equations (34), (39), and (40) this becomes

$$(\varepsilon_t + \varepsilon_r + \varepsilon_z)_{total} = \frac{\sqrt{3}(1-2\mu)Y_o n^2}{E\omega^2} \tag{52}$$

From assumption 2

$$\varepsilon_{z_{total}} = \varepsilon_z' = \frac{1}{E}\left[\sigma_z' - \mu(\sigma_r'+\sigma_t')\right]$$

$$= \frac{(1-2\mu)Y_o n^2}{\sqrt{3}E\omega^2}$$

If this is substituted into (52) together with the fact that

$$\varepsilon_t = \frac{u}{r} , \quad \varepsilon_r = \frac{du}{dr}$$

then

$$\frac{du}{dr} + \frac{u}{r} = \frac{2(1-2\mu)Y_o n^2}{\sqrt{3}E\omega^2} \tag{53}$$

Integration of this gives

$$ur = \frac{2(1-2\mu)Y_o n^2}{\sqrt{3}E\omega^2} \frac{r^2}{2} + C_1$$

so that

$$\varepsilon_t = \frac{u}{r} = \frac{(1-2\mu)Y_o n^2}{\sqrt{3}E\omega^2} + \frac{C_1}{r^2} \tag{54}$$

At the outside of the cylinder r = b

$$\varepsilon_t = \frac{1}{E}\left[\sigma_t' - \mu(\sigma_r'+\sigma_z')\right]$$

$$= \frac{(2-\mu)Y_o n^2}{\sqrt{3}E\omega^2} \tag{55}$$

If this is substituted into (54) then

$$C_1 = \frac{b^2(1+\mu)Y_o n^2}{\sqrt{3}E\omega^2}$$

so that

$$\varepsilon_t = \frac{Y_o n^2}{\sqrt{3}E\omega^2}\left[(1-2\mu) + (1+\mu)\frac{b^2}{r^2}\right] \tag{56}$$

The strain at the outside of the cylinder, r = b, for a typical value of μ = 0.3, is therefore

$$\varepsilon_t = \frac{1.7Y_o n^2}{\sqrt{3}E\omega^2} \tag{57}$$

Equation (57) gives the tangential strain at the outside diameter during the autofrettage process. Upon release of the pressure, the tube recovers elastically with the slope of the recovery line having the same slope as the pressure deformation line that occurred during the original pressurization before plastic flow began, provided reverse yielding does not occur during pressure release.

A basic computer program which provides the complete stress-strain history of a monobloc cylinder constructed of a perfectly elastic-plastic material is described in reference [3]. The program is based upon the equations developed in this chapter and includes reverse-yielding effects.

Reference [4] describes measurements made during the autofrettage of a cylinder having a wall ratio of 2. The cylinder was constructed of AlS1 4340 steel which had an average stress-strain curve from multiple specimens as shown in Figure 10. This stress-strain curve was approximated, as shown in the figure, with two perfectly elastic-plastic stress-strain curves, one of which had a yield of 120,000 psi (~0.83 GPa) and the other a yield of 138,000 psi (~ 0.95 GPa). The equations of this chapter were then used to predict the external strain as a function of the pressure during autofrettage to a pressure of 95,000 psi (~0.66 GPa) and compared to the experimentally measured values. The results are shown in Figure 11.

It is seen that the cylinder, as is to be expected, is exhibiting work hardening effects and the measured values of external strain lie in between the calculated strain values for the two yield strengths

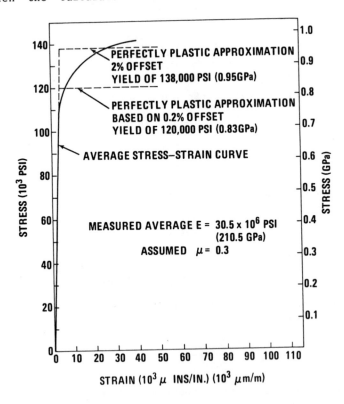

FIG. 10. Average stress-strain relation and approximations to a perfectly plastic material.

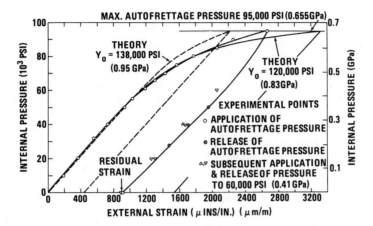

FIG. 11.   External strain versus internal pressure.
Comparison of theory and experiment.

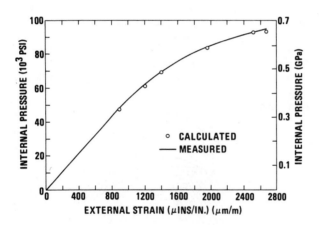

FIG. 12.   Comparison of calculated strain and measured
external strain during autofrettage accounting for work hardening
of the material.

assumed. Subsequent calculations were made to take account of work
hardening. The approach used was based on a method suggested by Nadai
[5]. The calculations are described in detail in reference [6]. Figure 12, taken from this last reference, shows a comparison of the calculated and measured strain values during the autofrettage pressure.
As can be seen the calculated strain values agree quite well with the
measured curve.

II.   SEGMENTED HIGH PRESSURE CHAMBERS

Figure 2 summarizes the elastic operating pressure capability
that can be achieved with autofrettaged monobloc or two piece shrink
fit cylinders. For the autofrettaged cylinder the limit with a large
wall ratio is of the order of 160,000 psi (~1.10 GPa) with a nominal
yield strength of 140,000 psi (~0.97 GPa). This is, of course, much
lower than the theoretical burst pressure but assumes that reverse
yielding does not occur.

Reference [7] describes various vessel designs that have been
studied to increase the restricted pressure range available to investigators. Poulter [8] patented a chamber design similar to that shown
in Figure 13. It consists of a thin inner elastic liner which of itself is not strong enough to carry even a small portion of the bursting forces generated by internal pressure. The bursting forces are
contained by a segmented ring which, in turn, is supported by an integral shell. The use of the segmented ring essentially eliminates the
tangential stresses that limit the pressure capability of an integral
cylinder. Also the pressure force at the bore is transferred through
the segments to the integral ring but the pressure is thereby reduced
by the diameter ratio. If the segmented pieces are made of a material
that can withstand high compressive stresses on the inner surfaces (a
material like tungsten carbide), then the chamber can handle substantially higher internal pressures than a conventional cylinder and
still operate elastically.

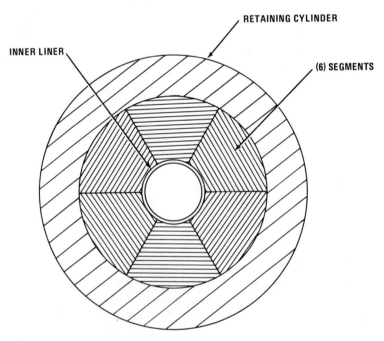

FIG. 13.   Segmented chamber design.

When internal pressure, p_i, is applied, a pressure, p, is developed between the segments and jacket so that

$$p_i d_i = p d_s \qquad \text{where } d_i \text{ is the I.D. of the liner and } d_s \text{ is the O.D. of the segments}$$

or

$$p = p_i \frac{d_i}{d_s} = \frac{p_i}{\omega_s} \tag{58}$$

This pressure must be contained by the integral jacket which begins to yield when

$$p = \frac{Y_o \left[\left[\frac{d_o}{d_s}\right]^2 - 1 \right]}{\sqrt{3}\left[\frac{d_o}{d_s}\right]^2} \qquad \text{where } d_o \text{ is the outer diameter of the cylinder}$$

so that

$$p_i = \frac{Y_o (\omega_o^2 - \omega_s^2) \omega_s}{\sqrt{3} \; \omega_o^2} \tag{59}$$

For a given value of ω_o the jacket and segments should have certain dimensions to provide maximum pressure capability. These dimensions can be determined by differentiating equation (59) with respect to ω_s and setting the result equal to zero. Thus

$$\frac{dp_i}{d\omega_s} = \frac{Y_o}{\sqrt{3}\omega_o^2} \left[(\omega_o^2 - \omega_s^2) + \omega_s^2 (- 2\omega_s) \right] = 0$$

$$\omega_o^2 - 3\omega_s^2 = 0 \tag{60}$$

When this value of ω_s is substituted into equation (59)

$$p_i^{max} = \frac{Y_o}{\sqrt{3}} \frac{(\omega_o^2 - \omega_o^2/3)}{\omega_o^2} \frac{\omega_o}{\sqrt{3}}$$

$$= \frac{2}{9} \omega_o Y_o$$

$$= 0.222 \; \omega_o Y_o \tag{61}$$

A plot of this result is shown in Figure 14 and shows that such a chamber is considerably stronger than a conventional nomobloc cylinder. Its use is limited to a minimum wall ratio of $\sqrt{3}$ as given by equation (60). Below this value the segmented chamber is less strong than the monobloc non-autofrettaged construction. For $\omega = \sqrt{3}$ to $\omega = 5$ the Poulter design is stronger than the monobloc non-autofrettaged

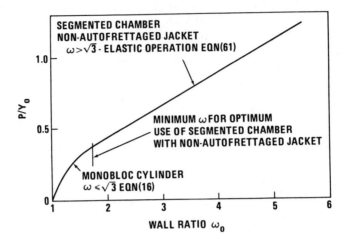

FIG. 14. P/Y_O versus wall ratio (segmented chamber non-autofrettaged jacket).

cylinder and for $\omega > 5$ it is stronger than the monobloc autofrettaged cylinder when the latter is limited to elastic operation, i.e., does not reverse yield when pressure is released.

It is apparent that the segmented chamber design of Poulter can be improved if the outer integral jacket is autofrettaged prior to assembly of the chamber. Dawson and Seigel [9] patented such a chamber design. For that type of construction, the maximum autofrettage pressure is given by equations (42) and (51)

$$\omega \leq 2.22 \qquad p = \frac{2Y_o}{\sqrt{3}} \ln \omega \qquad (42)$$

$$\omega \geq 2.22 \qquad p = \frac{2Y_o}{\sqrt{3}} \frac{\omega^2 - 1}{\omega^2} \qquad (51)$$

When an autofrettaged jacket is used with the segmented chamber construction, the optimum jacket-wall ratio is 2.22 so that

$$\frac{d_o}{d_s} = \frac{\omega_o}{\omega_s} = 2.22$$

To autofrettage a cylinder with this wall ratio requires a pressure as given by equation (42) of

$$p = 0.92 \ Y_o$$

Since

$$p = \frac{p_i d_i}{d_s} = 0.92 \ Y_o$$

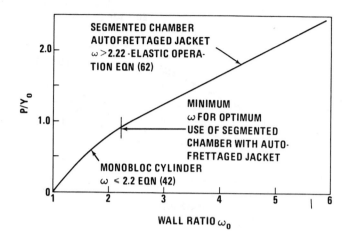

FIG. 15. p/Y_o vs wall ratio (segmented chamber autofrettaged jacket).

the internal pressure capability is

$$p_i = 0.92 \ Y_o \omega_s$$
$$= \frac{0.92 \ Y_o \omega_o}{2.22}$$
$$p_i = 0.415 \ Y_o \omega_o \qquad\qquad (62)$$

This result is plotted in Figure 15. As can be seen it provides a higher pressure capability than any of the systems considered when ω_o > 2.22. In fact when ω_o > 3.5 it has a higher elastic pressure capability than the theoretical burst pressure of a monobloc cylinder.

The design of the segmented elements is critical with respect to their material properties and shaping since very high stresses are generated where they contact each other on the inner chamber element. The segmented elements must be sealed by a thin liner of a ductile material. Radial displacement of the outer jacket can cause the opening of gaps at the bore of the segmented elements resulting in the liner extruding into the gaps. This problem is generally overcome by pre-shaping the segmented elements so that a residual compressive tangential pressure is generated between the faces of the segmented elements during the shrink fit. A chamber of this type was constructed and tested at the Naval Surface Weapons Center, White Oak, Maryland (formerly the Naval Ordnance Laboratory). The estimated pressure limit on the chamber was 750,000 psi (~5.2 GPa), which was the compressive strength of the tungsten carbide segments that were used. It is estimated that a pressure in excess of 450,000 psi (~3.1 GPa) was obtained before the experiment was terminated due to instrumentation difficulties. The cylinder had an inner bore diameter of 0.5" and an outside diameter of 12". References [9] and [10] contain further in-

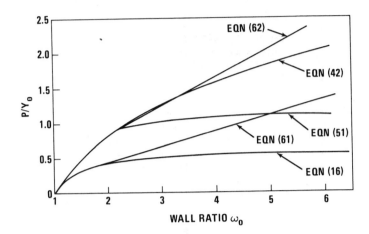

FIG. 16.   Summary plot p/Y_o vs wall ratio.

formation on the segmented chamber design and the subsequent tests.
     The limiting pressure curves for all the cylinders discussed in
this chapter are given in Figure 16.

III.   AUTOCLAVES, CREEP, AND STRESS RELAXATION

     Autoclaves are pressure vessels which are subjected to high tem-
peratures. During the heat-up process, when a sizable thermal gradient in the cylinder wall may occur, the thermal stresses that result
can be large and must be superimposed on the pressure stresses to determine yielding of the cylinder. Analysis of the thermal stress
equations [11] shows that they are beneficial, i.e., cause a lower resultant stress condition when superimposed on the pressure stresses,
if the thermal gradient is such that the bore is at a higher temperature than the outside surface. If the reverse occurs the resultant
stress condition is detrimental and yielding of the bore occurs at a
lower pressure than would normally be expected. Thus in autoclave design heating should be provided internally when possible, rather than
externally.
     Once the heating process is completed, the thermal gradient, and
hence the thermal stresses in the cylinder wall, are usually small.
Now the pressure stresses prevail and because of the temperature the
cylinder will be subject to creep if the temperature is high enough.
The effect of creep can be estimated from the calculated stresses if
creep data on the material are available.
     Most applications of autofrettaged cylinders have been at opera-
ting conditions near room temperature or where the cylinder is subjected to a heat pulse of extremely short duration as, e.g., in the
firing of a cannon. One of the reasons for this is the assumption
that heating of an autofrettaged cylinder will cause stress relief of
the residual stresses and may result in the eventual loss of the pressure capability created by the autofrettaging process. Obviously, if

autofrettaged vessels could be used, the pressure range of such devices could be increased or the wall ratio decreased for a given pressure.

Reference [4] reports on an experimental and analytical program
in which 0.5" (~12.7 mm) internal diameter autofrettaged cylinders of
wall ratio 2 were subjected to high temperatures for varying lengths
of time as indicated in the following table:

Table I

Specimen Number	Temperature	Time (hours)
1	Room (~25°C)	
2	400°F (204°C)	24
3	600°F (316°C)	24
4	850°F (454°C)	24
5	Room (~25°C)	
6	850°F (454°C)	1
7	850°F "	5
8	850°F "	72

Specimens 1 and 5 acted as control specimens of the original autofrettaged cylinders. The residual stresses were determined by means of
the Sach's boring-out technique [12]. From creep data that were
available on the steel used for the cylinders, it was possible to predict the tangential bore stress relaxation using either the strain-
hardening or time-hardening methods of doing this that are described
in reference [13]. Figure 17 shows a comparison between the measured
experimental values and the predicted curves for the cylinders heated
to 850°F (454° C).

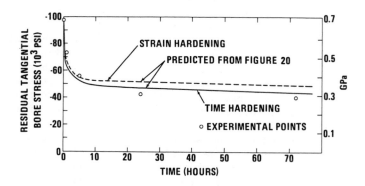

FIG. 17. Comparison of relaxation of residual, tangential,
bore-stress with predicted relaxation.

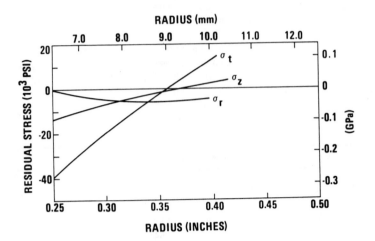

FIG. 18.   Experimental residual stresses versus radius for
specimen No. 8.

As expected the residual tangential bore-stress which, before any
heating of the specimen, was at a value of about 97,500 psi (~0.67GPa)
in compression, drops rapidly when the specimen is heated at 850°F
(454° C) for short periods of time. However, after about 5 hours at
temperature it becomes asymptotic to a value of about 37,500 psi
(~0.76 GPa). Thus, in spite of the temperature and time at temperature, the cylinder is left with a sizeable residual tangential compressive stress at the bore. The complete stress distribution in
specimen 8 is shown in Figure 18. The original residual tangential
stress distribution in the cylinders that resulted from an autofrettage pressure of 97,500 psi (~0.67 GPa) is shown in Figure 19, which
also indicates the experimental scatter band obtained from specimens 1
and 5. The solid lines represent the estimated residual stress pattern assuming a perfectly elastic plastic material with a yield of
120,000 psi (~0.83 GPa) and 138,000 psi (~0.95 GPa).

From the results of reference [4], it is concluded that autofrettaged autoclaves that are constructed of a creep-resistant material
can provide better performance, from the viewpoint of higher pressures
for a given wall ratio or lower wall ratios for a given pressure, than
a simple non-autofrettaged pressure. While the residual stresses will
relax in such vessels, this relaxation is amenable to prediction from
creep data on the material.

The design procedure and examples that are discussed below are
illustrative of the fact that the autofrettage technique can be used
to significantly extend present autoclave design methods. It is
cautioned that the procedure and examples are intended to demonstrate
the advantage of autofrettage but the reader must recognize that present design methods are regulated and dictated by the various ASME
pressure codes which must be adhered to (see Chapter 3).

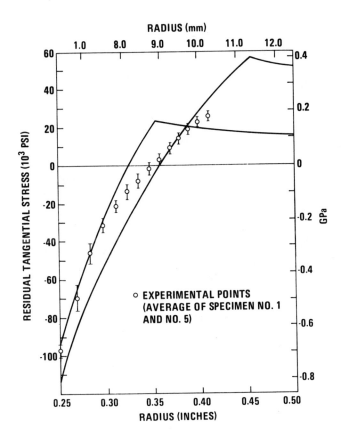

FIG. 19. Residual tangential stress versus radius
comparison of theory and experiment.

Suppose that creep data of the steel to be used in the design of
an autoclave are as shown in Figure 20. (These data are taken from
reference [4]).

Example I

Assume that an autoclave must withstand a pressure of 36,000 psi
(~0.25 GPa) internally applied at a temperature of 850°F (454°C).
Heat is applied after pressurization and cooling of the system occurs
prior to pressure release. The wall ratio of the cylinder is 2, and
failure is assumed to occur when the creep strain reaches a value of
2%. The material is 4340 steel with a yield strength of 85,500 psi at
850°F (~0.59 GPa at 454°C).

Reference [14] establishes a relationship between creep-rate components and stress components for three dimensional stress fields by
means of the distortion-energy yield condition. The applicable equations are

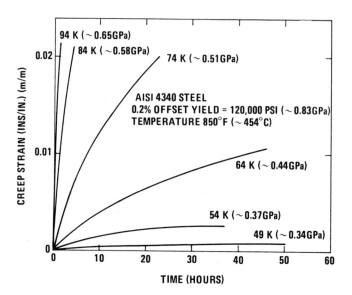

FIG. 20.   Creep data for 4340-steel.

$$\dot{\bar{\varepsilon}}_1 = \frac{\sigma_1 - \frac{1}{2}(\sigma_2 + \sigma_3)}{S} \, \bar{\varepsilon}$$

$$\dot{\bar{\varepsilon}}_2 = \frac{\sigma_2 - \frac{1}{2}(\sigma_3 + \sigma_1)}{S} \, \bar{\varepsilon} \left. \right\} \tag{63}$$

$$\dot{\bar{\varepsilon}}_3 = \frac{\sigma_3 - \frac{1}{2}(\sigma_2 + \sigma_1)}{S} \, \bar{\varepsilon}$$

The effective stress, S, is given by the equation

$$S = \frac{1}{\sqrt{3}} \sqrt{(\sigma_1 - \sigma_2)^2 + (\sigma_2 - \sigma_3)^2 + (\sigma_3 - \sigma_1)^2} \tag{64}$$

and $\bar{\varepsilon}$ is the effective creep strain rate. For a uniaxial stress application, S is equal to the applied stress, σ_1, and $\bar{\varepsilon}$ is equal to the creep strain rate, $\dot{\varepsilon}_1$, measured in a creep test. Thus the relationship between S and $\bar{\varepsilon}$ is obtained from tensile creep test results and is analogous to the concept of a combined stress yield theory using a yield stress obtained from a simple uniaxial tensile test.

In a closed end cylinder, where $\sigma_z = 1/2(\sigma_t + \sigma_r)$ substitution into the equations for $\dot{\bar{\varepsilon}}_1$, $\dot{\bar{\varepsilon}}_2$, and $\dot{\bar{\varepsilon}}_3$, with the subscripts changed to t, r, and z, respectively, yields

$$\dot{\bar{\varepsilon}}_t = \frac{\sqrt{3}}{2} \, \bar{\varepsilon}$$

$$\dot{\bar{\varepsilon}}_r = -\dot{\bar{\varepsilon}}_t \left. \right\} \tag{65}$$

$$\dot{\bar{\varepsilon}}_z = 0$$

Thus, for a closed-end cylinder, it is sufficient for comparison purposes to calculate the value of S and use Figure 20 to determine when
the creep strain reaches 2%.

If the cylinder is non-autofrettaged then with a wall ratio of 2
and a pressure of 36,000 psi (~0.25 GPa), the pressure of Lamé stresses at the bore are, according to the equations developed earlier in
this chapter,

$$\sigma_t = 60,000 \text{ psi} \quad (\sim0.41 \text{ GPa})$$

$$\sigma_r = -36,000 \text{ psi} \quad (-0.25 \text{ GPa})$$

$$\sigma_z = 12,000 \text{ psi} \quad (\sim0.08 \text{ GPa})$$

The effective stress, from equation (64) is

$$S = 83,000 \text{ psi} \quad (\sim0.57 \text{ GPa})$$

From Figure 20, with a stress of 84,000 psi (~0.58 GPa), 2% creep-
strain is reached in approximately four hours. For the conditions assumed in this example, then, the monobloc, non-autofrettaged autoclave would have a total operating life of approximately four hours.

Consider next the case when the cylinder to be used is first
autofrettaged to a pressure of 97,500 psi (~0.67 GPa). The residual
plus the pressure or Lamé stresses due to a subsequent pressure of
36,000 psi (~0.25 GPa) will be the following [4]:

$$\sigma_r = -36,000 \text{ psi} \quad (\sim-0.25 \text{ GPa})$$

$$\sigma_t = -37,500 \text{ psi} \quad (\sim-0.26 \text{ GPa})$$

$$\sigma_z = -15,200 \text{ psi} \quad (\sim-0.10 \text{ GPa})$$

From equation (64), S = 21,600 psi (~0.15 GPa) and from Figure 20 the
creep life is indefinite. By means of autofrettage the operating life
has been indefinitely extended as compared to the non-autofrettaged
cylinder.

Example II

If the monobloc cylinder of Example I has a given creep-strain
life for a given pressure, then the pressure that the autofrettaged
cylinder of Example I could withstand for the same creep-strain life
should be higher.

The distortion energy stress for the monobloc non-autofrettaged
cylinder is

$$S = \frac{\sqrt{3}p\omega^2}{\omega^2-1} = 2.31p \tag{66}$$

In order that the autofrettaged cylinder have the same creep-strain
rate as that of the non-autofrettaged cylinder the distortion energy
stress in the former must equal that of the latter during operation.

The autofrettaged specimen has the following bore stresses at
pressure and temperature:

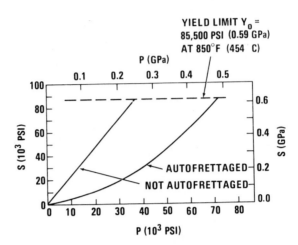

FIG. 21.   Pressure-stress relationships at 850°F (454°C).

$$
\left.\begin{array}{ll}
\text{tangential stress} & (5/3p - 97,500) \\[4pt]
\text{radial stress} & (-p) \\[4pt]
\text{longitudinal stress} & (p/3 - 27,500)
\end{array}\right\} \qquad (67)
$$

By assuming values of p, the stresses (67) can be determined and then S obtained from (66). Figure 21 results. For the same value of S (and therefore the same creep strain in a given time) the autofrettaged cylinder can withstand considerably higher pressure.

Example III

Consider the case where an autoclave is uniformly preheated for one hour at a temperature of 850°F (454°C) before pressurization.

For this case the autofrettage stress will relax during the preheat period. Thus, the number of cycles of operation is needed. It is apparent, from the relaxation data of Figure 17, that during the initial cycles, high residual stresses will relax rapidly. As the initial residual stresses become smaller, further cycling causes the relaxation curve to flatten out, and it becomes necessary to determine some design relaxation limit. This can be done from the creep data (or directly from relaxation data if available). However, for the purposes of this example, it is sufficient to note that there was no appreciable change in the residual longitudinal stress during heating of the specimens. It may reasonably be assumed, therefore, that the relaxation limit for this material is approximately 20,000 psi (~0.14 GPa). Thus, the design, residual, bore-stresses in the autofrettaged cylinder would be

tangential stress   (-20,000 psi)   (~-0.14 GPa)

radial stress   (0 psi)

longitudinal stress   (-20,000 psi)   (~-0.14 GPa)

At a pressure of 36,000 psi (~0.25 GPa) with a wall ratio of 2 this cylinder has a value of S = 66,500 psi (~0.46 GPa). From Figure 20, the time required to reach 2% creep strain at a stress of 66,500 psi (~0.46 GPa) is over 150 hours as against four hours for the non-autofrettaged cylinder of Example I which would be unaffected essentially by cycling.

This example shows that even relatively low residual stresses, properly oriented, can extend the creep life of a pressure vessel. It also indicates the importance of the order of application of temperature and pressure. In Example I, the pressure was applied first so that full advantage could be taken of the residual stresses.

In the third example the residual stresses could relax during the initial cycling periods so that the final residual stress levels had to be predicted on the basis of continued application of the time-hardening principle of reference [13].

These examples dramatically illustrate the importance of the order in which pressure and temperature are applied. Examples I and II show that the creep life for the autofrettaged cylinder is indefinite when the pressure is applied before temperature. Example III shows that the creep life is reduced to finite time (150 hours) when temperature application precedes pressurization.

The designer of autoclaves employing the autofrettage technique must therefore carefully analyze the particular conditions of temperature, pressure, time and cycle rate, the importance of each parameter having been indicated quantitatively in the examples above.

The conditions under which heat is applied to the cylinder as well as the manner in which it is cooled after pressurization are also a matter of concern just as they are in present design using non-auto-frettaged autoclaves. It was assumed in the examples above that heating and cooling were done slowly so that a uniform condition of temperature occurred throughout the wall thickness. This ensured that the thermal stresses would be small. Any rapid temperature change would naturally require the designer to consider thermal stress effects which would have to be superimposed on the Lamé and residual stress pattern.

As a final example of design procedure it is desirable to consider the effect of the cooling of an autoclave if pressure is released before the temperature is lowered. Suppose, for example, that the cycle is such that the pressure is applied to the cylinder which is then heated slowly to 850°F (454°C) as in Example I. After a period of time at pressure and temperature, the pressure is released and the cylinder, beginning at 850°F (454°C) is cooled at a slow uniform rate such that the thermal stresses developed are small.

In this case, it is obvious that during the pressurization and heating cycle the problem is essentially one of creep considerations as in Example I. During the cooling cycle, however, since only the residual stresses remain, the problem is one of relaxation at varying temperatures. Based upon the cooling cycle this can be handled by assuming that the cylinder is at a uniform temperature T_1 for a given length of time t_1, at which time drops to a new lower temperature T_2 where it remains for a time t_2 and so on until the final temperature is reached. Creep data at each assumed temperature used in approximating the cooling curve are necessary. Using the initial residual stress distribution, the creep data at temperature T_1 are used to determine a relaxation curve. At time t_1 the residual stress distribution can be determined from this curve.

On the basis of this new distribution a new relaxation curve is constructed from the creep data at temperature T_2 giving a new stress distribution at time t_2. This process is repeated until a limiting residual stress distribution is obtained. It is then necessary to repeat the entire process beginning with the final stress distribution obtained subjected to the higher temperatures in order to account for further cycling. The process is continued until additional cycling produces no change in the residual stress distribution. This final stress distribution is then used to determine the creep-rate during the first part of the cycle as in Example I.

In all of the foregoing analysis, thermal stresses were not considered since only slow uniform heating and cooling and small thermal gradients were assumed. The following example indicates the manner in which a significant steady state thermal gradient would affect the results.

Example IV

As in Example I, assume that an autoclave, with $\omega = 2$, O.D. = 8" (~203 mm), is pressurized internally to 36,000 psi (~0.25 GPa) and then heated. The creep life will be calculated for the following steady state conditions:

    1.   Bore temperature - 850°F (~454°C)

           Outside surface temperature - 650°F (~343°C)

    2.   Bore temperature - 850°F (~454°C)

           Outside surface temperature - 1050°F (~566°C)

This gradient (200°F) (~93°C) is considerably larger than would normally occur under steady state for a horizontal cylinder cooled by free convection and radiation. A temperature gradient of 200°F (~93°C) was chosen (instead of the more realistic value of 55°F (~12.8°C)) in order to develop relatively high thermal stresses.

From reference [11] the thermal stresses at the bore of a thick-walled cylinder under steady-state conditions are

$$\sigma_r = 0$$

$$\sigma_t = \frac{E\alpha(T_i - T_o)}{2(1-\mu)\ln\omega}\left[1 - \frac{2\omega^2}{\omega^2 - 1}\ln\omega\right]$$

$$\sigma_z = \frac{E\alpha(T_i - T_o)}{2(1-\mu)\ln\omega}\left[1 - \frac{2\omega^2}{\omega^2 - 1}\ln\omega\right]$$

where α is the thermal coefficient of expansion, T_i is the bore temperature, T_o is the outside surface temperature, μ is Poisson's ratio, E is Young's modulus and ω is the wall ratio.

Case a: For the two cylinders considered in Example I, with $T_i - T_o = 200°F$, $\alpha = 6 \times 10^{-6}/°F$, the following results are obtained for a creep strain of 2%: ($T_i - T_o = 111°C$, $\alpha = 10.8 \times 10^{-6}/°C$)

Non-Autofrettaged Cylinder

Stress	Lame Stress	Residual Stress	Thermal Stress	Resultant Stress	Distortion Energy Stress	Creep Life
σ_t	60,000 (0.41)	0	-31,600 (-0.22)	28,400 (0.20)		
σ_r	-36,000 (-0.25)	0	0	-36,000 (-0.25)	60,500 (0.42)	230 hrs.
σ_z	12,000 (0.08)	0	-31,600 (-0.22)	-19,600 (-0.14)		

Numbers without parenthesis - psi; with - GPa

Autofrettaged Cylinder

Stress	Lame Stress	Residual Stress	Thermal Stress	Resultant Stress	Distortion Energy Stress	Creep Life
σ_t	60,000 (0.4)	-97,500 (-0.67)	-31,600 (-0.22)	-69,100 (-0.48)		
σ_r	-36,000 (-0.25)	0	0	-36,000 (-0.25)	29,300 (0.20)	indefinite
σ_z	12,000 (0.08)	-27,200 (-0.19)	-31,600 (-0.22)	-46,800 (-0.32)		

Case b: $T_i - T_o = -200°F*$ (-111°C)

Non-Autofrettaged Cylinder

Stress	Lame Stress	Residual Stress	Thermal Stress	Resultant Stress	Distortion Energy Stress	Creep Life
σ_t	60,000 (0.41)	0	31,600 (0.22)	91,600 (0.63)		
σ_r	-36,000 (-0.25)	0	0	-36,000 (-0.25)	112,000 (0.77)	cylinder yields
σ_z	12,000 (0.08)	0	31,600 (0.22)	43,600 (0.30)		

Autofrettaged Cylinder

Stress	Lame Stress	Residual Stress	Thermal Stress	Resultant Stress	Distortion Energy Stress	Creep Life
σ_t	60,000 (0.41)	-97,500 (-0.67)	31,600 (0.22)	- 5,900 (-0.04)		
σ_r	-36,000 (-0.25)	0	0	-36,000 (-0.25)	45,500 (0.31)	7,000 hrs.
σ_z	12,000 (0.08)	-27,200 (-0.19)	31,600 (0.22)	16,400 (0.11)		

It is necessary to inject again words of caution. To the knowledge of the writer, no controlled experiments have been made on autofrettaged cylinders under simultaneous pressure and temperature application. Thus, the effect of plastic creep-strain on the residual
stress distribution is believed to be unknown. Obviously, this problem can be mitigated by developing a stress distribution which, under
pressure, develops a low distortion energy stress, S, and therefore a
low creep-rate. As calculated above, however, due to the order in
which pressure and temperature may be applied, this may not always be
possible (e.g. Example III).

The examples chosen above show the methods of approach that can
be considered for use in the design of autofrettaged autoclaves. They
are representative only of the fact that autofrettage may, under controlled circumstances, provide lower resultant stresses during operation of the autoclave. Since any autoclave design is a complex analysis of the pressure, temperature and type of cycle that the cylinder
will be subjected to, it is important that the designer conduct a
thorough study of these parameters for each specific autoclave design.

IV.   FATIGUE DESIGN PROCEDURE FOR
HIGH PRESSURE VESSELS

The data available on fatigue of high pressure vessels are limited and oftentimes conflicting. Reference [15] contains a good summary of the work of Morrison et al at the University of Bristol.
Tests conducted on small scale cylindrical specimens, which were machined and then honed to surface finishes of 0.1μ or better before
vacuum stress-relieving, indicated that the shear stress should be
used for design and the endurance limit in shear was about one-third
of the ultimate tensile strength.

On the other hand, experiments conducted at the Watervliet Arsen-
al by Davidson et al [16] with actual rifled and unrifled gun barrels
of 105 mm and 175 mm constructed of SAE 4330 type steel of higher
strength than the steels used by the Bristol investigators led the ex-
perimenters to conclude that the hoop stress at the bore provided a
better parameter than the maximum shear stress for fatigue design.
Davidson, Eisenstadt and Reiner [16] noted the discrepancy between the
Watervliet and Bristol data and pointed out that there were three substantial differences between the two studies, namely,

1.   The Watervliet tests were conducted under open-end
conditions versus closed-end conditions in the
Bristol experiments.
2.   The cyclic pressure rate was 6 per minute at
Watervliet versus 1000 per minute at Bristol.
3.   The bores of the Bristol specimens were more highly
finished than those of the Watervliet specimens.

It is likely that the latter condition has the most effect in explain-
ing the differences between the two sets of data. Both teams agreed
that autofrettaging of a cylinder had a beneficial effect as far as
fatigue is concerned. The Bristol experiments showed that autofrettaging a cylinder having a small, radial, side-hole provided a marked
improvement in its fatigue life.

It can be concluded that the existing data are conflicting and provide limited information at this time that can be used for design. In the absence of more definitive data and testing the approach recommended herein is based upon the techniques and data provided in reference [17]. An excellent summary of this method of approach is given in reference [18].

In the absence of actual fatigue strength data the designer can predict the mean endurance limit (S_e) of ferritic rotating beam specimens in terms of the ultimate tensile strength, S_{ut} as:

$$S_e' = 0.5\ S_{ut} \qquad S_{ut} < 200\ \text{kpsi} \qquad (\sim1.38\ \text{GPa})$$

$$= 100\ \text{kpsi} \qquad S_{ut} > 200\ \text{kpsi} \qquad (\sim1.38\ \text{GPa})$$
$$(\ 0.69\ \text{GPa})$$

This endurance limit applies to the type of specimen normally used in an R.R. Moore rotating beam machine, i.e., one of 0.30" ($\sim$7.6 mm) diameter subjected to pure bending with a highly polished surface. Thus it is necessary to correct this value for various factors to account for the actual design application. The estimated log S- log N curve where N is the number of cycles that would represent a minimum strength curve based upon a large compilation of rotation-beam experiments would consist of a straight line joining the point $0.9\ S_{ut}$ at 10^3 cycles to S_e at 10^6 cycles. Such a plot is shown in Figure 22. The next step in the procedure is to adjust this curve for the actual design application. This is done by applying to S_e various modifying factors by means of the equation

$$S_e = k_a k_b k_c k_d k_e k_f\ S_e'$$

where S_e = endurance limit of the mechanical element, (psi or GPa)

S_e' = endurance limit of the rotating beam speciman, (psi or GPa)

k_a = surface factor

k_b = size factor

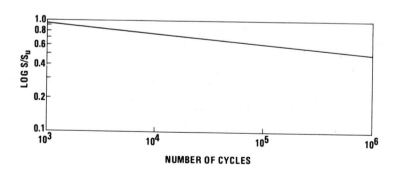

FIG. 22. Estimated S-N curve based on data obtained with standard, reversed-bending, fatigue specimens.

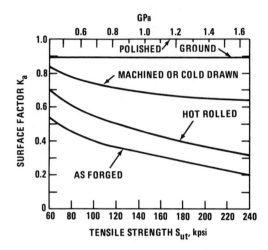

FIG. 23.   Surface-finish modification factors for steel.

k_c = reliability factor

k_d = temperature factor

k_e = modifying factor for stress concentration

k_f = miscellaneous effects factor

Figure 23 provides the value of k_a and adjusts the endurance limit from the highly polished surface that exists on a rotating beam specimen to the actual designed finish of the element. To account for size in the case of cylindrical pressure vessels, the value of k_b can be taken as

$$k_b = \begin{cases} 1 & \text{when} & d \leq 0.30 \text{ in} & (7.62 \text{ mm}) \\ 0.85 & \text{when } 0.30 < d \leq 2 & \text{in} & (50.8 \text{ mm}) \\ 0.75 & \text{when} & d > 2 & \text{in} & (50.8 \text{ mm}) \end{cases}$$

where d represents the wall thickness of the cylinder. The reliability factor, k_c, can be found from the following table:

Table II

Reliability (%)	Reliability factor, k_c
50	1.000
90	0.897
95	0.868
99	0.814
99.9	0.752
99.99	0.704

It should be noted that this Table is based upon a standard deviation of the endurance limit of 8% as noted from a large compilation of rotating beam tests.

The effect of temperature for steels can be estimated from the equation

$$k_d = \frac{620}{460+T} \quad (T \text{ in } °F) \quad T > 160°F$$

$$= \frac{344}{273+T} \quad (T \text{ in } °C) \quad T > 71°C$$

$$k_d = 1 \qquad\qquad T \leq 160°F(71°C)$$

A fatigue failure almost always originates at a discontinuity of some type, e.g., a radial side-hole into the bore of a cylindrical pressure vessel or a tool-mark on the bore. It is known that some materials are much more sensitive to notches than others so that it is necessary to adjust the geometric or theoretical stress concentration factor, K_t, obtained photoelastically or analytically, for this effect. The resulting factor is defined as the fatigue stress-concentration factor, K_f. The modifying factor for stress concentration is then given as

$$k_e = \frac{1}{K_f}$$

and

$$K_f = 1 + q \ (K_t - 1)$$

The notch-sensitivity factor, q, for steels subjected to reversed tension is given in Figure 24. The geometric or theoretical stress

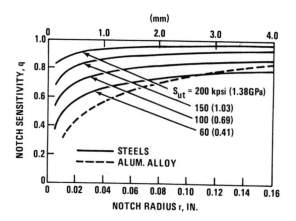

FIG. 24. Notch-sensitivity charts for steels and 2024-T wrought aluminum alloys subjected to reversed bending or reversed axial loads. Use the values for r = 0.16 in. for notch radii larger than 0.16 in. (Reproduced by permission from "Metal Fatigue", (George Sines and J. L. Waisman, eds., pp. 296, 298, McGraw-Hill Book Co., New York, 1959.)

concentration factor, K_t, is obtained from references [19] and/or [1]. The latter reference is particularly useful when pressure vessel design is involved. The factor k_f is intended to account for the reduction of the endurance limit due to all other effects, e.g., corrosion, plating, etc. Actual values of k_f are not available and depend to a large extent on the experience of the designer. For example, metallic coatings can seriously degrade the endurance limit to an extent that they may have to be eliminated.

Example I

A pressure vessel having and I.D. of 2" (~51 mm) and wall ratio of 3 is to be machined from AISI 4340 steel tempered and drawn at 1000°F (~538°C). The cylinder has a 1/8" (3.2 mm) radial side hole used for a pressure tap. It is desired to estimate the log S- log N curve for these vent conditions where it is assumed that the vessel will operate at room temperature with a reliability of 90%.

   Solution: For the material heat-treated as indicated

$$S_{ut} = 182 \text{ kpsi } (1.26 \text{ GPa})$$

Thus $S_e' = 0.5 S_{ut} = 91 \text{ kpsi } (0.63 \text{ GPa})$

From Figure 23 for a machined surface with a steel having an ultimate strength of 182 kpsi (~1.26 GPa), $k_a = 0.65$. Since the wall thickness of the cylinder is 2" (~51 mm), $k_b = 0.85$. For a reliability of 90%, from Table II, $k_c = 0.897$. Since the cylinder is to operate at room temperature, $k_d = 1$.

   From reference [1], for a heavy-walled cylinder with a side hole

$$K_t = \frac{1 + 4 \text{ b/a}}{2}$$

where b and a account for ellipticity of the side hole. For a circular hole a = b so that $K_t = 2.5$. This represents the geometric stress concentration factor and must be corrected from Figure 24 for the notch sensitivity, q, of the material. From the latter figure q = 0.92 with $S_{ut} = 182 \text{ kpsi }$ (~1.26 GPa) and r = 0.0625. Thus $K_f = 1+q$ $(K_t-1) = 1+0.92(2.5-1) = 2.38$ and $k_e = 1/K_f = 0.42$. In the absence of further information k_f is assumed to be 1. The actual endurance limit for this application is therefore

$$S_e = k_a k_b k_c k_d k_e k_f S_e'$$
$$= (0.65)(0.85)(0.897)(1.0)(0.42)(1.0)(91000)$$

$$= 18940 \text{ psi } (\sim 0.13 \text{ GPa})$$

The resulting log S- log N curve is drawn as shown in Figure 25 where

$$S_e = 0.9 S_{ut} = 163800 \text{ psi } (\sim 1.13 \text{ GPa}) \text{ at } 10^3 \text{ cycles}$$

$$S_e = 18940 \text{ psi } (\sim 0.13 \text{ GPa}) \text{ at } 10^6 \text{ cycles.}$$

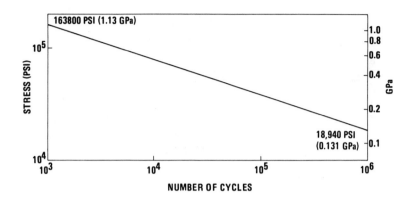

FIG. 25.   Predicted S-N curve.

Figure 25 represents the fatigue curve for an element subjected to reversed tension and must now be interpreted as it applies to a pressurized cylinder which may or may not be autofrettaged. Consider that the cylinder is to be non-autofrettaged with a closed end and subjected to a cyclic internal pressure, p. It will therefore be subjected to combined stresses which are not completely reversed so that the log S- log N curve constructed is not directly applicable.

The bore will represent the worst-stress condition and the Lamé stresses are

$$\sigma_t = p \frac{\omega^2+1}{\omega^2-1}$$

$$\sigma_r = -p$$

$$\sigma_z = \frac{p}{\omega^2-1} = \frac{\sigma_t+\sigma_r}{2}$$

These stresses are in phase and vary from 0 when p = 0 to the values shown above when the pressure cycles to its maximum value. They can be combined into a single alternating stress and mean stress by using the distortion-energy theory. Thus for the alternating combined stress

$$\sigma_{t_a} = \frac{1}{2}\,\sigma_t$$

$$\sigma_{r_a} = \frac{1}{2}\,\sigma_r$$

$$\sigma_{z_a} = \frac{1}{2}\,\sigma_z$$

and $S_{\substack{\text{combined}\\\text{alternating}}} = \frac{1}{\sqrt{2}} = \sqrt{(\sigma_{ta}-\sigma_{ra})^2 + (\sigma_{ra}-\sigma_{za})^2 + (\sigma_{za}-\sigma_{ta})^2}$

$$= \frac{\sqrt{3}\ p\omega^2}{2\,(\omega^2-1)} \tag{68}$$

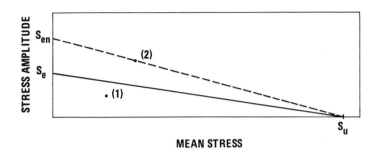

FIG. 26. Goodman diagram.

The mean combined stress for this case will have the same value, i.e.,

$$S_{\substack{combined \\ mean}} = \frac{\sqrt{3}\ p\omega^2}{2(\omega^2-1)} \tag{69}$$

These combined alternating and mean stresses can now be used with a modified Goodman diagram and the equivalent reversed tension stress derived to determine the cyclic fatigue life of the vessel. For example, Figure 26 shows a Goodman diagram [17] in which the endurance limit is plotted on the ordinate which represents the combined alternating stress component and the ultimate strength is plotted on the abscissa which represents the combined mean stress component. If the plot of equations (68) and (69) is represented by a point, say (1), which falls inside the line connecting S_e and S_{ut}, then the cylinder is safe and has an indefinite fatigue life. On the other hand, if the plotted point, say (2), falls outside of the line, the cylinder will have a finite fatigue life. Its cyclic life can be estimated by extending a straight line from S_{ut} through the plotted point to the abscissa where a value S_{en} is read off as shown in Figure 26. This value of S_{en} is then used in Figure 25 to determine the number of cycles to failure.

Example II

For the fatigue curve generated in the previous example, determine the safe operating pressure of the cylinder for an indefinite fatigue life.

Solution: Figure 27 represents the Goodman diagram constructed from Figure 25. Since the mean and alternating combined stresses are equal, the intersection of a 45° line from the origin with the line representing the indefinite cycle life represents the maximum safe operating pressure capability. This intersection is represented by the point labeled (1). Since $S_{ca} = S_{cm} = \dfrac{\sqrt{3}\ p\omega^2}{2(\omega^2-1)}$

$$S_{ca} = \frac{\sqrt{3}\ p\omega^2}{2(\omega^2-1)} = \frac{S_e\ S_{ut}}{S_e + S_{ut}}$$

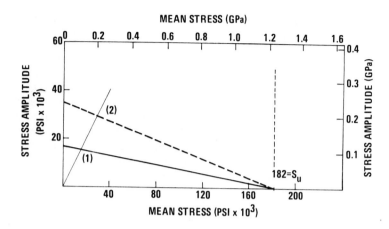

FIG. 27. Modified Goodman diagram.

and the safe operating pressure is

$$p = \frac{2S_e S_{ut}}{(S_e + S_{ut})} \frac{(\omega^2 - 1)}{\sqrt{3}\,\omega^2}$$

For the conditions of this example

$$S_e = 18{,}940 \text{ psi} \quad (\sim.130 \text{ GPa})$$

$$S_{ut} = 182{,}000 \text{ psi} \quad (\sim 1.26 \text{ GPa})$$

$$\omega = 3$$

and

$$p = \frac{2(18940)(182000)}{(18940 + 182000)} \frac{8}{\sqrt{39}}$$

$$= 17{,}600 \text{ psi} \quad (\sim.12 \text{ GPa})$$

Example III

Suppose now that it is desired to operate this cylinder at 30,000 psi (~.20 GPa). The fatigue life will now be finite and it is neces- sary to estimate how many cycles of operation can be expected.
Solution:

$$S_{ca} = S_{cm} = \frac{\sqrt{3}\ (30000)(9)}{2(8)} = 29{,}200 \text{ psi} \quad (\sim.20\text{GPa})$$

Point (2) on Figure 27 represents the operating condition. A line drawn from S_{ut} through point (2) and extended to the abscissa gives a value of $S_{en} = 34{,}800$ psi (~.240 GPa) With this value of S_{en} the number of cycles of operation as read from Figure 25 is 140,000.
These examples indicate the method of approach that can be used to make fatigue estimates in a given design application. Several things are to be noted from these examples, namely,
1. The fatigue curve shown in Figure 25 was adjusted by calcu-

lating S_e at 10^6 cycles. The 10^3 cycle value of 0.9 S_{ut} was not adjusted. This is in line with fatigue data results that indicate the low cycle values are not materially affected by the modifying factors used to adjust S_e.

2.   The stress-concentration factor due to the radial side-hole was used as a strength reduction factor to adjust the fatigue curve and then the normal Lamé stresses were used in the calculation. An alternate procedure would be to let $k_e = 1$, i.e., not reduce the strength due to the stress concentration, but then calculate the Lamé stresses by including the stress concentration factor as a multiplier. The latter approach might be preferred when the stress concentration factor is applied to only one of the Lamé stresses.

3.   For very low cycle fatigue, $N < 10^3$ cycles, the endurance limit should be taken as 0.9 S_{ut}.

4.   The recommended approach outlined is useful in providing some indication of the fatigue life that can be expected in a given design. The method can also be used, for example, in the analysis of threaded closures. However, it is not to be considered as absolute and the designer should be aware of the ASME pressure vessel code estimates of the fatigue life of a given design where safety and/or legal constraints require a coded vessel.

The advantages of autofrettage are apparent in designing against fatigue. Since the autofrettage causes compressive residual stresses at the bore, the Lamé stresses, when the cylinder is later pressurized, must be decreased by the amount of the residual stress. This means that the mean stress component is reduced and a higher fatigue life will result. For example, suppose that as a result of autofrettage a pressure vessel has compressive residual stresses induced tangentially and longitudinally of absolute magnitude σ_t^* and σ_z^*. When the cylinder is subsequently pressurized to a pressure p, the stresses at the bore vary from

$$\sigma_t = \sigma_t^*$$

$$\sigma_r = 0$$

$$\sigma_z = -\sigma_z^*$$

to

$$\sigma_t = p \frac{\omega^2 + 1}{\omega^2 - 1} - \sigma_t^*$$

$$\sigma_r = -p$$

$$\sigma_z = \frac{p}{\omega^2 - 1} - \sigma_z^*$$

Thus the alternating stress components are

$$\sigma_{ta} = 1/2 \left[p \frac{\omega^2 + 1}{\omega^2 - 1} - \sigma_t^* + \sigma_t^* \right]$$

$$= 1/2 \; p \left(\frac{\omega^2 + 1}{\omega^2 - 1} \right)$$

$$\sigma_{ra} = - p/2$$

$$\sigma_{za} = \frac{p}{2(\omega^2-1)}$$

and the mean stress components are

$$\sigma_{tm} = \frac{p}{2} \frac{\omega^2+1}{\omega^2-1} - \sigma_t^*$$

$$\sigma_{rm} = - p/2$$

$$\sigma_{zm} = \frac{p}{2(\omega^2-1)} - \sigma_z^*$$

The combined alternating and mean stresses are therefore

$$S_{\substack{combined\\alternating}} = \frac{\sqrt{3}p}{2} \frac{\omega^2}{(\omega^2-1)} \qquad (70)$$

$$S_{\substack{combined\\mean}} = \frac{1}{\sqrt{2}} \left[\left(\frac{p\omega^2}{\omega^2-1} - \sigma_t^* \right)^2 + \left(\sigma_z^* - \frac{p\omega^2}{2(\omega^2-1)} \right)^2 + \left(\sigma_t^* - \sigma_z - \frac{p\omega^2}{2(\omega^2-1)} \right)^2 \right]^{\frac{1}{2}}$$

$$(71)$$

Example IV

Suppose that the cylinder used in the previous example is auto-frettaged such that the calculated residual stresses have absolute values of

$$\sigma_t^* = 30000 \text{ psi} \big)^{\dagger} \quad (0.21 \text{ GPa})$$

$$\sigma_r^* = 15000 \text{ psi} \big) \quad (0.10 \text{ GPa})$$

It is desired to calculate the number of cycles of operation that can be expected if the cylinder is cyclically pressurized to 30000 psi.
    Solution: From equations (70) and (71),

$$S_{\substack{combined\\alternating}} = 29,200 \text{ psi} \quad (\sim.20 \text{ GPa})$$

$$S_{\substack{combined\\mean}} = 3250 \text{ psi} \quad (\sim.02 \text{ GPa})$$

From Figure 27, S_{en} = 29,800 psi ($\sim$.20 GPa). With this value of S_{en}, from Figure 25, the cyclic life is 230,000 cycles or an increase of 90,000 cycles compared to the non-autofrettaged cylinder.

---

† For the material used in the previous example, the yield stress is 162 kpsi. Thus the yield pressure is 83 kpsi and the autofrettage pressure to obtain these residual stresses is 96.5 kpsi.

V.   PRESSURE VESSEL CLOSURES

Figure 28a shows the conventional design used with many high pressure vessels. Basically it consists of an Acme thread machined into the wall of the vessel, an undercut to provide tool runout for the thread cutting operation, and one or more radial holes drilled through the vessel wall at the undercut region to prevent pressure buildup over the entire face area of the closure if the seal system fails.

Estimates of the compressive bearing stress and shear stress on the threads can be made using elementary stress equations but, at best, these values can only represent averages over the length of the closure threads.

The actual stress distribution in the threads is complicated by the following factors:

1.   the section modulus at the root of the thread is limited so that bending stresses occur on load application;

2.   the first thread picks up the load initially and as it deforms elastically it transfers the load to the next thread and so on until the entire thread system is loaded;

3.   in the usual closure, illustrated in Figure 28a, the tangential stress generated by the internal pressure may not be completely

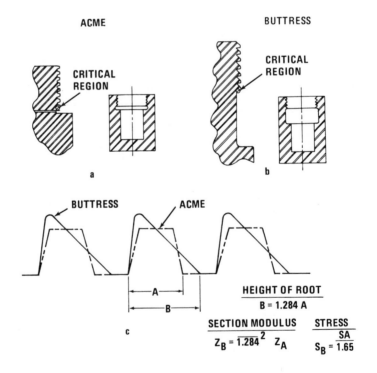

FIG. 28.   Comparison of Acme and Buttress threads.

attenuated near the first few threads because of the closeness of
these threads to the pressure sealing area. The result is a complicated stress condition involving axial, tangential, shear, and bearing
stresses;

4.   stress concentration factors exist due to the change in section and re-entrant corner between the thread section and the undercut, the change in section and re-entrant corner between the undercut
and the vessel inside diameter, and the effect of the penetration made
for pressure relief. The combined stress concentration factor will be
the product of the three individual factors due to the close proximity
of these configurations.

Figure 29 represents the actual load distribution in the type of
thread closure illustrated in Figure 28. As indicated above, the
first few threads carry the majority of the load. Under cyclic pressure loading it is easy to see why most thread failures occur in the
vicinity of the first thread.

For preliminary fatigue design and analysis, the approach outlined earlier in this chapter can be used provided the average stress
values calculated with elementary equations are increased by a factor
of 3.5 to 4 to account for the uneven load distribution. Generous
factors of safety should also be used whenever possible.

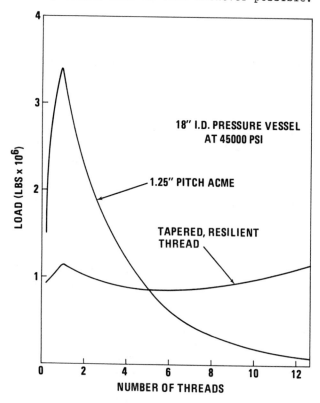

FIG. 29. Load distribution in conventional thread closure
and in tapered resilient thread.

The conventional closure can be improved so as to provide a more uniform load distribution over the threads. This can be achieved by minimizing the effect of any or all of the four factors listed previously. For example, the load on the first thread can be reduced by providing resiliency in the design so that the load is more equally distributed over all the threads. Modification of the thread form can be used to reduce the bending stress at the root of the first thread, Figure 28c. An increase in the length of the undercut region can be used to separate the longitudinal stresses caused by end loading from the tangential stresses caused by internal pressure, Figure 28c. Finally, elimination of the radial vent holes in favor of axial vent holes through the main body of the closure and careful attention to finish machine operations can be used to minimize stress concentration effects.

Reference [20] describes a resilient thread closure for which it is claimed that a load distribution such as indicated in Figure 29 is obtained. The closure consists of a tapered main nut which is machined with a semi-circular helical groove that mates with a semi-circular helical groove in the vessel (see Figure 5, Chapter 2). The resilient thread consists of special solid-wound spring sections with an inner core of steel rod which is wound around the grooves in the nut or vessel. The equivalent of a thread is thereby formed and the main nut can then be engaged with the body as with any other threaded closure. The use of a tapered closure aids in spreading the stress and providing a quick opening design.

Figure 28b represents a design described in reference [21], which tends to reduce all of the factors listed previously. The use of a buttress thread (Figure 28b and c) provides a greater section modulus at the thread root thereby reducing the bending stress. The long undercut region is of sufficient length so that the tangential stresses due to the internal pressure in the vessel are attenuated to a negligible level at the critical first thread region. Stress concentrations effects are also separated and the long undercut allows more generous radii and smoother transitions between sections. Radial vent holes have been eliminated by providing an axial vent between the cover and nut of the closure. Finally all critical surfaces are smoothed and polished to eliminate potential machining stress raisers.

VI.   SUMMARY

This chapter has been concerned with the design of cylindrical pressure vessels which provide high pressure containment as well as the potential for large volumes. The basic stress-strain equations were derived for simple cylinders and compound cylinders and the principles of the autofrettage process, together with the relevant equations applying to this technique, were discussed and derived. In addition, examples of segmented chambers providing higher pressure capability than conventional chambers of the same wall ratio were given.

Methods of approach that can be used to design against creep and fatigue in vessels constructed of ductile steels were presented. It is again noted that these techniques are essentially untried and should be used by the designer with caution. They are useful in pointing out where qualitative improvements in the design can be achieved but lack sufficient experimental backup at this time to allow for complete reliance on the quantitative results obtained. In the United States the ASME Pressure Vessel Code is the established and legal standard.

A brief discussion was included in the conclusion to this chapter on some of the problems associated with pressure vessel closure design.

In addition to the references mentioned in the body of this chapter, references [22, 23] are recommended at this point to the reader; [23] in particular is an interesting paper on pressure vessel design and contains 204 references which might well serve as the unreferenced bibliography for this chapter.

REFERENCES

1. Joseph H. Faupel, "Engineering Design", John Wiley and Sons, Inc., New York, New York (1964).
2. V.C.D. Dawson and A. E. Seigel, "Reversed Yielding of Fully Autofrettaged Tube of Large Wall Ratio", NOLTR 63-123, U.S. Naval Ordnance Laboratory, White Oak, Maryland (1963).
3. V.C.D. Dawson, "Computer Program for a Monobloc Hollow, Closed-End Cylinder Subjected to Internal Pressure", NOLTR 70-41, U. S. Naval Ordnance Laboratory, White Oak, Maryland (1970).
4. V.C.D. Dawson, "Investigation of the Relaxation of the Residual Stresses in Autofrettaged Cylinders", University of Maryland, Ph. D. Thesis (1963).
5. A. Nadai, "Plasticity", McGraw-Hill Book Co., New York, New York (1931).
6. V.C.D. Dawson, "Residual Bore Stress in an Autofrettaged Cylinder Constructed of a Strain Hardening Material", NOLTR 64-25, U.S. Naval Ordnance Laboratory, White Oak, Maryland (1964).
7. High Pressure Technology, Mechanical Engineering, December 1965.
8. T. C. Poulter, Patent No. 2,554,499, May 29, 1951.
9. V.C. Dawson and A. E. Seigel, "High Pressure Chamber Design", NOLTR 67-121, U. S. Naval Ordnance Laboratory, White Oak, Maryland (1967).
10. Unpublished correspondence with Dr. J. E. Goeller of the Naval Surface Weapons Center, White Oak Laboratory, Silver Spring, Maryland.
11. S. Timoshenko, "Strength of Materials", Vol. II, D. Van Nostrand Co., Inc., New York, New York.
12. G. Sachs and G. H. Campbell, "Residual Stress in SAE 4130 Steel Tubing", Welding Journal, November (1941).
13. J. H. Gieske, "Stress Relaxation in Aeronautical Fasteners", Progress Report I, National Bureau of Standards, August (1962).
14. Creep Design and Nuclear Reactor Applications, Proceedings for Short Course - Materials Engineering Design for High Temperatures, Penn. State University (1960).

15. W.R.D. Manning and S. Labrow, "High Pressure Engineering",
     Leonard Hill, London (1971).
16. T. E. Davidson, R. Eisenstadt, and A. N. Reiner, "Fatigue
     Characteristics of Open-End Thick Walled Cylinders Under
     Cyclic Internal Pressure", WVT-R1-6216, Watervliet Arsenal,
     Watervliet, New York (1962).
17. R. C. Juvinall, "Engineering Considerations of Stress, Strain,
     and Strength, McGraw-Hill Book Co., New York, New York (1967).
18. J. E. Shigley, "Mechanical Engineering Design", 2nd edition,
     McGraw-Hill Book Co., New York, New York (1972).
19. R. E. Peterson, "Stress Concentration Design Factors",
     John Wiley and Sons, Inc., New York, New York (1953).
20. Bulletin - Autoclave Engineers, New Flexing Resilient
     Thread Closure, Bulletin No. 320, Autoclave Engineers Inc.,
     Erie, Pa.
21. "Design of Closures for High Pressure Vessels", D. E. Witkin,
     Engineering Manager, Pressure Systems Division, National Forge
     Co., Irvine, Pa.
22. J. F. Harvey, "Theory and Design of Modern Pressure Vessels",
     Van Nostrand Reinhold Company, New York (1974).
23. K. V. Raghavan, "Pressure Vessel Design", Chemical and Process
     Engineering, October 1970.

Chapter 8

THE MEASUREMENT OF PRESSURE AND TEMPERATURE
IN HIGH PRESSURE SYSTEMS

Ian L. Spain

Laboratory for High Pressure Science and
Engineering Materials Program
Department of Chemical Engineering
University of Maryland
College Park, Maryland 20742

Jac Paauwe

Naval Surface Weapons Center
White Oak Laboratory
Silver Spring, Maryland 20910
and
Laboratory for High Pressure Science
University of Maryland
College Park, Maryland 20742

I.   INTRODUCTION

The success of many high pressure processes depends on   an   accurate characterization of the environment or of certain physical properties. In this Chapter a detailed discussion will be   given   of   the
measurement of the two primary quantities of interest, temperature and
pressure. Other measurements are dealt with in other   chapters,   with
a Bibliography of general works at the end of this chapter.

II.   MEASUREMENT OF PRESSURE

A.   Primary and Secondary Standards

Methods   of pressure measurement can be conveniently divided into
two classes -- primary and secondary. Primary   scales   are   based   on
fundamental equations relating pressure to other physical quantities.
Two examples are the fundamental equation relating pressure (P) to the
force (F) acting over area (a):

$$P = F/a \tag{1}$$

or the thermodynamic formula:

$$P = - \left(\frac{\partial A}{\partial V}\right)_T \tag{2}$$

where A is the Helmholtz function (A=U−TS) (see   Chapter   12),   V   the
volume and T the absolute temperature of the system. The first equation is the basis for the operation of the mercury manometer and   also
the pressure balance, and these instruments may be referred to as primary devices.

Some analogies may be drawn here with the measurement of temperature. The Second Law of Thermodynamics allows temperature to   be   defined and the properties of ideal (very low pressure) gases allow temperature to be measured absolutely. In practice, corrections have   to
be   made   since gases are not exactly ideal in their behavior.   Useful
temperature-measuring devices, however,   (e.g.   thermocouples, resistance thermometers,   expansion   devices) are secondary   instruments requiring calibration against primary standards or fixed points.

In high pressure measurement,   secondary devices are also most
widely used in practical situations. Most practical devices are
based   on   the   change of strain with pressure (strain gauges, Bourdon
gauges) or   electrical   resistance   of   a   material (e.g.   manganin).
Ideally the property that is used for this secondary measurement
should have a large pressure coefficient but be independent   of   other
measurement conditions (e.g. temperature). It should be reproducible,
free of hysteresis effects, be based on a material which   is   easy   to
handle, chemically inert and nontoxic. The property should be one for
which a detailed theory has been worked out for the specific   material
in question, and which does not depend on sample purity or microstructure. In practice these conditions are never   met. Most commonly,
temperature effects are important.

In practical temperature measurement, the pressure of a fixed-point calibrant is usually held constant, for example at P = 1 bar. Triple points may also be used (e.g. ice-water-water vapor) in which both P and T are fixed thermodynamically. In pressure calibration against fixed points, however, the temperature may be conveniently used to vary the transition pressure -- i.e. phase lines are more commonly used. An example is the melting line of mercury. The control of temperature is a very important prerequisite for the use of such phase lines for pressure measurement. For instance the slope of the P(T) curve for the mercury melting line is 197 bar/°C at 0°C.

Two reviews of pressure measurement techniques have been published recently (Decker, Bassett, Merritt, Hall, Barnett [1], Liu, Ishizaki, Paauwe and Spain [2]). A symposium on the Accurate Characterization of the High Pressure Environment (ACHPE) [3] was held at the National Bureau of Standards, Gaithersburg, Md. (1968). Working committees reviewed key areas and submitted recommendations (published as an appendix to the conference proceedings). It is interesting to note that whereas the First International Temperature Scale was accepted in 1927 and has been revised on several occasions since then (most recently in 1968) (see for example ref. 4) a universal set of pressure standards has still to be adopted.

In the International System of Units (SI) (for a review see ref. 5) the fundamental unit of pressure is the Pascal (1 Pa $\equiv$ 1 Newton m^{-2}). Here, the Newton (N) is the fundamental unit of force, defined as that force required to accelerate a mass of one kilogram at the rate of 1 $m\ sec^{-2}$. For high pressure measurement, this unit is small (1 bar $\equiv 10^5$ Pa), however the MPa (10^6 Pa) is a convenient unit for most purposes. In the immediate future it is likely that both Pascals and bars will be used. In this volume SI units have been given wherever other pressure units have been used (e.g. psi, $kg_f cm^{-2}$, etc.), but the bar has been considered an acceptable alternative. Conversions between different pressure units are given in Table 1.

In this Chapter, stress will be laid on practical means of measuring pressure in the laboratory or plant, rather than on specialized pieces of research equipment, or that maintained by Standards Laboratories. The pressure range will be limited to 0-30 kb. The subject of ultrahigh pressure apparatus and techniques is dealt with in Chapter 11. Furthermore, the chapter does not attempt to give an exhaustive list of references. Detailed references to earlier work can be obtained in ref. 1.

B.   Primary Measurement Standards

1.   Mercury Columns

Mercury columns are useful for pressure measurement for pressures up to about 10 bars, although specialized pieces of equipment have been constructed with column lengths up to ~30m (0-40 bars) and in some cases up to 300 m. For instance, Amagat constructed a column in the Eifel Tower, while Cailletet built one in a mineshaft to reduce temperature variations. As early as 1894 Stratton [6] proposed the use of a multiple-tube manometer with alternate tubes containing a

TABLE 1

Principal units of pressure and conversion factors

	bar	$N\ m^{-2}$ (Pa)	$kg\ cm^{-2}$	atm	$lb_f\ in^{-2}$	in. Hg	mm Hg
1 bar	1	10^5	1.01971	0.98692	14.5038	29.530	750.059
1 $N\ m^{-2}$ (Pa)	10^{-5}	1	1.0197×10^{-5}	0.98692×10^{-5}	14.5038×10^{-5}	29.530×10^{-5}	750.059×10^{-5}
1 $kg\ cm^{-2}$	0.98067	0.98067×10^5	1	0.96784	14.2234	28.959	735.56
1 atm	1.01325	1.01325×10^5	1.03322	1	14.6960	29.921	760.0
1 $lb_f\ in^{-2}$	0.068947	6.8947×10^3	0.070306	0.068045	1	2.0360	51.716
1 inHg	3.3864×10^{-3}	3.3864×10^3	0.034531	0.033421	0.49116	1	25.40
1 mmHg	1.332×10^{-3}	1.3332×10^2	1.3595×10^{-3}	1.3158×10^{-3}	0.019337	0.03937	1

1 bar = 10^6 dynes cm^{-2}

The pascal (1 Pa = 1 $N\ m^{-2}$) has been adopted as the unit of pressure in the international system of units (SI), Paris, 1960.

light liquid. This concept was developed by several workers. The pressure difference ΔP between two mercury levels h_1 and h_2 is:

$$\Delta P = \int_{h_1}^{h_2} \rho\,(h,T)\,\frac{g}{g_c}\,dh \qquad\qquad (3)$$

where ρ is the mercury density, which depends on the pressure (i.e. position in the column) and temperature, g is the acceleration due to gravity and g_c a unit factor (g_c is identically unity and dimensionless in the SI). Very accurate pressure measurement may be made with such columns (e.g. <1 in 10^5 for $P \lesssim 40$ bar [7, 8]). Temperature control of the column is the most stringent experimental requirement. Above about 10 bar, the pressure balance is a more convenient apparatus.

2.   The Pressure Balance

Without doubt the pressure balance is the most widely-used primary instrument for pressure measurement and is the accepted primary scale in the range ~0-30 kb. The principle of the balance is the loading of a piston with accurately known area, a, with a force, F, which exactly balances that exerted by the pressure on the piston (Eqn. 1). Earlier reviews have been given by Cross [9] and Meyers and Jessup [10], while a recent review has been given by Heydemann and Welch [11].

The fundamental principle is illustrated in Figure 1. The basic difficulties of the method are that there should be negligible fric-

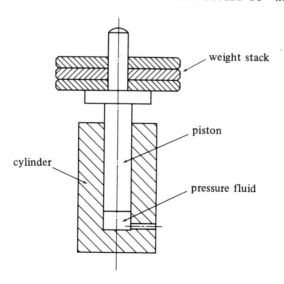

FIG. 1.   Schematic diagram showing the principle of the top-loaded pressure balance.

tion between cylinder and piston while at the same time there should be negligible leak of the fluid past the piston. These and other conflicting requirements limit the effective range to ~30 kb.

Friction may be almost completely eliminated at low pressure by rotating the piston [12, 13] above a certain critical speed. A thin oil layer between piston and cylinder acts as a real lubricant under these conditions. During measurement, the piston assembly is allowed to "coast" so that no vertical force is applied by the drive mechaniam. A pulley and belt system is normally used to drive the system, but other methods have been successfully tested (e.g. air-drive, magnetic drive). Below the critical speed, the film of lubricant breaks, and friction increases, thus reducing the accuracy of measurement and increasing the likelihood of damage to the walls. Corrections to the pressure due to leakage flow have been analyzed by Bennett and Vodar [14].

At higher pressure, the cylinder expands and the leak rate increases. Several solutions have been used to solve this problem.

First, pistons may be made with an interference fit. The piston is shrunk onto the cylinder, but at high pressure, the piston is freed by the expansion of the cylinder and is then capable of rotation. Measurement may be made over a range of pressures limited by excessive leak at high pressure and excessive friction at low pressure. Skill in construction and great care in operation are required with this design.

Second, a re-entrant cylinder was first used by Bridgman [15, 16] (Figure 2). The pressure on the re-entrant cylinder reduces its radial expansion. These designs, however, do not lead to simple calculations of the effective area of the balance based on analysis of elastic distortions.

A third solution to the problem is the controlled-clearance design of Johnson and Newhall [17] (Figure 3) in which a counterpressure acting on the outside of the measuring cylinder "exactly" compensates for the elastic deformation of the chamber. This is a very attractive solution to the problem and commercial equipment is available to ~14 kb (Figure 4).

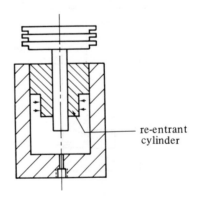

re-entrant
cylinder

FIG. 2. Schematic diagram of the piston-cylinder assembly of a top-loaded pressure balance with re-entrant cylinder.

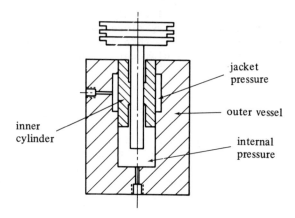

FIG. 3. The principle of the controlled clearance design of
Harwood Eng. Co.: by carefully adjusting the jacket pressure the
clearance between the piston and the inner liner may be controlled.

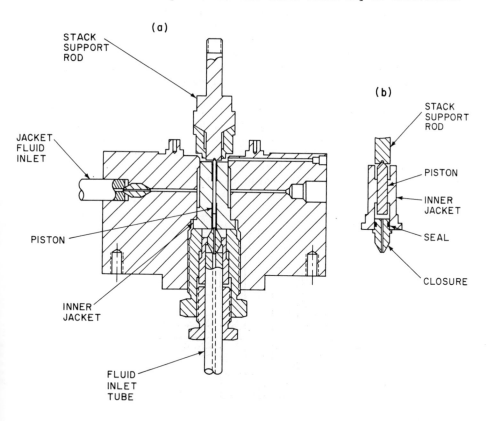

FIG. 4. (a) Detailed diagram of the Harwood controlled-clearance
pressure balance [17] for use to pressures up to 15 kb. The piston
shown is for operation up to high pressure. (b) Detail of a piston and
jacket for a low pressure range. (Figure drawn from prints supplied
by Harwood Engineering Co., Walpole, Mass.)

The effective area of a piston-cylinder assembly is the arithmetic mean of the cross-sectional areas of cylinder and piston. If the cylinder area can be brought very close to that of the piston, as in the controlled-clearance device, then the calculation of the effective area is much simplified. The diameter of the piston, d, varies with pressure, P, approximately as

$$d(P) = d(0)[1 + \frac{3\mu - 1}{E} P]$$

where μ is Poisson's ratio and E is Young's modulus. It may be assumed that this represents the variation of the effective area of the balance in the controlled clearance design, but it must be realized that the cross-sectional areas of cylinder and piston will vary along their length, so that further corrections should be necessary for work of the highest precision.

The design of the weight-stack and piston becomes more complicated at higher pressures. The top-loaded piston becomes an unacceptable design for large weights. Yoke mounting of the weights can be used (Figure 5). Michels [12, 13] devised the differential piston design, shown schematically in Figures 6, 7 and 8 to circumvent these difficulties. The weights can be hung from the bottom, and pistons with large diameter can be used with small differential cross-section. For instance, if the two piston diameters are 11.34 and 11.28mm, the differential area is ~1 mm^2. It would be impractical to design a conventional piston with this area. Several commercial designs use this principle (see Figures 7 and 8 for illustrations).

The pressure balance may be directly calibrated against the mercury manometer up to ~40 bars. Dadson [7, 8] has shown that pressures measured with the mercury column and balance can be obtained within one part in 10^5 of each other. At higher pressures the mercury column can be used as a differential manometer [19], allowing the pressure balance to be successfully calibrated to higher and higher pressures.

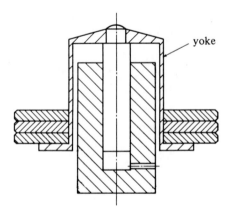

FIG. 5. Schematic diagram of a pressure balance with weights mounted on a yoke.

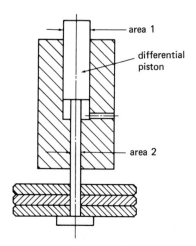

FIG. 6.   Schematic diagram of a differential-piston and cylinder arrangement.

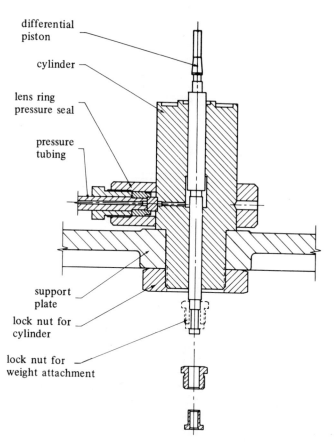

FIG. 7.   Scale drawing of the Michels differential-piston and cylinder assembly.

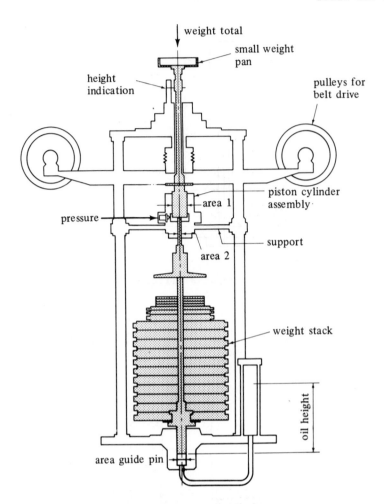

weight total

small weight pan

height indication

pulleys for belt drive

piston cylinder assembly

area 1

pressure

area 2

support

weight stack

oil height

area guide pin

FIG. 8.   Scale drawing of a Michels differential-piston pressure balance. Shaded regions indicate rotating parts.

Two such systems have been used up to ~3 kb at the Van der Waals Laboratory, Netherlands [13], and Imperial College, Great Britain [20, 21]. In these experiments one limiting factor is the knowledge of the density of mercury as a function of pressure.

Dadson [7, 8] and Dadson and Greig [22], at the National Physical Laboratory, Great Britain have discussed in detail the corrections necessary for the computation of the pressure in a balance, using two identical gauges constructed of different materials (similarity method). They claim that the uncertainty in pressure at 10 kb can be reduced to $\pm$ 1 bar. Their work can be considered the most detailed and accurate in this pressure range. Similar accuracies at 10 kb have been reported with the controlled-clearance device.

Johnson and Heydemann [23] have published results on a controlled-clearance device for use to 26 kb, at which pressure the error was stated to be $\pm$ 25 bars (see Chapter 11). It should be possible to reduce this error with better control of the clearance. Konyaev has also published work up to 25 kb [24].

Commercial balances may be purchased for use to ~6 kb from a number of sources and the controlled-clearance device for work up to 15 kb (Figure 9). Used with care as a calibration facility in common situations, accuracies of several parts in 10^4 can be achieved, and with great care approximately one part in 10^4. For such work the acceleration due to gravity must be known accurately, and corrections made for such things as the buoyancy of the air, thermal expansion of the piston and cylinder, possible buoyancy effects of lubricating oil in the lower bearings (Figure 8) etc.

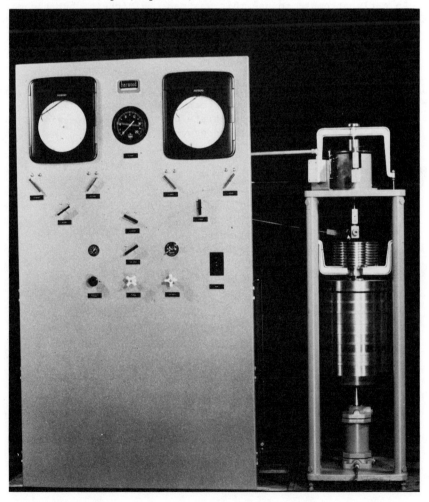

FIG. 9. Photograph of a pressure balance using the controlled-clearance principle. (Figure kindly supplied by Harwood Engineering Co., Walpole, Mass.)

3.   Comparison of Primary Gauges

A comparison of primary gauges is given in Table 2.   This table
includes a new type of primary instrument based on the simultaneous
measurement of volume and compressibility (κ).   This system is still
in an early stage of development and is only included in the Table
for completeness.   It is based on the fundamental equation:

$$P_2 - P_1 = \int_{P_1}^{P_2} dP = -\int_{v_1}^{v_2} \frac{1}{\kappa v} dv \qquad (4)$$

An early attempt to use this was made by Michels, but the method
has been revived and made much more precise [25, 26].

C.   Measurement of Pressure Based on Secondary Scales

In practice, measurements are mostly made with secondary scales,
based on the variation of a physical property with pressure.   A number of different properties have been used for special applications,
but in this Chapter we will concentrate on the two most widely used
-- strain and electrical-resistance gauges.

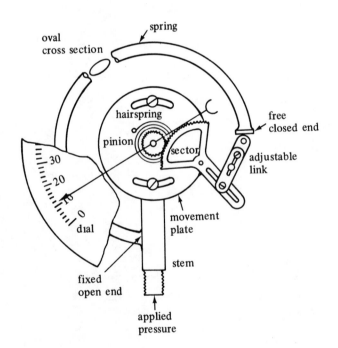

FIG. 10.   Schematic diagram of a Bourdon gauge with mechanical
linkage (figure redrawn from ref. 27).

Table 2. Primary pressure measurement methods; their range, sensitivity, and accuracy.

Type of gauge or principle	Pressure range (kbar)	Sensitivity (bar)	Possible accuracy	Limiting factor in temperature effect	Temperature control (K) for measurement to 1 in 10^4	Other remarks
Mercury manometer	0-0.04	$<10^{-6}$	>1 in 10^5	thermal expansion of mercury, container, and scale	~0.5	
Differential mercury manometer and pressure balance	0-3	$<10^{-6}$	1 in 10^5	thermal expansion of mercury, container, scale, piston, and cylinder	~0.5	equation of state of mercury is the limiting factor in the accuracy
Pressure balance	0-26					commercial units working to 15 kbar are available
3 kbar -------------	-------------	0.01-------	0.1bar	thermal expansion of piston and cylinder	~0.5	
10 kbar ------------		0.01-------	1 bar			
26 kbar ------------		0.1 -------	25 bar			
Simultaneous measurement of volume and compressibility	not stated but should be useable to >20kbar and to 80 kbar in modified form	not stated but <0.1 bar	±1.6 bar at 7.5 kbar	thermal expansion of specimen	5×10^{-3}	Ruoff et al. [25,26]; this gauge is new and further work is necessary before its adoption as a primary measurement system

293

1. Bourdon and Other Hollow-Tube Gauges

Without doubt the most widely used gauge in commercial applications is the Bourdon Gauge, discussed in a number of reports (see for
example Benedict, ref. 27). The gauge is normally in the form of a
hollow tube bent into a spiral, one end of which is fixed, while the
other is free to move. Indication of the pressure is usually made
via a mechanical linkage (see Figure 10) which includes amplification
of the motion. The element is normally surrounded by air, so correction is needed for atmospheric pressure.

Bourdon gauges can be obtained commercially to measure pressures
up to ~15 kb. In the pressure range up to about 6 kb gauges can be
supplied with certified pressure calibration to $\pm$ 0.1% of full-scale
deflection (see later comments). Some gauges have compensation for
temperature changes built into the mechanical system (e.g. temperature changes within 5°C of 25°C do not change calibration by more
than 0.2%). The gauge may be built in other geometries than a tubular coil, such as a flat tube twisted about its axis, straight tube
with eccentric bore, spiral, etc. For pressures below about 40 bars
a diaphragm arrangement similar to that used for barometers is commonly encountered. The movement of the coil may be sensed by electrical or optical means (Figures 11 and 12). A diagram showing the
principle of a commercial gauge [28] which measures the axial strain
of a straight tube is shown in Figure 13.

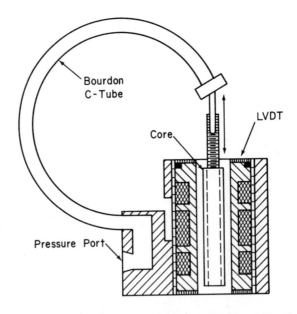

FIG. 11. Schematic diagram of a Bourdon Gauge with displacement measured with a linear, variable, differential trans-
former (LVDT). Movement of the magnetic core produces a change
in inductance of the LVDT windings. (Figure redrawn from
Schaevitz Engineering Co. Handbook, Camden, N.J.)

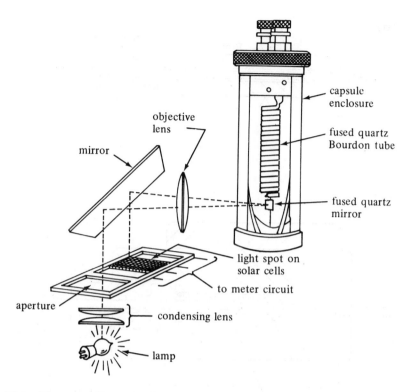

capsule
enclosure

fused quartz
Bourdon tube

fused quartz
mirror

objective
lens

mirror

light spot on
solar cells

to meter circuit

aperture

condensing lens

lamp

FIG. 12. Bourdon Gauge with spiral tube of quartz for
extremely precise measurement. The optical system senses the
pressure-induced torque in the coil (figure redrawn from ref. 29).

Below about 40 bar, many designs have been developed, principally for application in the aerospace industry. Deflection of a diaphragm can be sensed in a number of ways, including the use of strain-
gauges (see following section); capacitance, with the diaphragm as a
moving capacitance plate; variable differential transformers etc.
Such products can be readily found in standard listings of products
and companies.
It is very much easier to clean the bore of the Bourdon tube if
the closed end is fitted with a removable closure. Some gauges are
fitted with mechanical stops at the zero. It is prudent to remove
them, since an accidental, precipitate loss of pressure can cause the
pointer to hit the stop with excessive force, resulting in bending.
The mechanical linkage inevitable has some backlash, the effects of
which can be minimized by gently vibrating the gauge. Simple elec-
tromechanical vibrators are readily obtainable and are particularly
suitable for applications where the equipment is located behind protective screens. For readout in such situations, resistive transducers may be obtained to give a signal proportional to the angular
rotation of the pointer, which can be useful for operating electro-
pneumatic equipment, such as safety vent valves.

(a)

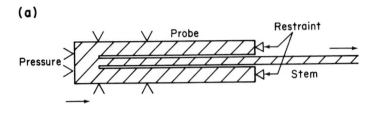

(b)

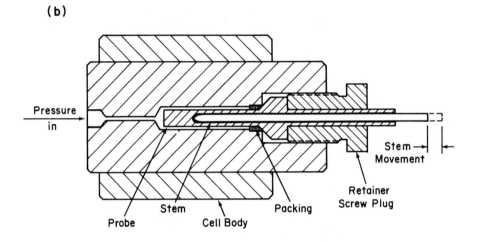

FIG. 13. (a) Principle of a bulk modulus cell in which axial strain of a straight tube is sensed; (b) schematic of the commercial gauge using this principle (figure redrawn courtesy of Harwood Engineering Co., Walpole, Mass.).

The selection of material for the sensor tube is a prime consideration. The gauge must operate within its elastic limits and not be susceptible to creep, which would result in a gradual (upward) change in the reading of the gauge if subjected to high pressure for an extended length of time. Suitable materials are fairly brittle. Beryllium-copper is chosen for many applications. Many workers recommend that a gauge should only be used to 50-75% of its stated full range, to avoid problems associated with creep or non-elastic behavior of the element. This is perhaps excessively cautious as a general rule, but it is unwise to operate gauges at their limit for long periods of time without checking their calibration. In the present author's experience, gauge calibration should be periodically checked even if used to only relatively low pressure, since problems can arise with the mechanical linkage. Checks are particularly important after transportation.

For work of the highest accuracy, quartz may be used for the tube material. However, the available pressure range is ~0.3 kb . An example of a system produced commercially [29] is shown in Figure 12. The optical system used to detect the rotation of the spiral avoids any problems with mechanical linkage connected to it. However, robust mechanical linkage from the optical system can be used to give a visual readout on a digital counter.

Units of this kind can be supplied with quartz spiral and mirror capsules for a wide range of pressures up to 1000 psi (~64 bar) with temperature control. Resolution and repeatability are 1 part in 10^5, and accuracy .01% (depending on standard used for calibration). Spiral, metallic Bourdon capsules are also available (e.g., Ni-Span C and Inconel) for pressures up to 10,000 psi (~640 bar), but here the repeatability and accuracy is only 0.05%.

Effects of creep and hysteresis, while not as obvious with conventional dial-face guages, become very apparent with the type of stable, high resolution, optical readout described above. A typical response curve with time is illustrated in Figure 14. The manufacturer recommends that these gauges be cycled several times to within 10% of full scale deflection before use if the gauge has not been used for several hours or more.

If the gauge is pressurized to its maximum value, then pressure is lowered to 50% of this value, the reading may differ by as much as

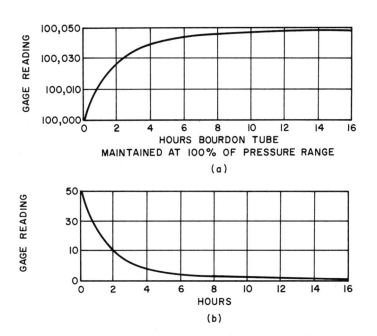

FIG. 14.   Typical response curves for a Bourdon gauge subjected to: (a) constant pressure of 100% of its range; (b) after depressurization from extended period at full scale. These curves are applicable to metallic, not quartz, spiral elements. (Figure redrawn by permission of Texas Instruments Company, Inc.)

0.5% from the true value. If pressure is then lowered to zero, then
brought back to this 50% value, the reading will be correct to 0.05%.
Thus, the manufacturer also recommends that the gauge be calibrated
both with increasing and decreasing pressure. Both the calibration
and subsequent use of the gauge with decreasing pressure should be
carried out from the same highest point (e.g., 75% of max pressure).
The hysteresis of Bourdon gauges was recognized by Bridgman as early
as 1909 [15], yet many manufacturers do not include details of the
effect in their gauge specifications.

     These points illustrate the fact that there are fundamental limi-
tations to the use of Bourdon Gauges for pressure measurement. Hysteresis and transient effects are presumably important in all such
gauges, although most manufacturers ignore them. One is led to conclude that most conventional Bourdon gauges are accurate to only ~1%
unless calibrated thoroughly and used in a specific manner consistent
with the results of the calibration. For the relatively low pressure
range 0-1000 psi (0-~64 bars) the quartz spiral gauge is a possible
alternative to the pressure balance or mercury column as a general
laboratory standard.

     A general note is in order concerning the safe use of Bourdon
gauges. The safest gauges are constructed with a solid front sepa-
rating the high pressure parts and the front dial, while a light, deformable dust cover is at the rear. Such gauges should never be mounted directly onto a rigid wall, which prevents the dust cover from releasing pressure, even if liquid is used as a pressure transmitting
medium (gas may be trapped in the gauge, or later workers may decide
to change to a gas pressure transmitting medium). A 50 mm (2") gap
between gauge and wall is usually satisfactory, although mounting in
a hole in the wall is preferred.

     If the gauge does not have a strong solid front then it should
always be used with a rigid, clear reinforced plastic about 50 mm
(~2") in front of the gauge. This does not impede reading accuracy,
but prevents splinters from flying into the observer's face in case of
an accidental explosion. Wherever possible, gauges should be   pro-
tected with a safety blow-off valve, thus protecting observer and
gauge. Glass cover plates should be replaced with clear plastic (e.g.
lucite).

2.   Strain-Gauges

     Strain-gauges can be used to sense the strain in a vessel or mem-
ber, and thus measure the internal pressure. The active elements are
usually in an electrical bridge, arranged so that pressure is read as
an out-of-balance signal. Strain-gauges have been reviewed by a number of authors (see for instance refs. 30-32). Commercial strain-
gauge systems are available to 15 kb [33] (Figure 15). In high pressure technology they are more usually employed for monitoring strain
at critical points in vessels or autoclaves.

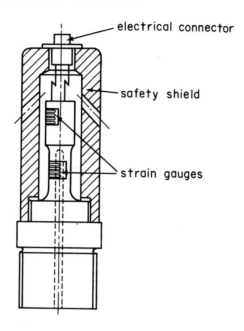

electrical connector

safety shield

strain gauges

FIG. 15. A strain gauge for use to 15 kb. Use of two elements
reduces effects of temperature changes (figure redrawn from ref. 33).

3.   Electrical Resistance Gauges

Similar in some ways to the strain-gauge is the electrical resistance gauge in which a resistance element is located inside the
pressure vessel. Again, pressure-induced changes in this resistance
are most conveniently read as an-out-of-balance signal from a bridge.
Manganin wire has been used extensively for this purpose since Bridgman's pioneering studies in 1911 [16]. Since then only two reasonable alternatives have been proposed (Table 3). The main problem is
to find a material whose pressure coefficient of resistance, $\frac{1}{R}\left(\frac{\partial R}{\partial P}\right)_T$,
is large compared with its temperature coefficient, $\frac{1}{R}\left(\frac{\partial R}{\partial T}\right)_P$. For manganin, the resistance-temperature curve reaches a broad maximum at am-
bient temperature and low pressure (Figure 16). To some extent temperature effects can be reduced by employing a secondary coil of the
same resistance as the primary coil at ambient pressure, attached to
the pressure vessel. This ensures that changes of laboratory temperature over a long time period can be compensated for. However, internal changes of temperature, due to compression of the fluid for example, are not compensated for.
Darling and Newhall [34] have proposed that an alloy of gold and
chromium (Au + 2.1%Cr) is useful for this purpose also. Its temperature-coefficient of resistance is smaller than that of manganin, but
so also is its pressure coefficient (Table 3). Studies also indicate
[35, 36] that the temperature-coefficient of resistance increases
strongly with pressure (Figure 17). This effect is less pronounced

TABLE 3

Temperature and Pressure Coefficients of Resistance
For Three Resistance-Gauge Materials

Material	$\frac{1}{R}\left(\frac{\partial R}{\partial T}\right)_P$ (K^{-1})			$\frac{1}{R}\left(\frac{\partial R}{\partial P}\right)_T$ (bar^{-1})
	0°C	20°C	40°C	
Manganin	−15–20 $\times 10^{-6}$	$^+_-5 \times 10^{-6}$	$10 - 16$ $\times 10^{-6}$	$2.4 - 2.5 \times 10^{-6}$
Au+2.1%Cr	-1.5×10^{-6}	$<1 \times 10^{-6}$	1.5×10^{-6}	$0.99 - 1.05 \times 10^{-6}$
Zeranin	$<10^{-6}$	$<10^{-6}$	$<10^{-6}$	1.6×10^{-6}

Data from ref. 33 and 34.

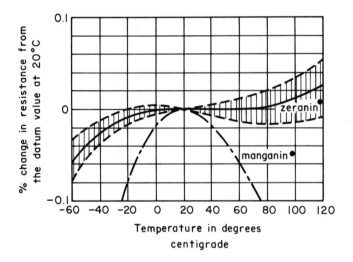

FIG. 16. Electrical resistance of zeranin and manganin in
relationship to temperature. (Figure redrawn from ref. 33.)

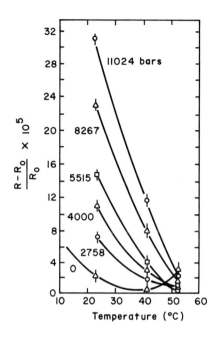

FIG. 17. Resistance vs. temperature characteristics of Au-Cr pressure gauge determined at several pressures. The alloy, insensitive to temperature at room pressure, becomes more temperature sensitive at high pressure. R_o is chosen arbitrarily. (Figure redrawn from ref. 36.)

for manganin [37] but is also troublesome. Also, Au-Cr gauges are susceptable to drift and show hysteresis effects on pressure cycling [35, 36]. An alternative material, zeranin, has a pressure coefficient about two thirds that of manganin and a much smaller temperature coefficient over a wide range of temperature [33] (Figure 16). Although used in a commercial gauge for use to 15 kb, no details are known of hysteresis or long-term stability.

Before use, manganin pressure sensors must be seasoned, otherwise hysteresis effects and gradual drift of the resistance with time are serious. The usual seasoning procedure is to anneal at ~140°C followed by a quench at liquid nitrogen temperature (77°K) [38, 39]. This procedure is normally repeated several times. Alekseev et al. [40] have reported an alternative method in which a large current through the coil flash-heats it to ~500°C. Coils of wire should be loosely wound and insulated.

Water in the pressure transmitting medium can cause resistance-shorting effects which become apparent as an abrupt change in resistance when the water freezes out ("water-kick"). Dissolved hydrogen can affect both manganin and gold-chromium gauges [41].

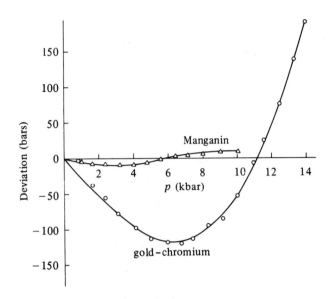

FIG. 18. Deviations from linearity of Manganin and gold-2.1
wt.% chromium resistance gauges. (Figure from ref. 35.)

Above a certain pressure the gauge response is not linear with
pressure. The response is normally fitted for pressures up to ~25 kb
to an equation of the form:

$$P = \alpha\Delta R + \beta(\Delta R)^2 \qquad\qquad (5)$$

where α and β are constants. Experiments to obtain and   are best
done above 4 kb to avoid erratic behavior in the low pressure region
(possibly related to inelastic effects). Typical curves of deviations
from linearity for manganin and gold-chromium are given in Figure 18.
     If manganin gauges are calibrated at P = 0 and 7569 bars (Hg
melting point at 0°C) then maximum deviations from linearity (ΔP_M)
can be compared for different coils prepared by different workers
(Table 4). It is not known whether these differences arise from dis-
similar material properties, winding and seasoning procedures, or calibration accuracies. Atanov and Ivanova [42, 43] have reported results on seventy-two gauges prepared from the same batch, showing that
average values for α and β in eqn. 5 obtained between 4 and 15 kb
could be used to obtain pressure to $\pm$ 0.5% up to 30 kb. This shows
the usefulness of manganin gauges for pressure measurement.
     A diagram of a commercially available gauge is shown in Figure
19. If a Mueller bridge is used to detect resistance changes, sense-
tivities of 0.01 bar can be achieved with coil resistances between 10
and 100Ω (see for example refs. 44, 45). Simple bridge circuits can
be devised giving 1 bar sensitivity, which is adequate for measurements to 15 kb, the limit for the gauge shown.

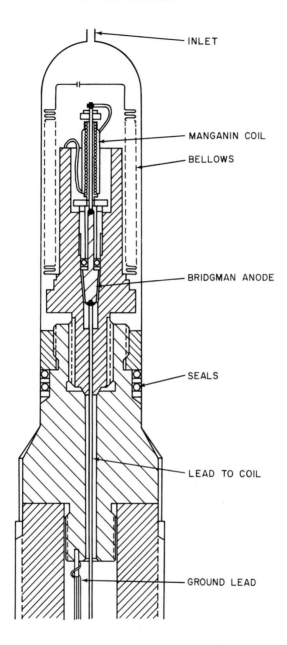

FIG. 19. A schematic diagram of the Harwood manganin pressure gauge for use up to 15 kb. Internals only are shown. The bellows need only be used with polar fluids. (Figure redrawn from prints supplied by Harwood Engineering Company, Walpole, Mass.)

TABLE 4

Maximum deviation of the Manganin gauge from linearity, ΔP_m.
The freezing point of mercury (7569 bars) and P = 0 bar are
used as fixed points. The maximum deviation occurs about
midway between the two points (Figure 16) (Data adapted
from ref. 25, 26).

ΔP_M(bar)	Source
-10	D. H. Newhall, private communication
-11	D. P. Johnson, private communication
-21	Boren et al. ref. 35
-21	Zeto and Vanfleet ref. 46
-11.6	Ruoff et al. ref. 25, 26

Resistance gauges can be used successfully as secondary devices
in many systems. Reported use has been even to 60 kb in fluid media
[46] but spurious effects arise in solid media [47]. There is still
scope for development of new materials with smaller temperature coefficients and larger pressure coefficients of resistance.

4.   Comparison of Secondary Gauges

There are many other gauge principles that have been proposed in
the literature, utilizing the variation of a number of different physical properties with pressure. A list of some of the main types is
given in Table 5. In all types of gauges, a major factor in their accuracy is the magnitude of the temperature effect. This is estimated
in the Table. Rough estimates of the sensitivity and accuracy of the
system are also given.

For general measurements in a plant or laboratory, a Bourdon
gauge usually offers the simplest solution. For more accurate measurements, or for higher pressure conditions, the gauge chosen usually
depends on subsidiary factors, such as availability of equipment, convenience to the particular system to be used, factors related to corrosive fluids etc. There is not a best solution applicable to all
cases.

5.   Use of Phase Lines for Pressure Measurement

In the pressure region below 30 kb, there are a number of transitions that can be used for pressure measurement, but since other
means exist for measuring pressure, their usefulness is limited. A
list of useful transitions is given in ref. 1, 2. In general, phase
lines are of much greater use in the region above 30 kb (see Chapter
11).

Table 5 Secondary pressure measurement methods; their range, sensitivity, and accuracy.

Type of gauge or principle	Pressure range (kbar)	Sensitivity (bar)	Possible accuracy	Limiting factor in temperature effect	Temperature control (K) required for measurement	Subsidiary comments
Melting line of mercury	0-15	<0.5	~1 bar	slope of melting line	<0.01 for ±1 bar	provisional pressure scale 0-15 kbar
Strain of elastic element (e.g. Bourdon tube)	0-15	$<10^{-6}$	~1 in 10^3	thermal expansion of element and reference plate	~0.5 for typical gauge for ±1 bar	most widely used type of gauge
Resistance element (e.g. Manganin or gold-chromium alloy, (or zeranin))	0-30	$\sim10^{-2}$	~1 in 10^3	temperature coefficient of resistivity	~0.1 (depends on pressure) for ±1 bar	probably the most useful secondary gauge from 10 to 30 kbar
Capacitance gauge (e.g. CaF_2)	0-2.5 reported	$\sim10^{-1}$	~1 in 10^4	temperature coefficient of capacitance	~0.01 for ±1 bar	new and promising type of gauge [ref 48]
Equation of state of solids (volume-pressure) (See Chapter-13)	0-300	~100	~1% below ~30 kbar	thermal expansion coefficient	~20 for ±100 bar	volume determined from x-ray lattice parameter [see ref 49]
Ultrasonic velocity measurement	0-20	<1	±2bar for P<20kb [50]	coefficients of thermal expansion and compliance	~0.5 for ±1 bar	gauge has also been developed for use to ~50 kb [refs 51-53]
Optical: Shift of ruby R lines with pressure	0->500	~500	precision greater than accuracy of present scale	shift of line with temperature	~5 for 1 kbar	For details see Chapter 11 [refs. 54,55]

The most useful phase line is that of the melting of mercury.
The transition can be readily detected by electrical resistance measurements and it is sharp, reproducible, and with negligible hysteresis. A large number of measurements have been made (see refs. 1 and
2 for a review) but the accepted equation describing the variation of
melting pressure with temperature is [56]

$$P(bars) = 38227 \left\{ \left(\frac{T(K)}{234.29}\right)^{1.1772} -1 \right\}$$

(6)

giving 7569 bars at 0°C.

The other most useful transition in the range below 30 kb is the
BiI-II (solid-solid) transition. (This transition is discussed in
greater detail in Chapter 4, Vol. II. The accepted value for the
transition pressure is $25.5\overset{+}{-}.06$ kb [3, 56, 57].

III.   THE MEASUREMENT OF TEMPERATURE IN HIGH PRESSURE SYSTEMS

Temperature measurement is straightforward in a high pressure
system if the sensor can be kept at ambient pressure. This presents
no difficulties below 300°C, but above it, problems become severe.
Only in the temperature range up to ~500°C, P<6 kb can the vessel be
heated to the same temperature as the pressurized contents. Usually
in such cases the seals of the vessel are kept cool ("cold-seal vessel") to avoid problems of leakage.

In cases where the thermometer must be exposed to the pressure
inside the chamber, thermo-couples are invariably used. They are
small, rugged, relatively reliable, and pressure corrections for some
combinations of thermocouple materials are relatively small.

It must be stressed immediately that there is not a uniform correction that can be applied to all cases. Consider for example the
two extreme cases shown in Figure 20. In case A, which approximately,
corresponds to the case normally encountered in practice, the entire
temperature drop occurs within the high pressure region. In case B,
the temperature drop along the thermocouple wires occurs outside the
high pressure region. The voltage developed between the end contacts
of the thermocouple is, for all cases:

$$V = \oint S dT = \oint S \frac{dT}{dx} dx$$

(7)

where x represents a spatial coordinate along the thermocouple wires,
$\frac{dT}{dx}$ the temperature gradient at any point x, and S is the thermoelectric power or Seebeck coefficient of the wire, which depends on temperature and pressure. The sign $\oint$ denotes an integral around the
complete circuit.

$$\oint S dT = \int_{T_1}^{T_2} S_I dT + \int_{T_2}^{T_2} S_{II} dT = \int_{T_1}^{T_2} (S_I - S_{II}) dT.$$

(8)

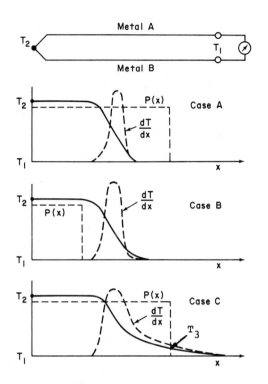

FIG. 20.   Schematic diagram of the thermocouple arrangement
in a high pressure vessel:
    Case A: An internal furnace produces the temperature
              gradient entirely within the high pressure region.
    Case B: The entire temperature gradient occurs at ambient
              pressure.
    Case C: Typical situation encountered in practice with an
              internal furnace. A small temperature gradient
              exists through the anode region of the pressure
              vessel.

where T_1 and T_2 are the junction temperatures for the two thermocouple wires of thermopower S_I and S_{II} respectively.
    Relationship (7) above shows clearly that the voltage between the
output terminals is produced <u>along</u> the wire <u>in the regions of temperature gradient</u>. Thus, in case B, there will be no effect of pressure
on the thermocouple output since the thermal gradients exist at ambient pressure. This is clearly not the case in A, where all the output voltage is developed in the high pressure environment.
    It is not possible to obtain conditions approximating case B in
practice except in special circumstances. Case C represents a typical
practical situation in which part of the temperature gradient is at
high pressure, part at ambient pressure. Ideally the temperature difference $T_3 - T_1$ is made as small as possible compared to $T_2 - T_3$ so that
corrections measured in an experiment corresponding to case A can be
applied with little error.

The effect of pressure on the output of several technologically important thermocouples has been measured by many groups. Results in general differ considerably for a given thermocouple. A particularly useful summary of work up to 1968 has been given by Hanneman, Strong, Bundy [58]. Other informative papers on this subject appear in the same conference proceedings [59-61].

The most accurately-characterized thermocouples at high pressure are Chromel - Alumel (C/A) and Platinum - Platinum/10% Rhodium (P/P10R), and for these couples the most accurate work has been performed by Lazarus and co-workers [62, 63] (Figure 21). Their data for P/P10R is in general agreement with that of Getting and Kennedy [61, 64] but in disagreement with that of Hanneman and Strong [58, 65, 66]. The effect of pressure on C/A thermocouples has generally been reported as small (e.g. $\frac{\Delta T}{\Delta P} \leq 0.1$ °C kb^{-1}, where ΔT is the temperature correction to be applied). However, at high temperature Cheng, Allen and Lazarus [63] found a relatively large effect ($\frac{\Delta T}{\Delta P} \geq 0.4$ °C kb^{-1} ≥ 950°C) (see Figure 21). It is important to note that corrections for C/A and P/P10R are of opposite sign. ΔT must be added to the measured temperature using calibration curves at 1 bar for P/P10R, but subtracted for C/A.

In the pressure range up to 20-30 kb high temperatures are normally generated in furnaces within the vessel and thermocouple leads are taken out through Bridgman anodes (see Chapter 5). While the cones must be constructed of hardened steel, it is important that the leads to both sides of these anodes be made of the correct thermocouple material. This is because temperature gradients invariably exist in this area. The reader is warned that commercial equipment is sold with anodes for thermocouples attached via long lengths of steel or other tough metals to the furnace and external wires. Errors in temperature measurement result.

Another potential source of error lies in the possibility of wires in the high pressure region becoming locally work-hardened through "kinking" or "pinching". This can cause the thermoelectric power , S, to be locally changed from its value in an unstrained wire. If this local region lies in a temperature gradient, then spurious voltages can be produced.

An alternative means of measuring temperature at high pressure has been proposed [67-70], using the thermal, or "Johnson", noise generated in a resistor, R. The average value of the square of the voltage fluctuation, $\overline{E^2}$, due to thermal agitation is:

$$\overline{E^2} = 4kTR\Delta f$$

where k is Boltzmann's constant, and Δf the frequency bandwith in which measurements are being made. Apart from the change in resistance with pressure that can be monitored, this noise is independent of pressure. The method has not been used commercially to the present author's knowledge.

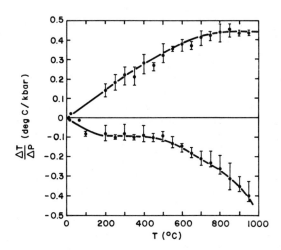

FIG. 21. Pressure correction term for platinum/platinum-10%
rhodium (P/P10R) and chromel/alumel (C/A) thermocouples up to 1000°C.
The correction term is additive for P/P10R but subtractive for C/A.
(Figure redrawn from ref. 62.)

IV.   MEASUREMENT OF OTHER PROPERTIES AT HIGH PRESSURE

There is a large literature devoted to techniques of measuring
different properties at high pressure. Certain techniques are discussed briefly in other Chapters. It is not possible to discuss this
subject in the limited space available in the present volume. At the
end of the chapter a bibliography is given, which lists several general articles on experimental techniques and the reader is referred
to these for further details.

REFERENCES

1. D. L. Decker, W. A. Bassett, H. T. Merrill, H. T. Hall,
    J. D. Barnett, J. Phys. Chem. Ref. Data 1, 773 (1972).
2. C. Y. Liu, K. Ishizaki, J. Paauwe, I. L. Spain, High Temp-
    High Pressure, 5, 359 (1973).
3. E. C. Lloyd (ed.), Proceedings of the Symposium on the
    Accurate Characterization of the High Pressure Environment,
    NBS, Gaithersburg (1968), NBS Spec. Publ. 326 (1971).
4. J. G. Hust, Cryogenics, 9, 443 (1969).
5. "Quantities, Units and Symbols" (2nd. Ed.), written by the
    Symbols Committee of the Royal Society, London. (London:
    Royal Society, 1975).
6. S. W. Stratton, Phil. Mag., 38, 160 (1894).

7. R. S. Dadson, Nature 176, 188 (1955).
8. R. S. Dadson, Proc. of the Joint Conference on Thermodynamics
    and Transport Properties of Fluids (Inst. Mech. Eng. London)
    p. 37 (1958).
9. J. L. Cross, NBS Monograph, 65 (1964).
10. C. H. Meyers, R. S. Jessup, J. Res. Nat. Bur. Stand., 6,
    1061 (1931).
11. P. L. M. Heydemann and B. E. Welch, "Piston Gauges", Chapter 4,
    Part 3 in "Experimental Thermodynamics of Non-Reacting Liquids"
    (B. Le Neindre and B. Vodar, eds.) (Butterworths, 1974).
12. A. Michels, Ann. Phys., 72, 285 (1923).
13. A. Michels, Ann. Phys., 73, 577 (1924).
14. C. O. Bennett and B. Vodar, High Pressure Measurement (A. A.
    Giardini and E. C. Lloyd, eds.) (Butterworths, London), p. 115
    (1963).
15. P. W. Bridgman, Proc. Am. Acad. Arts & Sci., 44, 201 (1909).
16. P. W. Bridgman, Proc. Am. Acad. Arts & Sci., 47, 321 (1911).
17. D. P. Johnson, D. H. Newhall, Trans. ASME, 75, 301 (1953).
18. D. P. Johnson, J. L. Cross, J. D. Hill, H. A. Bowman,
    Ind. Eng. Chem., 49, 2046 (1957).
19. L. Holborn & H. Schultze, Ann. Phys. (Leipzig) 47, 1089 (1915).
20. K. E. Bett, P. F. Hayes, D. M. Newitt, Phil. Trans. Roy. Soc.
    (London) 247, 59 (1954).
21. K. E. Bett, D. M. Newitt, "Physics and Chemistry of High
    Pressures" (Soc. of Chem. Ind., London) p. 99 (1963).
22. R. S. Dadson, R. G. P. Greig, Brit. J. Appl. Phys., 16, 1711
    (1965).
23. D. P. Johnson and P. L. M. Heydemann, Rev. Sci. Instr., 38,
    1294 (1967).
24. U. S. Konyaev, Instrum. Exp. Tech. (USSR) 1961, 728 (1961).
25. A. L. Ruoff, R. C. Lincoln, Y. C. Chen, Appl. Phys. Let. 22,
    310 (1973).
26. A. L. Ruoff, R. C. Lincoln, Y. C. Chen, J. Phys. (London)
    D6, 1295 (1973).
27. R. P. Benedict, Electro-Technology, p. 71, (Oct. 1967).
28. D. H. Newhall and L. H. Abbot, Measurements and Data,
    March - April (1970).
29. J. B. Damrel, Instr. of Control Syst., 36, 87 (1963).
30. M. Dean, "Semiconductor and Conventional Strain Gauges"
    (Academic Press, N.Y., 1962).
31. W. M. Murray, P. K. Stein, "Strain-Gauge Techniques" (Stein
    Eng. Systems, Inc., Phoenix, Ariz., 1964).
32. C. C. Parry and H. R. Lissner, "The Strain-Gauge Primer"
    (McGraw Hill, 1962).
33. A. W. Birks, C. A. Gall, Strain, p. 1, April, 1973.
34. H. E. Darling and D. H. Newhall, Trans. ASME 75, 311 (1953).
35. M. D. Boren, S. E. Babb and G. J. Scott, Rev. Sci. Instr. 36,
    1456 (1965).
36. L. A. Davis, R. B. Gordon, Rev. Sci. Instr., 38, 371 (1967).
37. C. Y. Yang, Rev. Sci. Instr., 38, 24 (1967).
38. L. H. Adams, R. W. Goranson, R. E. Gibson, Rev. Sci. Instr.
    8, 230 (1937).

39. P. W. Bridgman, "The Physics of High Pressure", (Bell, London, 1958).

40. K. A. Alekseev, Y. A. Atanov, L. L. Burova, Proc. of the Committee of the Inst. of Standards (U.S.S.R.) 75, 44 (1964).

41. R. Wisniewski, Rev. Sci. Instr., 42, 1226 (1971).

42. Y. A. Atanov, E. M. Ivanova, Izmitel. Tekhn., 2, 46 (1971).

43. Y. A. Atanov and E. M. Ivanova, p. 49, NBS Spec. Publ. 326 (1971).

44. D. M. Warschauer and W. Paul, Rev. Sci. Instr., 42, 1266 (1971).

45. W. Wilson and D. Bradley, Deep Sea Research, 15, 355 (1968).

46. R. J. Zeto, H. B. Van Fleet, J. Appl. Phys., 40, 2227 (1969).

47. J. Lees, High Temp-High Pressure, 1, 477 (1969).

48. C. Andeen, J. Fontanella, D. Schuele, Rev. Sci. Instr., 42, 495 (1971).

49. M. D. Banus, High Temp-High Pressure, 1, 483 (1969).

50. P. L. M. Heydemann, J. Basic Eng., 89, 551 (1967).

51. T. J. Ahrens and S. Katz, J. Geophys. Res., 68, 529 (1963).

52. P. L. M. Heydemann, J. C. Houck, J. Appl. Phys., 40, 1609 (1969).

53. P. L. M. Heydemann and J. C. Houck, NBS Spec. Publ., 326, 11 (1971).

54. R. A. Forman, G. J. Piermarini, J. D. Barnett, S. Block, Science, 176, 284 (1972).

55. J. D. Barnett, S. Block, G. J. Piermarini, Rev. Sci Instr. 44, 1 (1973).

56. E. C. Lloyd, C. W. Beckett, F. R. Boyd, Science, 164, 860 (1969).

57. P. L. M. Heydemann, J. Appl. Phys., 38, 2640 (1967).

58. R. E. Hanneman, H. M. Strong, F. P. Bundy, NBS Spec. Publ. 326 (E. C. Lloyd, ed.) p. 53 (1971).

59. P. M. Bell, J. L. England, F. R. Boyd, NBS Spec. Publ. 326 (E. C. Lloyd, ed.) p. 63 (1971).

60. P. J. Freud and P. N. LaMori, NBS Spec. Publ. 326 (E. C. Lloyd, ed.) p. 67 (1971).

61. I. C. Getting and G. C. Kennedy, NBS Spec. Publ. 326 (E. C. Lloyd, ed.) p. 77 (1971).

62. D. Lazarus, R. N. Jeffery and J. D. Weiss, Appl. Phys. Lett., 19, 371 (1971).

63. V. M. Cheng, P. C. Allen and D. Lazarus, Appl. Phys. Lett., 26, 6 (1975).

64. J. C. Getting and G. C. Kennedy, J. Appl. Phys., 41, 4554 (1970).

65. R. E. Hanneman and J. Strong, J. Appl. Phys., 36, 523 (1965).

66. R. E. Hanneman and J. Strong, J. Appl. Phys., 37, 612 (1966).

67. J. B. Garrison and A. W. Lawson, Rev. Sci. Instr., 20, 785 (1949).

68. R. Aumont, J. Romand, B. Vodar, Comptes Rendus Acad. Sci. (Paris) 238, 1293 (1954).

69. R. H. Wentorf, NBS Spec. Publ. 326 (E. C. Lloyd, ed.) p.81 (1971).

70. J. Fujishoro, H. Mii, M. Senoo, J. Satoo, Proc. IVth Int. Conf. on High Pressure (Kyoto) p. 818 (J. Osugi, ed.-in-chief) (1975).

BIBLIOGRAPHY

R. H. Perry and C. H. Chilton, eds., "Chemical Engineer's Handbook", (McGraw-Hill Book Company, New York, 5th edition, 1973). (Section 22 contains information on process measurements, including temperature, pressure, flow, fluid level.)

W. Paul and D. M. Warschauer, eds., "Solids Under Pressure", McGraw-Hill Book Co., Inc., New York (1963). (This volume contains a valuable Bibliography of High Pressure Techniques up to ~1963. Specific areas listed include Measurement of Temperature, Electrical and Microwave Techniques, Radiation Windows, Combined Pressure and High Temperature, Low Temperature Apparatus, Magnetic Techniques, Microwave Techniques, Optical Techniques, X-ray Techniques).

G. C. Ulmer, ed., "Research Techniques for High Pressure and High Temperature". (Springer Verlag, Berlin, 1971.)

A. A. Giordini and E. G. Lloyd, "High Pressure Measurement". (Papers presented at the High Pressure Measurement Symposium, New York City, 1962) (Butterworths, London, 1963).

C. C. Bradley, "High Pressure Methods in Solid State Research", (Plenum Press, New York, 1969).

D. S. Tsiklis, "Handbook of Techniques in High Pressure Research and Engineering", (A. Bobrowsky, transl. & ed.) (Plenum Press, New York, 1968). (This volume contains information on measurement of pressure, flow, temperature, optical, x-ray and electrical measurements, with chapters on methods of investigating phase equilibria at high pressure, surface tension, compressibility of gases and liquids.)

J. C. Jamieson and A. W. Lawson, "Solid State Studies Under High Pressure". Chapter 6, volume 6A, "Methods of Experimental Physics", p. 407 (Academic Press, New York, 1959).

E. C. Lloyd (ed.), "Accurate Characterization of the High Pressure Environment" (Proceedings of Symposium held at the National Bureau of Standards, 1968) (NBS Special Monograph 326 (1971)).

"Advances in High Pressure Research" (Academic Press, London, New York). Several chapters in this series include details of techniques, as follows:

Volume 1, R. S. Bradley, ed.
  J. Lees, "The Design and Performance of U.H.P. Equipment. An Interim Report on the Tetrahedral-Anvil Apparatus", Chap. 1, p. 2.

Volume 1 (continued)
E. Whalley, "Effect of Pressure on the Refractive and Dielectric Properties of Solids and Liquids", Chap. 3, p. 143.

L. S. Whatley and A. Van Valkenburg, "High Pressure Optics", Chap. 6, p. 327.

Volume 2 (1969), R. S. Bradley, ed.
J. S. Dugdale, "Some Aspects of High Pressures at Low Temperatures", Chap. 2, p. 101.

R. A. Horne, "The Effect of Pressure on the Electrical Conductivity of Aqueous Solutions", Chap. 3, p. 169.

G. J. Hills, "Pressure Coefficients of Electrode Processes", Chap. 4, p. 226.

Volume 3 (1969), R. D. Bradley, ed.
D. Bloch and A. S. Pavlovic, "Magnetically Ordered Materials at High Pressures", Chap. 2, p. 41.

P. G. Menon, "Absorption of Gases at High Pressures", Chap. 5, p. 313.

Volume 4 (1974), R. H. Wentorf, Jr., ed.
W. A. Bassett & T. Takahashi, "X-Ray Diffraction Studies Up to 300 kbar", Chap. 2, p. 165.

Chapter 9

MATERIALS FOR USE IN HIGH PRESSURE EQUIPMENT

P. Bolsaitis

Instituto Venezolano de Investigaciones Cientificas (IVIC)
Center of Engineering and Computation
Caracas, Venezuela

I.   INTRODUCTION

One of the first stages in the design of apparatus and instruments for use in conjunction with high pressure is, generally, the selection of adequate materials. This becomes a particularly acute problem when the pressures reach above the 10-20 kbar range and are compounded with the requirements imposed by cryogenic temperatures (as is frequently the case with experimental conditions for measurement of physical properties of materials) or very high temperatures required in many process type applications.

Although the primary concern may be the search for adequate structural materials to withstand extremely severe conditions, the selection of packing and insulating materials, lubricants, and pressure-

transmitting media are also of great importance in the proper design
of high-pressure equipment. These are dealt with in Chapters 5 and 6.
This chapter will discuss the main concern in the selection of structural materials - their mechanical properties and thermal and chemical
stability. A review of the effects of pressure on other properties
(e.g. electrical, magnetic, etc.) is the topic of Chapter 13.

II.   PROPERTIES OF STRUCTURAL MATERIALS

As is well known, the theoretical strength of materials, i.e.
that calculated on the basis of the cohesion between atoms, is very
high - an order of magnitude, or more, higher than that of practical
materials. The reason for this is that most mechanical properties are
controlled by the presence and movement of defects (cracks, scratches,
dislocations, vacancies) rather than the strength of interatomic
bonds. Defects also complicate considerably the analytic and quantitative interpretation of the macroscopic properties in terms of structural features. High-strength alloys are particularly complex systems
since they usually consist of a large number of components and several
phases.

The ultimate criterion in the selection of materials must rest
with test data at conditions that best resemble the severest to be encountered in actual use. However, the understanding of the basic concepts that account for the behavior of materials can be of great assistance in the preselection of materials and as a guideline for the
design of better new materials for a given application.

Figure 1 illustrates the main parameters measured in a normal
tensile test on materials. This type of data: elastic modulus, yield
strength, ultimate tensile strength, elongation - together with hardness, which is related to the yield strength of the material, constitute the bulk of mechanical data available on standard alloys. Many

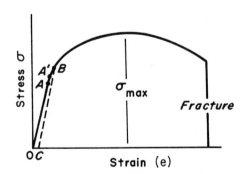

FIG. 1.   Typical tension stress-strain diagram, $e = \ell/\ell_0$
(engineering strain), $\sigma = F/A_0$ (engineering stress). True
strain defined as $\varepsilon = \ell_n (\ell/\ell_0)$, true stress as $\sigma_t = F/A$.
The stress at B, where a line parallel to the Hooke's law re-
gion intersects the stress-strain curve for $OC = 0.2\%$ is called
the 0.2% yield-stress. σ_{max} is the ultimate tensile strength.

alloys, including most of the high-strength alloys, are "heat-treat-able", i.e. by exposing the material to given temperatures for certain periods of time various phases in different amounts may be precipita-ted inside the material. The presence of such phases may drastically alter the properties of the material. Hence the properties of such alloys are not only a function of their composition but also of their previous thermal history or heat-treatment.

A similar criterion applies to properties in terms of the history of mechanical processing. Most engineering alloys come into existence through some refining or alloying process which takes place in the liquid phase, and thus the initial structure of the solid material (grain size and orientation, segregation of phases, impurity distri-bution) is controlled by the conditions imposed by the solidification process. For materials which solidify to their final form (castings) further modification of their structure is possible only by heat-treatment. However, for materials produced in the form of bars, sheet, strip, or other standard forms, the initial structure produced in the solidification step is later modified by the forging, rolling, drawing, or other mechanical deformation processes used to produce the final shape.

Because of the highly inhomogeneous deformations occurring in these processes, a high degree of anisotropy may be introduced in the material, mainly in the form of preferential grain orientation, in addition to the structure and shape of second phase particles in the material. These inhomogeneities are frequently of concern in applica-tions using heavy sections of material since the inner regions of such sections may not be accessible to the heat-treatment necessary to effect the required transformations.

It is for the above reasons that data on some crucial mechanical properties (e.g. fracture toughness) of the thousands of commercially available alloys is so scarce. Numerous microstructures can be ob-tained in an alloy of a given composition and properties may also de-pend on the fabrication defects (blowholes, inclusions) that any one piece of material may contain.

Special, severe conditions, i.e. very high or very low tempera-tures, corrosive environments, cyclical loading etc, impose additional criteria to those expressed by standard tensile and hardness tests, as will be discussed below. Cryogenic temperatures can dramatically change the mechanical behavior of some materials. At very high tem-peratures the deformation of materials by "creep", which may be lik-ened to a viscous-flow process, becomes operative and leads to perman-ent deformation at stresses below the yield stress measured at normal conditions. Vibrations or cyclic loading require the evaluation of fatigue properties of materials, since under these conditions fracture may occur at stresses far below the usual fracture stress.

Finally, the interaction of a material with the environment can have a profound effect on its mechanical strength and change its prop-erties with time. Such interactions range from ordinary corrosion processes which gradually chew up a material, to a rapid penetration of gases or liquid metals into grain boundaries leading to catastroph-ic failure. High pressure generally increases the rate of kinetic processes and hence any deleterious environmental effect will be in-

tensified as the pressure is increased. The disastrous effects of
high pressure hydrogen and mercury on pressure vessels made of certain steels are well known (see Chapter 4).

In the subsequent sections the effect of alloy structure on the
following material properties will be discussed briefly:

    a)   Yield strength
    b)   Plastic deformation (ductility)
    c)   Fracture strength and toughness
    d)   High-temperature strength
    e)   Environmental effects on strength of materials

A.   Yield Strength

The permanent deformation that is observed when materials are
subjected to stresses above their yield strength takes place by two
basic mechanisms: slipping and twinning. The first of these is best
described in terms of the movement of dislocations, while deformation
twinning consists of shearing of atomic planes over each other.
Twinning is not a significant deformation mechanism for cubic materials except at very low temperatures and high strain rates. It is more
important in hcp (hexagonal close-packed) crystals where, by reorien-
ting sections of the crystals, it helps to activate slip systems.
The processes by which deformation takes place can be described accurately only for idealized systems such as high purity single crystals, yet this description serves as a valuable reference point for
the criteria necessary to design and select engineering alloys.

The most fundamental concept for the movement of dislocations is
the Peierls-Nabarro stress. It defines the stress which must be applied to an isolated dislocation to move it from an equilibrium position in a high purity crystal. This stress is calculated as [1]

$$\sigma_p = \frac{2G}{1-\nu} e^{-4\pi\varepsilon/b} \tag{1}$$

where G is the shear modulus of the metal, ν its Poisson's ratio, b
the Burger's vector and ε the "dislocation width" defined for an edge
dislocation as $\varepsilon = a/2(1-\nu)$, where a is the distance between glide
planes. The movement of a large number of dislocations along a given
plane produces deformation by slip and, hence, yielding of the material.

The changes in yield strength in going from high purity single
crystals to alloys may be interpreted in terms of the changes in the
forces needed to move dislocations, and therefore of their mobilities,
as the structure of the material through which they move becomes increasingly complex. Work-hardening (i.e. the increase in hardness of
plastically deformed materials) can thus be related directly to this
decrease in dislocation mobility. It has been shown [2, 3] that the
yield stress of metals increases with the square-root of the dislocation density, as is illustrated in Figure 2.

An effect similar to the interaction between dislocations results from the introduction into the lattice of solute elements,
either substitutionally or interstitially. Such alloying invariably
increases the strength of fcc (face-centered cubic) and hcp metals,

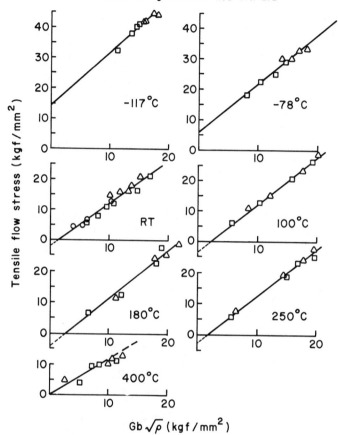

△ Points for grain size 0.025 mm dia
□ Points for grain size 0.2 mm dia
○ Points for grain size 2.0 mm dia

Tensile flow stress (kgf/mm^2)

Gb $\sqrt{\rho}$ (kgf/mm^2)

FIG. 2.   Variation of tensile flow stress with dislocation density and temperature in iron (after Dingley and McLean [3]). G = shear modulus, b = Burgers vector, ρ = dislocation density; 1 Kg/mm^2 = 9.81 MPa.

and of most bcc (body-centered cubic) metals (some substitutional solutes when added in amounts of less than 5% reduce the yield strength of some bcc matrices, leading to "solid-solution softening"). The solution-hardening effect is due to several mechanisms, the principal one being the segregation of solute atoms towards dislocations forming an "atmosphere" around them and decreasing their mobility.

This mechanism, proposed and elaborated by Cottrell and co-workers [4-6] explains the observation that very small quantities of solute atoms are sufficient to produce a significant change in the yield strength of a material. The effect of solute atoms on the

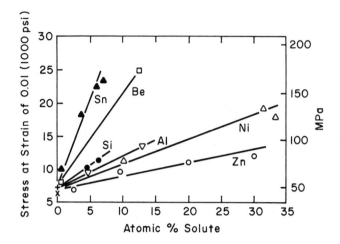

FIG. 3.   Effect of solute concentration on the yield
strength of copper alloys (after French and Hibbard [7]).

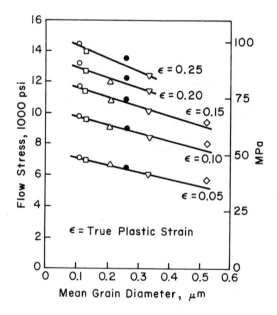

FIG. 4.   Effect of grain size on the flow stress of
aluminum at 194K, (after Dorn, Pietrokowsky and Tietz [12]).

yield strength derives from the interaction of the stress fields resulting from the impurity atoms with the stress fields of the disloc-
cations.   Since interstitial impurity atoms are usually associated
with shear force fields as well as hydrostatic distortion of the lattice, they are more effective strengtheners than substitutional impurities which only cause hydrostatic force fields, (e.g. the introduction of, say, 0.5% of carbon in interstitial sites of the iron
crystal lattice causes a much larger increase in the yield strength
than the introduction of an equal percentage of nickel atoms substitutionally).

The effect of each alloying element is different and depends on
the nature of the matrix as well as of the alloying material, as is
illustrated in Figure 3. These effects can be correlated principally
with differences in size and elastic constants of the constituent materials [8, 9].

Other effects that decrease the mobility of dislocations are
short-range-order hardening in concentrated solid solutions [10] and
the bending of dislocations between highly stressed areas in the lattice [11]. Also grain boundaries are barriers to the movement of dislocations and, consequently, a decrease in grain size contributes to
increasing the yield strength of a material, as is illustrated for the
case of aluminum in Figure 4.

It should be noted, however, that at high temperatures and slow
strain rates, deformation is localized at the grain boundaries and
hence a decrease in grain size may lead to higher rates of deformation
and accelerated failure under these conditions.

As the concentration of impurity atoms in a material is increased
beyond their solubility limit the formation of a second phase becomes
possible. If the volumetric fraction of the second phase is low, its
presence may cause stress-concentration areas or interphase boundaries
in the material which, in a manner similar to the previous effects,
serve as barriers to dislocation motion. Thus the introduction of a
second phase within a matrix is generally associated with an increase
in the yield strength of a material. However, a brittle precipitate
forming at the grain boundaries in a ductile matrix can cause the material to become very brittle [13].

As the volume fraction of the second phase increases the macroscopic properties of the material become a function of the properties
of both phases. The relationships are, however, very complex and
correlations between the properties of individual phases of a poly-
phase material and its overall properties are at best qualitative.
For a material consisting of two ductile phases two basic criteria of
deformation have been proposed. One theory proposes that the strains
in both phases should be equal, hence the average stress in the alloy
would be

$$\sigma = \chi_v' \sigma' + \chi_v'' \sigma'' \qquad\qquad (2)$$

where χ_v' and χ_v'' are the volume fractions of the two phases, and σ'
and σ'' the stresses in the phases. In this case the stress for a
given strain would increase linearly with the volume fraction of the

stronger phase. Another hypothesis formulates equal stress in both phases, hence the average strain (ε) in the two-phase material would be given by

$$\varepsilon = \chi_V' \, \varepsilon' + \chi_V'' \, \varepsilon'' \tag{3}$$

Where ε' and ε'' are the strains in each of the two phases.

Most experimental evidence shows the actual behaviour to fall somewhere between the above extreme assumptions.

The mechanical properties of poly-phase alloys consisting of brittle and ductile phases depend on the microstructure, (i.e. part-icle size and distribution), degree of coherency and interfacial bond-ing between the phases, and the properties of the individual phases. The numerous variables that enter into the description of such sys-tems make a systematic correlation impossible.

Practically all high-strength alloys depend on dispersion-or pre-cipitation-hardening to achieve their strength - including materials such as cemented carbides which consist of hard, carbide particles imbedded in, and held together by, a more ductile, metallic matrix. Other classic examples of this type of hardening are the age-hardening of aluminum, maraging steels, and dispersions of carbides (NbC, VC, etc.) in steels. Figure 5 illustrates the effect of the presence of Fe_3C particles in various ferritic steels as a function of the disper-sion of such particles (distance between particles). Figure 6 shows the effect of the volume fraction of WC in Carbolloy on its ductility and tensile strength.

In designing with conventional, low strength, ductile materials the yield strength is the fundamental criterion for the selection of a material. Barring catastrophic effects due to unusual environmental conditions, the plastic deformation of such materials provides the

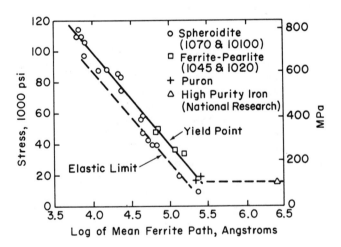

FIG. 5. Effect of fineness of dispersion of Fe_3C on the yield strength of steels (after Dorn and Starr [14]).

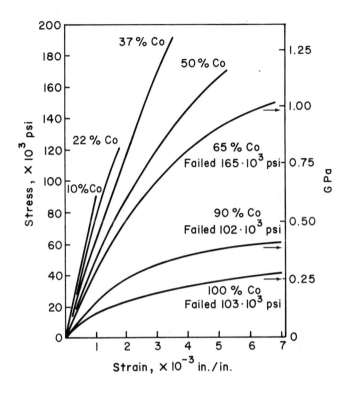

FIG. 6.   Tensile deformation of a composite material:
WC-Co alloy (after Nishimatsu and Gurland [15]).

necessary safety margin against fracture, should the design loads be
exceeded.   For high strength materials, i.e. those having very high
yield strengths, the fracture toughness is generally low (as will be
discussed below) hence the fracture stress rather than the yield
stress becomes the basic design criterion.

B.   Plastic Deformation and Ductility

Once a material is stressed past its yield point it will deform
plastically.   The amount of such deformation that may take place before fracture occurs is a measure of the ductility of the material.
A material is termed brittle when the fracture stress is very near the
yield point (i.e. no plastic deformation is observed before fracture).
The area under the stress-strain curve (Figure 1) is proportional to
the energy that the material absorbs before fracture and is called the
fracture toughness of the material. Evidently, increases in either
the yield strength, ductility, or fracture strength, with the other
two remaining constant, would lead to an increase in the fracture-
toughness of the material.   The fracture toughness is not only the
most crucial property for materials subjected to impact loading, but
is also related to the rate of crack propagation under static loading
conditions and hence to the eventual failure of materials.

Adequate ductility is necessary for the redistribution of localized stresses caused by notches or other accidentally formed stress-
concentration points, such as inclusions, microcracks or other internal defects that may exist in a material. The ability of a material
to deform plastically provides a safeguard in operations involving
tensile stress and certain elongation ductility is a basic requirement
for materials used in the design of high-pressure vessels. The required ductility of typical components in a high-pressure apparatus
will vary widely depending on the type of loading. For instance, the
body and end-plugs of pressure vessels are subjected to high tensile
stresses and require use of materials with elongation preferably above
15%; certain anvils and back-up rings subjected to compressive stresses are better fabricated from low ductility materials with very high
compressive strengths (e.g. tungsten carbide, hardened steel ($\geq$50R$_c$).

The two basic factors which determine the stress-strain curve
past the yield point are the strain-hardening behaviour of the material and the mechanisms that lead to the nucleation and propagation of
cracks. In the present section we shall review briefly the effect of
metallurgical variables on the ductility of materials.

As the yield point is associated with the force necessary to permanently displace dislocations from their equilibrium positions, the
ductility may be associated with the process of the continuing motion
of these dislocations, and the changes in the structure of the material resulting from these displacements until such time when the growth of a crack ultimately leads to fracture. The correlation of ductility and plastic deformation with metallurgical parameters is particularly difficult because of the competition of two deformation
modes: plastic flow and fracture, and a resulting high sensitivity on
the stress configuration and rate of loading.

For a high-purity crystal of a ductile material the plastic flow
is determined by the movement of dislocations and their interactions;
hence it is very sensitive to the level of impurities [16, 17], the
orientation of the crystal [18], and temperature [19]. The presence
of interstitial impurities in bcc metals leads to the congregation of
impurity atoms around dislocations and causes the yield point phenomenon observed in these materials [5] (a discontinuity in the stress-
strain curve at the onset of plastic flow, observed in mild steels and
some other alloys). The effect of substitutional impurities on ductility (as measured by % elongation) varies with the nature and concentration of the alloying elements, as is illustrated in Figure 7.
The slope of the stress-strain curve, which may be represented analytically by the empirical stress-plastic strain relation

$$\sigma = A\varepsilon^n \tag{4}$$

is usually, but not always, increased by the addition of solute elements to a pure metal, but such changes are generally not very large
[21].

Materials that undergo a transition from ductile-to-brittle failure at low temperatures may lose their ductility in the presence of
small amounts of impurities. The body-centered cubic transition metals (Fe,Mo,Cr,W) are the foremost examples of such behaviour. In a

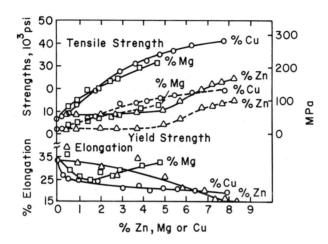

FIG. 7.   Influence of alloying elements on the tensile
and yield strength and elongation of aluminum (after F. N.
Rhines [20]).

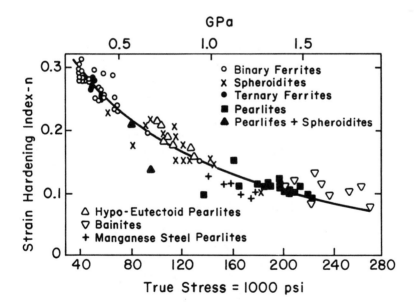

FIG. 8.   Strain hardening index 'n' in terms of deformation
stress at 0.2% strain for various steels (after Gensamer [23]).

high-purity state these materials are ductile to fairly low temperatures yet the addition of 10 to 100ppm of interstitial impurities
(C,N,O) or a percent, or so, of substitutional impurities may embrit-
tile them at room temperature.

The grain boundaries present in polycrystalline materials act as
barriers against which "dislocation pile-ups" occur, producing back
stresses and a resulting hardening of the material. Also the continuity condition within a polycrystalline material requires hetero-
geneous deformation across some boundaries and hence the activation
of some higher energy slip systems. The result is a larger rate of
strain-hardening in fine-grained materials. However, at larger total
strains the grain-size effect diminishes [22].

The ductility and plastic deformation of multiphase alloys, such
as are of primary interest in engineering applications, depend on the
nature of the phases present, their relative amounts and configuration. The equal-stress and equal-strain hypotheses for mixtures of
ductile phases were mentioned in the previous section. However, more
generally, engineering alloys contain some hard and brittle phases in
varying amounts and microstructural forms, depending on the previous
heat treatment. Within groups of similar alloys a correlation between
strength (as measured by stress at a given strain) and the strain
hardening index "n" of equation (4) can frequently be found. (This is
also equivalent to a correlation between "A" and "n" of equation (4)).
Figure 8 illustrates such correlation for a number of different
steels. Such correlations have been shown to exist for the same steel
after different levels of heat-treatment, as is shown for plain carbon
steels in Figure 9, where the different stresses correspond to changes
in strength due to quenching and tempering at different temperatures.

Similar effects have been found in age-hardened and precipit -
tion-hardened alloys. Age-hardened aluminum, for instance, shows a
relatively low rate of strain-hardening when in a highly hardened
state due to the presence of Guinier-Preston zones. With overaging,
as the coherency between the matrix and the precipitate is lost, the
decrease in yield strength is accompanied by a higher rate of strain
hardening [25-27]. This phenomenon has been explained on the basis
that particles of partly coherent precipitates are strong obstacles to
dislocation movement and thus lead to higher rates of work-hardening
[28].

C.   Fracture Strength and Toughness

The fracture properties of materials rest primarily with the
mechanisms by which fracture may take place in various materials. The
yield strength of a material generally defines the limits on the design of any structural part. However, as one increases the strength
of alloys by manipulating various metallurgical parameters (alloys
with yield strengths in excess of 300,000 psi (~2GPa) are presently
available), such materials also become increasingly sensitive to flaws
and hence to sudden failure. The elastic energy ($\sigma^2/2E$) that may be
stored in a material of high yield point becomes very large and any
stress-concentration point in the material may be the focal point for
the rapid propagation of a fracture. In brittle materials such fail-

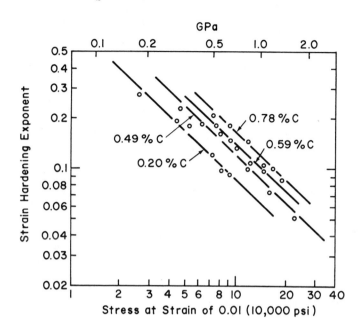

FIG. 9. Variation of strain hardening index 'n' witn strength for steels of several carbon contents (after Hollomon [24]).

ures may take place at a high kinetic rate, while in materials retaining some ductility some of the kinetic energy is dissipated in plastic work. Consequently, utmost attention concerning the fracture of materials goes into characterizing the conditions (temperature, environment) and structural features that may lead to the embrittlement of materials. From this point of view the most significant parameter for characterizing high strength materials is their fracture toughness.

The fracture toughness (G_c) is related to the fracture stress σ_F (critical stress required for spontaneous crack growth) by the equation [29]

$$G_c = \frac{\sigma_F^2}{E} \alpha \pi c \tag{5a}$$

or, using a stress-intensity parameter, K_c, instead:

$$K_c^2 = (\sigma_F/\sqrt{\alpha\pi c})^2 = EG_c \tag{5b}$$

where c represents the depth of a surface flaw (or one half of the length of an internal flaw), E is Young's modulus and α is a parameter that represents the size of the plastic zone near the crack tip (if the crack is small relative to the dimensions of the sample and $\sigma_F \ll \sigma_y$, i.e. fracture takes place before reaching the yield point of the material, then $\alpha \simeq 1$).

The quantitative characterization of fracture toughness of materials is usually made in terms of so-called "notch-toughness tests",
which are designed to measure the resistance to impact loading of a
material in the presence of a stress concentrator such as a notch or a
crack. The notch toughness of a material should be differentiated
from the more general concept of toughness which frequently refers to
the area under the stress-strain curve obtained in a normal tensile
test. While most materials that are brittle will show a low notch
toughness in a normal tensile test, the converse is not always true.
A number of materials, such as heat-treated alloy steels, while relatively ductile in a normal tensile test, may still show a low notch
toughness.

Of the notch-toughness tests the ones used most frequently are
the Izod and Charpy tests - the former of wider useage in Britain, the
latter in the United States. Both tests are based on striking with a
pendulum of calibrated weight and predetermined velocity a bar of the
material to be tested of standard dimensions and precut notch. While
these tests do not give a full assessment of all the complexities associated with toughness and fracture characteristics of materials,
they do provide a scale for comparison of various materials. For further details of the notch-toughness tests the interested reader is referred to the ample literature available on the subject (e.g. references [30-32]).

Figure 10 illustrates the general effect of yield strength level
on plane-strain fracture toughness for a number of high- and medium-
strength steels. The behaviour of the maraging steels included in
this Figure may be approximated by the analytical equation (K_{Ic} is
equivalent to K_c of equation 5(b) under plane-strain conditions,
$K_{Ic}^2 = K_c^2 (1 - \nu^2)$)

$$K_{Ic} \simeq 363 - 1.1\sigma_y \text{ (Kpsi) for } (180 \leq \sigma_y \leq 330) \qquad (6)$$

$$\simeq 2.47 - 1.1\sigma_y \text{ (GPa) for } (1.2 \leq \sigma_y \leq 2.25)$$

A relation similar to that found for steels and shown in Figure
10 has also been found for aluminum alloys (Figure 11) and other materials [33].

Relations such as this, together with equations (5), may be used
to estimate the maximum flaw size (with an adequate safety factor)
permissible in a material intended for specified service conditions.
It should be noted that under cyclic loading conditions, or in corrosive environments, there will be a tendency for flaws to grow in size
leading to failure after some time in service. Methods for calculating such rates of growth have been discussed by several authors [35-
37].

In considering the effect of metallurgical parameters on fracture
the most important aspect is their effect on the ductile-to-brittle
fracture transition. In this regard it is convenient to separate
metallic systems of the three common crystal types into two groups:
the fcc materials, which remain ductile at all temperatures and the
bcc and hcp metals, where most of the former and some of the latter
undergo a ductile-to-brittle transition at low temperatures. This

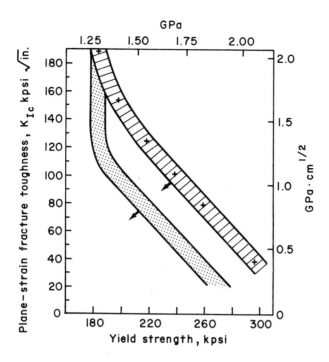

FIG. 10. Effect of yield strength level on plane-strain
fracture toughness for various medium-and high-strength steels
(after Pellini [33]). The upper band refers to "new" steels
such as maraging and H.P. 9-4; the lower band to "old" steels
e.g. 4330, 4335, H 11 etc. (Arrow indicates decrease with increasing thickness or in presence of moisture (stress-corrosion
cracking)).

transition implies a change in fracture mode from transgranular shear
to crystallographic cleavage.

Interstitial solutes in bcc metals play a very important role in
determining the temperature at which the ductile-to-brittle transition
occurs. In moderately pure iron cleavage fracture occurs in the -195°
to -250° range; however, zone-refined molybdenum and iron are ductile
to 4 K [38, 39] leading to the hypothesis that all bcc transition metals may be ductile at all temperatures if sufficiently pure. For example, the effect of small concentrations of oxygen on the toughness
of ferrite (as measured by the Charpy impact energy) is illustrated in
Figure 12. As can be noted, the temperature of the ductile to brittle
transition increases with increasing impurity concentration. Similar
effects are observed for other types of interstitials and matrices
[41-43].

Carbon, nitrogen and oxygen, if present in excess of their solubility limits, may lead to the formation of carbides, nitrides and oxides of the host metals. The distribution and form of these precipitates may lead to further modifications of the properties of the al-

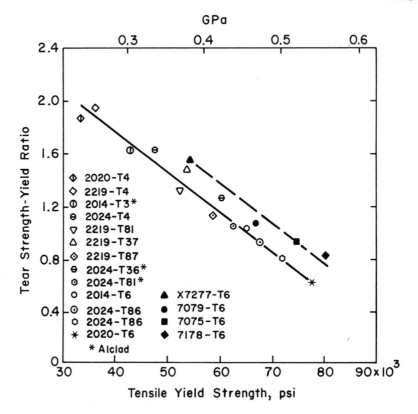

FIG. 11. Tear strength-to-yield strength ratio in terms of tensile yield strength for commercial aluminum alloys (after Kaufman and Hunsicker [34]).

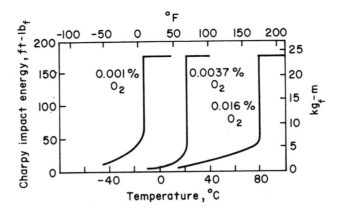

FIG. 12. Effect of oxygen content on the Charpy impact energy of ferrite (after Biggs [40]).

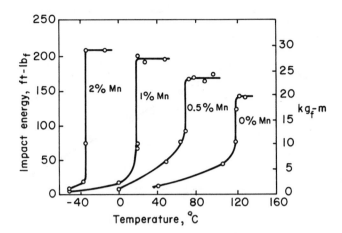

FIG. 13.   Effect of manganese on the impact energy of
iron (after Allen et. al. [44]).

loy. Especially undesirable is the precipitation of these compounds
along grain boundaries, which may lead to the embrittlement of the material.

The effect of substitutional alloying elements is twofold: by
their own effect on the properties of the host lattice and, more im-
portantly, by their interaction with the interstitial impurities in
the redistribution of carbides and nitrides, and by their association
with the residual sulfur and phosphorus present in steels, both of
which have a marked deleterious effect on mechanical properties of
steels.

Manganese and nickel improve the low-temperature ductility of
iron and are technologically important alloying elements in steels for
this reason. The effect of manganese is to redistribute grain-boundary carbides to intragranular sites and to spheroidize embrittling
grain-boundary sulfides. The effect of manganese additions on the
toughness of iron is illustrated in Figure 13. Some other alloying
elements such as the platinum-group elements [45] also lower the transition temperature, while most other alloying elements have the opposite effect [46].

Pure fcc metals and their solid-solution alloys generally retain
ductility at all temperatures. However, embrittlement may occur by
segregation of impurities at grain boundaries, or by increased sensitivity of alloys (as compared to pure metals) to the effects of corrosive environments [47, 48].

The tensile transition temperature of the bcc and hcp metals that
undergo a ductile-brittle transition is lower for smaller grain size
materials, as is illustrated by the example shown in Figure 14. On
this basis some solute elements may improve the fracture resistance of
metals by acting as grain refiners.

The fracture behaviour of precipitation-or dispersion-strength-
ened (i.e. multi-phase) alloys is very complex since, depending on the

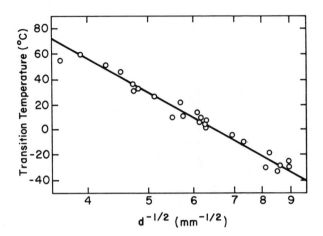

FIG. 14.   Transition temperature of 0.11%C mild steel as a
function of grain size (i.e. mean grain diameter d), (after
Petch [49]).

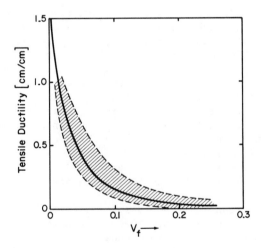

FIG. 15.   Effect of volume fraction (V_f) of hard particles
on the tensile ductility of two-phase copper alloys (after Edelson
and Baldwin [51]).   Figure includes data of copper-iron-molybdenum,
copper-holes (i.e. porosity), copper-chromium, copper-alumina,
copper-iron, copper-molybdenum, copper-alumina, copper-silica.

nature of the precipitate and its form and distribution within the matrix, many different effects may be obtained. Crack nucleation at matrix-particle interfaces, stress concentrations at inclusions, retardation of grain growth, increase in yield strength, are but a few of the effects associated with the presence of second phases. The formation of precipitates along grain boundaries is one of the most common mechanisms for the embrittlement of alloys, yet properly dispersed particles are an essential feature of many high-strength materials, such as maraging steels [50]. Where second-phase particles form on grain boundaries, or by their size and lack of coherency with the matrix act as fracture-initiation sites, a decrease in fracture toughness with the amount of the second phase may be expected, as is illustrated in Figure 15. On the other hand, dispersions that are associated with pronounced strengthening effects are distinguished by their fineness and uniform distribution throughout the matrix. Such fine and uniform dispersions are obtained by creating, within the matrix and prior to precipitation, a high density of active sites where precipitation may take place, for example by introducing a high dislocation density. In maraging steels the high dislocation density is the result of the martensitic transformation before ageing [50]. Similarly the dislocation structure produced by ausforming alloy steels (strain-hardening before converting to martensite) insures a uniform distribution of subsequently precipitated alloy carbides [52, 53].

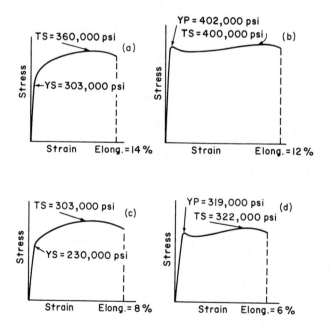

FIG. 16. Effect of ausforming and strain ageing on stress-strain curves of H-11 steel (after Zackay and Parker [53]): a) Ausform H-11 steel, quenched and tempered; b) Ausform H-11 steel, quenched, tempered and strain-aged; c) Conventional H-11 steel, quenched and tempered; d) Conventional H-11 steel, quenched, tempered and strain-aged.

The above are just two examples of the effects of mechanisms of transformation on the final properties of a material. It must be strongly emphasized that the proper mechanical and thermal processing of alloys is crucial in determining their final structure and resulting fracture toughness.

An illustration of how different mechanical and thermal processing parameters may affect the stress-strain curve of an alloy (the high strength H-11 alloy) is shown in Figure 16.

D.   High-temperature Strength

The main feature of the mechanical behaviour of materials at elevated temperatures, apart from the physical stability of the phases, is the strain-rate dependence of deformation and fracture. While the results obtained from short-time mechanical tests, such as tensile, impact, hardness tests, may not show major changes with temperature, it is found that under prolonged static loading at elevated temperatures materials do not behave elastically but, rather, flow or "creep" under loads much lower than their yield strengths measured in short-time tests.

The concept of elevated temperature is, of course, relative, in view of the wide range of melting temperatures for various materials. A rule of thumb criterion is that a temperature of 0.5 T_m (where T_m is the absolute melting temperature) sets the limit for the mechanical usefullness of materials. On a more absolute scale, temperatures above 1000 K (1340°F) are considered as high temperatures. The number of materials whose melting points are above 2000 K are not numerous. Apart from some platinum-group metals, only zirconium, hafnium, vanadium, niobium, tantalum, chromium, molybdenum and tungsten fall in this category and hence are referred to as "refractory metals". Many ceramics (oxides, carbides) have elevated melting points - the highest known is that of HfC (4160°C) - but, of course, the limitations of brittleness and the associated difficulties in fabrication remain even at high temperatures.

In terms of microstructure, the difference between low-and high-temperature deformation is the fact that at high temperatures other deformation mechanisms become operative. One mechanism that becomes important at high temperatures is grain boundary sliding; furthermore, recovery processes (i.e. annealing) operate simultaneously with deformation. At elevated temperatures slip remains a primary mode of deformation; however, dislocations may overcome barriers by a "climb mechanism" which involves the diffusion of vacancies to dislocations. Since diffusion is a thermally activated process, such dislocation climb, which is related to the rate of deformation, may be expected to be sensitive to temperatures. On the basis of this type of analysis a relationship between the strain rate $\dot{\gamma}$, applied stress (σ), and temperature may be derived:

$$\dot{\gamma} = B (\sigma)e^{-Q}/kT \qquad (7)$$

where B(σ) is some function of the applied stress and Q an activation energy, which at high temperature is generally the same as that for self-diffusion in the matrix under consideration.

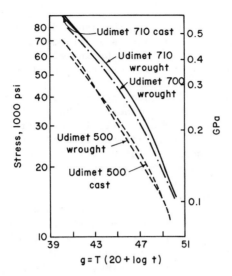

FIG. 17. Larson-Miller plot for some Ni-Cr-Co superalloys
(T in degrees Rankine), (after Ruoff [55]).

Equation (7) is the basis for other relationships used to predict
service lifetime of high-temperature materials. One integrated form
of Equation (7) is the Larson-Miller [54] relation

$$T(C + \log t) = g(\sigma) \qquad (8)$$

which relates temperature (T), service life (t) and applied stress
(σ). An illustrative diagram of this type, for several high-temperature alloys, is shown in Figure 17. The Larson-Miller relation is
found to be best applicable at high stress levels, while at lower
stresses other, similar equations, such as those proposed by Orr,
Sherby and Dorn [56] and Barret and Sherby [57] are found to give better correlations.

A more condensed form of representation of the stress-tempera-
ture-time relation and one which lends itself to tabulation, is the
rupture strength for a given temperature in terms of 10,000 or 1000
hours of static loading.

Fracture of metals and alloys at elevated temperatures is inter-
granular, indicating that under these conditions the grain boundaries
are weaker than the crystallites. Intercrystalline cracking is noted
to increase with decreasing strain rate, increasing temperature and
decreasing grain size.

Grain size affects the rupture life of materials in two ways. As
sources and sinks for vacancies, grain boundaries have a moderate
effect on the strain or creep rate under a steady load, where the
creep rate decreases with increasing grain size. More important, however, is the effect due to nucleation of cracks at grain boundaries,
especially those in planes normal to an applied tensile stress. An

illustrative example of this effect concerns recently developed materials for turbine blades which, through a directional solidification process, contain no grain boundaries in planes perpendicular to the blade axis. This results in very substantial increase in service life as compared to blades of the same material produced by conventional processes [58].

In terms of crystallographic structure it is found that close-packed structures (fcc,hcp) generally have better creep resistance than more open structures (bcc). Alloying elements and precipitates are the main structural modifications used to increment the high temperature mechanical properties of materials. Solutes added to restrain creep operate mainly through their effect on recovery processes, i.e. the presence of solute atoms in a lattice makes dislocation climb and cross-slip more difficult.

The most important method for imparting high-temperature strength to alloys is the formation of some hard, fine, stable and uniformly distributed precipitate, which is effective in inhibiting dislocation climb and grain boundary sliding. Such precipitates are most generally oxides, carbides, or some intermetallic compounds. Many metals with internally dispersed oxide particles such as $Al + Al_2O_3$ [59], $Cu + Al_2O_3$ [60], $Cu + SiO_2$ [61] and $Ni + Al_2O_3$ [62] have been shown to have considerably better thermal stabilities than the corresponding pure metals.

The high-temperature strength of cobalt-based superalloys depends primarily on precipitation along the grain boundaries of finely dispersed carbides, while nickel-base superalloys depend primarily on the formation of a coherent aluminum and titanium compound (γ'-$Ni_3Al(Ti)$) for high temperature strength [52]. The marked increase in high temperature strength with the content of aluminum plus titanium of nickel base superalloys is shown in Figure 18.

Ceramic materials also exhibit creep deformation under static loading at high temperatures. Since most engineering ceramics are sintered materials, they exhibit some degree of porosity which affects their creep behaviour; e.g. it has been shown that the shear-stress necessary to maintain a given creep rate decreases with increasing porosity [63]. The mechanisms operative in the high temperature deformation of ceramic materials are similar to those operative in metals and alloys. The much longer rupture life of single-crystal alumina as compared to polycrystalline alumina, shown in Figure 19, is an illustration of this fact.

E.   Environmental Effects

All structural materials are susceptible to deterioration caused by some environments, through various mechanisms. Such environmental effects may be of particular concern for high-strength materials since there is a general tendency for an increase in susceptibility to environmental effects as the stress on the material is increased. Furthermore, substances contained at high pressures are likely to exhibit an enhanced activity in terms of electrochemical and oxidation reactions, solubility, and other mechanisms by which materials interact with the surrounding environment.

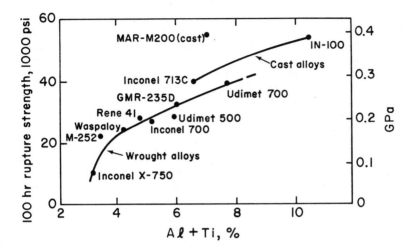

FIG. 18.   The effect of Ti and Al content on the 100 hour
rupture strength of nickel base alloys at 1600F (1144 K),
(after Tetelman and McEvily [29]).

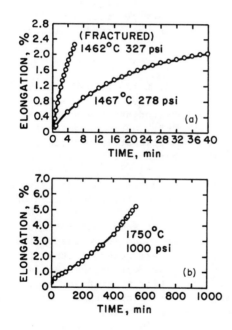

FIG. 19.   Creep of Al_2O_3:   a) polycrystalline sample;
b) single crystal,   (after Chang [64]).

   The usual corrosive attack which leads to failure of materials
through a reduction of cross section with time is of no special interest here, since it falls under the normal criteria of selection and
protection of materials for corrosion resistance. Of particular interest for materials in high pressure applications is, however, local
attack - i.e. cracks which form and grow as a result of the synergic
effect of stress and corrosive environment, which may lead to failure
of materials at stress levels considerably below their normal yield
strengths. This type of failure is called "stress-corrosion cracking" and, together with hydrogen damage, constitute the primary problems in regard to environmental effects on high-strength materials.
   The exact mechanism of stress-corrosion cracking is not well understood. It shows a very pronounced specificity of combination of
materials and environments and no evident relation to normal corro-
sivity. Pure metals are generally immune to stress-corrosion cracking
while most alloys are susceptible to such failure in certain environments, and trace amounts of certain impurities may have a very pronounced effect on the susceptibility of particular alloys. In aqueous
solutions the presence of certain ionic species may have a strong inhibiting effect. Unfortunately no general correlation of resistance
to stress-corrosion cracking to alloy composition and environment has
been found.
   The tensile stresses causing failure by stress-corrosion cracking
may be externally applied or residual. For a large number of alloys
in environments causing stress-corrosion cracking a relationship between applied stress and service life, similar to the Larson-Miller
relation (see Equation 8), has been found suitable:

$$\sigma = -a \log t + b \tag{9}$$

where σ is the applied stress, t the time to failure and a and b are
constants that depend on the nature of the alloy and the environment.
This relationship has been tested and the constants 'a' and 'b' estimated for aluminum alloys [65], carbon steels [66], austenitic stainless steels [67, 68], martensitic stainless steels [69] and other materials. By virtue of increased residual stresses, cold-work has a
negative effect on the resistance to stress-corrosion cracking. It
has also been found that an increase in grain size of an alloy generally decreases its resistance to stress-corrosion cracking in a given
environment [70, 71].
   Table 1, shows a listing of some combinations of materials and
environments that have been found to be susceptible to stress-corrosion cracking. Such listings are however rather incomplete in view of
the high sensitivity of stress-corrosion cracking to residual impurities, the presence of inhibiting substances, etc.
   Recently a large number of studies of stress-corrosion of high-
strength materials have been published, e.g. of precipitation-hardening stainless steels [72], high strength aluminum alloys [73], maraging steels [74-76], Fe-Cr-Ni base alloys [77]. However, the great
diversity of environments to be found in practice may require the previous testing of materials for particular high pressure applications
in potentially corrosive environments.

TABLE 1[78]

Environments Conducive to Stress-Corrosion Cracking

Material	Environments
Aluminum Alloys	NaCl solutions, seawater, water vapor
Copper Alloys	Ammonia Vapors and solutions, amines, water, water vapor
Nickel Alloys	Caustic-soda solutions
Carbon Steels	Nitrate solutions, acidic H_2S solutions, NaOH solutions, seawater
Stainless Steels	Acidic-chloride solutions, seawater, H_2S
Titanium Alloys	Fuming HNO_3, seawater, N_2O_4

Hydrogen has a deleterious effect on a number of different engineering alloys by various mechanisms such as hydrogen-blistering, hydrogen-embrittlement and decarburization. The active species in these
processes is atomic hydrogen which has a high diffusional mobility in
the relatively open bcc structures. For this reason it is the bcc
steels, titanium, zirconium and the bcc refractory metals (vanadium,
chromium, molybdenum etc.) which are susceptible to hydrogen damage
while the fcc austenitic steels (e.g. 300 series stainless) are not.
The sources of atomic hydrogen may be corrosion processes, electroly-
sis (e.g. electroplating), hydrogen sulfide, chemical reactions, or
just moist atmospheres. There is some indication, for instance, that
the embrittlement of ultra-high-strength steels by water vapor results
from the presence of dissociated hydrogen [79].

The mechanism by which hydrogen-induced damage takes place is
different in different materials. Generally, embrittlement results
from hydrogen contents greater than the solubility limit and the common forms of hydrogen precipitation are as molecular hydrogen (near
precipitate particles or in blow-holes as is commonly observed in
steels), or as hydrides (as in titanium and zirconium alloys). Decarburization results from the reaction of hydrogen with carbon contained
in steels and occurs at high temperatures in the presence of moist hydrogen.

III.   SELECTION OF HIGH-STRENGTH MATERIALS

In this section a few selected properties of typical high-
strength materials are shown and some specific applications discussed.
The number of materials available is, of course, very large and the
listings here are intended as a guide to selection rather than a substitute for the appropriate handbooks and data sheets. Listed at the
end of the chapter are a number of handbooks, reports and other publications which may be useful for the final selection of materials.

For the purpose of the present discussion the structural materials will be grouped as follows:
   a.   Ultra-high strength steels
   b.   High-strength aluminum alloys
   c.   High-strength titanium alloys
   d.   Other structural materials
   e.   Materials for cryogenic service
   f.   High-temperature materials

Because of the special effect of extreme temperature conditions
on the properties of materials, the properties of structural materials
for use at cryogenic and high-temperature applications are grouped
separately.

A.   Ultra-high-strength Steels

The classification of steels on a strength scale is rather arbitrary and confusing. The main reason for this is the rapid advance of
steel technology over the past decades, producing ever stronger materials so that those designated as "high-strength steels" some years
ago have been surpassed by "extra-high-strength steels" which in turn
have been outdone by "ultra-high-strength steels". For the purpose of
the present discussion we shall designate as high-strength steels
those of yield strengths in the range of 50,000 psi to 80,000 (~0.34 -
0.55 GPa); those in the yield strength range of 80,000 psi to 115,000
psi (~0.55 - 0.89 Pa) as extra-high-strength steels; those with a
yield strength above 115,000 psi (~0.89 Pa) as ultra-high-strength
steels.

Steels are, of course, the work horse of structural materials and
a very large number of standard and proprietary carbon and alloy
steels are available commercially. Most of these may be used for high
pressure applications in the appropriate pressure range and factors
such as cost, corrosion resistance, fatigue resistance, weldability,
machinability, etc. determine the ultimate selection of a particular
steel. Since the properties of the more conventional steels are amply
documented in standard references (see listings at the end of this
chapter) the discussion here will be limited to steels in the highest
strength bracket.

The yield strengths of ultra-high-strength steels commonly fall
in the 250,000 to 300,000 psi range (~1.7 - 2.0 GPa) hence they are
usually formed or joined in the soft condition and then heat-treated
to full strength. The ultra-high-strength steels may be grouped in
the following categories, characterized by the type of mechanism and
microstructure causing the strengthening effect:
1.   Ultra-high-strength alloy steels:
   a.   Low-alloy hardenable steels
   b.   Medium-alloy hardenable steels
   c.   High-alloy hardenable steels
2.   Maraging steels
3.   Ultra-high-strength stainless steels:
   a.   Martensitic stainless steels
   b.   Semi-austenitic stainless steels
   c.   Austenitic stainless steels.

The low-alloy hardenable steels are characterized by the widely used AISI 4130 (chromium-molybdenum) and AISI 4340 (chromium-nickel-molybdenum) types, of which most other steels in this category may be considered as descendants. These steels contain 0.35 to 0.45% carbon and thus may be hardened to great strengths (240,000 - 280,000 psi) (~1.6 - 1.9 GPa) by means of the martensitic transformation. The alloying elements, besides increasing the hardenability of the alloy (by favoring the martensitic transformation), contribute also with some solid-solution strengthening. Vanadium added to some of these alloys acts as a grain refiner and increases toughness. These alloys have good weldability and machinability characteristics and are used for structural components in aircraft, solid-propellant rocket-motor cases, high pressure gas-storage bottles, etc. The number of steels in this category, of AISI grades as well as proprietary, is very large and Appendix Table 1 lists the compositions of only a few typical materials.

The medium-alloy ultra-high-strength steels can be grouped in two categories: the 5% Cr types (e.g. 5CrMoV, AISI H11, AISI HB) and the 5% Ni types (e.g. HY 130, HY 150). The 5% Cr types are generally air-hardening and temper at high temperatures. Yield strengths of these materials range up to 250,000 psi (~1.7 GPa) and they retain their strength to high temperatures. They are used primarily as hot-work tool steels, for structural components, and also for pump components, actuating cylinders, etc. The yield strengths of the 5% Ni types are lower (about 150,000 psi; ~1.0 GPa) but they are very tough and have good weldability. They are used extensively for pressure vessels, deep submersibles, and nuclear reactor components. Appendix Table 1 lists a few typical steels in this category.

The high-alloy steels, a few examples of which are given in Appendix Table 1, are similar to the medium-alloy steels but have somewhat higher yield strengths. Having good toughness and weldability they are used for pressure vessels, deep submersibles, and armor.

The maraging steels are one of the more recent additions to the ultra-high-strength steel group. They are characterized by very low carbon contents (<0.03%) and high nickel contents (17 to 25%). The maximum strength in these alloys is developed by first forming martensite and subsequently ageing the alloy at temperatures around 480°C (900°F), to induce precipitation-hardening, which is due to the presence in the alloy of titanium, cobalt and molybdenum. The hardening can be done in a single step heat-treatment that causes virtually no distortion nor dimensional change. The fracture-toughness of these alloys is another of their outstanding properties. Besides the usual applications requiring high strength and toughness, these alloys are especially suitable for precision machine components. Some typical maraging steel compositions are shown in Appendix Table 2.

Martensitic stainless steels derive from hardenable chromium-type stainless steels, which are traditionally used for cutlery and surgical instruments (e.g. AISI-420, AISI-431). Additions of various other alloying elements have given rise to a large number of proprietary alloys of this type, characterized by ultra-high strength and a moderate corrosion resistance (better than low- and medium-alloy UHS steels). These steels are hardened by quenching and tempering, although some also undergo a precipitation-hardening process which further enhances their strength. Martensitic stainless steels are susceptible to stress-corrosion cracking (e.g. in the presence of water) and hydrogen-embrittlement, which should be taken into account if they are used for high pressure equipment. In their annealed form martensitic stainless steels are readily cut, formed, machined, and welded.

Semi-austenitic stainless steels derive their name from the fact that, by proper heat treatment, they can be made austenitic, hence soft and pliable, or martensitic, hence strong and hard. The alloying elements in these steels represent a careful balance between those that stabilize austenite and those that do not. If the steel is cooled from a solution-treatment temperature, 1030°C-1070°C(1950-1950° F), where all alloying elements are in solution, it retains an austenitic structure. However, on reheating to a temperature where some of the carbon precipitates as chromium carbide, or by severely cold-working the material, the austenitic phase transforms to martensite. Some of these alloys can be further strengthened by precipitation-hardening. These steels have a good corrosion resistance and are widely used for nuclear reactor components, compressor disks, storage and pressure tanks, etc.

The ultra-high-strength austenitic stainless steels depend largely on cold-work for their final properties and are the only ones in the ultra-high strength category designed to be used in the cold-rolled condition. Consequently they are only available in the form of sheet, foil, strip, and tube. As would be expected, the alloying elements are primarily of the austenite stabilizing group (i.e. Mn,Ni,C, N). These steels are resistant to atmospheric corrosion and, in spite of the heavily cold-worked condition in which they are used, have an excellent resistance to stress-corrosion cracking (e.g. 316 s/s is frequently used for high pressure tubing). Low carbon grade austenitic stainless steels show an improved weldability. Also the fact that austenitic stainless steels are nonmagnetic makes them desirable for the use in equipment and instrumentation where such properties are required. These steels primarily are used for tanks and containers for oxidizing materials, fuels, foods, and other chemicals. Some typical ultra-high-strength stainless steels are listed in Appendix Table 4.

B.   High-strength Aluminum Alloys

Even though aluminum alloys do not compare with steels in terms of absolute strength (i.e. the maximum yield strength of heat-treatable aluminum alloys is about 80 to 90,000 psi ($\sim$0.55 - 0.6 GPa)), the nearly three-to-one density ratio between steels and aluminum makes some aluminum alloys comparable to ultra-high-strength steels on a per-unit-weight basis.

Aluminum alloys offer some other advantages: good electrical and heat conductivity, corrosion resistance and, generally, an improvement in mechanical properties with decreasing temperature. For these reasons, and others, aluminum alloys find wide useage in a number of moderately high pressure applications.

Aluminum alloys may be classified into two broad categories: "non-heat-treatable alloys" and "heat-treatable alloys". The "non-

heat-treatable alloys" depend on solid-solution hardening and work-
hardening for the enhancement of mechanical properties. Manganese,
silicon, iron and magnesium are the common alloying elements, and most
of these alloys fall in the series designated 1000,3000,4000 and 5000.
Although these alloys do not reach the strength of some heat-treatable
alloys, (yield strengths in the 20 - 60,000 psi range (~0.14 - 0.4
GPa) being typical), they possess good weldability, corrosion resistance, and good low temperature strength and toughness.

In heat-treatable alloys the main strengthening effect is derived from the precipitation of second phases, as the alloy becomes
supersaturated on quenching from an elevated temperature. Copper,
magnesium, zinc, and silicon are the most common alloying elements in
these materials and most of them fall in the 2000, 6000 and 7000 series designation. By various combinations of solution heat-treatment,
cold-working, quenching, and ageing, various "tempers", and hence
properties, may be obtained. (The label Hn or Tn, where n is a number
following the alloy designation, refers to the "temper" of the material).

The higher strength levels attainable with heat-treatable aluminum alloys are accompanied by the deterioration of some other properties, e.g. 2000 and 7000 series alloys generally have a lower corrosion resistance and poorer weldability than other aluminum alloys.

As in the case of steels, the fracture toughness of aluminum alloys decreases with strength level (see Figure 11). This is particularly evident for precipitation-hardened alloys where segregation of
particles near grain boundaries may promote intergranular fractures as
well as fatigue and stress-corrosion cracking.

Furthermore, for materials used in the "as-fabricated" form, or
after mild heat-treatment, the preferential grain orientation may
cause a considerable anisotropy in properties. This effect is particularly noticeable in extruded stock, where lower ductility and intergranular fracture are observed in tensile test specimens cut in a
direction transverse to the extrusion direction.

A special type of aluminum-base materials are the "clad alloys".
Some of the heat-treatable alloys containing copper and zinc as the
major alloying constituents show a marked decrease in corrosion resistance. For this reason, sheet and plate of such alloys is often
clad with a layer (2 to 5% of total thickness) of high-purity aluminum
or of a low concentration alloy. The cladding protects by its own
corrosion resistance as well as by a galvanic effect.

Compositions and sample properties of some typical aluminum alloys that have been used for pressure vessels, hydraulic cylinders,
storage tanks, submersible hulls etc. are listed in Appendix Table 3.
For more complete listings of alloys and properties the reader is referred to the bibliography at the end of the chapter.

C. High-strength Titanium Alloys

The properties of titanium alloys are controlled, primarily, by
an allotropic transformation from a low-temperature hcp phase (α-
phase) to a high-temperature bcc phase (β-phase) that takes place, in
pure titanium, at 883°C (1625°F). The effect of various alloying ele-

ments is to stabilize one or the other phase and to precipitate some
age-hardening second phases. Thus a number of different types of
microstructures may be obtained in these alloys. Single-phase alloys
are weldable and exhibit good ductility. Two-phase alloys, on the
other hand, can be strengthened by various heat-treatments. Beta-alloys are usually hardened by heat-treatments leading to the precipitation of some alpha-phase or some intermetallic compound (e.g. $TiCr_2$).
Yield strengths in excess of 200,000 psi (~1.4 GPa) can be achieved
by proper heat-treatment of certain beta-alloys.

Titanium alloys find use principally because of their high
strength/weight ratio and their corrosion resistance at temperatures
up to 540°C (1000°F). At low temperatures interstitial elements such
as carbon and oxygen have an adverse effect on the toughness of titanium, hence alloy grades labeled ELI (extra low interstitial concentration) are used for low temperature applications. In the ELI grade,
several titanium alloys (e.g. 5Al, 2.5V and 6Al, 4V) have excellent
low temperature properties and are used for cryogenic, high-pressure
vessels.

Hydrogen, past a certain concentration level (125 to 250 ppm),
causes embrittlement of titanium alloys through the formation of hydrides.

Titanium alloys have found application in chemical process equipment, especially heat-exchange equipment, because they are relatively
resistant to fouling and buildup of corrosion products on the surface.
Some typical high-strength titanium alloys are listed in Appendix
Table 5.

D.   Other Structural Materials

1. Alloys.   Besides steels and aluminum and titanium alloys, a
number of structural materials based on other matrices are widely
used. Nickel- and cobalt-based alloys, known as "super-alloys", as
well as alloys based on refractory metals such as niobium, molybdenum,
tungsten and tantalum exhibit good mechanical properties at elevated
temperatures. We have reserved a special section for a discussion of
these materials. Of other alloys, manganese and copper alloys are the
most frequently encountered in structural applications.

Copper alloys, although more costly than most structural alloys,
offer the advantage of excellent machinability and formability characteristics, excellent corrosion resistance, and high electrical and
thermal conductivity, and are generally non-magnetic. A number of
copper alloys can be precipitation-, or age-hardened and thus, by the
appropriate heat-treatment, considerable strength levels may be
achieved. Foremost among these heat treatable alloys are beryllium-
copper alloys, aluminum bronzes and manganese-bronze alloys. Some
typical copper-base alloys and their mechanical properties are given
in Appendix Table 6. Magnesium, for engineering applications, is usually alloyed with aluminum, zinc, and manganese. Such alloys are age-
hardenable but the strength achieved is not very high. These alloys
are also susceptible to stress-corrosion cracking. The outstanding
property of magnesium and its alloys is their low density; for this
reason most of their structural applications are in the aerospace

area. Most magnesium alloys exhibit good mechanical properties at low temperatures. A few examples of the strongest magnesium alloys and their mechanical properties are listed in Appendix Table 6.

2. Ceramics and glasses. In this category may be grouped a large number of materials that are nonmetallic and inorganic in nature, with ionic or covalent type bonding. Included in this category are graphite and diamond, various oxides, nitrides, carbides, silicates and other more complex compounds, and a number of multi-phase or composite materials such as cement, concrete, and cemented carbides or oxides ('cermets'). As may be expected, such a wide variety of materials offers a wide range of properties. By the nature of the bonding, however, some common features may be established.

Materials in this group are characterized by high hardness and stiffness, good compressive strength, lack of plasticity and a tensile strength that is generally far below theoretical values. This latter effect is related to the lack of plasticity and makes the tensile strength of the material extremely sensitive to surface flaws. As a result, the tensile strength of these materials is generally very sensitive to the condition of the surface, environment, size of sample, and rate of loading. Other results of the bonding type of these materials are their low thermal and electrical conductivity, and relative chemical stability (high corrosion resistance). The fundamental criteria relating microstructure to mechanical properties are essentially the same as discussed previously for metallic materials. For those materials consisting of two admixed phases (e.g. cermets, concrete) the wetting of the imbedded phase by the matrix and the strength of bonding between the phases are additional important factors for the overall mechanical properties of such materials.

Included in this group is concrete, the most widely used structural engineering material, mainly for large structures including, for instance, high pressure vessels for nuclear reactors.

Diamond has long been used for manufacturing ultra-high pressure anvils (see Chap. 11) and cermets (e.g. WC/Co) are increasingly being used in various high-pressure applications. Many refractory oxides have long been used as high temperature materials; however, the inherent brittleness of these materials and the ensuing poor machinability place some limits on their applicability. Graphite, on the other hand, is more readily machinable and retains its strength to about 2500°C [80]. However, at temperatures above 2000°C the creep rates become appreciable in graphite and should be taken into consideration when designing with graphite for high temperature - high pressure applications [80]. Another problem in using graphite at temperatures above 500°C in oxygen containing atmospheres is that of oxidation. For this reason, special, glass-impregnated graphite materials have been developed by introducing into the pores of graphite a vitreous substance which is partly fluid at the operating temperature and which mechanically reduces the rate of the oxidation reaction [81]. It should also be noted that graphite can be produced in many different grades of widely varying properties. Grain orientation, particle size, density, types of pitch and coke used, as well as temperature of graphitization used, all affect the physical and mechanical properties of the material.

Appendix Table 7 lists some typical high-strength ceramic and other materials together with some of their mechanical properties.

Glasses are similar to other ceramics in most properties, being different insofar that they are largely amorphous in structure. The single most important property of glass is its transparency to light which makes it irreplaceable for some applications. The tensile strength of glasses is extremely sensitive to surface flaws and hence very difficult to evaluate quantitatively. Small diameter (0.0002") glass fibers have been measured to have strengths of $1.5.10^6$ psi (~10 GPa) if tested in vacuum, yet strength drops to 20-30% of this value if the diameter of the fiber is increased to 0.0010", or if the test if performed in wet air. Fracture stresses for ordinary glasses fall in the range of 3,000 to 20,000 psi (0.02 - 0.14 GPa). It is of interest to note that the compressive strength of glass is much higher than its tensile strength, being of the order of 100 - 200 kpsi (~0.7 - 1.4 GPa).

It is a generally observed phenomenon, and of particular import-ance for naturally brittle materials, that the ductility and strength of materials increases if they are subjected to a hydrostatic pressure in addition to the deforming load. It has been shown, for instance, that under a combination of hydrostatic pressure and shear silicate glasses show a yield strength of the order of 600,000 psi [82] (~4 GPa). This observation suggests the interesting possibility that in certain high pressure applications, through proper design, the effect of uniform hydrostatic pressure may be used to enhance the properties of otherwise brittle materials. (See also Chapter 15).

3. Plastics. Although plastics are rarely considered for appli-cations requiring high strength, some grades, especially reinforced ones, have strengths that make them comparable to some light-weight metallic alloys. Thus, for the purpose of comparison, the properties of some of the tougher polymeric materials are shown in Appendix Table 8. Most mechanical properties of plastics are evaluated by tests sim-ilar to those used for metals; however, tensile test data must be in-terpreted cautiously because plastics are not always elastic at room temperature, but frequently exhibit viscoelastic behaviour. The me-chanical properties of plastics generally change much more rapidly with temperature than those of metals and are also more strain-rate dependent. At low temperatures a "glass transition" takes place which results in embrittlement of these materials, while the nature of the bonding places upper limits on the useful temperature range. The properties of polymers vary with such internal structural parameters as average molecular weight and degree of crosslinking.

E.   Materials for Cryogenic Service

The mechanical properties of structural materials at cryogenic temperatures are of interest in two types of applications: first, some technological applications such as containers for cryogenic liq-uids (e.g. liquid air, fuels for missiles and rockets, coolants for superconductors, etc.) require materials with satisfactory properties at low-temperature, high-pressure conditions; and second, the study

of properties of materials at low temperature, where thermal vibrations are reduced to a minimum, is of great interest for the understanding of fundamental aspects of the structure and properties of materials, and equipment necessary for the study of such properties requires adequate structural materials. The combination of high pressures and low temperatures is a particularly demanding one in terms of
material performance.

As has been discussed previously, low-temperature embrittlement
is the central issue when considering materials for use at cryogenic
temperature. The yield strength and tensile strength of most metals
increases with decreasing temperature, which may be explained in terms
of a larger force required to move dislocations when thermal activation is decreased. On the other hand, the ductile-to-brittle transition that many metals undergo at low temperature limits their usefulness for cryogenic applications. Crystalline structure appears as the
dominant parameter in this regard since the toughness of bcc and most
hcp metals tends to undergo a ductile-to-brittle transition. The alloys generally used for high-strength applications at cryogenic temperatures are based on matrices having fcc structure, e.g. aluminum,
copper, and titanium alloys, austenistic stainless steels, and nickel-
based superalloys.

Mechanical properties of some of the most widely used materials
for high-strength, low-temperature applications are shown in Appendix
Table 9.

F.   High-temperature Materials

The selection of suitable materials for use at high temperatures
(e.g. above 1500°C) must consider, in addition to an adequate
strength, creep resistance (i.e. dimensional stability), the oxidation
and corrosion resistance at the temperature of use. For these reasons
the quality of materials for use at high temperatures is frequently
rated in terms of service life at a given temperature and loading.

Materials especially designed for use at high temperatures fall
into three general categories: the superalloys, the refractory-metal
alloys, and ceramics.

The superalloys are nickel-, cobalt- and/or iron-based alloys,
generally of the fcc austenitic structure, which are able to retain
good tensile strength, creep properties, and oxidation resistance at
elevated temperatures, i.e. in the 700-1100°C (1300 - 2000°F) range.
These alloys are strengthened by precipitation of intermetallic compounds, carbides, and by solid-solution strengthening. The use of
powder-metallurgical techniques also permits the fabrication of superalloys with internally dispersed oxides. Oxidation resistance is a
prime requisite of superalloys and this is achieved through the addition of alloying elements (chromium, aluminum) that help to form a
tight, continuous, surface-oxide layer that acts as a barrier to further oxidation [83].

The main market for super alloys is that related to gas turbines,
rocket engines and steam power plants, but they are also being used
increasingly in petrochemical plants, cryogenic equipment, submarines,
etc. The numerous proprietary alloys of this type may be grouped ac-

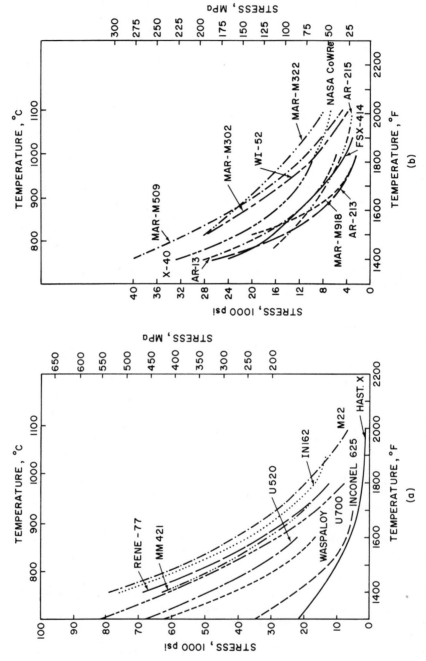

FIG. 20. Stress versus temperature curves for rupture in 100 hours for typical superalloys:
a) Nickel-base alloys; b) Cobalt-base alloys (Reference [84], Appendix B).

cording to the predominant metal as nickel-base, iron-base, and cobalt-base superalloys. Appendix Table 10 shows the compositions of a
few typical superalloys while Figure 20 shows the stress-temperature
curves for a number of superalloys.

For certain high-temperature applications the refractory metals
and their alloys are most adequate, since some of these materials retain strength at temperatures above 2000°C. The term "refractory metals" generally identifies the elements V, Nb, and Ta (Group V-a) and
Cr, Mo, and W (Group VI-a in the Periodic Table). Their melting temperatures range from 1875°C for chromium to 3410°C for tungsten.
There are other metals such as rhenium, osmium, iridium, rhodium, and
ruthenium which have comparable melting points, yet, because of their
scarcity are considered "precious metals" and used only occasionally
as alloying elements in engineering alloys. One of the main limitations of the refractory metals is their susceptibility to oxidation,
hence they are generally used in neutral or reducing atmospheres or
vacuum. The design of refractory metal alloys is primarily concerned
with the ductile-to-brittle transition that these materials (with the
exception of tantalum) undergo at low temperatures and embrittlement
through oxidation or absorption of other interstitial impurities (N,H)
at high temperatures [85]. Most engineering alloys based on refractory metals have a niobium, molybdenum, tantalum, or tungsten matrix. A
number of tungsten alloys are strengthened by dispersed oxides (e.g.
ThO_2), such materials being fabricated by powder-metallurgical methods. Some illustrative examples of refractory metal alloys and their
high-temperature strength are shown in Appendix Table 11.

The third alternative in the selection of materials for use at
high temperatures rests with ceramic materials or graphite. Although
these materials offer high temperature stability their brittleness
limits their applicability where large tensile stresses are involved.
The mechanical properties of ceramics are sensitive to the fabrication
methods (e.g. porosity, purity, shape of part, etc.).

REFERENCES

1. F. R. N. Nabarro, Proc. Phys. Soc. (London) 59, 256 91947).
2. J. D. Livingstone, Acta. Met. 10, 229 (1962).
3. D. J. Dingley and D. McLean, Acta. Met. 15, 855 (1967).
4. A. H. Cottrell in "Report of the Conference on Strength of
    Solids", (Physical Society, London, 1948).
5. A. H. Cottrell and B. A. Bilby, Proc. Phys. Soc. (London)
    62A, 49 (1949).
6. A. H. Cottrell and M. A. Jaswon, Proc. Phys. Soc. (London)
    199, 104 (1949).
7. R. S. French and W. R. Hibbard Jr., "Tensile Deformation
    of Copper", Trans. AIME 188, 53 (1950).
8. R. Fleischer in "The Strengthening of Metals", p. 93,
    (Reinhold Press, New York 1964).
9. J. C. Warner and J. O. Verhoeven, Metall. Trans. 4, 1255
    (1973).
10. J. C. Fisher, Acta. Met. 2, 9, (1954).

11. N. F. Mott and F. R. N. Nabarro in "Report of the Conference on Strength of Solids", (Physical Society, London 1948).

12. J. E. Dorn, P. Pietrokowsky and T. E. Tietz, Trans. AIME 188, 933 (1950).

13. L. E. Samuels, J. Inst. Metals 76, 91 (1949).

14. J. E. Dorn and C. D. Starr in "Relations of Properties to Microstructure" (ASM., 1954).

15. C. Nishimatsu and J. Gurland, Trans. ASM 52, 469 (1960).

16. K. Lucke and H. Lange, Z. Metallk. 43, 55 (1952).

17. F. D. Rosi, Trans. AIME 200, 1009 (1954).

18. J. Diehl, Z. Metallk. 47, 331 (1956).

19. E. N. Andrade and D. A. Aboav, Proc. Roy. Soc. A240, 304 (1957).

20. F. N. Rhines in "Effect of Residual Elements on the Properties of Materials" (ASM, Ohio, 1957).

21. E. R. Parker and T. H. Hazleh in "Relation of Properties to Microstructure" ASM, Ohio (1954).

22. D. McLean, "Mechanical Properties of Materials", p. 122ff, (J. Wiley, New York, 1962).

23. M. Gensamer, Trans. ASM Vol. 36, 30 (1946).

24. J. H. Hollomon, Trans. AIME 162, 268 (1945).

25. G. Greetham and R. W. K. Honeycombe, J. Inst. Met. 89, 13 (1960).

26. D. Dew-Hughes and W. D. Robertson, Acta. Met. 8, 147 (1960).

27. A. Kelly and R. B. Nicholson, Prog. Mat. Sci. 10, 151 (1963).

28. R. W. K. Honeycombe, "The Plastic Deformation of Metals" (St. Martin's Press, New York, 1968).

29. A. S. Tetelman and A. J. McEvily, "Fracture of Structural Materials" (J. Wiley, New York, 1967).

30. J. F. Baker and C. F. Tipper, J. Mech. Eng. (London) 170, 65 (1956).

31. "Fracture Toughness Testing and its Applications" (ASTM Symposium, Chicago, June 1964) (ASTM, 1965).

32. J. D. Lubahn in "Fracturing of Metals", (ASM, 1948).

33. W. S. Pellini et al., Naval Research Laboratory Report #6300, Washington, D.C. (1965).

34. J. G. Kaufman and H. Y. Hunsicker in "Symposium on Fracture Toughness Testing and Its Applications". STP 381, 290, (1965) (ASTM, Philadelphia, Pa.).

35. A. J. McEvily and A. P. Bond, J. Electrochem. Soc. 112, 131 (1964).

36. P. C. Paris in "Fatigue-An Interdisciplinary Approach". (Syracuse University Press, Syracuse, N.Y.)

37. B. F. Brown and C. D. Beachem, Corrosion Sci. 5, 745 (1965).

38. A. Lawley, J. Van den Sype and R. Maddin, J. Inst. Metals 91, 23 (1962).

39. R. L. Smith and J. L. Rutherford, Trans. AIME 209, 857 (1957).

40. W. D. Biggs, "Brittle Fracture of Steel" (MacDonald & Evans, London, 1960).

41. B. E. Hopkins and H. R. Tipler, J. Iron Steel Inst. (London) 177, 110 91954).

42. B. A. Loomis and O. N. Carlson in "Reactive Metals", (Wiley Interscience, New York, 1959).

43. G. T. Hahn, A. Gilbert and R. I. Jaffee in "Refractory Metals and Alloys" (Wiley Interscience, N.Y., 1963).

44. N. P. Allen, B. E. Hopkins and H. R. Tipler, J. Iron and Steel Inst. (London) 174, 108 (1953).

45. E. A. Loria, Trans. ASM, 58, 221 (1965).

46. J. A. Rinebolt and W. J. Harris, Trans. ASM. 43, 1175 (1951).

47. D. Tromans and J. Nutting, Corrosion 21, 143 (1965).

48. T. L. Johnston, R. G. Davies and N. S. Stoloff, Phil. Mag. 12, 305 (1965).

49. N. J. Petch in "Fracture", (B. L. Averbach et. al. eds.) (J. Wiley, New York, 1959).

50. R. F. Decker, J. T. Eash and A. J. Goldman, ASM. Trans. Quart. 55, 59 (1962).

51. B. Edelson and W. Baldwin, Trans. ASM 55, 230 (1962).

52. G. Thomas, D. Schmatz and W. Gerberich in "High Strength Materials" (V. F. Zackay ed.) (J. Wiley, New York, 1965).

53. V. F. Zackay and E. F. Parker in "High Strength Materials" (V. F. Zackay ed.) (J. Wiley, New York, 1965).

54. F. R. Larson and J. Miller, Trans. ASME 74, 765 (1952).

55. A. L. Ruoff, "Materials Science" (Prentice Hall, Englewood Cliffs, N.J., p. 622 1973).

56. R. L. Orr, O. D. Sherby, and J. E. Dorn, Trans. ASM 46, 113 (1954).

57. C. R. Barret and O. Sherby, Trans. AIME 233, 116 (1965).

58. B. G. Piearcey and F. L. Versnyder, Pratt and Whitney Report No. 65-007, April 1965.

59. G. S. Ansell and F. V. Lenel, Trans AIME 221, 452 (1961).

60. O. Preston and N. J. Grant, Trans AIME 221, 164 (1961).

61. L. J. Bonis and N. J. Grant, Trans AIME 218, 877 (1960).

62. H. S. Cross, Met. Prog. 87, 67 (1965).

63. R. L. Coble and W. D. Kingery, J. Am. Ceram. Soc. 39, 377 (1956).

64. R. Chang, "Creep and Anelastic Studies of Al_2O_3", USAEC Research and Development Report NAA-SR-2770, Sept. 1958.

65. H. Farmery and U. R. Evans, J. Inst. Met. 84, 413 (1956).

66. R. Parkins, J. Iron Steel Inst. (London) 172, 149 (1952).

67. M. Scheil in "Corrosion Handbook" (H. Uhlig ed.) (J. Wiley, New York, 1948).

68. T. Haar and J. Hines, J. Iron Steel Inst. (London) 182, 124 (1956).

69. J. Truman, R. Perry and G. Chapman, J. Iron Steel Inst. (London) 202, 745 (1964).

70. E. Coleman, D. Weinstein and W. Rostocker, Acta. Met. 9, 491 (1961).

71. R. Parkins, Met. Rev. 9, 209 (1964).

72. C. S. Carter, D. G. Farwick, A. M. Ross and J. M. Uckide, Corrosion 27, 190 (1971).

73. M. V. Hyatt, Corrosion 26, 487 (1970); 27, 49 (1971).

74. H. P. Leckie in "Proceedings of the Conference on
     Fundamental Aspects of Stress Corrosion Cracking" (R.W.
     Staehle ed.) (National Association of Corrosion Engineers,
     Houston, 1969).
75. B. C. Syrett, Corrosion 27, 270 (1971).
76. A. Gallacio and M. A. Pelensay in "Stress Corrosion Testing"
     (ASTM. Philadelphia, Pa., 1967).
77. R. W. Staele et al., Corrosion 26, 451 (1970).
78. M. G. Fontana and N. D. Greene, "Corrosion Engineering"
     (McGraw-Hill, New York, 1967).
79. G. L. Hanna, A. R. Troiano and E. A. Steigerwald, Trans.
     ASM 57, 658 (1964).
80. C. Malmström, R. Keen and L. Green Jr., J. Appl. Phys.
     22 593 (1951).
81. E. I. Shobert II, in "Modern Materials", (B. W. Gonser
     and H. H. Hausner, eds.) 4, 1 (1964).
82. P. W. Bridgman and I. Simon, J. Appl. Phys. 24, 4 (1953).
83. G. E. Wasielewski and R. A. Rapp, "High Temperature
     Oxidation", J. Wiley Sons, New York (1972).
84. The Super Alloys" (C. T. Sims and W. C. Hagel eds.),
     Appendix B; J. Wiley & Sons, New York (1972).
85. W. D. Wilkinson "Properties of Refractory Metals",
     Gordon and Breach, New York (1969).

APPENDIX

The data contained in the Tables of this Chapter are fragmentary
and intended primarily for guidance and illustration. The following
list of references pertains to handbooks, manuals, and other data compilations that provide more ample information of the types of materials available and their properties.

A.   Data on Metals and Alloys

A1. "Metals Handbook", T. Lyman Ed., 8th ed. v. 1, (American
     Society of Metals, Metals Park, Ohio, 1961).
A2. "Metal Progress Databook", (American Society of Metals,
     Metals Park, Ohio 1968).
A3. "Metal Progress Databook", Metal Progress 106, 1, (1974).
A4. "Engineering Alloys" 5th ed., (N. E. Woldman and R. C.
     Gibbons eds.) (Van Nostrand-Reinhold Co., New York, 1973).
A5. "Metal Data", (S. L. Hoyte ed.), (Reinhold Publ. Co.,
     New York, 1952).
A6. "Metals Reference Book" 3rd ed., C. J. Smithells, (Butter-
     worths, London, 1962).
A7. "Alloy Digest" (Data Sheets), (Engineering Alloy Digest
     Inc., Upper Montclair, New Jersey, 1952 -).

B.   Data on Solid Materials in General

A8. "Materials Data Book for Engineers and Scientists",
     E. Parker, (McGraw-Hill, New York, 1967).
A9. "Materials Selector Issue of Materials Engineering"
     (Chapman-Reinhold, New York (published yearly)).
A10. "Handbook of Engineering Materials", (D. F. Miner and
     J. B. Seastone eds.) (J. Wiley, New York, 1955).
A11. "Engineering Materials Handbook" (C. L. Mantell ed.)
     (McGraw-Hill, New York, 1958).
A12. "The Encyclopedia of Engineering Materials and Processes"
     (H. R. Clauser ed.) (Reinhold Publ. Co., New York, 1963).
A13. "Materials Handbook", G. S. Brady (McGraw-Hill, 1971).

C.   Data on Special Materials

A14. Super-alloy Data, Appendix B in "The Superalloys" (C. T.
     Sims and W. C. Hagel eds.) (Wiley Interscience, New York,
     1972).
A15. "Cryogenic Materials Data Handbook", F. R. Schwartzenberg,
     S. H. Osgood, R. D. Keys and T. F. Kiefer, Air Force Mate-
     rials Laboratory ML-TDR-64-280 Wright Patterson AFB Ohio
     (1964).
A16. "Aluminum Standards and Data", 1st ed. (The Aluminum
     Association, New York, 1968). (See also "Alcoa Aluminum
     Handbook", Aluminum Company of America, Pittsburgh, Penna.).
A17. "Ceramic Data Book" (Cahners Publ. Co., Chicago (published
     yearly)).
A18. "Modern Plastics Encyclopedia" (McGraw-Hill, New York
     (published yearly)).
A19. "Rare Metals Handbook" (C. A. Hampel ed.) (Reinhold Publ. Co.
     New York, 1961).
A20. "Thermophysical Properties of High Temperature Solid
     Materials" (6v), (Y. S. Touloukian ed.) (The Macmillan Co.,
     New York, 1966).
A21. "Plenum Press Handbook of High Temperature Materials" No. 1
     Materials Index (T. B. Shaffer ed.), No. 2 Properties Index
     (C. V. Somsonov ed.) (Plenum Press, New York, 1964).
A22. "Engineering Properties of Selected Ceramic Materials"
     (J. Lynch, C. Ruderer and W. Duckworth, eds.) (American
     Ceramic Society, Columbus, 1966).
A23. "Handbook of Engineering Data" (National Carbon Company,
     Division of Union Carbide Corporation, New York, 1962).

D.   Data on Corrosion-Resistance

A24. "Corrosion Guide", E. Rabald (Elsevier Publ. Co., Amsterdam,
     1968).
A25. "Corrosion-Resistant Materials Handbook", I. Mellan (Noyes
     Data Corp., New York, 1971).
A26. "Corrosion Handbook" (H. H. Uhlig ed.) (J. Wiley, New York,
     1948).

Appendix – Table 1

Typical Ultra-High – Strength Alloy Steels

DESIGNATION	COMPOSITION (weight percent)	MAX. YIELD STRENGTH 10^3 psi (kbar)	MAX. TENSILE STRENGTH 10^3 psi (kbar)	ELONGATION (%)	HARDNESS
Low Alloy					
AISI 3140	0.40C, 0.50Mn, 0.26Si, 1.28Ni, 0.95Cr	240 (16.9)	270 (19.0)	5 – 10	40 – 44 Rc
AISI 4140	0.40C, 0.90Mn, 0.30Si, 0.95Cr, 0.20Mo	250 (17.6)	290 (20.4)	10	45 – 55 Rc
AISI 4340	0.40C, 0.70Mn, 0.30Si, 0.80Cr, 1.80Ni, 0.20Mo	230 (16.2)	280 (19.7)	6 – 10	40 – 55 Rc
Crucible D-6 (Crucible Spec. Metals)	0.46C, 0.75Mn, 0.22Si, 1.0Cr, 0.55Ni, 0.08V	210 (14.8)	300 (21.1)	8 – 18	19 – 61 Rc
Hy-Tuf (Crucible Spec. Metals)	0.25C, 1.3Mn, 1.5Si, 1.8Ni, 0.4Mo	195 (13.7)	235 (16.5)	13	50 Rc
USS-Strux	0.43C, 0.9Mn, 0.9Cr, 0.75Ni, 0.55Mo, 0.9B, 0.05V		280 (19.7)		
Tricent (Crucible Spec. Metals)	0.43C, 0.80Mn, 1.6Si, 0.85Cr, 1.8Ni, 0.38Mo, 0.08V	250 (17.6)	290 (20.4)	10	36 Rc
Bearcat (Bethlehem Steel)	0.5C, 0.7Mn, 0.3Si, 3.25Cr, 1.4Mo	205 (14.4)	340 (23.9)	4 – 20	57 Rc

Material	Composition				
Medium Alloy					
AISI H11	0.35C, 0.30Mn, 1.00Si, 5.20Cr, 1.5Mo, 0.4V	310 (21.8)	240 (16.9)	4 – 10	
5Cr-Mo-V (H11 die steel)	0.40C, 0.30Mn 1.00Si, 5.00Cr, 1.3Mo, 0.5V				
HY 130 (USS)	0.12C, 0.6–0.9 Mn, 0.4–0.7Cr, 0.3–0.65Mo, 0.02Ti, 0.05– 0.10V 0.14Cu	130 (9.2)		15	50 Rc
Hy 150(USS)	0.16–0.20C, 0.4–0.6Mn, 3.5–4.0Ni, 0.3–0.5Mo 0.07–0.12V	180 (12.7)			
High Alloy					
HP 9-4-20	0.20C, 0.30Mn, 0.10Si, 0.75Cr, 9.0Ni, 0.75Mo, 4.50Co, 0.10 V	180 (12.7)			
HP 9.4.30	0.30C, 0.20Mn, 0.10Si, 1.00Cr, 7.5Ni, 1.00Mo, 4.50Co, 0.10 V	220 (15.5)			
HY-180 (USS)	0.10C, 2.0Cr, 10.9Ni, 1.0Mo, 8.0Co.	180 (12.7)			

Appendix - Table 2
Typical Maraging Steels

Designation	Composition (weight percent)	Max. Yield Strength 10³ psi (kbar)	Max Tensile Strength 10³ psi (kbar)	Elongation %	Hardness
Gr. 200 (USS)	0.03 max.C, 17-19Ni, 7.0-8.5Co, 4.0-4.5Mo, 0.10-0.30Ti, 0.05-0.15Aℓ,B,Zr,Ca	215 (15.1)	225 (15.8)	12	48 Rc
Gr. 250 (USS)	0.03max.C, 17-19Ni, 7.0-8.5Co, 4.6-5.1Mo, 0.10-0.30Ti, 0.05-0.15 Aℓ,B,Zr,Ca	245 (17.2)	255 (17.9)	10	52 Rc
Gr. 300 (USS)	0.03max.C, 17-19Ni, 8.0-9.5Co, 4.6-5.1Mo, 0.6-0.8Ti, 0.05-0.15Aℓ, B,Zr,Ca	290 (20.4)	295 (20.7)	8	54 Rc
20 Ni (International Nickel Co.)	0.03max.C, 18-20Ni, 1.3-1.6Ti, 0.15-0.35 Aℓ, 0.3-0.5Cb	250 (17.6)	265 (18.6)	5-12	
25 Ni (International Nickel Co.)	0.03max.C, 25-26Ni, 1.3-1.6Ti, 0.15-0.35Aℓ, 0.3-0.5Cb	260 (18.3)	285 (20.0)	5-12	

Appendix-Table 3

Typical Aluminum Alloys For High-Pressure

Or High-Strength Applications*

Designation	Nominal Composition (weight percent)	Max. Yield Strength 10³ psi (kbar)	Max. Tensile Strength 10³ psi (kbar)	Elongation %	Hardness Bhn
Non Heat-Treatable					
3003-H18	0.12Cu, 1.2Mn	27 (1.9)	29 (2.0)	4-10	
3004-H38	1.2Mn, 1.0Mg	36 (2.5)	41 (2.9)	5-6	77
5154-H38	3.5Mg, 0.25Cr	39 (2.7)	48 (3.4)	10	80
5454-H34	0.8Mn, 2.7Mg, 0.12Cr	35 (2.5)	44 (3.1)	10	81
5456-H321	0.8Mn, 5.1Mg, 0.12Cr	37 (2.6)	51 (2.6)	16	90
5056-H18	0.12Mn, 5.1Mg, 0.12Cr	59 (4.2)	60 (4.2)	10	105
Heat-Treatable					
2014-T6	0.8Si, 4.4Cu, 0.8Mn, 0.5Mg	60 (4.2)	70 (4.9)	13	135
2218-T72	4.0Cu, 1.5Mg, 2.0Ni	37 (2.6)	48 (3.4)	11	95
6061-T6	0.6Si, 0.27Cu, 1.0Mg, 0.2Cr	40 (2.8)	45 (3.2)	12-17	95
6262-T9	0.6Si, 0.27Cu, 1.0Mg, 0.55Pb, 0.55Bi, 0.1Cr	55 (3.9)	58 (4.1)	10	120
7001-T6	7.4Zn, 3.0Mg, 2.1Cu, 0.30Cr	91 (6.4)	98 (6.9)	9	160
7079-T6	0.6Cu, 0.2Mn, 3.3Mg, 0.2Cr, 4.3Zn	68 (4.8)	78 (5.5)	14	145
7178-T6	2.0Cu, 2.7Mg, 0.30Cr, 6.8Zn	78 (5.5)	88 (6.2)	10	145

*From reference A2 (Appendix)

Appendix-Table 4

Typical Ultra-High - Strength Stainless Steels

Designation	Composition (weight percent)	Max. Yield[+] Strength 10³ psi* (kbar)	Max.Tensile[+] Strength 10³ psi* (kbar)	Elongation[+] %	Hardness[+]
Martensitic AISI - 410	0.15min.C, 1.0max.Mn, 1.0max Si, 11.5 to 13.0 Cr, 2.0Ni	87 (6.1) [R.T.] 148(10.4)[70K]	110 (7.7)[R.T.] 158(11.1)[70K]	20[RT] 10[70K]	92 Rb
AISI - 431	0.2max.C, 1.0max.Mn, 1.0max.Si, 16.0Cr, 2.0Ni	95 (6.7)	125 (8.8)	20	
Custom 455 (Carpenter Technology Corp.)	0.03C, 11.75Cr, 8.5Ni, 1.2Ti, 0.3Cb, 2.25Cu,	245(17.2)	330(23.2)	13	50 Rc
Crucible 422M (Crucible Spec.Metals)	0.28C, 0.9Mn, 0.2Si, 12Cr, 0.2Ni,0.5V, 1.7W, 2.2Mo	180(12.7)	255 (17.9)	11	36 Rc
ARMCO 17-4PH	0.05C, 17Cr, 4Ni, 4Cu	200(14.1)	210(14.8)	6-15	48 Rc
ARMCO PH 13-8Mo	0.05max C, 13Cr, 8 Ni, 2Mo, 1 Al	205(14.4)	225(15.8)	12	
Stainless W(USS)	0.07C, 17Cr, 7Ni, 0.7Ti, 0.2Al	210(14.8)	225(15.8)	3-20	48 Rc
Semi-Austenitic ARMCO 17-7PH	0.09C, 16-18Cr, 6.5-7.75Ni, 0.75-1.25Al	185-260 (13.0-18.3)	200-265 (14.1-18.6)	2-9	
ARMCO 15-7Mo	0.09max C, 14-16Cr, 6.5-7.75Ni, 2.0-3.0Mo, 0.75-1.25Al	210-260 (14.8-18.3)	220-265 (15.5-18.6)	2-7	50 Rc

	Composition				
AM-357 (Allegheny-Ludlum)	0.21-0.26C, 0.5-1.25Mn, 13.5-14.5Cr, 4.0-5.0Ni, 2.5-3.25Mo, 0.07-0.13N	305 (21.5)	310 (21.8)	4	
Austenitic AISI 202	0.15maxC, 7.5-10Mn, 17-19Cr, 4.6Ni	55 (3.9) [RT] 170 (12.0)[20K]	100 (7.0) [RT] 220 (15.5)[20K]	55 [RT] 5 [20K]	90 Rb
AISI 301	0.10C, 2.0 max Mn, 1.0max Si, 17.0Cr, 7.0Ni, 1.0max Si	40 (2.8) [RT] 75 (5.3) [20K]	105 (7) [RT] 275 (19.4)[20K]	60 [RT] 30 [20K]	85 Rb
AISI 316	0.08C, 2.0Mn, 1.0Si, 16-18Cr, 10-14Ni, 2.0-3.0Mo	37 (2.6) [RT] 84 (5.9) [20K]	85 (6.0)[RT] 210 (14.8)[20K]	65[RT] 52[20]	78 Rb
17-7MnV (USS)	0.11C, 13.5Mn, 16.2Cr, 4.6Ni, 0.92V, 0.36N	260 (18.3)	270 (19.0)	2	
Micromach (Washington Steel Co.)	0.08-0.12C, 2.0max Mn, 1.0max Si, 17.3Cr, 6.2Ni	180 (12.7)	200 (14.1)		
USS Tenelon	0.10C, 14.5Mn, 17Cr, 0.4N	210-240 (14.8-16.9)	225-255 (15.8-17.9)	3-9	

+ Properties quoted at room temperature [RT] unless otherwise indicated with temperature in brackets, e.g. [20K]

* Values given in kpsi followed by kbars in brackets

Appendix – Table 5

Typical Titanium Alloys (*)

Alloy (weight percent)	Condition	Average Yield Strength 10³psi (kbar)	Average Tensile Strength 10³ psi (kbar)	Elongation	Charpy Impact strength Ft-lb. (Kg-m)
99.0Ti	Annealed	85 (6.0)	96 (6.7)	25	15 (2.07)
α-Alloys					
5Aℓ-2.5Sn	Annealed	117 (8.2)	125 (8.8)	18	19 (2.63)
α-β-Alloys					
6Aℓ-4V	Annealed	134 (9.4)	144 (10.1)	35	14 (1.93)
6Aℓ-6V-2Sn	Solution + Aging	170 (12.0)	185 (13.0)	28	
β-Alloys					
3Aℓ, 8V					
6Cr, 4Mo,					
4Zr	Solution + Aging	200 (14.1)	210 (14.8)	20	7.5 (1.04)
11.5Mo, 6Zr,					
4.5Sn	Solution + Aging	191 (13.4)	201 (14.1)	16	

(*)Reference A2 (Appendix), Source Titanium Metals Corp. of America and RMI Co.

Appendix - Table 6

Typical High Strength Copper and Magnesium Alloys

Designation	Nominal Composition (weight percent)	Max. Yield Strength 10³psi (kbar)	Max. Tensile Strength 10³psi (kbar)	Elongation %	Hardness
Copper Alloys:					
ASTM-B194	1.9Be, 0.2Ni	150(10.5)	200(14.1)		Rc 42
Berylco 25 (Berylium Corp. of America)	2.0Be, 0.3Co	170(12.0)	200(14.1)	3-35	Rc 43
ASTM-B150	9.5Al, 5Ni, 2.5Fe, 1Mn	65-95(4.6-6.7)	100-150(7.0-10.5)	1-30	
ASTM-B97	3Si	70(4.9)	145(10.2)		
Magnesium Alloys					
AZ31B	3Al, 1 Zn, 0.2Mn	25(1.8)	37(2.6)	6-12	
AZ80A	8.5Al, 0.5Zn	33(2.3)	47(3.3)	2-5	
HM31A	3Th, 1.2Mn	26(1.8)	37(2.6)	~4	

Appendix - Table 7

Typical Ceramic And Cermet Materials

Material	Youngs Modulus 10^{6}psi (Mbar)	Tensile Strength 10^{3}psi (kbar)	Bending Strength 10^{3}psi (kbar)	Compressive Strength 10^{3}psi (kbar)
Single Phase Ceramics				
Al$_2$O$_3$ 85%	32(2.2)	~20 (1.4)	~40 (2.8)	~250 (17.6)
95%	40(2.8)	~30(2.1)	~45 (3.2)	~300 (21.1)
99$^+$%	50(3.5)	~40 (2.8)	~47 (3.3)	~300-500(21.1-35.2)
SiO$_2$(polycryst. quartz glass)	~10(0.70)	14.3(1.0)		
BeO	~50(3.5)	~25(1.8)		~220 (15.5)
TiC	45-55(3.2-3.9)	120-240(8.4-16.9)	25-135(1.8-9.5)	250-450(17.6-31.6)
WC	60-95(4.2-6.7)	175-450(12.3-31.6)	130+(9.1+)	520-800(36.6-56.2)
Si$_3$N$_4$	~35(2.5)	~16(1.1)		~90(6.3)
Multiphase Materials				
3Al$_2$O$_3$, 2SiO$_2$ (Mullite)	22(1.5)	15(1.1)	27(1.9)	75-80(5.3-5.6)
Cermet(90%WC, 10%Co)	(5.9)	~230(16.1)		~650(45.7)
Cermet(54%TiC, 40%Ni)		~140(9.8)		
Concrete: hydrated cement, sand, ballast	2.0-5.0 (0.14-0.35)	~0.4(0.028)		35~60(2.5-4.2)
Carbon base materials				
Diamond	135 (9.5)	600-1200(42-84)		~1,500(100)
Graphite fibers		45 (3.2)		
40% Phenolic resin-60% graphite fabric composite		17 (1.2)	27 (1.9)	40(2.8)
High density molded graphite	23 (0.16)	5.7 (0.40)		

Appendix - Table 8

Typical High-Strength Plastics

Material	Tensile Strength 10³psi (kbar)	Compressive Strength 10³psi (kbar)	Flexural Strength 10³psi (kbar)	Tensile Elongation (%)	Hardness
Single phase					
Polymethylmethacrylate	8-11(0.56-0.77)	11-19(0.77-1.34)	12-17(0.84-1.20)	2-7	M80-M100
Nylon 6-6	9-12(0.63-0.84)	7-13(0.49-0.91)	8-14(0.56-0.98)	25-300	R110
Polystyrene-Acrylonitrile Copolymer	9-12(0.63-0.84)	14-17(0.98-1.20)	14-19(0.98-1.34)	1-3	M85
Molding Reinforced Plastics					
Polyester(65% glass cloth)	30-70(2.1-4.9)	25-50(1.8-3.5)	40-90(2.8-6.3)		
Epoxy(+70% glass cloth)	20-70(1.4-4.9)	50-70(3.5-4.9)	70-100(4.9-7.0)		
Phenolic(+65% Asbestos mat)	40-60(2.8-4.2)	45-55(3.2-3.9)	50-90(3.5-6.3)		

Appendix - Table 9

Mechanical Properties of Typical Metals For Cryogenic Service

(Values are given in units of 10^3 psi followed in brackets by kbar.)

Material & Condition	Yield Strength				Tensile Strength				Elongation %			
	70°F 294K	-100°F 200K	-320°F 77K	-423°F 20K	70°F 294K	-100°F 200K	-320°F 77K	-423°F 20K	70°F 294K	-100°F 200K	-320°F 77K	-423°F 20K
Aluminum base alloys												
2021-T8151	67 (4.7)	70 (4.9)	79 (5.6)	86 (6.1)	74 (5.2)	79 (5.6)	90 (6.3)	102 (7.2)	8	9.5	11	12.5
5083-H321	34 (2.4)		40 (2.8)	42 (3.4)	49 (3.4)		66 (4.6)	90 (6.3)	15		31.5	30
7005-T5351	55 (3.9)	58 (4.1)	68 (4.8)	73 (5.1)	62 (4.4)	67 (4.7)	84 (5.9)	103 (7.2)	15	14	17	18.5
7075-T6	73 (5.1)	80 (5.6)	89 (6.3)	99 (7.0)	83 (5.8)	90 (6.3)	101 (7.1)	114 (8.0)	11	10	8	8
Titanium base alloys												
Ti-5Al-2.5Sn EL1	100 (7.0)		175 (12.3)	215 (15.1)	110 (7.7)		185 (13.0)	225 (15.8)	17		14	12
Ti-6Al-4V (EL1)	125 (8.8)		205 (14.4)	250 (17.6)	140 (9.8)		220 (18.3)	260 (18.3)	13		12	5
Austenitic Stainless Steels												
301 (XFH)	195 (13.7)		235 (16.5)	285 (20.0)	225 (15.8)		315 (27.1)	330 (23.2)	12		40	20

Table 9 (cont.)

304 (annealed)	40 (2.8)		60 (4.2)	70 (4.9)	95 (6.7)		220 (15.5)	250 (17.6)	40		20	20
310 (75% coldworked)	160 (11.2)		225 (15.8)	260 (18.3)	190 (13.4)		260 (18.3)	300 (21.1)	4		11	10
Superalloys												
Elgiloy	220 (15.5)		255 (17.9)	280 (19.7)	250 (17.6)		325 (22.8)	365 (25.7)	12		14	7
Inconel 718	170 (12.0)		215 (15.1)	230 (16.2)	195 (13.7)		240 (16.9)	275 (19.3)	10		15	15
Rene 41	110 (7.7)	130 (9.1)	140 (9.8)	160 (11.2)	175 (12.3)	185 (13.0)	210 (14.8)	230 (16.2)	28	28	26	22
Copper base alloys												
Berylco 25	135 (9.5)	150 (10.5)	175 (12.3)	190 (13.4)	165 (11.6)	180 (12.7)	210 (14.8)	220 (15.5)	5	6	8	10

Appendix – Table 10

Compositions of Typical Superalloys (weight percent)

Alloy	Fe	Ni	Co	C	Mn	Si	Cr	Mo	W	Nb	Ti	Al	B	Zr	Other
Nickel Base Alloys															
TRW-NASA VI A	–	61	7.5	0.13	–	–	6	2.0	5.8	0.5	1.0	5.4	0.02	0.13	9.0Ta, 0.5Rc, 0.43Hf
Rene 41	–	55	11	0.09	–	–	19	10	–	–	3.1	1.5	0.01	–	–
Inconel 600	7.2	77	–	0.04	0.20	0.20	15.8	–	–	–	–	–	–	–	–
M-22	–	71	–	0.13	–	–	5.7	2.0	11.0	–	–	6.3	–	0.60	30Ta
Waspalloy	–	58.3	13.5	0.08	–	–	19.5	4.3	–	–	3.0	1.3	0.006	0.06	–
Hastelloy-X	18.5	47.3	1.5	0.10	0.50	0.50	22.0	9.0	0.6	–	–	–	–	–	–
Cobalt Base Alloys															
Mar-M322	–	–	61.0	1.0	–	–	21.5	–	9.0	–	0.75	–	–	2.25	4.5Ta
X-40	[	10.5	54.0	0.5	0.75	0.75	25.5	–	7.5	–	–	–	–	–	–
AR-213	–	–	66.0	0.18	–	–	19.0	–	4.7	–	–	3.5	–		6.5Ta
Iron Base Alloys															
Alloy 901	36.0	42.5	–	0.05	0.10	0.10	12.5	5.7	–	–	2.8	0.2	0.015	–	–
Discalloy	54.3	26.0	–	0.04	0.90	0.80	13.5	2.7	–	–	1.7	0.1	0.005	–	–

Appendix - Table 11

Composition and High-Temperature Strength of Some Refractory-Metal Alloys

Alloy Designation	Composition %	Ultimate Tensile Strength (10^3 psi)*		Rupture Strength After 100 hr. (10^3 psi)*	
		2000°F (1093K)	2400°F (1316K)	2000°F (1093K)	2400°F (1316K)
Niobium Base					
Nb 753	93.75Nb,5V,1.25Zr.	41.5(2.92)	18.6(1.31)	14(0.98)	—
D 43	88.9Nb,10W,1Zr,0.1C	47(3.3)	27 (1.9)	26(1.8)	6.5 (0.46)
Nb 752	87.5Nb,10W,2.5Zr.	44(3.1)	25 (1.8)	18(1.3)	8 (0.56)
AS 30	79Nb,20W,1Zr,0.08C	85 (6.0)	—	45(3.2)	—
Molybdemum Base					
Mo-0.5Ti	99.5Mo,0.5Ti,0.02-0.05C	79 (5.6)	31 (2.2)	36(2.5)	10 (0.70)
TZM	99,4Mo,0.5Ti,0.08Zr,0.015C	73 (5.4)	54 (3.8)	53(3.7)	14 (0.90)
TZC	98.3Mo,1.25Ti,0.3Zr,0.15C	91 (6.4)	60 (4.2)	55(3.9)	21 (1.5)
Tantalum Base					
GE-473	90 Ta,7W,3Re	32 (2.2)	—	—	—
T 111	90 Ta,8W,2Hf	37 (2.6)	13 (0.91)	—	—
T 222	88 Ta,9.6W,2.4Hf,0.01C	54 (3.8)	14 (0.98)	28(2.0)	5 (0.35)
Tungsten Base					
W-2ThO2	98W,2ThO2	29.5 (2.07)	17.5(1.23)	23(1.6)	10.5(0.73)
W-15Mo.	85W,15Mo.	36 (2.5)	5.6(0.39)		7.3(0.51)
W-25Re	75W,25Re	32.6 (2.29)	11.3(0.79)		5.8(0.41)

*Values given in kpsi followed in brackets by bars.

Appendix - Table 12

High-Temperature Properties of Some Refractory Ceramics

Material	Temperature °F (K)	Elastic Modulus 10⁶ psi (Mbar)	Bending Strength 10³ psi (kbar)	Tensile Strength 10³ psi (kbar)	Compressive Strength 10³ psi (kbar)
Al_2O_3	70 (294)	50 (3.5)	20-38(1.4-2.7)	25-35(1.8-2.5)	400(28.1)
	1000 (811)	47 (3.3)	20-35(1.4-2.7)	20-35(1.4-2.5)	200(14.1)
	2000 (1366)	42 (3.0)	10-25(0.70-1.8)	15-30(1.1-2.1)	100(7.0)
	3000 (1922)	30-35(2.1-2.5)	1-3 (0.07-0.21)	2-3(0.14-0.21)	1(0.07)
BeO	70 (294)	40-55(2.8-3.9)	40 (2.8)	12-18(0.84-1.3)	110-250(7.7-17.6)
	1000 (811)	38-50(2.7-3.5)		10-16(0.70-1.1)	30-80(2.1-5.6)
	2000 (1366)	20-44(1.4-3.1)	17 (1.2)	3-10(0.21-0.70)	20-50(1.4-3.5)
MgO	70 (294)	40 (2.8)	22 (1.5)	15 (1.1)	200(14.1)
	1000 (811)	38 (2.7)	28 (2.0)	15 (1.1)	___
	2000 (1366)	34 (2.4)	15 (1.1)	10 (0.70)	
ZrO	70 (294)	24 (1.7)	26 (1.8)	20 (1.4)	300(21.1)
	1000 (811)	21 (1.5)		17 (1.2)	200(14.1)
	2000 (1366)	18 (1.3)	___	13 (0.91)	120(8.4)
	3000 (1922)	-	___	1 (0.07)	1(0.07)

Chapter 10

NONDESTRUCTIVE TESTING

Edward L. Criscuolo

Naval Surface Weapons Center
White Oak Laboratory
Silver Spring, Maryland 20910

I.   INTRODUCTION

Nondestructive test methods are used to detect discontinuities without impairing the usefulness of the material or part. The demand for high quality materials in critical applications has stimulated the development and use of nondestructive testing (NDT). Industries, such as nuclear, aerospace, utilities and petroleum rely upon nondestructive testing to assure material quality for improved safety and reliability. In addition, a cost savings can be achieved by preventing faulty material from being fabricated into parts or structures. If properly applied as an in-service detection technique, catastrophic failure of a system can be prevented.

By indirect means nondestructive testing can detect voids, cracks, porosity, laminations, shrinkage, lack of fusion, etc. Each test method varies in its ability to reveal a particular type of defect; therefore, more than one method may be needed to completely inspect an item. Used properly, the methods complement each other.

High pressure systems require a high degree of reliability and safety, because a failure could result in extreme damage and even loss of life. Nondestructive testing applied to components of a system can help assure reliability and safety. Components such as pumps, valves, piping, pressure vessels, etc. are made from castings, weldments, pipe and forgings and are suitable for NDT. An analysis must be made of each component to determine which types of flaw are acceptable and what should be their maximum size. With this information the proper nondestructive tests and acceptance standards can be selected. Codes have been issued by a number of societies which should be followed when they apply [5, 9, 10]. In some cases they are mandatory by law. There are two important elements in specifying NDT: 1. requirements for the specific method; 2. acceptance levels. A method specification gives details on how to conduct the inspection, and an acceptance specification defines the flaw size allowable.

The five most common NDT methods are Radiography, Ultrasonics, Magnetic Particle, Liquid Penetrant, and Visual. Other test methods include Acoustical Emission, Holography, Neutron Radiography, and Eddy Current. These latter methods are used in special applications. The success of these methods depends greatly upon proper application and interpretation of the results. Only qualified personnel should apply the tests and interpret the results.

II.   RADIOGRAPHY

Radiography is one of the oldest nondestructive test methods used for the detection of flaws. Industrially it has been applied for material inspection since the early 1920's and currently it is widely used on all types of materials. As X-rays or gamma-rays pass through material they are attenuated exponentially as a function of material thickness, density, and atomic number. Because of this characteristic, a void in a block of material will permit more penetrating radiation to pass than the surrounding region. Since X-rays cause a photographic film to be exposed, an image of the void can be obtained by placing a film behind the object. A typical radiographic exposure is shown in Figure 1.

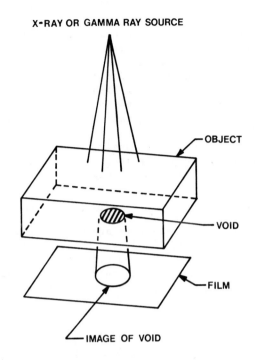

FIG. 1. Typical radiographic exposure.

A.   Radiation Sources

The penetrating ability of radiation is dependent upon the wavelength, the shorter the wavelength the greater the penetrating ability
of the radiation. Figure 2 illustrates the electromagnetic spectrum
and the position of X- and gamma-rays. X-rays refer to machine-produced radiation and cover the wavelength region from .005 to 10 $\overset{o}{A}$
(0.0005- 1nm). Gamma-rays refer to emissions from natural, or artifi-
cially produced isotopes and they produce radiation at discrete wavelengths. Four isotopes, Cobalt 60 (Co^{60}), Cesium 137 (Cs^{137}), Iridium

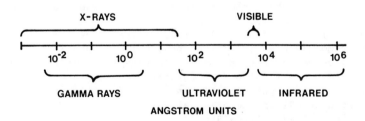

FIG. 2. Electromagnetic spectrum showing the range of
wavelengths of different types of radiation.

192 (Ir^{192}), and Thulium 170 (Th^{170}), are generally used for industrial radiography.

Elements with the same atomic number but with different atomic weights are called isotopes. Some isotopes are stable; others are unstable or radioactive. A radioactive isotope is one in which the nuclei of the atoms disintegrate. The disintegration (decay) of the nuclei proceeds with the emission of alpha- or beta-particles; accompanying this decay, generally with the beta-particle, is a gamma-ray. It is those isotopes that emit gamma-rays that are of value in radiography.

The decay of an isotope is purely random, and the number of disintegrations per second is proportional to the amount of radioactive material present. This leads to a decay formula

$$N = N_o e^{-t/\tau} \tag{1}$$

where N is the number of radioactive atoms at time t, N_o the number of radioactive atoms present at t = 0, and e is the natural logarithmic constant (e = 2.718...). τ is the decay time of the isotope or the time after which the number of radioactive atoms has decayed to 1/e the original number, as indicated in equation (1).

The activity of a radioactive source is measured by its disintegration rate.

$$\frac{dN}{dt} = \frac{-N_o e^{-t/\tau}}{\tau} \tag{2}$$

The curie is the unit of measurement of source activity and is defined as the quantity of any radioactive material that has a disintegration rate of 3.7×10^{10} disintegration/sec.

Radiation is measured in roentgens per time, and one method of measuring source strength is by specifying the radiation output in roentgens per hour at one meter (Rhm). The Roentgen (R) is that quantity of X- or gamma-radiation such that the associated corpuscular emission per 0.001293 gm of air (i.e. that mass of air occupying 1mℓ at 0°C, and 760 mmHg) produces, in air, ions carrying 1 e.s.u. of quantity of electricity of either sign [11].

Clearly, a source with a given activity (e.g. 1 Curie) will produce a certain radiation (Rhm). However, the specific amount of radiation will depend on the particular type of radiation emitted by the source (e.g. soft or hard X-rays or gamma-rays), so that there is not a general relationship between source activity and radiation, but only one for each radioactive material.

Table I gives the half-lives and the radiation outputs of the isotopes commonly used in radiography, while Figure 3 is a graph of the decay curves of the isotopes. An example illustrating the use of Figure 3 and Table I follows: Consider a two-curie source of Cobalt 60 which has an output of 2.64 Rhm; three years later the source will have decayed to 67% of its original value, and its radiation output will be 2.64 x 0.67 = 1.769 Rhm.

Aside from radioactive isotopes, electronically produced X-rays are used extensively for industrial radiography. X-rays are produced by bombarding a metallic target (e.g. tungsten) with high energy elec-

TABLE I

CHARACTERISTICS OF ISOTOPES USED IN RADIOGRAPHY

	Tm^{170}	Ir^{192}	Cs^{137}	Co^{60}
Average Energy (Mev)	.083	.28	.66	1.25
Production Process	Pile	Pile	Fission	Pile
Production Process	Pile Produced	Pile Produced	Fission Product	Pile Produced
Half Life	125 days	74 days	37 years	5.3 years
Radiation output Rhm/curie	45×10^{-3}	.55	.39	1.32
HVL inches	0.42*	0.52	0.67	0.75
(mild steel) mm	10.7	13.2	1.70	1.90

* measured beyond the fifth HVL

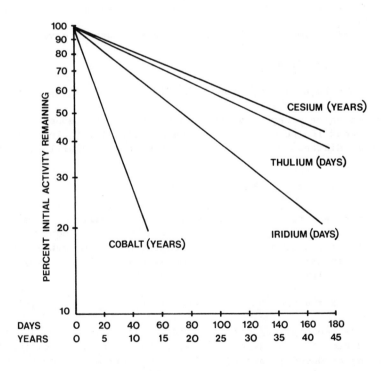

FIG. 3.   Decay curves for cesium, thulium, iridium and cobalt. Curves are labelled days and years and refer to different scales on the x-axis (abscissa).

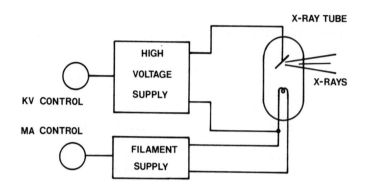

FIG. 4. Diagram showing principle of an X-ray source.

trons. These rays emanate from the target in straight lines.   The
electron current controls the X-ray intensity and the high voltage applied to the tube controls the wavelength and intensity of the radiation.   Figure 4 shows a schematic diagram of an X-ray machine. X-rays
can be produced from a few kilovolts to over 30 million volts. The
very high energy X-rays are used to penetrate massive and very thick
objects and very low energy X-rays are used on very thin or low density subjects.

The relationship between voltage and wavelength is given by

$$\lambda = \frac{12.34}{V} \tag{3}$$

where λ is wavelength in angstroms ($1\overset{o}{A} \equiv 10nm$) and V is the high voltage in kilovolts.

When X-rays or γ-rays pass through any material, absorption occurs.   If a beam is incident normally on a material with intensity I_o,
then, at some distance x in the material the intensity will have fal-
len to:

$$I(x) = I_o e^{-\mu x} \tag{4}$$

where μ is the absorption coefficient. More normally, the mass absorption coefficient is specified (μ/ρ), where ρ is the density of the
material. The depth of material at which the beam has been attenuated
to half its original value is called the half-value layer (HVL) thickness. From (4), $I(x) = I_o/2$ when

$$x = 0.6931/\mu = \frac{0.6931}{\rho}/(\mu/\rho) \tag{5}$$

The mass absorption coefficient, and hence HVL, depends on the
material and also the energy of the radiation. For instance, for typ-
ical machine-produced X-rays (E = 20kV, λ = 0.617 $\overset{o}{A}$ (6.17nm)) μ/ρ for
Aλ is 3.3 cm^2/gm (0.33m^2/kg) whereas at 1.25 MeV appropriate to gamma-radiation from Co^{60} it is ~0.055 cm^2/gm (0.0055 m^2/kg).   Further

information, including graphical data of cross-sections as a function
of photon energy, interpolation formulae, etc. can be found in refs.
12-14.

B.   Film Characteristics

X-ray film consists of a silver halide emulsion placed upon both
sides of a clear safety-base sheet. Radiation absorbed by the emulsion makes the silver halide grains developable. When a film is developed, those areas that absorbed radiation are dark, while those
that absorbed less radiation are lighter. The "darkness" of a film is
measured in units of density. Density is defined as the logarithm of
the ratio of the incident light on the film to the emergent light

$$D = \log_{10} \frac{I_o}{I_1} \qquad\qquad (6)$$

For example, a density of one indicates that the incident light on a
film is ten times more intense than the emergent light, while a density of 1.3 indicates that the incident light is 20 times more intense
than the emergent light.

The density of a film is dependent on the exposure time. A plot
of the density versus log exposure, which gives the characteristic
curve of the exposed film, is shown in Figure 5. From curves of this
type, one can determine the change in exposure time to get a given
change in density, determine the relative speeds between different
films, and, by the slope of the curve, determine which film will give
the highest contrast at a given density. The slope of the straight
line portion of a characteristic curve is called the film gradient.
Satisfactory industrial radiographs vary in density from about 1.5 to
4.0, although the best sensitivity is obtained at a density of about
2.5.

The speed of a film is the reciprocal of the time required to get
an arbitrary density on the film. In general, each film manufacturer
selects one of his films as speed 100 and bases the speed of his other
types of film relative to the arbitrary film.

C.   Radiographic Techniques

The radiographic procedure consists of four steps -- setup, exposure, development, and interpretation. Each of these steps will be
considered separately.

1.   Setup

The setup includes all the preliminary steps before actually making the exposure, such as the selection of film and screen combination, source, source-to-film distance, filter and scatter precautions.
The selection of film and screens usually depends upon the resolution
required in the radiograph and the length of exposure that can be tol-
erated.

The primary function of the intensifying screen is to reduce the
exposure time. The screens are placed on each side of the film inside

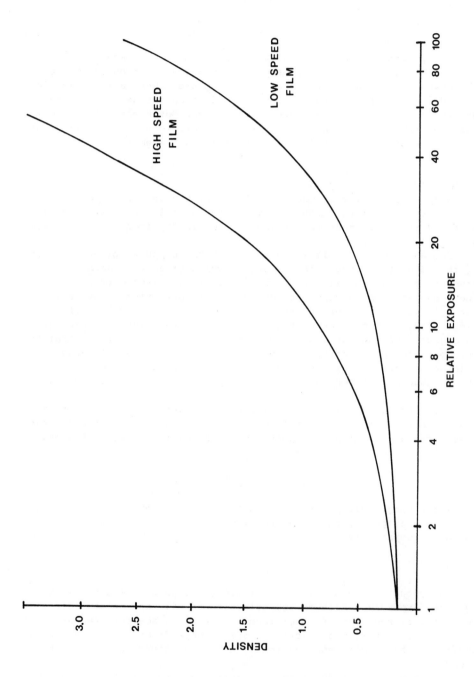

FIG. 5. Characteristic curves for X-ray films. The density and relative exposure are defined in the text.

the holder or cassette. When irradiated, lead screens eject photo-electrons which are captured by the film, thus producing increased blackening of the developed film. For radiography with isotopes, lead screen thickness of .005 inch ($\sim$.127 mm) front and .010 inch ($\sim$.254 mm) back are commonly used.

The selection of the proper radioactive isotope and X-ray energy is important in producing a good radiograph. A good rule for the selection of an isotope for inspection of a given thickness of material is to choose a source whose half-value layer (HVL) thickness is 1/3 to 1/6 times the section thickness. The HVL for steel with Co^{60} is about 0.75 inch ($\sim$19 mm) so that a layer 3 inches ($\sim$76.2 mm) thick represents 4 half-value layers; therefore, Co^{60} is a satisfactory source for this thickness of steel.

The source-to-film distance is another factor to consider. Since the image on the film is formed by geometrical projection, the source size and distances will determine the lack of sharpness due to geometrical factors, (u_g). A large distance will result in a very long exposure time; a too-short distance will not produce a sharp image. Figure 6 shows the geometry involved in the proper setup. The source size, distance of the object from the film, and the source-film distance are related to u_g in the following manner:

$$u_g = \frac{\text{object-to-film distance x source size}}{\text{source-to object distance}} \qquad (7)$$

For optimum resolution, the u_g should be equal to or less than the lack of sharpness due to film factors (u_f).

For example, a radiographic setup is to be made with Co^{60} to inspect 2 inches ($\sim$50 mm) of steel at a distance of 2 feet ($\sim$300 mm). The source size is a 1/8 inch ($\sim$3 mm) cube. u_f is .003 inch ($\sim$.077 mm). Is the geometrical resolution satisfactory in this setup? Using (5), u_g is 0.014 inches ($\sim$.264 mm). Since the $u_f < u_g$, the source-to-film distance should be increased for improved resolution.

Another consideration is scatter, a complicated subject of which only a brief discussion will be given here. Scatter originates primarily from several sources -- from the room, benches, and setup (room-scatter) and from the object (forward scatter). Scatter tends to fog the film, thus reducing contrast. In almost any setup, there is a certain amount of radiation being scattered from the wall, ceiling, and floor of a room. Room scatter can be minimized (a) by irradiating only the pertinent area so that little radiation reflects off the wall, and (b) by shielding the film (placing lead behind it) to avoid back scatter from the floor.

Scatter from the object can be prevented from reaching the film by placing a lead filter of the proper thickness over the film which will differentially filter the scattered radiation and allow the primary radiation to penetrate. A .030 inch ($\sim$0.774 mm) lead filter for Ir^{192} and a 1/8 inch ($\sim$3.17 mm) or 1/4 inch ($\sim$6.34 mm) for Co^{60} are commonly used. A penetrameter is placed on the source side of the object; this device is an image quality indicator. It consists of a shim of material (same composition as the object containing three drilled holes). The ASTM Penetrameter is shown in Figure 7. The visibility of the holes on a radiograph indicate the film quality. Us-

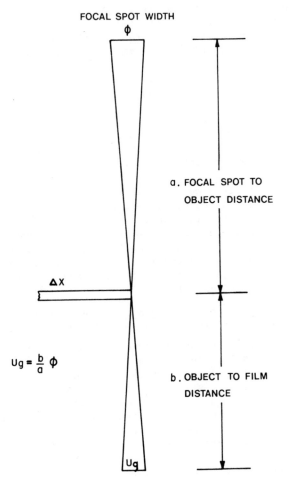

FIG. 6. Line diagram showing the geometry for a radiographic set-up.

ually the shim thickness is 2% of the object thickness. Many code bodies such as AWS, ASTM, ASME, etc. give specific requirements to be met.

2.   Estimation of Exposure Time

An accurate estimation of exposure time is very important with isotopes since the time involved is usually long compared to exposures made with X-ray equipment. The opportunity for a second exposure may not always occur. There are approximately three ways of estimating exposures: (a) by technique curves, (b) by calculation, and (3) by

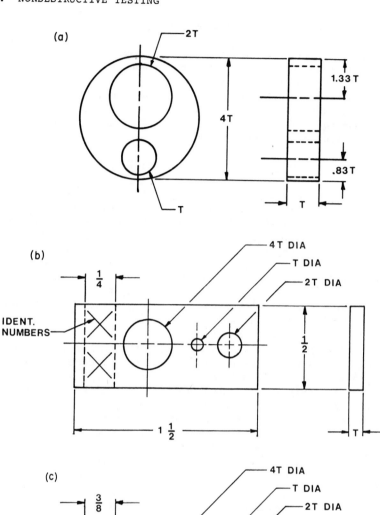

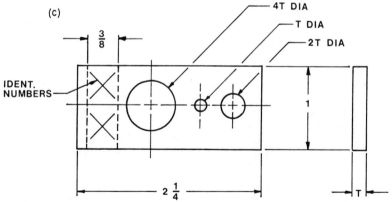

FIG. 7. Penetrameter Designs: (a) for thickness of 0.180 in.
(~4.57 mm) and over; (b) for thickness from 0.005 in. (~.127 mm)
and including 0.050 in (~1.27 mm); (c) for thickness from 0.060 in.
(~1.52 mm) to and including 0.160 in. (~4.06 mm). Dimensions given
in figures are in inches (1 in. = ~25.4 mm).

radiation measurements. Exposure calculators are also available and eliminate the need for calculation. They are a convenient form of technique curves. Details can be found in a number of texts.

In estimating the exposure time, the shape of the object must be considered. If the object is a flat plate of known and uniform thickness, it is simple to estimate the exposure time. An object with varying thickness poses greater problems because the exposure for each thickness must be considered. Some complex objects will require two or more exposures for complete coverage.

D.   Development Procedures

Film should be developed according to standard procedures as published by ASTM in E94-, Recommended Practice for Radiographic Testing, or according to manufacturers' recommendations. Reasonable care in processing of film will ensure the highest quality radiograph.

The processing cycle for radiographs includes X-ray developer, rinse, stop, fixer, and wash. A typical cycle might be to develop for 5 min. at 68°F (20°C) (8 min. for maximum speed), rinse 15 sec. in clear water, place for 30 sec. in the stop bath, fix for 5-10 min., and then wash in running water for about 30 minutes. After washing, the film may be placed in an emulsion hardening bath for about 1 min. Film should be dried by a forced draft of air.

It is during the developing cycle while the emulsion is soft that care should be taken to see that the films do not contact each other and that handling of the film is at a minimum. Poor procedure during developing will result in film blemishes that may appear as defects or may mask actual defects in the material radiographed.

Automatic processors are available that give a start to finish time of about 7 minutes. This equipment will produce very good results provided the manufacturers instructions are followed. Control of such items as temperature, replenishment rate and maintenance must be exercised.

E.   Interpretation of Radiographs

1.   General

Interpretation is an important step of radiographic inspection. The interpreter must be given a set of standards and/or reference radiographs by which an evaluation of the object can be made. Familiarity with the material inspected, radiographic procedure, and service requirements are necessary before an intelligent evaluation can be made.

a.   Film-Viewing Procedure.   An important piece of equipment used by an interpreter is the viewer. Two types are necessary -- a large screen and a high-brightness spot viewer. Some manufacturers make these combined into one unit with a dial to control the intensity of the spot light. Another device which is useful in the viewing room is a small magnifier of approximately 7X. This is made in several styles in which the reticule contains scales, lines, and/or different size circles for ease in measurement.

A few procedures that should be followed when viewing film are:
i.   View film in a darkened room. A dim sidelight for
     note-taking is permissible.
ii. Illuminate the film area only (avoid glare).
iii. Use spot viewer for dense areas.
iv. Keep films clean.
The first thing that an interpreter looks for on a radiograph is the
penetrameter. This device will indicate if the technique is satisfactory. Secondly, a comparison is made with the radiographic standard
to accept or reject the specimen. A record of the findings is made
for future reference.

     b.   Standards.   Numerous standards have been developed during
the past twenty years for castings and weldments of aluminum, magnesium, steel, and bronze. In these standards, a discontinuity is illustrated in varying degrees of severity. The product specification usually defines which degree is acceptable. In practice, the interpreter makes a comparison to the standard and makes a decision to accept
or reject the specimen part.
     Many times the film interpreter will find defects that will cause
rejection or necessitate repair of completed work. At times he will
be called upon to substantiate his findings. The use of film standards and ASTM terminology give the film interpreter a solid founda-
tion for his decisions.
     The value of radiographic inspection is not only to reject defec-
tive parts but also to assist in the production of higher quality material. For example, in the inspection of pipe and pressure vessel
welds, the radiographer may find the first indication of poor quality
by noting porosity or other discontinuities in the welds. A conference between radiographer and welder should be held, and this "defect"
can be corrected by the welder. Occasional conferences between welder and radiographer should assure the welder that radiography is used
to assist him. With or without radiography, welders today produce
high quality welds most of the time. It is the function of radiography to detect any faults, determine whether they are of significance for the specific application, and if so to recommend corrective
action.

                    F.   Radiation Safety

     The characteristics of X-rays and gamma-rays that make them use-
ful for industrial inspection are the same characteristics that make
them dangerous to the human body. The ability of radiation to penetrate large masses of material and to ionize matter can result in damage to the body. This problem of radiation injury has been recognized
since the early days of X-rays. Recent advances in nuclear technology
have emphasized and enlarged the problem so that considerable work has
been done on the subject of radiation safety. Groups such as the National Bureau of Standards, the Atomic Energy Commission, the Public
Health Service, the National Committee on Radiation Protection, and
many others have published very valuable data on radiation safety.
These factors are considered in greater detail in Chapter 4.

III.   ULTRASONIC TESTING

Ultrasonic testing uses high frequency, low energy sound to detect and locate flaws in material. A short pulse of ultra-sound is directed into a material by a piezoelectric transducer which is in contact with the surface. The sound beam is fairly well defined and the divergence seldom exceeds 10 to 15 degrees. High frequency sound waves behave in many respects like other wave phenomena such as light and audible sound. They travel in a straight line until they reach an acoustic boundary where reflection occurs. A flaw such as a gashole in the path of the sound beam presents an acoustical boundary and sends an echo back to the transducer. The back surface will also return an echo.

The signal from the transducer is displayed by a cathode-ray tube where signal intensity as a function of time is displayed. A simplified sketch of the ultrasonic technique is shown in Figure 8. The position of the flaw signal will indicate the relative depth of the flaw from the test surface of the object. The amplitude of the signal indicates the amount of sound energy reflected by the flaw. This does not necessarily indicate the size of the flaw since the returned signal depends upon many factors such as flaw shape, size, acoustical coupling, etc. An experienced operator can give accurate estimates of flaw size and shape if the flaw is probed from various angles and the operator is familiar with the geometry of the test object.

Ultrasonics is almost a universal tool for materials that transmit acoustical energy provided that the shapes of the materials are not too complex. It is applied to wrought products, weldments, steel castings and forged materials.

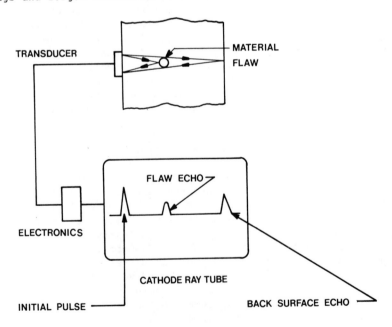

Fig. 8.   Schematic of the ultrasonic test method.

A.   Fundamentals of Ultrasonics

The transmission of ultra-sound through a medium depends upon particle displacement as the energy travels through the medium. Ultrasonic waves are classified into several types depending upon mode of the particle vibration. If particles of the transmitting medium travel in the same direction as the wave propagation, they are commonly called longitudinal waves. In the transverse or shear wave the particles of the transmitting medium vibrate at right angles to the direction of wave propagation. Figure 9 illustrates longitudinal and transverse waves. These two wave types are most commonly used for nondestructive testing. Surface waves are divided into three classes, Rayleigh, Love, and Stonely waves. These waves are roughly analogous to water waves. Since these waves are used only for special inspection problems they will not be discussed here in any detail. The interested reader may find further discussion in references [1, 2].

The velocities of longitudinal (V_t) and shear (V_s) waves in an isotropic medium can be calculated by the following equations (see also Chapter 13 and ref. [1]).

$$V_L = \left[\frac{E(1 - \sigma)}{\rho(1 + \sigma)(1 - 2\sigma)} \right]^{1/2} \tag{8a}$$

$$V_T = \left[\frac{E}{\rho} \frac{1}{2(1 - \sigma)} \right]^{1/2} \tag{8b}$$

where E is Young's modulus, ρ is the density, σ is Poisson's ratio. Typical values are given in Table 2.

The ultrasonic frequencies used for material inspection range from 0.5 to 10 MHz. The frequency and wavelength are related by

$$f\lambda = V \tag{9}$$

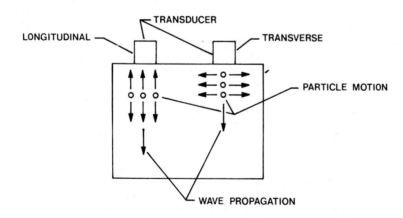

FIG. 9. Longitudinal and transverse waves in a transmitting medium.

TABLE 2

VELOCITY OF ULTRASONIC WAVES IN MATERIALS

Material	Velocity $[(m./sec.) \times 10^3]$	
	V_L	V_T
Metals		
Aluminum	6.35	3.10
Brass (Naval)	4.43	2.12
Copper	4.66	2.26
Steel	5.85	3.23
Nonmetals		
Air	0.33	--
Water	1.49	--

where f is frequency, λ is wavelength and V is the velocity of sound. The velocity of sound is a function of material and when the sound travels from one material to another, there will be a change of wavelength.

B.   Propagation of Ultrasound

As ultrasound propagates through any material it experiences a loss of energy. This loss can be attributed to four mechanisms, heat conduction, viscous friction, elastic hysteresis, and scattering. These losses result in attenuation of the beam. In most engineering materials and alloys the attenuation is small and thereby permits the inspection of material several feet thick. For some materials such as rubber, or components and materials with large grain structure, the sound is greatly attenuated.

Ultrasonic waves are highly directional and travel in straight lines. The beam spread is a function of frequency; for a 1 inch (~25.4 mm) diameter transducer the spread is approximately 4° at 5 MHz and 9° at 1 MHz [2].

As an ultrasonic wave traverses an acoustical interface, presented by two different media, part of the energy is reflected and part is transmitted. This property is useful for the detection and location of flaws. A steel-to-air interface reflects almost 100% of the energy; thereby a void is very easy to detect by the pulse-echo technique. References 1 and 2 should be consulted for more detailed treatment of this subject.

C.   Ultrasonic Test Equipment

The basic ultrasonic test equipment consists of a transducer and
electronic circuits to generate, receive, and display the electronic
signal.   The transducer is selected based upon frequency and requirements of the test. Therefore, it is necessary to have a wide selection of transducers available. Test instruments are available in several forms, such as portable, modular and laboratory style; many instruments have built-in circuits to assist the operator's interpretation of the electronic signal. Some of these are an alarm gate which
produces an audio or visual alarm to indicate that the signal amplitude has exceeded a preset level and timing marks across the face of
the scope which can be preset to correspond to thickness of material
under test. An accurate measurement of the flaw depth can be obtained.

Figure 10 shows a block diagram of an ultrasonic flaw detector.
The clock controls the proper timing and synchronization of the pulser, receiver, mark generator and cathode-ray display tube. A high
voltage pulse is provided to the probe by the pulser. The probe converts the electrical energy to ultrasound. The same or another probe
can be used to receive the ultrasonic echo from a flaw or back surface. The receiver amplifies the signal and displays it on the scope.
A timing mark generator produces a calibrated scale for measuring the
distance between signals. Laboratory instruments have provisions to
accept special modules to do such things as compensate for attenuation, gate signals, convert to a decibel scale, etc.

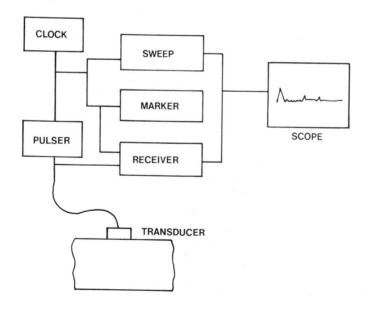

FIG. 10.   Block diagram of pulse-echo flaw detector.

D.   Ultrasonic Testing Procedure

In order to obtain a satisfactory ultrasonic test it is necessary to establish an acceptable procedure. A number of factors must be considered: 1) the shape of the part, 2) the type of material, 3) the surface conditions, 4) defect orientation and type, and 5) calibration.

From the shape of the part, the probe and wave type is selected. This is illustrated in Figure 11(a) where a uniform object is to be inspected. Longitudinal waves produced from a straight transducer should give satisfactory results. Figure 11(b) illustrates how a weld is inspected. The sound wave is introduced into the weldment by a probe and a lucite wedge. The sound beam is reflected from the upper and lower surfaces until it traverses the weld metal. A flaw will reflect some of the acoustical energy back to the probe. For some complicated geometries it may be necessary to inspect the weldment from several positions. A scan pattern for the probe must be established in order to completely cover the object.

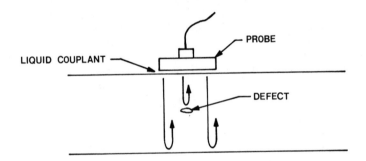

(a) INSPECTION OF AN UNIFORM OBJECT

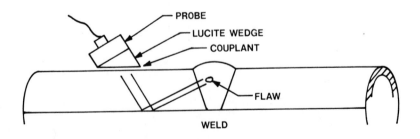

(b) WELD INSPECTION OF PIPE OR VESSEL

FIG. 11. Ultrasonic inspection methods for different cases.

As discussed earlier the attenuation varies from one material to another. For example, cast iron is difficult to inspect because of the coarse grain size. Also, the surface condition is very important in order to couple sound into the part. Rough surfaces may require grinding before an ultrasonic test can be performed. If a back surface is used as a reflector, it must be uniform also.

Since reflectors normal to the direction of wave propagation produce the maximum signal, laminar defects such as cracks, lack of fusion, unbonded joints, etc. are easier to detect than fine porosity or defects that are parallel to the sound beam. An object should be probed from many angles in order to fully inspect the object.

Reference blocks made of the same material as the object to be inspected are used to calibrate the ultrasonic system. The block contains drilled holes that are used as reflectors. The signal amplitude on the scope is adjusted as required by specifications. Standard block designs have been established by ASTM, ASME, IIW and other standard groups.

IV.   LIQUID PENETRANT INSPECTION

The liquid penetrant method of nondestructive testing is used to detect fine cracks or pores that are open to the surface. This method is based upon the ability of certain types of liquids to enter into narrow voids by capillary action, and to remain there after the excess is removed. Subsequently the penetrant is drawn out of the crevice by a developer powder thus making the void visible. This method is applicable to magnetic and nonmagnetic materials. It is particularly useful for inspecting nonmagnetic materials where the magnetic particle method cannot be used. The method when properly carried out is reliable, sensitive and of low cost. It can be applied to castings, welds and mill products. In many parts of a high pressure system this may be one of the few methods that can be applied to nondestructively test a part, and in many cases will be the first method used, since apparatus requirements are minimal.

A.   Fundamentals of Penetrant Testing

Penetrant inspection techniques are divided in two basic methods, Fluorescent and Visible-Dye. Fluorescent penetrant inspection makes use of a penetrant that fluoresces under black light. Exceptionally good visibility of an indication is obtained with this method. The black light is near ultra-violet with a wavelength of approximately 3300 to 3900 Å (330-390 nm). Lamps producing the necessary radiation are commercially available. Inspection using visible-dye technique is based on the use of a penetrant containing a vivid red dye which is well contrasted against a background of a white developer. The procedure can be broken down into seven basic steps as follows:

i.   Clean the surface to be inspected.
ii. Apply the penetrant.
iii. Allow sufficient penetrant dwell time
iv. Remove excess penetrant.
v.   Apply the developer.

vi.   Examine and inspect the test surface.

vii. Post-clean the surface.

The above steps apply irrespective of which type of test is selected -- either visible-dye or fluorescent-penetrant method.

B.   Fluorescent-Penetrant Method

The fluorescent penetrant method differs from the visible dye
method only in the type of penetrant used and the technique for viewing the indications. Fluorescent-penetrants are available as water-
washable for easy removal of excess penetrant, or can be made water-
washable by the application of an emulsifier to the surface of the
penetrant. The direct water-washable type may contain additives that
reduce the penetrating effectiveness. The purpose of a developer is
to draw out the penetrant from a discontinuity where the fluorescent
penetrant can be made visible by black light. Developers are manufactured in the following types, dry powder, water-suspension, or non-
aqueous suspension. The following steps should be used in fluorescent
penetrant inspection.

1.   Precleaning

Precleaning is an important step in the process since dirt, scale
grease or other surface coatings could prevent the penetrant from entering a discontinuity.

Scale, slag, grease, oil, paint, water or other surface coatings
must be removed. Cleaning with solvents or mechanical means are effective.   Care should be taken not to use such methods as sand blasting or other mechanical methods that might close a surface-opening of
a defect. Sufficient time should be allowed for the evaporation of a
volatile cleaner.

2.   Application of Penetrants

Penetrant is applied to the part by spraying, brushing or dipping.   The surface should be kept well coated during the dwell time.
On very large objects where only a small section is to be inspected it
may be necessary to mask the object to prevent the penetrant from
getting into unwanted areas.

3.   Dwell Time

The dwell or soaking time is the period of time during which the
penetrant is allowed to remain on the test part. Sufficient dwell
time must be allowed for the penetrant to enter the discontinuity.
The dwell time varies from 10 to 20 minutes. Each manufacturer of
penetrant material has some specific recommendations which should be
followed.

4.   Penetrant Removal

The water-washable penetrant can be removed from the part with a water spray. The temperature of the water should not be in excess of 110°F (~43.3°C) since hot water tends to remove the penetrant from the void.

If a post-emulsifiable type penetrant has been used, an emulsifier must be applied to render the excess penetrant water-washable. It is applied by dipping, spraying or flowing on. Manufacturers of such materials give specific recommendation on the dwell time. After emulsification, the washing technique is the same as an emulsified penetrant-removal technique. In order to ensure that the removal of penetrant is complete, a black light is used to check the surface.

5.   Application of Developer

A dry developer is dusted onto a part that has been completely dried after the excess penetrant has been removed. The developer draws the penetrant from the discontinuity. Wet developers are made up of dry powder in a water suspension. These are applied directly to the part immediately after the rinse for removal of excess developer. The part is air-dried by means of a warm air stream. Excessively thick coating of developer should be avoided because of the possibility of masking small indications.

Non-aqueous wet developers are applied to a dry part. All wet developers may be applied by spraying or brushing.

6.   Inspection

Inspection is performed after the developer has been on the surface of the part for at least one half the penetrant dwell time. The surface is examined in a darkened room under black light. Presence of voids will be indicated by brillant-fluorescence under the black-light excitation.

7.   Final Cleaning

The final step in the process is cleaning. Solvents are usually used for cleaning.

C.   Visible-Dye Penetrant Method

Visible-dye penetrants are available in water-washable, post-emulsible or non-water washable types. The non-water washable type is removable by solvents. Most of the steps to be followed are the same as in the fluorescent-penetrant method except for the inspection. Black light is not used on the part. The examination is performed with the part illuminated with ordinary light. Discontinuities are shown as a red indication against a white background.

V.   MAGNETIC PARTICLE INSPECTION

Magnetic particle inspection utilizes leakage magnetic fields to locate surface defects in magnetic materials. This method is especially useful for detecting fine cracks which may not be visible to

the naked eye. The magnetic method of inspection is applicable onl
to ferromagnetic materials and cannot be used to inspect non-ferrous
material or austenitic steels.

A defect must be sufficiently large to interrupt or distort the
magnetic field to cause external leakage. A fine, long defect such as
a crack that lies in the direction of the magnetic field will cause
little leakage. The most desirable position of a defect is with the
long direction perpendicular to the magnetic field. Since the direction of defect is not usually known it is necessary to inspect by mag-
netizing the part in more than one direction.

The method is used in conjunction with other nondestructive tests
that can locate internal defects.

A.   Fundamentals of Magnetic Particle Inspection

The basic principles of magnetic particle inspection are illustrated in Figure 12 where the magnetic flux about a magnetized object
containing a crack is shown. The leakage flux in the region of the
crack will have a stronger attraction for the magnetic particles
placed upon the surface and therefore the crack will be indicated.
The part can be magnetized by conducting a large current through it or
by placing a current-carrying coil around it. The area to be inspected is covered by finely-divided magnetic particles which react to the

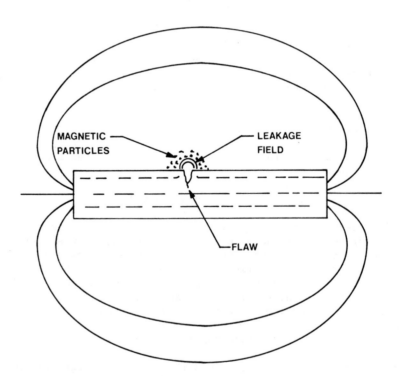

FIG. 12.   Illustration of the magnetic particle inspection
technique.

leakage flux produced by a flaw. These magnetic particles will form
an indication on the surface approximately the same shape as the flaw.

B.   Methods of Magnetization

1.   Circular Magnetization

Figure 13 shows the method of producing circular magnetization by
passing current directly through the part. The current produces a
field in the circumferential direction of the part which locates flaws
in the long direction of the test piece. Circular magnetization can
be obtained in hollow parts such as cylinders by placing a current-
carrying conductor through the center. This permits the inspection of
inner surfaces of large pipes or cylinders.

2.   Longitudinal Magnetization

A coil or wire wrapped around the part produces a field in the
longitudinal direction of the part, Figure 14. Defects in the circum-

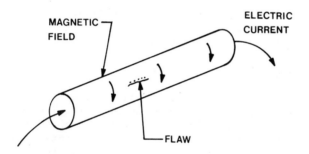

(a) ELECTRIC CURRENT IS CONDUCTED THROUGH PART

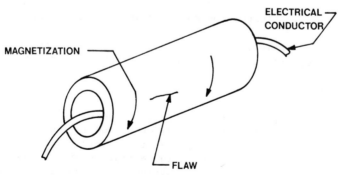

(b) MAGNETIZATION OF CYLINDER

FIG. 13. Methods for obtaining circular magnetization of
cylindrical parts.

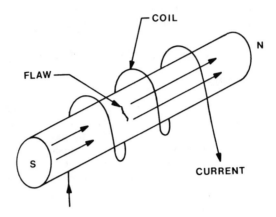

FIG. 14. Method for obtaining longitudinal magnetization of a long part.

ferential direction are located by this technique. Longitudinal magnetization can also be obtained by using the part under test as a link in the magnetic circuit of an electromagnet.

3. Prod Magnetization

For parts that are too large to magnetize as a whole, prods are used to pass current in a local region of the part. Prods are handheld on the surface while a low voltage, high current is passed through the part, as shown in Figure 15. The magnetization is such that defects along a line between prods are indicated. Care must be used to not produce an arc at the surface. The current is applied after the prods are in contact with the surface.

C. Inspection Media (Magnetic Particles)

Magnetic particles are available in various colors for different surfaces. They are applied by one of two methods, dry or wet.

1. Dry Method

Dry magnetic particles are applied by means of a dusting bag, atomizer or spray gun. The dry powder method is easy to use on rough surfaces and is portable. The powder is lightly applied in a low velocity cloud to the region being inspected. Excess powder is removed by a light air stream. The indications are observed with normal lighting conditions.

2. Wet Method

Particles for the wet method are smaller than those used in the dry method. The particles are suspended in a batch of light oil or

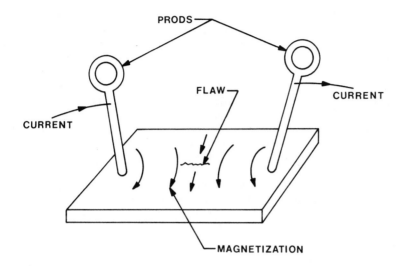

FIG. 15. Method for obtaining local magnetization with prods.

water. Because of the smaller particle size the technique is capable of indicating smaller defects. Since the particles are in suspension the bath must be continuously agitated to prevent settling. The mixture is either flowed or sprayed over the surface of the part, or the part can be immersed in the suspension. The indications are read in the same manner as with the dry method. Particles are avilable with red, black or fluorescent material coatings. Particles coated with fluorescent material require black light to fluoresce the particles and give greater sensitivity than with other techniques.

D. Demagnetization

Parts that have been magnetized retain some magnetization and tend to pick up filings, chips, grindings, etc. These unwanted particles could cause problems during the life of the part, therefore, parts should be demagnetized by placing them in a field of an alternating current solenoid. There are a number of other demagnetization techniques that can be satisfactorily used [1].

VI. OTHER METHODS OF NONDESTRUCTIVE TESTING

There are numerous nondestructive tests for inspecting high pressure systems that have not been discussed in detail in this chapter. Some of these are Visual, Leak Testing, Acoustical Emission, Holography and Neutron Radiography.

Visual inspection is often given little attention since it is one of the easier tests to apply. Aids such as magnifiers or borescopes are often used. It is important to establish a procedure which an inspection must follow.

Leak-testing is another important nondestructive test; it is a form of proof-testing where defects are revealed by the leakage of gas or liquid through the defect. The test can be as simple as a bubble test or as complicated as a mass spectrometer detector. The proper leak detection method should be selected based upon the application of the system.

Acoustic emission is a very new method where emission of sound is recorded while the system is loaded. This method offers the possibilities of early detection of fatigue type flaws [6].

Holographic interferometry is another recent development for observing small deflections of a structure while under partial load.

Neutron radiography utilizes thermal neutrons as a source of radiation. The process is similar to X-radiography but it offers unique detection capabilities. Neutrons can penetrate lead easily but are stopped by hydrogenous materials such as wax, explosives, etc. It is therefore possible to inspect materials which are difficult by X-ray methods [8].

REFERENCES

1. R. C. McMaster, Nondestructive Testing Handbook, Vol. 1, The Ronald Press Co., New York (1959).

2. W. J. McGonnagle, Nondestructive Testing, Second Edition, Gordon and Breach Publ., New York (1961).

3. E. A. Fenton, Welding Inspection, Am. Welding Society, Inc., New York (1968).

4. R. Halmshaw, Physics of Industrial Radiology, Am. Elsevier Publ. Co., Inc., New York (1966).

5. American Society for Testing and Materials, Annual Book of ASTM Standards, Part 3, p. 504 (1971).

6. ASTM, Acoustic Emission - STP 505, Publication Code Number 04-505000-22 (1972).

7. ASTM, Monitoring Structural Integrity by Acoustic Emission - STP 571, Publication Code Number 04-571000-22 (1975).

8. H. Berger, Neutron Radiography, Elsevier Publ., New York (1965).

9. Am. Society of Mechanical Engineers, ASME Boiler and Pressure Vessel Code, Latest Edition.

10. Am. Welding Society, Structural Welding Code D1. 1-75.

11. American Institute of Physics Handbook, McGraw-Hill Book Company Inc. (1963).

12. R. D. Evans, "The Atomic Nucleus", McGraw-Hill Book Company Inc. (1955).

13. U. Fano, L. V. Spencer, M. J. Berger, Penetration and Diffusions of X-rays in "Handbuch der Physik", Vol. XXXVIII/2 (S. Flüge, ed.) Springer Verlag, Berlin (1959).

14. G. W. Grodstein, "X-ray Attenuation Coefficients from 10 kev to 10 MeV". Natl. Bur. Standards Circular 583 (1957).

Chapter 11

ULTRAHIGH PRESSURE APPARATUS AND TECHNOLOGY

Ian L. Spain

Laboratory for High Pressure Science and
Engineering Materials Program
University of Maryland
College Park, Maryland 20742

I.   INTRODUCTION

Up to approximately 20kb, compressors, containment vessels,
valves, fittings and tubing can be designed and constructed using conventional techniques discussed in earlier chapters, provided that care
is taken in selecting the correct materials. Commercial equipment of
this type is available up to at least 15kb. A comparison of the basic
capabilities of different designs above ~20kb is given in Table 1.

Above about 20kb special techniques have to be developed to compress and contain materials. The most obvious factor is the lack of
high pressure tubing to convey the pressurizing fluid from the com-

TABLE 1

Principal Designs Used for Reaching High Pressures

Type of Apparatus	Pressure Range (k bar)	References
1. Piston-Cylinder		
a. Unsupported piston (multiple-ring cylinder, or variable radial support on inner cylinder)	0-50	6-9, 11
b. Supported piston (multiple-ring cylinder with axial support on cylinder and radial support on piston	0-80	10
c. Inner piston-cylinder inside outer piston-cylinder	0-100	6, 12
2. Opposed Anvils		
a. Unsupported Bridgman anvils (tungsten carbide)	0-180	1, 7, 19
b. Unsupported Bridgman anvils (single crystal diamond (diamond anvil cell))	0-1000?	4, 31-37
c. Supported Bridgman (tungsten carbide)	0-250	20-29
d. Supported Bridgman (tungsten carbide anvils tipped with sintered diamond)	0-350	5
e. Bridgman anvils with variable lateral support	0-350	30
f. Belt and girdle designs	0-120 (0-80*)	13, 48 - 54
3. Multiple Anvils		
a. Tetrahedral Press	0-120 (0-80*)	55-58
b. Cubic Press	0-110 (0-80*)	59-63
c. Split-Sphere	0-300?	65-67
d. Multiple Anvil, Sliding Systems (MASS)	0-300?	69-73

* Denotes max pressure for repeated use in commercial applications.

pressor to other parts of the system. This considerable restriction implies that the material to be processed must be directly compressed, that it must be heated within the confining volume, and that any pressure-or temperature-measuring devices must also be included there.

Although cylindrical vessels can be used up to 60kb or even 80kb (Table 1), parts that are highly stressed must be supported with stresses that increase the compressive components and decrease the deviatoric components. A simple way of achieving this goal was developed by Bridgman - the opposed-anvil apparatus, which utilized his "principle of massive support" [1]. Many newer designs use the "principle of active support" in which controlled compressive stresses are used to support the most highly stressed members of the apparatus. As a rough criterion the deviatoric stress, or maximum stress difference between principal components, should not exceed the shear strength of the material.

For most applications, the medium to be compressed is solid. However, a 4:1 mixture by volume of methanol: ethanol has been found to remain fluid-like up to pressures of over 100 kb [2] at ambient temperature. This is a significant improvement over the 1:1 mixture of pentane-isopentane used for many years, which remains fluid up to about 65kb [3].

An important factor in ultra-high pressure designs is the choice of materials with the correct, or best, mechanical properties for gaskets, which must deform in such a way that the sample material to be compressed remains in a quasi-hydrostatic environment. Also in some systems, the gasket must flow in such a way as to maximize the lateral support of anvils. The amount of flow is critical: if not sufficient, then local stress concentrations lead to anvil failure; if too great, then the efficiency of the system is reduced. In some cases the gasket material must also withstand high temperature with low heat loss, or be electrically insulating.

Another problem is the measurement of pressure. Usually the pressure is calibrated in a commercial apparatus with the use of fixed points and a pressure versus applied-load curve is developed. The pressure obtained with such a chart can at best be only approximate and depends upon the exact maintenance of anvil and gasket geometry, and also the gasket material. As an example of the difficulties inherent in pressure measurement, the pressure scale above 400 kb has recently been revised downwards by almost a factor of two [4, 5]!

Another major consequence of the high stress level is that the volume of material subjected to high pressure is relatively small. Thus, only materials of great inherent value can be produced or processed economically (e.g. diamond, BN, Chapter 8, Vol. II). It would therefore be inappropriate in the present volume to dwell at length on this topic. The present chapter will simply serve as a guide to further study. Attention will be focused on designs built commercially, while a brief discussion will be given of research apparatus for the ultra-high pressure region. A bibliography of general works in this field is given at the end of the chapter.

It would be inappropriate to write an introduction to this field without stressing the great contribution made by P. W. Bridgman. He introduced most of the fundamental concepts, built and tested many of the basic types of apparatus. He was the first worker in the pressure region above 20kb, and the first to reach 100 kb. Most workers who have contributed to this field have openly expressed their debt to his pioneering work. Even today a valuable beginning is afforded by study of Bridgman's book and collected works (Ref. 1, 2 in the bibliography).

II.   PISTON-CYLINDER DEVICES

Piston and cylinder devices can be operated up to about 80kb (Table 1). The two main considerations are the design of the cylinder and piston. Pistons are generally constructed of cemented tungsten carbide which fracture at ~50kb if unsupported. Lateral support can be provided in a number of ways. Cylinders are generally of multi-ring design, and for the highest pressures, employ variable radial and axial support on inner rings.

A commercially available apparatus for the compression of fluids up to 50kb is shown in Figure 1 [6]. It is based on an original design of Bridgman [7] for use up to 30kb which was later modified by Birch [8]. In one version a 3/4" (~19.0mm) piston is used (30kb) while a higher pressure (50kb) version uses a 1/2" piston (~12.7mm). With the piston at the top of its stroke, fluid is introduced past the lower piston (Fig. 1b) and pre-compressed to ~2kb if a gas. Thereafter pressure is increased by advancing the piston past the entry port. Variable support for the inner vessel with conical outer plan is obtained by thrusting it against the outer vessel which bears against the platen. Lubrication between the outer and inner vessels is afforded by a thin sheet of lead, lubricated with a glycerine/graphite powder paste.

Several Bridgman electrodes (see Chapter 5) can be brought in through the lower end-plug. In the author's experience such an apparatus can be successfully used with even leak-prone gases such as helium after minor modifications have been made. The apparatus maintains a strictly hydrostatic environment and can be used for many studies. It is relatively easy to incorporate an internal heater, for instance, for temperature studies up to at least 1200°C. Pressure measurement is readily made with a manganin gauge. A discussion of the wide capabilities of this design is given in Refs. 6 and 8.

Several designs capable of compressing solids (or liquids) up to ~60-80kb have been published. Two basic designs of Boyd and England [9, 10] for example are capable of reaching ~50kb with an unsupported piston and ~80kb in a two-stage configuration. In this latter design, the advance of the piston compresses a soft solid such as KBr (potassium bromide) so that it supports the piston. The design of Kennedy and LaMori [11] is similar to the single stage design of Boyd and England, but allows the piston to be rotated for more precise determination of pressure from the applied load (see section IV). Several Russian designs are described in Tsiklis' book (Bibliography, ref. 6).

Piston-cylinder apparatus for use to pressures above 50kb have been commercially developed, but their use is so restricted that none are available at the present time, to this author's knowledge. For commercial applications in this pressure range either the belt-apparatus or multi-anvil systems are invariably used.

(a)

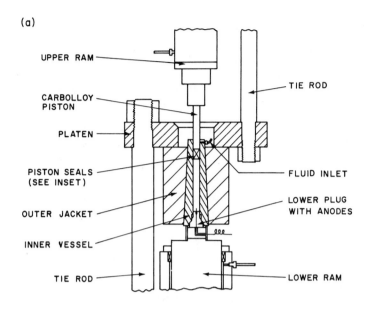

UPPER RAM

TIE ROD

CARBOLLOY
PISTON

PLATEN

PISTON SEALS
(SEE INSET)

FLUID INLET

OUTER JACKET

LOWER PLUG
WITH ANODES

INNER VESSEL

TIE ROD

LOWER RAM

(b)

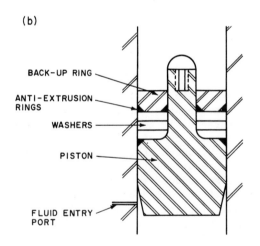

BACK-UP RING

ANTI-EXTRUSION
RINGS

WASHERS

PISTON

FLUID ENTRY
PORT

FIG. 1. (a) A sketch of the Harwood 30/50 kb apparatus [6] showing the tapered-support principle; (b) Insert showing Bridgman piston seal. Fluid enters through the side-port, is precompressed if gaseous, then the piston is driven down, sealing fluid below it. (Figures courtesy of Harwood Engineering Co., Walpole, Mass.)

In principle, by constructing one piston-cylinder device inside an outer stage, pressures in excess of 100kb can be reached. This design was first realized by Bridgman [12] who reached 100kb in this way. An alternative design is given by Tsiklis (ref. 6 in bibliography), while the Harwood piston-cylinder apparatus described earlier can be adapted for this purpose [6]. However, their operation is extremely complicated, and even the failure of a small component in an inner stage can cause expensive replacements.

The concept of "multistaging" is of general application to ultra-high-pressure generation. The concept can be used to describe widely differing applications. The apparatus shown in Figure 1, for example, could be described as two-stage in concept (driving piston plus lower piston providing radial and axial support to the cylinder). A three-stage system would result if another piston-cylinder stage were incorporated. Alternatively, if the piston were supported, as in Boyd and England's later design [10], the apparatus could also be described as three-stage.

III.   OPPOSED ANVIL DEVICES

In order to achieve higher pressures, the idea of a single-stage cylindrical pressure vessel and piston must be abandoned. Several classes of design can be envisaged. In the simplest case, Bridgman [1, 7] suggested the use of two opposed-anvils. This design has led to the "Belt-apparatus" of Hall [13] which in some ways incorporates ideas of the piston-cylinder and opposed-anvil devices. Further developments employ several anvils -- usually four (tetrahedral) or six (cubic). Commercial applications in this pressure range all utilize variants of these basic types.

A.   Bridgman Anvils [1]

The basic design of the anvils is illustrated in Figure 2. The truncated anvils of cemented tungsten carbide are supported by steel binding rings with an interference-fit to apply inward-acting radial stresses. The sample is in the form of a thin disk surrounded by a gasket, normally of pipestone (grade-A lava, or wonderstone) (see ref. 14 for discussion of this material). When the anvils compress the sample, radial extrusion may be reduced by coating the anvils with a diamond paste, or by optimizing the proportions of sample to gasket, having regard to the compressibilities of sample and gasket.

This design uses Bridgman's principle of "massive support". The principle uses the fact that the yield stress of a flat, semi-infinite plate subjected to an indenter is several times higher than the compressive yield-stress of the plate material. This factor is reduced for a truncated cone (Figure 3), falling to unity at $\Theta=0$, where Θ is the cone semi-angle. The factor has been theoretically calculated [15] and experimentally determined for copper [15], steel [16] and carbide [17, 18]. Clearly a large angle is favored. In the normal, opposed-angle device the half-angle is taken as ~80° (for a review, see ref. 14).

With a large cone-angle, some improvement also results if the anvil is supported by a steel ring [1, 20]. Using a strengthening factor of ~3 (Figure 3) for a supported-carbide anvil of half-angle 80°, and a compressive strength of ~56kb, the ultimate pressure capability should be ~170kb. Some improvement of the design results if

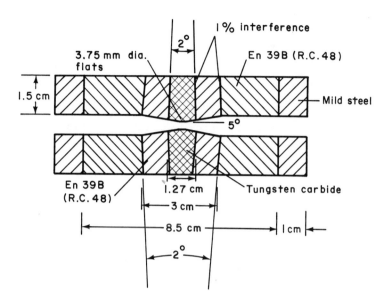

FIG. 2.   Schematic of Bridgman-anvils for high pressure use.
(Figure adapted from Bradley, ref. 7 in bibliography.)

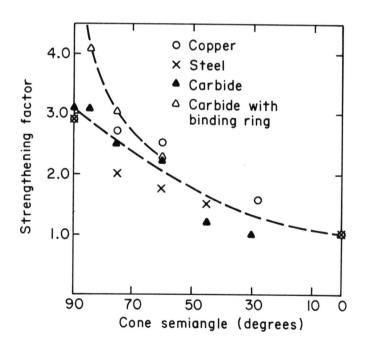

FIG. 3.   The strengthening factor in anvils from the massive
support principle.   (Figure adapted from Lees, ref 14.)

the cemented carbide anvil is slightly tapered on its outside so that
the applied ram force increases the lateral support (Figure 2).
Bridgman anvils of cemented tungsten carbide can be readily used up to
about 100kb and with care up to 180kb [19]. (See also later remarks
about the diamond-anvil high pressure cell in this section, and general comments about the ultimate pressure capability in opposed-anvil
devices in Section V.)

The main disadvantage of this design is the thinness of the sample. Apart from its small volume, an associated difficulty arises
when heating the sample, which cannot be effected without seriously
weakening the anvils. However, the simplicity of the design makes it
attractive for a number of exploratory studies.

Balchan and Drickamer [21] developed an improved design using
"supported" Bridgman anvils (Figure 4). Drickamer subsequently proposed versions suitable for measurements of electrical resistance [21,
22], optical absorption [23, 24], x-ray diffraction [25], nuclear magnetic resonance [26] and Mössbauer resonance [27]. A review of earlier work is given in ref. 28. Support of the anvils is provided by
the pyrophyllite pellet, which is compressed as the anvils advance.
Although in early papers pressures as high as 600 kb were claimed,
Drickamer revised his fixed-point-scale in 1971 [29] (see section IV-
C). A more recent revision of the pressure scale places his upper
pressure limit at ~250 kb [4, 5] (see sections IV and V).

Bundy [5] has recently extended the upper pressure limit of the
Drickamer design to over 400 kb using carboloy pistons tipped with
dense, sintered diamond (Figure 4) (see also remarks in section IV and
V). An alternative design using sintered tungsted carbide anvils has
been developed by Kendall, Dembowski and Davidson [30] in which variable lateral support is provided to the anvils. Pressures up to 300
kb have been reached.

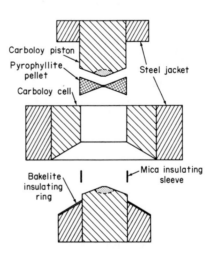

FIG. 4. Schematic of supported-anvil device of Balchan and
Drickamer [21]. The dashed line represents the limit of sintered
diamond in Bundy's apparatus [5] based on this design.

The use of single-crystals of diamond in an opposed-anvil, Bridgman geometry was developed independently by Jamieson and Lawson [31,
32] for x-ray diffraction experiments and by Weir, Lippincott, Van
Valkenburg and Bunting [33-35] for optical experiments. However, it
is the geometry pioneered by the latter group at the National Bureau
of Standards that has been adopted by succeeding workers. They used
incident and exit optical, or x-ray beams parallel to the axis of the
anvils, rather than perpendicular to it, as in the design of Jamieson
and Lawson. A diagram of a recent cell which has been used to 500 kb
[4] is shown in Figure 5.

The basic operation is very simple. A gasket of thin metal (e.g.
0.01" (0.25mm) hardened Be/Cu alloy, Duranickel or Inconel 750X) contains the sample in a drilled hole (~0.01" Dia (0.25mm)). Sample and
gasket are then compressed between the opposed diamond anvils. Essential to the ultra-high pressure operation of the system is the alignment of the anvils, both initially and during compression. The diamonds are mounted on rockers (either cylindrical [36] or hemispherical
[4]) and initial angular alignment can be obtained easily using optical fringes. A gasket is not absolutely essential to the operation of
the instrument [36], but pressure gradients are severe across the sample width [2, 37] if a gasket is not used.

A wide range of studies can be carried out easily because the
diamonds are transparent to photons of a wide range of energies.
Cells suitable for optical [33, 35, 38, 39] (including Raman Scattering [40]), x-ray diffraction [41, 42] (single and polycrystal), superconducting studies [43] have been described in the literature. Using
continuous-wave high energy lasers, the sample can be heated to very
high temperatures [44]. The use of the pressure scale based on the
ruby R-line fluorescence shift [45, 2, 38, 46] makes pressure measurement very simple and rapid. Pressure is also very conveniently measured using the NaCl scale (see section IV).[†]

Although the small sample size in the diamond cell limit its application, its versatility, economy and convenience make it the most
exciting research tool for ultrahigh pressure studies. Its ultimate
pressure capability is not known, but pressures of 1 Mb based on the
extrapolated Ruby R-line fluorescence scale (see section IV-B) have
recently been reported [47]. The next few years will see its use
greatly increased in a diversity of solid state and liquid studies,
and it may revolutionize exploratory studies relating to synthesis of
materials at high temperature and pressure.

[†] Note added in proof: A design has been published for measuring
strain and strength of materials at pressures up to 300kb. (G. L.
Kinsland and W. A. Bassett, Rev. Sci. Instr. 47, 130 (1976)). Another
report describes the use of the diamond cell for measuring elastic
constants, (C. H. Whitfield, E. M. Brody and W. A. Bassett, Rev. Sci.
Instr. 47, 8 (1976)), while an optical spectroscopic system is described by B. Welber (Rev. Sci. Instr. 47, 183 (1976)). X-ray studies
at low temperature (4-300K) have been described by E. F. Skelton, I.
L. Spain, S. C. Yu, C. Y. Liu, E. R. Carpenter (Rev. Sci. Instr. 48, 7
(1977).

(a)

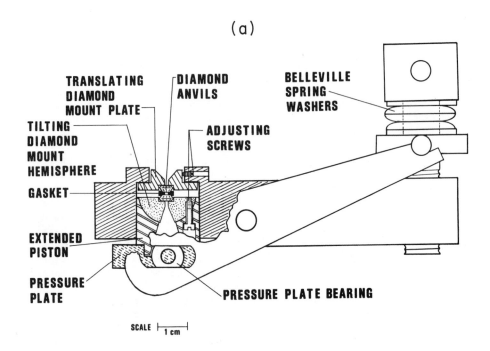

TRANSLATING DIAMOND MOUNT PLATE

DIAMOND ANVILS

BELLEVILLE SPRING WASHERS

TILTING DIAMOND MOUNT HEMISPHERE

GASKET

ADJUSTING SCREWS

EXTENDED PISTON

PRESSURE PLATE

PRESSURE PLATE BEARING

SCALE ⊢──┤
1 cm

(b)

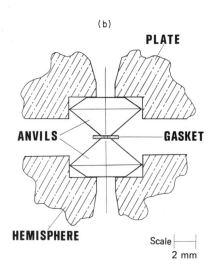

PLATE

ANVILS

GASKET

HEMISPHERE

Scale ⊢──┤
2 mm

FIG. 5.   (a)   Cross-section of a diamond cell.   Force is
applied by tightening the screw onto the spring washers.
(b)   Magnified cross-section of the diamonds, seats,
gasket and sample.   (Figure from Piermarini and Block, ref. 4
kindly supplied by G. J. Piermarini.)

B.   The Belt and Girdle Designs

The belt apparatus [13] is a logical extension of the opposed-anvil concept, giving greater volume and high-temperature capability. Its development has been outlined by Bundy [48]. Bundy describes the apparatus in Chapter 8, vol. II (see his Figs. 2 & 3), particularly its use at high temperature for the production of diamond. A photograph of a production unit is shown in Figure 6 of this chapter. Apparently most synthetic diamond and cubic boron nitride is produced in presses of this type. A high compression version of the apparatus which can be used to ~160 kb (new scale, see section IV-C) has been devised by Bundy [49].

FIG. 6. A photograph of a belt apparatus used for the production of diamond. Dr. F. P. Bundy is shown in the photograph, giving an idea of the large scale of the equipment. (Photograph kindly supplied by H. P. Bovenkerk, General Electric Company, Worthington, Ohio.)

Very similar in design to the "belt" apparatus is the "girdle" configuration of Wilson [50]. Alternative designs have been proposed by Daniels and Jones [51] (girdle & compression gaskets), Young et al [52] (girdle & sandwich gaskets), Lorent [53] (ultrahigh pressure girdle apparatus) and Fukunaga [54] (large-volume girdle apparatus).

Apparatus using heating arrangements similar to those described by Bundy in Chapter 8, vol. II can be used to temperatures ≥2,500°C at ~80 kb, even though pyrophyllite melts at ~1500°C at atmospheric pressure [14]. For many commercial uses, an apparatus of this type is to be preferred because of its relative simplicity and large volume.

C. Tetrahedral, Cubic and Other Multiple-Anvil Presses

The basic concept of Hall's tetrahedral apparatus [55] is indicated in Figures 7 and 8. Many versions of this apparatus have been constructed, and commercially-built equipment is avilable. A thorough review of its design and operation has been given by Lees [14].

A central, tetrahedrally-shaped sample is compressed simultaneously by four anvils driven independently by four external rams. The anvils are in the form of frustra of triangular pyramids, changing to cylindrical shapes on their base (Figure 7). Initially the sample tetrahedron is oversized so that, on compression, material flows into the gaps between the anvils, forming gaskets.

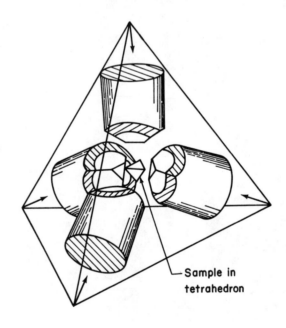

Sample in
tetrahedron

FIG. 7. Figure showing the geometry of a tetrahedral press (guide pins not shown).

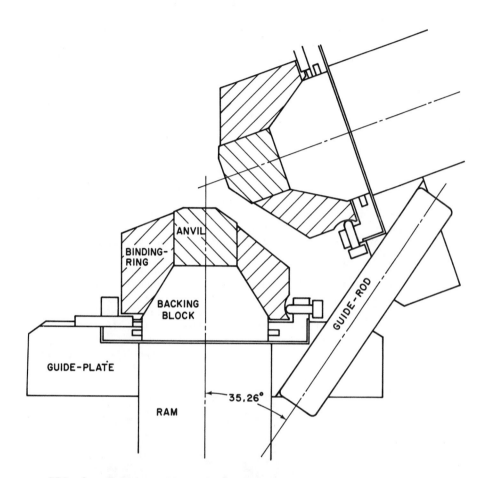

FIG. 8.   Cross-section of a tetrahedral unit, showing guide
pins (figure kindly supplied by Dr. H. Tracy Hall).

The successful operation of this and other devices such as the
belt apparatus depends on the correct choice for the gasket material.
It is usually chosen to be pyrophyllite although a magnesia-resin mixture is claimed to be superior (for a discussion of gaskets, see ref.
14). It must be capable of deforming when an increased load is applied, and yet support the large stress difference across the gasket.
    Another critical factor in the operation of the unit is the init-
ial alignment and subsequent motion of the anvils for symmetric posi-
tioning. In later models, Hall [56] devised a set of guide-pins so
that all four external rams could be operated at the same pressure,
while the guide pins maintained relative alignment of the anvils within tolerable limits (Figure 8).
    Another design of the tetrahedral press has been constructed by
workers at the National Bureau of Standards [57]. In this design, the
three lower anvils are nested in a steel cone. The fourth anvil is

then driven down with a single ram. In order that the anvils maintain alignment the lower three anvils must slide on the conical surface towards each other. Thus, the cone angle (ideally 19°28' in a frictionless device) must be chosen to allow for frictional forces between the anvils and the cone. Frictional forces are reduced by using teflon (PTFE) sheets for the sliding surfaces.

Operation of this type of tetrahedral press was simpler than the earlier versions developed by Hall [55], but not significantly so compared with his devices utilizing guide pins. A major factor influencing one's choice of design must be the availability or otherwise of a press which could be used with the NBS design. Another factor is the specific operation. For instance, if x-ray diffraction experiments are to be performed during mineral synthesis [58], then the open design of Hall's apparatus may make this experiment easier.

Similar in concept to the tetrahedral apparatus is the cubic press [59, 60]. Again there are many different designs, some using six independent rams [60], while others use only a single, vertical ram [59, 61]. Another design uses linkage-type anvil supports and a single ram [62, 63]. A photograph of a cubic press used by H. Tracy Hall for development of super-hard materials is shown in Figure 9.

According to the data shown in Figure 3, the tetrahedral press (maximum cone semi-angle 71°, minimum 55°) should be capable of use to higher pressure with tungsten carbide anvils (mean strengthening factor ~2.1, max pressure ~120 kb) than the cubic design (maximum semiangle 55°, minimum 45°, mean strengthening factor 1.8, maximum pressure ~105 kb). In practice, both types are limited to normal-life,

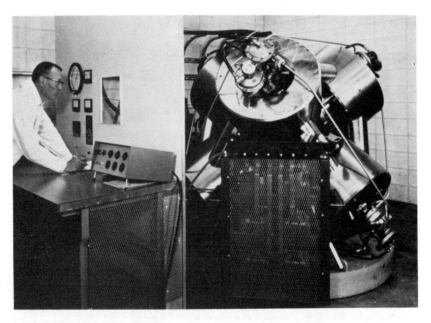

FIG. 9. Photograph of a cubic press used for development of ultrahard materials. Dr. H. Tracy Hall is shown at the controls. (Photo kindly supplied by H. Tracy Hall, Megadiamond Corporation, Provo, Utah.)

repeat-loading pressures of ~60 kb. It is interesting to note that in
these systems, anvil breakage invariably occurs when the rams are being retracted.

Lees [14] has published a particularly interesting review article
on the mechanical factors of importance in the performance of tetrahedral presses. His comments are of relevance also to cubic presses.
Amongst other factors he discusses the optimum size of the tetrahedron
to give the best gasket characteristics; calibration procedures; analysis of pressure and strength distribution within the tetrahedral sample and gaskets; effects of load cycling etc.   Some further remarks
pertinent to these types of apparatus will be made in sections V and
VI. Another useful review is given by Bradley (ref. 7 in bibliography).

In all apparatus of the type described so far, a key element is
the design of the rams (for a review of ram design, see Hall [64]).
Rams with 1000 metric tonnes force or more (1 metric tonne = 1000 kg
weight = ~2163 lb ~9.8 kN) are large and costly. A press capable of
producing a load of ~50,000 tonnes has been constructed in the U.S.S.
R., according to recent reports, and will be used for ultra-high pressure research.

A form of the cubic press devised by von Platen [65] circumvented
the need for rams. In this apparatus six spherical segments, separated by air gaps, were pushed inwards onto a centrally located sample
by surrounding the resultant sphere with a deformable membrane, and
immersing it in fluid which could be compressed to high pressure
(Figure 10 (a)). Diamond was possibly synthesized for the first time
in von Platen's press, although this type of equipment is no longer
used commercially.

The split-sphere has been adapted by Kawai [66, 67] for ultra-
high pressure measurements. Later designs [67] used two stages (see

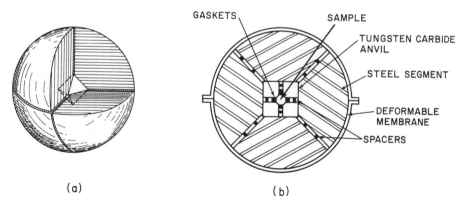

FIG. 10. (a) Sketch showing the principle of the split-
sphere apparatus. In operation the sphere is surrounded by a
leak-tight deformable membrane . When immersed in a hydrostatic
fluid the segments are forced inwards, compressing the sample.
(b) Two-stage version of the split-sphere apparatus.
Outer segments of hard steel, inner segments of tungsten carbide.
(Figure 10 (a) adapted from ref. 66; figure 10 (b) from ref. 67.)

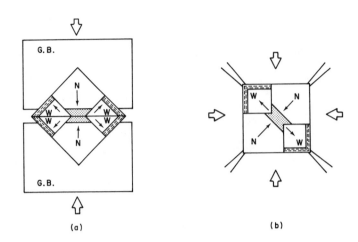

FIG. 11.   Two representative designs of multiple anvil
sliding systems (MASS).   Anvils (N) are driven forward by the
rams, while wedge anvils (W) are driven back onto deformable
pads.   Ideally the sample can be reduced to zero volume.
(Figure adapted from ref. 71.)

Figure 10 (b)).   Outer segments (6) are of hardened steel while inner
segments (8) are of tungsten carbide or sintered alumina.   Although
early estimates exceeded 1 Mb for the ultimate pressure capability in
Kawai's presses, the recent revision of the pressure scale (see next
section) throws these estimates in doubt.   Reliable fixed points in
the range 300 - 1000 kb are badly needed. A variant of the split-
sphere apparatus was proposed by Newhall [6] and various types are
under development in the U.S.A. at the present time.

Of greater relevance to technological applications are Multiple
Anvil Sliding Systems (MASS) [69-73]. Two representative systems are
sketched in Figure 11.   The fundamental idea is to allow the anvils
to support each other so that the stress deviatoric is kept below the
strength of the anvil (see remarks in section V) while allowing large
volume reduction to generate high pressure. At the same time the
compressed volume can be relatively large.

At the present time, test apparatus is in operation in France
[69] and Japan [73].   In both instances pressures of ∼180 kb have
been reached.   Some practical aspects are discussed in ref. 71 and
results of tests in ref. 73.

IV.   PRESSURE STANDARDS AND CALIBRATION

The measurement of pressure in an ultra-high pressure system is
particularly difficult. For commercial applications, pressure is
usually estimated from the end loads; but, as will be discussed below, this is not a reliable method. Initially, the apparatus must be
calibrated against standards and this aspect will be discussed first.

A. Primary Standards

A true primary standard of the pressure balance design is only available up to ~26 kb [74]. The design developed jointly by the National Bureau of Standards and Harwood Engineering Company is illustrated in Figure 12. It employs the controlled-clearance principle (see Chapter 8) and at its maximum pressure the stated accuracy is $^{+}_{-}25$ bars (~0.1%). Further improvements should be possible when the clearance is controlled more precisely.

The piston-cylinder apparatus with solid compressing-medium is considered by some to be a primary instrument [75]. However, there must be serious questions as to the applicability of this title because of the non-hydrostatic nature of the stresses within even soft solids compressed in it. Even when the piston is rotated, the hyster-

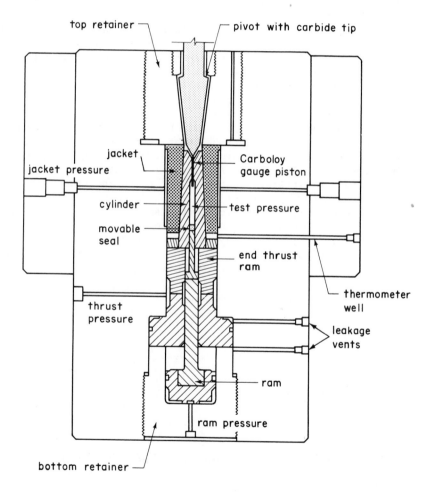

FIG. 12. Schematic of the NBS pressure balance for use to 30 kb, designed and constructed by Harwood Engineering Company. (Figure adapted from ref. 74.)

esis between up- and down-stroke load-pressure curves is as high as 3
kb. (See for example Fig. 7-2, ref. 76.) At its upper pressure limit
(~80 kb) uncertainties are probably as high as $^+_-3$ kb and at 50 kb of
the order of $^+_-0.5$ kb.

B.   Secondary Pressure Standards

In fluid media the manganin pressure gauge can be successfully
used up to at least 60 kb [77] as a secondary pressure standard.
However, when embedded in even a relatively soft solid medium such as
pyrophyllite, spurious effects preclude its use [78, 79].

X-ray diffraction studies can be conveniently made in several
ultra-high pressure cells (for a review, see ref. 80) thereby afford-
ing a useful way of measuring pressure through its effect on the lattice spacing of simple cubic solids. Decker [81-83] has theoretically estimated the variation of the lattice spacing of NaCl and CsCl
with pressure up to ~400 kb. Up to its transition pressure to the
CsCl structure at ~290 kb [84], the NaCl scale appears to be reliable,
although there are small discrepancies between the NaCl and CsCl
scales up to 30 kb [85] and more serious discrepancies up to 300 kb
[86]. The influence of non-hydrostatic stress in the NaCl on the accuracy of the scale has been considered [87, 88].

It is possible that a similar scale based on the lattice compression of metals such as Cu, Al may prove of use in the future since
equations of state have been calculated to pressures of the order of
1 Mb [89, 90]. Also shock-studies indicate no evidence of phase
transitions to similar pressures [91]. The interested reader may find
several papers of interest on the use of these and other materials in
ref. 8 of the general bibliography.

A new secondary scale based on the shift of the ruby R-fluores-
cence lines with pressure has been developed at the National Bureau of
Standards [45, 46] (Figures 13 and 14). The method is particularly

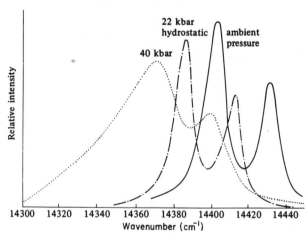

FIG. 13. A graph showing the Ruby R-line fluorescence
doubled and its shift with pressure. At 40 kb the peaks are
broadened due to non-hydrostatic stresses within the sample
volume. (Figure adapted from Forman, Piermarini, Barnett and
Block, ref. 45.)

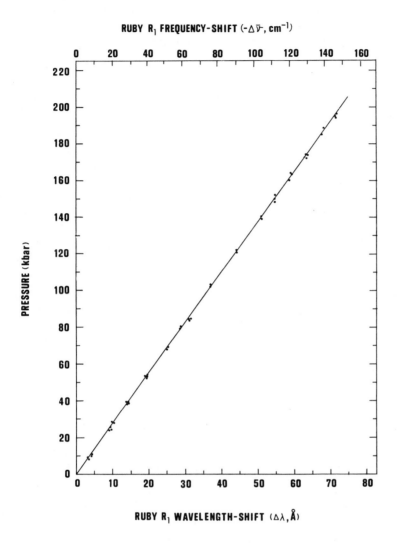

RUBY R₁ FREQUENCY-SHIFT $(-\Delta\bar{\nu}, cm^{-1})$

PRESSURE (kbar)

RUBY R₁ WAVELENGTH-SHIFT $(\Delta\lambda, \AA)$

FIG. 14. The pressure dependence at 25°C of the ruby R₁
fluorescence line (6942 Å at P = 1 bar) as a function of wavelength (Å) and wavenumber (cm⁻¹). Pressure values are based on
Decker's equation of state for NaCl [81-83]. (Figure from
Piermarini, Block, Barnett & Forman (ref. 46), kindly supplied
by G. J. Piermarini.)

suited for use with the diamond-anvil cell. Even with its small sample volume, relatively small crystals of ruby can be incorporated in
the cell allowing the pressure-displacement profile to be obtained.
This scale can give reproducibility of ± 1/2 kb up to 100 kb -- and is
thus more precise than the accuracy of present standards. It is capable of use up to at least 600 kb and possibly to much higher pres-

sures [4, 47]. Although of direct use only as a research tool, this scale is revolutionizing our knowledge of fixed point pressures, which are of direct use in commercial equipment.

C.   Fixed Points

A number of fixed points are used in the pressure region above 20 kb. Examples are given in Table 2. Usually the transition is detected from resistance measurements, so that transitions of the insulator-to-metal type are most useful. Resistance changes between metallic phases (e.g. Ba, Fe-Co alloys, Pb, etc.) are difficult to observe because of the smallness of the resistivity (normally $\lesssim 50$ $\mu\Omega$m). Transitions in the semi-metal Bi are somewhat easier to detect. Care must be taken to reduce to a minimum the resistance in the electrical path to the specimen. A four-point resistance technique is favored [92]. Even changes in phase of non-conducting phases (e.g. KCl (NaCl phase)-KCl(CsCl phase) can be detected electrically by using the change in resistance of a metal wire embedded in it [93]. In all cases in Table 2, values of high pressure transitions are quoted on the up-stroke.

The pressure assigned to fixed points above about 50 kb has changed several times in the last few years, so that comparison values are given in Table 2. In the pressure range between about 30-100 kbar, pressure standards were largely based on Bridgman's [3] fixed

TABLE 2

Fixed Point Pressures For Use Above 20 Bars

Substance	1975	1971-1975	Up to 1971
Bi (I-II)	25		
Ba (I-II)	53		
Bi (IV-VI)*	74	$77^{\pm}3$	88
Fe ($\alpha-\varepsilon$)	112	110-115	133
Ba (II-III)	120	118-122	144
Pb (I-II)	130	128-132	160
Fe$_{.8}$Co$_{.2}$ ($\alpha-\varepsilon$)	190		
CdS (max)	190-200 $220^{\pm}1$	330-340	460
GaP (ins-met)	230-240	>400	
Fe$_{.6}$Co$_{.4}$ ($\alpha-\varepsilon$)	285-295		
NaCl	290-300		
EuO	~400		
ZnS	$150^{\pm}5$	185	240-245
ZnSe	$137^{\pm}3$	134	165
Si	$125^{\pm}5$	150	195

points observed by detecting discontinuities in electrical resistance. Unfortunately the pressures observed by him as volume discontinuities in the same substance were approximately 30% lower than those in the electrical resistance. Kennedy and La Mori [11] suggested that the transitions were of the same origin and that pressures obtained from the volume discontinuities were nearly correct. These suggestions have been substantiated by further work. A detailed discussion has been given by Hall [94].

Above about 100 kb, most of the fixed points have been suggested by Drickamer and co-workers. Following the revision of fixed-point pressures in the 30-100 kb range, Drickamer [29] reassigned pressures of his higher fixed points. However, these pressures have again been revised substantially downwards in the last two years [4, 5, 30]. Accordingly, comparison values are given for some of the transitions in Table 2.

The transition pressure of alloys based on some of these fixed points can be useful for "tailoring" a substance with transition pressure between the two constituents. Examples are; KBr-KCl [95] (17 - 19 kbar); RbCl-KCl [95] (5.3 - 19 kbar); KCl-NaCl [96] (19 - 300 kbar) [84]; InSb-GaSb [97] (23 - 62 kbar) (see also Fe-Co alloys [5] in Table 2.) It is possible that other group III-V semiconducting alloys may be of interest, in particular for the higher pressure range (P>100 kbar). Theoretically calculated transition pressures for these substances have been given by Van Vechten [98] and data for Ge, Si, GaP are included in Table 2.

For high temperature systems, melting lines may be of interest for calibrating pressure, e.g. the noble metals, Au, Ag, Cu [99], Alkali halides [100] or elemental semiconductors [101, 102].

More detailed information on fixed points can be obtained from refs. 103, 104 or from several papers in ref. 8 of the general bibliography.

D.   Calibration of Ultrahigh Pressure Apparatus

For most commercial applications, the pressure is estimated from the anvil load using a pressure-load graph. A detailed account for a tetrahedral press in given by Lees [14], which should be qualitatively useful for other devices. Some of the main factors affecting the calibration (and thus changing the calibration) are:

1. Oversize of the sample volume compared to the anvil. This controls the initial gasket width.
2. The materials used for the sample container and gaskets. In some cases, gaskets may be pre-formed of a different material than the sample container.
3. The anvil material also affects the calibration. This effect is related to the difference between elastic properties.
4. Friction effects can include (a) differences in surface treatment of anvils to control anvil-gasket-friction (e.g. jeweller's rouge or graphite coating); (b) differences in anvil/nest friction; (c) piston-cylinder-sample friction in piston-cylinder devices.

5. Sample geometry or material changes can produce appreciable differences in calibration, presumably due to changes in the flow-stress pattern in the sample container.

6. Loading history - calibration on the up- and down-stroke will differ, and also the calibration after successive up- and down-excursions.

7. Change of temperature of the sample container will change the calibration.

From the above it is clear that the calibration should be regarded as only approximate and relevant to only a particular geometry, sample, temperature etc. Great care should be taken to reproduce exactly the conditions of the calibration in the commercial process.

V.   ULTIMATE CAPABILITY OF ULTRAHIGH PRESSURE SYSTEMS

Some initial consideration of the strength of anvils has been given in section III. However, these considerations did not take into account the influence that gaskets can exert by reducing stress gradients across the anvil. Also, in assessing the ultimate pressure capability the possible strengthening effect of pressure on the anvil material should be considered as well as the possibility that plastic deformation can occur before fracture.

Normally, it would be advantageous to operate a system within elastic limits. However, initial runs exceeding the elastic limit can lead to a strengthening of the material. This effect has been used, of course, for over a century in the autofrettage of cylinders. Drickamer [28] has also discussed the strengthening effect on Bridgman anvils of bonded tungsten carbide.

Bundy [105] has recently considered the ultimate pressure capability of Drickamer (supported Bridgman) anvils within the elastic limit. As shown in Figure 15, he considers a geometry in which the flat anvil face has radius x_o and the gasket supports the anvil to a radius x_1. The stresses in the gasket, which support the anvil, vary radially.

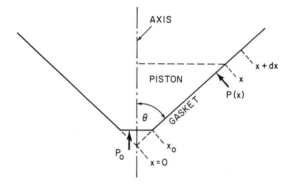

FIG. 15. A sketch showing the anvil geometry considered by Bundy [105]. (Figure redrawn from copy kindly supplied by F. P. Bundy [105].

The optimum stress distribution can be obtained readily. The condition at each point on the anvil face is:

$$\frac{dP(x)}{dx} < -\frac{2S}{x} \tag{1}$$

S is identified with the compressive strength measured on a test sample with diameter approximately equal to the height. If the pressure across the flat anvils is P_o then the gasket pressure can ideally be set at $P_o - S$ at $x = x_o$. Using this boundary condition and integrating (1) it is found that the maximum pressure obtainable is:

$$P_{o,max} = S[1 + 2\ln x_1/x_o] \tag{2}$$

The "efficiency" of the device can be found readily:

$$\text{Efficiency} = \frac{\text{Load Applied to Flat Anvils}}{\text{Total Load Applied to Anvil}} \tag{3a}$$

$$\equiv \frac{L_f}{L_t} = \frac{\pi P_o x_o^2}{\pi P_o x_o^2 + \pi \int_{x_o}^{x_1} P(x)x\,dx} \tag{3b}$$

A corresponding graph can be used to estimate the maximum pressure obtainable within these limits as a function of the inverse efficency L_t/L_f (Figure 16).

As noted by Bundy, the logarithmic function in (2) increases slowly with x_1/x_o. However, the maximum obtainable pressure depends directly on the compressive strength. For this reason he strengthened the anvil tips in his apparatus (Figure 4) with sintered diamond (S(sintered diamond) ~100 kb: S(sintered tungsten carbide) ~30 kb).

In Figure 16, representative data for several types of apparatus are given, showing efficiencies and maximum pressures attained. For the Drickamer apparatus, for example, x_1/x_o ~10, and if pressures of ~250 kb are reached, then S ~50 kb, indicating that considerable strengthening of the anvil has occurred due to "autofrettage".

Kumazawa [72] and Ruoff [106] have both stressed the importance of strengthening that occurs in anvil materials as the pressure increases. By including this effect considerably higher pressures are theoretically made possible. Also, these authors consider the system in the plastic limit. Both authors assume a relationship between critical stresses and pressure for the material, then derive estimates for the maximum pressure attainable in containers with simple geometry, such as hollow spheres. The value of the critical stress at P = 1 bar (τ_o) and its pressure derivative may be related to the elastic constants of the material [106].

Ruoff and Wanagel [107] have tested this idea using a Drickamer cell with anvils constructed of steel, for which τ_o is considerably lower than for tungsten carbide. They conclude that pressures higher

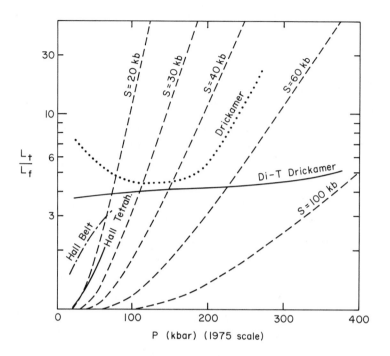

FIG. 16. A graph of the maximum pressure capability of anvils operating within elastic limits versus inverse efficiency [105]. Dashed curves are theoretically calculated from equations 2 and 3 for several compressive strengths. Data is also included for several operating designs including the tetrahedral [55], belt [13], Drickamer anvils [20] and Drickamer anvils tipped with sintered diamond [5]. (Figure redrawn from copy kindly supplied by F. B. Bundy.)

than four times the strength of the steel can be reached. It is concluded [72, 106] that pressures considerably higher than 1 Mb should be possible using anvils of tungsten carbide into the plastic region. Even higher pressures should be available if large pieces of high-quality, sintered or polycrystalline diamond are made available [108-110]. Considerable efforts appear to have been devoted to this end in the U.S.S.R. [108, 110] and an opposed-anvil apparatus producing (unsubstantiated) pressures above 1 Mb has been reported [111] (see also Bundy [5]).

The next few years may see many developments in the ultrahigh pressure range. The immediate objective will be to reach and exploit reasonably well-characterized pressures up to 1 Mbar [47]. One objective of possible technological importance is the metallization of hydrogen already mentioned in the introductory chapter.

BIBLIOGRAPHY OF GENERAL WORKS

1. "The Physics of High Pressure", P. W. Bridgman (G. Bell & Sons, 1952).
2. "Collected Experimental Papers", P. W. Bridgman (Harvard University Press, 1964).
3. "Progress in Very High Pressure Research", F. P. Bundy, W. R. Hibbard, Jr., H. M. Strong (eds.) (John Wiley, 1961).
4. "Modern Very High Pressure Techniques", R. H. Wentorf (ed.) (Butterworths, London, 1962).
5. "High Pressure Measurement", A. A. Giardini & E. C. Lloyd (eds.) (Proceedings of the Bolton Landing Conference, N.Y., June, 1960, published by Butterworths, Washington, 1963).
6. "Handbook of Techniques in High Pressure Research and Engineering", D. S. Tsiklis (translated from original Russian by A. Bobrowsky) (Plenum Press, New York, 1968).
7. "High Pressure Methods in Solid State Research", C. C. Bradley (Plenum Press, New York, 1969).
8. "The Accurate Characterization of the High Pressure Environment", NBS Special Monograph 326 (ed. E. C. Lloyd, 1971).

REFERENCES

1. P. W. Bridgman, J. Appl. Phys. $\underline{12}$, 461 (1941) (also Revs. Mod. Phys. $\underline{18}$, 1 (1946)).
2. G. J. Piermarini, S. Block and J. D. Barnett, J. Appl. Phys., $\underline{44}$, 5377 (1973).
3. P. W. Bridgman, "The Physics of High Pressure", (Bell & Sons, London, 1952).
4. G. J. Piermarini and S. Block, Rev. Sci. Instr., $\underline{46}$, 973 (1975).
5. F. P. Bundy, Rev. Sci. Instr., $\underline{46}$, 1318 (1975).
6. D. H. Newhall and L. H. Abbot, Proc. Instn. Mech. Engr. $\underline{182}$, 288 (1966-7).
7. P. W. Bridgman, Proc. Am. Acad. Sci., $\underline{72}$, 157 (1938) (see also Proc. Roy. Soc., London, $\underline{A203}$, 1 (1950)).
8. F. Birch, E. C. Robertson and S. P. Clark, Ind. Eng. Chem., $\underline{49}$, 1965 (1950).
9. F. R. Boyd and J. L. England, J. Geophys. Res., $\underline{65}$, 74 (1960).
10. F. R. Boyd, Chapter 8, p. 151 in "Modern Very High Pressure Techniques", ed. R. H. Wentorf (Butterworths, 1962).
11. G. C. Kennedy and P. N. LaMori, p. 304 in "Progress in Very High Pressure Research", (F. P. Bundy, W. R. Hibbard, H. M. Strong, eds.) (John Wiley, 1961).
12. P. W. Bridgman, Proc. Am. Acad. Arts. Sci., $\underline{74}$, 425 (1942).
13. H. T. Hall, Rev. Sci. Instr., $\underline{31}$, 125 (1960).
14. J. Lees, Chapter 1, "The Design and Performance of Ultrahigh Pressure Equipment. An Interim Report on the Tetrahedral Anvil Apparatus" in "Advances in High Pressure Research", Vol. $\underline{1}$ (R. S. Bradley, ed) (Academic Press, 1966).

15. D. Tabor, "The Hardness of Metals" (Oxford Univ. Press, 1951).
16. R. P. Levey and R. L. Huddleston, "Tooling Development for Very High Pressure Pressing", Report # Y-DA-470, Union Carbide Co., 1964.
17. G. Gerard, "Investigation of Massive Support Principle for Ultra-High Pressure Anvils", Air Force Final Report, Contract #AF19(504)-7438, June 1962.
18. S. Tsujii, M. Jinushi, Abstract of 11th Symposium on High Pressure, Japan, p. 15 (1969). (In Japanese)
19. F. Dachille and R. Roy, Chapter 9, "Opposed Anvil Pressure Devices" in "Modern Very High Pressure Techniques" (R. H. Wentorf, ed.) (Butterworths, 1962).
20. G. Jura, R. E. Harris, R. J. Vaisnys, H. Stromberg, p. 165, in "Progress in Very High Pressure Research" (F. P. Bundy, W. R. Hubbard, H. M. Strong, eds.) (Wiley, 1961).
21. A. S. Balchan and H. G. Drickamer, Rev. Sci. Instr., 32, 308 (1961).
22. K. F. Forsgren and H. G. Drickamer, Rev. Sci. Instr., 36, 1709 (1965).
23. R. A. Fitch, T. E. Slykhouse and H. G. Drickamer, J. Opt. Soc. Amer., 47 1015 (1957).
24. W. F. Sherman, J. Sci. Instr., 43, 462 (1966).
25. E. A. Perez-Albuerne, K. F. Forsgren and H. G. Drickamer, Rev. Sci. Instr., 35, 29 (1964).
26. V. Cleron, C. J. Coston and H. G. Drickamer, Rev. Sci. Instr., 37, 68 (1966).
27. D. N. Pipcorn, C. K. Edge, P. Debrunner, G. de Pasquali, H. G. Drickamer and H. Frauenfelder, Phys. Rev., 135, A1604 (1964).
28. H. G. Drickamer and A. S. Balchan, Chapter 2, "High Pressure Optical and Electrical Measurements" in "Modern Very High Pressure Techniques", (R. H. Wentorf, ed.) (Butterworths, 1962).
29. H. G. Drickamer, Rev. Sci. Instr. Notes, 42, 1667 (1971).
30. D. P. Kendall, P. V. Dembowski and T. E. Davidson, Rev. Sci. Instr., 46, 629 (1975).
31. J. C. Jamieson, A. W. Lawson, N. D. Nachtrieb, Rev. Sci. Instr. 30, 1016 (1959).
32. J. C. Jamieson and A. W. Lawson, Chapter 4, "Debye-Scherrer X-Ray Techniques for Very High Pressure Studies" in "Modern Very High Pressure Techniques", (R. H. Wentorf, ed.) (Butterworths, 1962).
33. C. E. Weir, E. R. Lippincott, A. Van Valkenburg and E. N. Bunting, J. Res. Nat. Bur. Stand., A63, 55 (1959).
34. E. R. Lippincott, C. E. Weir, A. Van Valkenburg and E. N. Bunting, Spectrochim Acta, 16, 59 (1960).
35. C. E. Weir, A. Van Valkenburg, E. R. Lippincott, Chapter 3, "Optical Studies at High Pressures Using Diamond Anvils" in "Modern Very High Pressure Techniques" (R. H. Wentorf, ed.) (Butterworths, 1962).
36. W. A. Bassett, T. Takahashi and P. W. Stook, Rev. Sci. Instr., 38, 37 (1967).

37. E. R. Lippincott and H. C. Duecker, Science 144, 1121 (1964).

38. J. D. Barnett, S. Block and G. J. Piermarini, Rev. Sci. Instr., 44, 1 (1973).

39. S. Block, G. J. Piermarini, High Temp-High Press, 5, 567 (1973).

40. B. A. Weinstein and G. J. Piermarini, Phys. Rev., B12, 1172 (1975).

41. C. E. Weir, G. J. Piermarini and S. Block, Rev. Sci. Instr., 40, 1133 (1969).

42. L. Merrill and W. A. Bassett, Rev. Sci. Instr., 45, 290 (1974).

43. A. W. Webb, D. U. Gubser, L. C. Towle, Rev. Sci. Instr., 47, 59 (1976).

44. L. Ming and W. A. Bassett, Rev. Sci. Instr., 45, 1115 (1974). See also P. M. Bell and H. K. Mao, Carnegie Inst., Wash., Yearbook 74, 399 (1975).

45. R. A. Forman, G. J. Piermarini, J. D. Barnett and S. Block, Science 176, 284 (1972).

46. G. J. Piermarini, S. Block, J. D. Barnett and R. A. Forman, J. Appl. Phys. 46, 2774 (1975).

47. H. K. Mao and P. M. Bell, Science, 191, 851 (1976).

48. F. P. Bundy, Chapter 1, "General Principles of High Pressure Apparatus Design" in "Modern Very High Pressure Techniques" R. H. Wentorf, ed.) (Butterworths, 1962).

49. F. P. Bundy, J. Chem. Phys., 38, 631 (1963).

50. W. B. Wilson, Rev. Sci. Instr., 31, 331 (1960).

51. W. B. Daniels and M. T. Jones, Rev. Sci. Instr., 32, 885 (1961).

52. A. P. Young, P. B. Robbins, C. M. Schwarz, p. 262 in "High Pressure Measurements" (A. A. Giardini and E. C. Lloyd, eds.) (Butterworths, 1963).

53. R. E. Lorent, Rev. Sci. Instr., 44, 1691 (1973).

54. O. Fukunaga, "Proceedings of the IVth International Conf. On High Pressure", p. 798 (J. Osugi, ed.-in-chief) Kawakita Print Co., Kyoto, 1975).

55. H. T. Hall, Rev. Sci. Instr., 29, 267 (1958).

56. H. T. Hall, Rev. Sci. Instr. Notes, 33, 1278 (1962).

57. E. C. Lloyd, U. O. Hutton and D. D. Johnson, J. Res. Nat. Bur. Stand., 63C, 59 (1959).

58. J. D. Barnett and H. T. Hall, Rev. Sci. Instr., 35, 175 (1964).

59. J. C. Houck and H. O. Hutton, p. 221, in "High Pressure Measurement" (A. A. Giardini and E. C. Lloyd, eds.) (Butterworths, 1963).

60. L. F. Vereschagin, p. 290, in "Progress in Very High Pressure Research" (F. P. Bundy, W. R. Hubbard, H. M. Strong, eds.) (Wiley, 1961).

61. J. Osugi, K. Shimizu, K. Inoue, Y. Yasunami, Rev. Phys. Chem., Japan, 34, 1 (1964).

62. M. Wakatsuki, K. Ichinose and T. Aoki, J. Appl. Phys., Japan, 10, 357 (1971).

63. S. Saito, A. Sawaoka, E. Tani, T. Mashimo, Y. Ozaki, Proc. IVth Int. Conf. High Pressure, Kyoto (1974) p. 786.

64. H. T. Hall, Rev. Sci. Instr., 37, 568 (1966).

65. B. Von Platen, Chapter 6, "A Multiple Piston, High Pressure, High Temperature Apparatus" in "Modern Very High Pressure Techniques", (R. H. Wentorf, ed.) (Butterworths, 1962).

66. N. Kawai, p. 45 in "Accurate Characterization of the High
     Pressure Environment" (E. C. Lloyd, ed.) (NBS Spec. Publ.
     326, 1971).

67. N. Kawai and S. Endo, Rev. Sci. Instr., 41, 1178 (1970).

68. N. Kawai and S. Mochizuki, Sol. St. Comm., 9, 1393 (1971).

69. R. Epain, C. Susse and B. Vodar, Comptes Rend. Acad. Sci.
     Paris, 265A, 323 (1967).

70. M. Kumazawa, High Temp-High Press., 3, 243 (1971).

71. M. Kumazawa, K. Masaki, H. Sawamoto, M. Kato, High Temp-
     High Press., 4, 293 (1972).

72. M. Kumazawa, High Temp-High Press., 5, 599 (1973).

73. K. Masaki, H. Sawamoto, E. Ohtani, M. Kumazawa, M. Machida,
     S. Mizukusa and N. Nakayama, Rev. Sci. Instr., 46, 84 (1975).

74. D. P. Johnson and P.L.M. Heydemann, Rev. Sci. Instr., 38,
     1294 (1967).

75. L. F. Vereschagin and E. V. Zubova, Dokl. Acad. Nauk USSR,
     169, 74 (1966).

76. G. C. Kennedy, R. C. Newton, Chapter 7, p. 163 in "Solids
     Under Pressure", (W. Paul and D. M. Warschauer, eds.)
     (McGraw Hill, 1963).

77. R. J. Zeto and H. B. Vanfleet, J. Appl. Phys., 40, 2227 (1969).

78. J. Lees, High Temp-High Press., 1, 477 (1969).

79. T. Kozuka and Y. Yamamoto, Proc. IVth Int. Conf. High
     Pressure, Kyoto (1974) p. 814.

80. M. D. Banus, High Temp-High Press., 1, 483 (1969).

81. D. L. Decker, J. Appl. Phys., 36, 157 (1965).

82. D. L. Decker, J. Appl. Phys., 37, 5012 (1966).

83. D. L. Decker, J. Appl. Phys., 42, 3239 (1971).

84. W. A. Bassett, T. Takahashi, H. Mao and J. S. Weaver, J. Appl.
     Phys., 39, 318 (1968).

85. D. L. Decker and T. G. Worlton, J. Appl. Phys., 43, 4799 (1972).

86. S. C. Yu, E. F. Skelton and I. L. Spain (unpublished).

87. J. C. Jamieson and B. Olinger, NBS Special Publ. 326 (E. C.
     Lloyd, ed.) (1971) p. 321.

88. A. K. Singh and G. C. Kennedy, J. Appl. Phys. 45, 4686 (1974).

89. D. J. Pastine and M. J. Carroll, p. 91, in "Accurate Character-
     ization of the High Pressure Environment" (E. C. Lloyd, ed.)
     NBS Spec. Publ. 32b (1971).

90. C. Friedli and N. W. Ashcroft, Phys. Rev. B12, 5552 (1975).

91. W. J. Carter, S. P. Marsh, J. N. Fritz, R. G. McQueen, p. 147,
     in "Accurate Characterization of the High Pressure Environ-
     ment" (E. C. Lloyd, ed.) NBS Spec. Publ. 326 (1971).

92. See for example R. C. Dunlap, Chapter 72, p. 32 in "Methods
     of Experimental Physics", Vol. 6B-Solid State Physics
     (K. Lark Horovitz and V. A. Johnson, eds.) (Academic Press,
     1959).

93. J. Lees, Nature 203, 965 (1964).

94. H. T. Hall, NBS Specl. Publ. 326, p. 303 (E. C. Lloyd, ed.)
     (1971).

95. A. J. Darnell, Bul. Am. Ceramic Soc., 44, 634 (1965).

96. J. C. Jamieson, p. 44 in "Physics of Solids at High Pressure"
     (C. T. Tomizuka and R. M. Emrick, ed.) (Academic Press, 1965).

97. C. Y. Liu, PhD Thesis, Univ. of Md. (1975) (Available from University Microfilms, Anne Arbor, Michigan).

98. J. A. Van Vechten, Phys. Rev. B7, 1479 (1973).

99. J. Akella and G. C. Kennedy, J. Geophys. Res., 76, 4969 (1971).

100. H. W. Schamp, J. N. Mundy, E. Rapoport, p. 355 in "High Pressure Measurement" (A. A. Giardini and E. C. Lloyd, eds.) (Butterworths, 1963).

101. S. N. Vaidya, J. Akella, G. C. Kennedy, J. Phys. Chem. Solids, 30, 1411 (1969).

102. J. Lees, High Temp-High Press., 1, 601 (1969).

103. D. L. Decker, W. A. Bassett, H. T. Merrill, H. T. Hall, J. D. Barrett, J. Phys. Chem. Ref. Data, 1, 773 (1972).

104. C. Y. Liu, K. Ishizaki, J. Paauwe and I. L. Spain, High Temp-High Press., 5, 359 (1973).

105. F. P. Bundy (private communication, and to be published). (A Summary of his findings was given at the Conference on Ultra High Pressure Technology, Rennselaerville, N.Y., June 1976).

106. A. L. Ruoff, Advances in Cryogenic Eng., 18, 435 (1973). (K. K. Timmerhaus, ed.)

107. A. L. Ruoff and J. Wanagel, Rev. Sci. Instr. 46, 1294 (1975).

108. L. F. Vereschagin, E. N. Yakovlev, T. D. Varfolomeeva, T. D. Seslarev, L. E. Shterenberg, Dokl. Acad. Nauk USSR, 185, 555 (1969) (English Trans. -Sov. Phys.-Dokl. 14, 248 (1969).)

109. H. T. Hall, Science 169, 868 (1970).

110. L. F. Vereschagin, E. N. Yakovlev, T. D. Varfolomeeva, V. N. Seslarev, L. E. Sterenberg, High Temp-High Press., 3, 239 (1971).

111. L. F. Vereschagin, E. N. Yakovlev, B. V. Vinogradov, G. N. Stepanov, K. Kh. Bibaev, T. I. Adaeva, V. P. Sakun, High Temp-High Press., 6, 499 (1974).

Chapter 12

PROPERTIES OF FLUIDS AT HIGH PRESSURES

William B. Streett

Science Research Laboratory
United States Military Academy
West Point, New York 10996

I.   INTRODUCTION

     Most industrial high pressure processes involve   fluids. As a
result, knowledge of the properties of fluids at high pressure is of
great importance in modern technology.

425

The growth of interest in fluids under pressure can be traced to the discovery by Andrews, in 1869, of the critical point of carbon dioxide [1]. Extensive experiments on the compressibilities of gases and liquids were carried out by Amagat, in France, during the latter part of the 19th century, culminating in the publication, in 1893, of tables of compressibilities of 13 fluids [2] at pressures up to 3 kbar*. This was soon followed by the experiments of P. W. Bridgman [3, 4], at Harvard University, whose early work included extensive systematic measurements of the compressibilities of gases and liquids. By 1932 he had measured the compressibilities of more than 50 fluids at pressures up to 15 kbar. This work, which is summarized in his book, The Physics of High Pressure [3], still provides one of the most useful and comprehensive introductions to fluids at high pressures.

In comparing the several states of matter, a clear distinction can be drawn between the crystalline solid state on the one hand and the fluid states, liquid and gas, on the other. The distinction between gas and liquid states for a pure substance exists only at pressures and temperatures between the triple point and the gas-liquid critical point. Inasmuch as the critical pressures of most ordinary liquids are of the order of 50 bars, this distinction will be of little interest here. In mixtures, two or more fluid phases have been observed to coexist in equilibrium at pressures as high as 15 kbar; however, in such cases no clear distinction between gas and liquid is possible, or even desirable. An example of the difficulties which can arise in attempting to draw this distinction is illustrated by the system carbon dioxide-water [5]. The critical temperature and pressure of carbon dioxide are 31.1°C and 73.8 bars, and for water 374.1°C and 221.3 bars. If liquid water is pressurized with carbon dioxide gas at 110°C, with suitable agitation to maintain equilibrium between the phases, the system behaves in a predictable manner for pressures up to about 1300 bars. At this pressure, the density of the gaseous phase, which is mainly carbon dioxide, becomes equal to that of the liquid phase, which is mainly water. Upon further increase of pressure the density of the "gas" exceeds that of the "liquid" and the two phases exchange positions within the pressure vessel. Similar density inversions have been observed in mixtures such as argon-ammonia, helium-hydrogen, and neon-methane. The explanation for this seemingly anomalous behavior is found in the relative molecular weights and volatilities of the two components in each mixture. In these systems the more volatile component (the one which condenses at the higher pressure at a given temperature) has the higher molecular weight. Consequently, at high pressures, where both phases have been compressed to liquid-like densities, the phase richer in the heavy component has the higher density.

Although some progress has been made in recent years in understanding the structure and behavior of fluids at the molecular level [6, 7], for present purposes the subject of fluids at high pressures is best approached through a survey of the known effects of pressure on macroscopic properties. Such a survey can logically be divided

_____

* 1 kbar = 1000 bars = 100 MPa

into two main parts: (1) the PVT equation of state and the equilibrium thermodynamic properties, and (2) transport properties. In section II the thermodynamic relations required to understand and interpret the PVT equation of state, and to calculate the equilibrium
properties from PVT and other types of experimental data, are developed. Section III is devoted to the PVT behavior of pure fluids and
methods of calculating thermodynamic properties, and section IV to a
survey of the effects of pressure on the transport properties. Section V is devoted to a survey of phase behavior in fluid mixtures at
high pressures. Finally, in section VI, a brief discussion is given
of expected future progress in the field of fluids at high pressures.

No attempt has been made to include a comprehensive and up-to-
date bibliography for each topic. Not only would such a bibliography
be unduly long and soon out of date, it is unnecessary, owing to the
appearance during the past decade of several outstanding bibliographies of high pressure research and thermodynamics. For high pressures, the most important of these is the Bibliography of High Pressure Research [8], prepared by the High Pressure Data Center at Brig-
ham Young University, Provo, Utah, under the direction of H. Tracy
Hall. The first part, published in 1970, covers the period 1900 to
1968. Subsequent bibliographies have been published in the form of a
bimonthly current awareness bulletin, entitled Bibliography on High
Pressure Research. Volumes I-IV in this series have been combined
and published in book form [9], covering the period 1968-72. Presumably this practice will continue, resulting in further editions in
book form at intervals of several years. The growth of high pressure
research is readily illustrated by the fact that part 2 (1968-72)
contains more than half the number of references in part 1 (1900-68).

A second important reference work, which is not limited to high
pressures, is the Thermophysical Property Data series prepared and
published at Purdue University, Lafayette, Indiana, under the super-
vision of Y. S. Touloukian [10]. Lide and Rossmassler [11] have compiled a useful list of data centers which specialize in the collec-
tion and dissemination of thermodynamic and thermochemical data.

II.   THERMODYNAMIC RELATIONS

A.   Systems of Fixed Composition

Experience tells us that only two independent intensive properties are required to describe the thermodynamic state of a system consisting of a single homogeneous phase of fixed composition. This is
in fact a statement of a specific case of the Gibbs phase rule, which
expresses the number of independent variables f as a function of the
number of components c and the number of phases p in a system at
equilibrium. The relation is f=c+2-p. For a single homogeneous
phase of fixed composition f=2. One interpretation of the phase rule
in this case is that any independent property can be expressed as a
function of any two others by equations of the form x=x(y,z). In
laboratory studies of fluids a convenient set of variables of measurement are pressure, temperature and volume. The laws of thermody-

namics provide expressions for calculating the pressure and volume
dependence of other thermodynamic properties, such as internal energy
U, entropy S, and enthalpy H, from the PVT equation of state. The
temperature dependence of these properties can be calculated from
knowledge of the heat capacity (C_p or C_V) as a function of temperature. The PVT equation can be written in any of the three forms
$T = T(P,V)$, $P = P(T,V)$, or $V = V(P,T)$, although the latter two forms
are most often used.

A combined statement of the first and second laws of thermodynamics can be written in the form

$$dU = TdS - PdV \tag{1}$$

where dU, dS and dV represent, respectively, small changes in the internal energy, entropy, and volume of a closed system (i.e., one
which does not exchange energy with its surroundings). Hence the
five fundamental thermodynamic properties of interest are P, V, T, U,
and S. In dealing with certain types of reactions and processes, it
is convenient to define three additional properties: the enthalpy H,
the Gibbs energy G, and the Helmholtz energy A, as follows

$$H \equiv U + PV \tag{2}$$

$$G \equiv H - TS \tag{3}$$

$$A \equiv U - TS \tag{4}$$

Differentiating these relations and substituting in equation (1)
yields three alternative forms of that equation:

$$dH = TdS + VdP, \tag{5}$$

$$dG = -SdT + VdP, \text{ and} \tag{6}$$

$$dA = -SdT - PdV. \tag{7}$$

Equations (1) and (5) - (7) are sometimes called the fundamental
equations of thermodynamics. The eight properties which appear in
these equations are all properties of state: that is, they are functions only of the thermodynamic state of the system. Changes in
these properties which accompany a change in state are dependent only
on the initial and final states of the system, and not on the sequence of intermediate states which characterizes the change. In
mathematical terms the properties are said to be independent of path,
and the fundamental equations are exact differentials. If we consider the dependence of the internal energy on entropy and volume, we
can write $U = U(S,V)$. For small changes in S and V, the total differential of U takes the form

$$dU = \left(\frac{\partial U}{\partial S}\right)_V dS + \left(\frac{\partial U}{\partial V}\right)_S dV. \tag{8}$$

Comparing this with equation (1) yields the identities

$$\left(\frac{\partial U}{\partial S}\right)_V = T, \text{ and} \tag{9}$$

$$\left(\frac{\partial U}{\partial V}\right)_S = -P. \tag{10}$$

Other useful identities can be derived in this way, including the following:

$$\left(\frac{\partial A}{\partial V}\right)_T = -P, \tag{11}$$

$$\left(\frac{\partial H}{\partial P}\right)_S = \left(\frac{\partial G}{\partial P}\right)_T = V, \tag{12}$$

$$\left(\frac{\partial H}{\partial S}\right)_P = T, \text{ and} \tag{13}$$

$$\left(\frac{\partial A}{\partial T}\right)_V = \left(\frac{\partial G}{\partial T}\right)_P = -S. \tag{14}$$

An important property of exact differentials of the form $dZ = MdX + NdY$ is that

$$\left(\frac{\partial M}{\partial Y}\right)_X = \left(\frac{\partial N}{\partial X}\right)_Y.$$

Applying this relation to the four fundamental equations yields the following equations, known as Maxwell's relations.

$$\left(\frac{\partial T}{\partial V}\right)_S = -\left(\frac{\partial P}{\partial S}\right)_V \tag{15}$$

$$\left(\frac{\partial T}{\partial P}\right)_S = \left(\frac{\partial V}{\partial S}\right)_P \tag{16}$$

$$\left(\frac{\partial P}{\partial T}\right)_V = \left(\frac{\partial S}{\partial V}\right)_T \tag{17}$$

$$\left(\frac{\partial V}{\partial T}\right)_P = -\left(\frac{\partial S}{\partial P}\right)_T \tag{18}$$

The heat capacities at constant volume and constant pressure are defined as:

$$C_V \equiv \left(\frac{\partial U}{\partial T}\right)_V, \text{ and} \tag{19}$$

$$C_P \equiv \left(\frac{\partial H}{\partial T}\right)_P \tag{20}$$

Combining equations (1), (14), and (19) leads to

$$C_V = \left(\frac{\partial U}{\partial T}\right)_V = T\left(\frac{\partial S}{\partial T}\right)_V = -T\left(\frac{\partial^2 A}{\partial T^2}\right)_V. \tag{21}$$

Similar expressions for C_P are derived from equations (5), (14) and (20):

$$C_P = \left(\frac{\partial H}{\partial T}\right)_P = T\left(\frac{\partial S}{\partial T}\right)_P = -T\left(\frac{\partial^2 G}{\partial T^2}\right)_P .$$

$$(22)$$

For calculating thermodynamic properties these can be rearranged to more useful forms. For example, the first equality in (21) can be re-written as

$$dU_V = C_V dT_V \tag{23}$$

where the subscripts on the differentials indicate that the change takes place at constant volume. Integrating gives

$$\Delta U_V = \int_{T_1}^{T_2} C_V dT_V . \tag{24}$$

Thus changes in internal energy with temperature, at constant volume, can be found if the temperature dependence of C_V is known. Other useful relations which follow from equations (20) and (21) are

$$\Delta H_P = \int_{T_1}^{T_2} C_P dT_P , \tag{25}$$

$$\Delta S_V = \int_{T_1}^{T_2} \frac{C_V}{T} dT_V \tag{26}$$

$$\Delta S_P = \int_{T_1}^{T_2} \frac{C_P}{T} dT_P . \tag{27}$$

Equations (24) - (27) can be used to calculate the temperature dependence of S, U, H, G and A. In practice it is only necessary to calculate ΔS_T and either ΔU_T or ΔH_T. For example, if ΔS_T and ΔU_T are known, the remaining differences can be found from equations (2)-(4). Since only properties of state are involved, one can write

$$\Delta H = \Delta(U+PV) = U_2 - U_1 + (P_2 V_2 - P_1 V_1) \tag{28}$$

where subscripts 1 and 2 denote initial and final states. Similar expressions follow for ΔG and ΔA. Equation (28) is not limited to changes at constant P or V, but applies to any change of state.

The fundamental equations and definitions can also be used to derive expressions for the pressure- and volume-dependence of thermodynamic properties. Rearranging equation (4) and differentiating with respect to V at constant T gives

$$\left(\frac{\partial U}{\partial V}\right)_T = \left(\frac{\partial A}{\partial V}\right)_T + T\left(\frac{\partial S}{\partial V}\right)_T . \tag{29}$$

Eliminating the derivatives on the right hand side by means of equations (11) and (17) gives

$$\left(\frac{\partial U}{\partial V}\right)_T = -P + T\left(\frac{\partial P}{\partial T}\right)_V.$$

(30)

Hence the volume-dependence of internal energy at constant temperature is given by

$$\Delta U_T = \int_{V_1}^{V_2}\left[-P + T\left(\frac{\partial P}{\partial T}\right)_V\right]dV_T.$$

(31)

Differentiating equation (3) and combining with (12) and (18) leads to a similar relation for isothermal changes in enthalpy as a function of pressure,

$$\Delta H_T = \int_{P_1}^{P_2}\left[V - T\left(\frac{\partial V}{\partial T}\right)_P\right]dP_T.$$

(32)

Expressions for the pressure and volume dependence of entropy at constant temperature follow from the Maxwell relations (17) and (18):

$$\Delta S_T = -\int_{P_1}^{P_2}\left(\frac{\partial V}{\partial T}\right)_P dP_T = \int_{V_1}^{V_2}\left(\frac{\partial P}{\partial T}\right)_V dV_T.$$

(33)

Equations (24) - (27) can be combined with (31) - (33) to yield the thermodynamic equations of state:

$$U_2 - U_1 = \int_{T_1}^{T_2}C_V dT + \int_{V_1}^{V_2}\left[T\left(\frac{\partial P}{\partial T}\right)_V - P\right]dV$$

(34)

$$H_2 - H_1 = \int_{T_1}^{T_2}C_P dT + \int_{P_1}^{P_2}\left[V - T\left(\frac{\partial V}{\partial T}\right)_P\right]dP$$

(35)

$$S_2 - S_1 = \int_{T_1}^{T_2}\frac{C_V}{T}dT + \int_{V_1}^{V_2}\left(\frac{\partial P}{\partial T}\right)_V dV = \int_{T_1}^{T_2}\frac{C_P}{T}dT + \int_{P_1}^{P_2}\left(\frac{\partial V}{\partial T}\right)_P dP.$$

(36)

In these equations the only variables which appear in the temperature integrals are temperature and heat capacity, and in the pressure and volume integrals the only variables are pressure, temperature, and volume. Thus the complete pressure, temperature, and volume dependence of internal energy, entropy, and enthalpy can be calculated from the PVT equation of state, together with knowledge of C_P as a function of T for a single isobar, or C_V as a function of volume for a single isochore.

Differentiating equation (30) with respect to T at constant V gives

$$\left(\frac{\partial^2 U}{\partial V \partial T}\right) = T \left(\frac{\partial^2 P}{\partial T^2}\right)_V. \tag{37}$$

Since the order of differentiation is unimportant, this can be re-written

$$\left(\frac{\partial C_V}{\partial V}\right)_T = - T\left(\frac{\partial^2 P}{\partial T^2}\right)_V. \tag{38}$$

The analogous expression for the pressure dependence of C_P is

$$\left(\frac{\partial C_P}{\partial P}\right)_T = -T\left(\frac{\partial^2 V}{\partial T^2}\right)_P. \tag{39}$$

Thus in principle the pressure and volume dependence of the heat capacities can be calculated from the PVT equation of state. However, as we shall see below, equations (38) and (39) can seldom be used effectively for fluids at high pressures because of the difficulty of obtaining accurate values of second derivatives of the PVT surface. Another useful relation between the heat capacities and the PVT surface can be derived from equations (2) and (30),

$$C_P - C_V = T\left(\frac{\partial P}{\partial T}\right)_V \left(\frac{\partial V}{\partial T}\right)_P. \tag{40}$$

Differentiating the expression $P = P(V,T)$ gives

$$dP = \left(\frac{\partial P}{\partial V}\right)_T dV + \left(\frac{\partial P}{\partial T}\right)_V dT. \tag{41}$$

Taking similar derivatives for dV and dT from $V = V(P,T)$ and $T = T(P,V)$, and eliminating dP, dV and dT gives the relation between the three mutual derivatives of the PVT equation of state

$$\left(\frac{\partial V}{\partial P}\right)_T \left(\frac{\partial T}{\partial V}\right)_P \left(\frac{\partial P}{\partial T}\right)_V = -1 . \tag{42}$$

The same result is obtained by cyclic differentiation of $V = V(P,T)$. These derivatives appear in definitions of the mechanical coefficients, commonly used in engineering practice. These are: the isothermal compressibility β_T, the thermal expansion coefficient α_P, and the thermal pressure coefficient γ_V, defined as

$$\beta_T \equiv - \frac{1}{V} \left(\frac{\partial V}{\partial P}\right)_T, \tag{43}$$

$$\alpha_P \equiv \frac{1}{V} \left(\frac{\partial V}{\partial T}\right)_P, \text{ and} \tag{44}$$

$$\gamma_V \equiv \left(\frac{\partial P}{\partial T}\right)_V. \tag{45}$$

The factor $1/V$ is included in (43) and (44) to make β_T and α_P intensive properties, and the minus sign is included in (43) to make β_T positive. Many of the earlier equations can be rewritten in terms of these coefficients. Among the more useful are the following:

$$\alpha_P = \beta_T \gamma_V, \tag{46}$$

$$\left(\frac{\partial S}{\partial P}\right)_T = -V\alpha_P, \tag{47}$$

$$\left(\frac{\partial U}{\partial V}\right)_T = -P + T\alpha_P/\beta_T = -P + T\gamma_V, \tag{48}$$

$$\left(\frac{\partial H}{\partial P}\right)_T = V - TV\alpha_P, \text{ and} \tag{49}$$

$$C_P - C_V = TV\alpha_P\gamma_V. \tag{50}$$

Another useful property is the adiabatic coefficient of compressibility, β_S, defined by

$$\beta_S \equiv -\frac{1}{V}\left(\frac{\partial V}{\partial T}\right)_S. \tag{51}$$

Its usefulness in deriving fluid properties arises from its relation to the velocity of sound W, through the equation

$$W^2 = \frac{V}{M\beta_S}, \tag{52}$$

where V is the molar volume and M the molecular weight. The velocity defined by equation (52) is a purely thermodynamic quantity, but is found to be equal to the measured velocity in fluids over a wide range of experimental conditions. The ratio of β_S to β_T can be derived from the relation between P, V, and S. For these three variables the analogue of equation (42) is

$$\left(\frac{\partial V}{\partial P}\right)_S = -\frac{\left(\frac{\partial S}{\partial P}\right)_V}{\left(\frac{\partial S}{\partial V}\right)_P}, \tag{53}$$

which can be rewritten, with the aid of equations (21) and (22) as

$$\left(\frac{\partial V}{\partial P}\right)_S = -\frac{\left(\frac{\partial S}{\partial T}\right)_V \left(\frac{\partial T}{\partial P}\right)_V}{\left(\frac{\partial S}{\partial T}\right)_P \left(\frac{\partial T}{\partial V}\right)_P} = \frac{C_V}{C_P}\left(\frac{\partial V}{\partial P}\right)_T. \tag{54}$$

Hence,

$$\frac{\beta_T}{\beta_S} = \frac{C_P}{C_V} \ . \tag{55}$$

Combining equations (50) and (55) gives

$$C_P = \frac{TV\alpha_P^2}{\beta_T - \beta_S} \ . \tag{56}$$

Thus, in the absence of experimental heat capacity data, measurements of the velocity of sound can be used with the PVT equation of state to derive the equilibrium properties, including heat capacities, from expressions which involve only first derivatives of the equation of state.

B.   Solutions

A solution is a homogeneous mixture of two or more components, and may be in the gas, liquid or solid phase. The extensive properties of a single phase solution are functions of two intensive properties and of the quantities of each individual component. If X is an extensive property of a single phase solution,

$$X = X \ (T, P, N_A, N_B, \ ...), \tag{57}$$

where N_A, N_B etc. represent the mass (or number of moles) of components A, B, etc., and temperature and pressure have been arbitrarily chosen as the two independent intensive properties. In treating properties of solutions the following expressions are particularly useful:

$$U = U \ (S, \ V, \ N_A, \ N_B \ ...), \tag{58}$$

$$H = H \ (S, \ P, \ N_A, \ N_B \ ...), \tag{59}$$

$$A = A \ (T, \ V, \ N_A, \ N_B \ ...), \ \text{and} \tag{60}$$

$$G = G \ (T, \ P, \ N_A, \ N_B \ ...). \tag{61}$$

Differentiating these expressions gives

$$dU = \left(\frac{\partial U}{\partial S}\right)_{V,N} dS + \left(\frac{\partial U}{\partial V}\right)_{S,N} dV + \sum^{i} \left(\frac{\partial U}{\partial N_i}\right)_{S,V} dN_i, \tag{62}$$

$$dH = \left(\frac{\partial H}{\partial S}\right)_{P,N} dS + \left(\frac{\partial H}{\partial P}\right)_{S,N} dP + \sum^{i} \left(\frac{\partial H}{\partial N_i}\right)_{S,P} dN_i, \tag{63}$$

$$dA = \left(\frac{\partial A}{\partial T}\right)_{V,N} dT + \left(\frac{\partial A}{\partial V}\right)_{T,N} dV + \sum^{i} \left(\frac{\partial A}{\partial N_i}\right)_{T,V} dN_i, \ \text{and} \tag{64}$$

$$dG = \left(\frac{\partial G}{\partial T}\right)_{P,N} dT + \left(\frac{\partial G}{\partial P}\right)_{T,N} dP + \sum^{i} \left(\frac{\partial G}{\partial N_i}\right)_{T,P} dN_i. \tag{65}$$

In each of these expressions the first two partial derivatives on the right are restricted to systems of fixed composition, hence the thermodynamic relations of the previous section apply. Substituting the expressions in equations (9) - (14) and transposing terms to the left side gives

$$dU - TdS + PdV = \sum^{i} \left(\frac{\partial U}{\partial N_i}\right)_{S,V} dN_i \tag{66}$$

$$dH - TdS - VdP = \sum^{i} \left(\frac{\partial H}{\partial N_i}\right)_{S,P} dN_i \tag{67}$$

$$dA + SdT + PdV = \sum^{i} \left(\frac{\partial A}{\partial N_i}\right)_{T,V} dN_i \tag{68}$$

$$dG + SdT - VdP = \sum^{i} \left(\frac{\partial G}{\partial N_i}\right)_{T,P} dN_i \tag{69}$$

From equations (1) - (4) it follows that the left sides of equations (66) - (69) are equal, therefore the summations on the right sides are equal. Therefore,

$$\left(\frac{\partial U}{\partial N_i}\right)_{S,V} = \left(\frac{\partial H}{\partial N_i}\right)_{S,P} = \left(\frac{\partial A}{\partial N_i}\right)_{T,V} = \left(\frac{\partial G}{\partial N_i}\right)_{T,P} \equiv \mu_i. \tag{70}$$

J. W. Gibbs was the first to recognize the importance of these equalities, and he called the partial derivatives in equation (70) "chemical potentials". They are usually represented by the symbol μ.

The partial molal property of a component in solution is defined as the differential change in that property with respect to a differential change in the amount of that component when the temperature, pressure, and amounts of the other components are held constant. From equation (70) it is evident that the chemical potential is equal to the partial molal Gibbs energy, $\mu_i = (\partial G/\partial N_i)_{T,P,N_i} = G_i$ (an upper case symbol with a subscript denotes a partial molal quantity).

A derived property of special thermodynamic importance is the fugacity, f, which is essentially a pseudo-pressure [59]. For a pure substance the fugacity is defined by the relation

$$(dG)_T = (RT\ d\ln f)_T. \tag{71}$$

From equation (6), $(dG)_T = (VdP)_T$, and for ideal gases this becomes

$$(dG)_T = (RT\ d\ln P)_T. \tag{72}$$

Thus for ideal gases fugacity is numerically equal to pressure. From equations (6) and (71) one obtains

$$(RT\ d\ln f)_T = (PV\ \frac{dP}{P})_T = (PV\ d\ln P)_T \tag{73}$$

from which it follows that

$$\frac{\partial \ln f}{\partial \ln P} = \frac{PV}{RT} = Z,$$

where Z is the compressibility factor. Since $Z \to 1$ as $P \to 0$, it follows that $f \to 0$ as $P \to 0$.

The fugacity of component i in solution is defined in terms of partial molal free energy:

$$dG_i = d\mu_i = RTd\ln(f_i)_T \tag{74}$$

For component i in solution, equation (6) becomes

$$dG_i = - \mathbf{S}i \ dT + \bar{V}_i \ dP. \tag{75}$$

At constant temperature,

$$\left(\frac{\partial G_i}{\partial P}\right)_T = \left(\frac{\partial \mu_i}{\partial P}\right)_T = V_i \tag{76}$$

Combining equations (76) and (74) gives the pressure-dependence of fugacity at constant temperature as

$$\left(\frac{\partial \ln f_i}{\partial P}\right)_T = \frac{V_i}{RT}. \tag{77}$$

The ratio of the fugacity of component i in any state to its fugacity in an arbitrarily chosen standard state is called the activity a_i; thus,

$$(a_i)_T = (f_i/f_i^\circ)_T, \tag{78}$$

where the "degree" symbol indicates standard state. Integrating equation (74) at constant temperature from the standard state to the present state gives

$$(G_i - G_i^\circ)_T = (\mu_i - \mu_i^\circ)_T = (RT \ln \frac{f_i}{f_i^\circ})_T = (RT \ln a_i)_T. \tag{79}$$

The departure of a solution from ideal behavior is described by the activity coefficient, γ, usually defined as

$$\gamma_i = \frac{a_i}{N_i}, = \frac{f_i}{f_i^\circ N_i}. \tag{80}$$

From equations (77) and (80) the pressure-dependence of the activity coefficient of component i at constant temperature and composition is given by,

$$\left(\frac{\partial \ln \gamma_i}{\partial P}\right)_{T,N_i} = \left(\frac{V_i - V_i^\circ}{RT}\right)_{T,N_i} \tag{81}$$

III.   THE PVT EQUATION OF STATE AND THE EQUILIBRIUM PROPERTIES

A detailed discussion of experimental methods for the study of
PVT behavior in fluids under pressure is beyond the scope of this
book. An excellent review, complete with many explanatory figures
and diagrams, can be found in the book by Tsiklis [12]. A review of
methods suited for studies at low temperature and high pressures has
been presented by Levelt-Sengers [13].

A.   The Equation of State

The effects of pressure on the volumes of several fluids are
shown in Figure 1. Bridgman's experiments on the compressibilities of
more than 50 liquids, at pressures up to 12 kbar, show a remarkable
uniformity in the compressibilities of substances which are liquids at
ordinary temperatures. Most organic liquids contract by 25 to 30 per
cent of their initial volume when compressed from 1 bar to 12 kbar.
Water behaves in a similar way, but mercury is far less compressible,
exhibiting a contraction of less than 4 per cent at 10 kbar.

The pressure dependence of β_T, the isothermal compressibility, is
obtained from P-V isotherms by taking the derivative $(\partial P/\partial V)_T$. In
general, the most useful method of deriving other thermodynamic properties from PVT data is to fit the data to an analytical equation of
the form $P = P(T,V)$ or $V = V(P,T)$, and calculate the properties using
the equations of Section II. No single equation of state for fluids
has proven to be entirely satisfactory, although a great many have
been proposed. Some representative equations are given in Table 1.

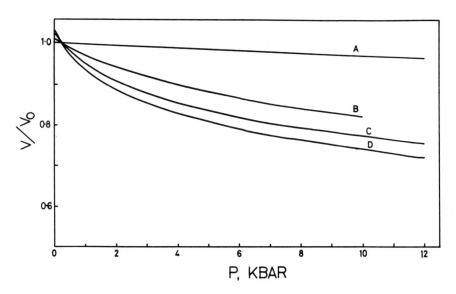

FIG. 1.   The relative volume V/V_o, of several liquids as a
function of pressure. Legend: A, mercury at 20°K; B, water at 50°C;
C, methyl alcohol at 20°C; D, ether at 20°C. V_o is the volume at 25
bars. Data from Bridgman [3].

Table 1.   Some Equations of State for Fluids

Name	Equation	Remarks
Perfect Gas Equation	$Pv = RT$	Applicable only to gases at low pressure
Reduced Coordinate Equation with Compressibility factor	$P_r v_r = ZRT_r$ $P_r = P/P_c, v_r = v/v_c, T_r = T/T_c$	Data for large numbers of fluids can be conveniently represented in plots of $Z(P_r T_r)$
Virial Equation (Inverse powers of molar volume)	$Z = 1 + B/v + C/v^2 + D/v^3 + \ldots$	Value of virial coefficients depends only on temperature
Virial Equation (Powers of pressure)	$Z = 1 + B'P + C'P^2 + D'P^3 + \ldots$ $B' = B/RT, \; C' = (C-B^2)/(RT)^2$ etc.	Value of virial coefficients depends only on temperature
Van der Waals Equation	$P = \dfrac{RT}{v-b} - \dfrac{a}{v^2}$ $v_c = 3b; \; P_c = \dfrac{a}{27b^2}; \; T_c = \dfrac{8a}{27Rb}$	Generally only useful for qualitative description of fluid behaviour
Redlich-Kwong Equation	$P = \dfrac{RT}{v-b} - \dfrac{a}{T^{1/2} \, v(v+b)}$	$a = \dfrac{0.4278 R^2 (T_c)^{2.5}}{P_c}$ $b = 0.0867 \, RT_c/P_c$
Beattie-Bridgman Equation	$P = \dfrac{RT}{v^2}\left[v + B_o\left(1 - \dfrac{b}{v}\right)\right]\left(1 - \dfrac{c}{vT^3}\right)$ $\qquad - \dfrac{A_o}{v^2}\left(1 - \dfrac{a}{v}\right)$	B_o, b, c, A_o, a are arbitrary constants fitted to experimental data
Benedict-Webb-Rubin Equation $(d \equiv v^{-1})$	$P = RTd + (B_o RT - A_o - C_o/T^2)d^2$ $\qquad + (bRT - a)d^3 + a\alpha d^6$ $\qquad + \dfrac{cd^3}{T^2}(1 + \gamma d^2)\exp{-\gamma d^2}$	$B_o, A_o, C_o, b, a, c, \alpha$ and γ are arbitrary constants fitted to experimental data. Gives good agreement with data at high density

A large body of data has been fitted to the Redlich-Kwong equation, but for applications at higher density the Benedict-Webb-Rubin equation is generally used.

Methods of developing equations of state can be broadly classified as theoretical or empirical. In the former, emphasis is placed on describing the behavior in terms of intermolecular forces and the size and shape of molecules, using the principles of statistical mechanics and kinetic theory; in the latter, an attempt is made to devise an empirical equation which describes the observed macroscopic behavior over a specific range of P, V and T. Because of slow progress in developing the molecular theory of dense fluids, presently available theoretical equations are of limited practical use in calculating thermodynamic properties. Important advances have taken place in this field during the past decade, however, resulting in accurate theoretical equations for the special case of simple spherical molecules (such as the noble gases), and work is now in progress to extend these methods to complex molecules [14].

Martin [15], Ott, et al., [16] and McDonald [57] have reviewed some of the more frequently used empirical equations. With modern digital computers, equations containing large numbers of adjustable constants (typically 10 to 30, but more in some cases) can be easily handled, and can provide accurate representations of the PVT surface over extended ranges of pressure and temperature. A general least squares method for fitting PVT data to empirical equations containing any desired number of constants has been described by Hust and McCarty [17].

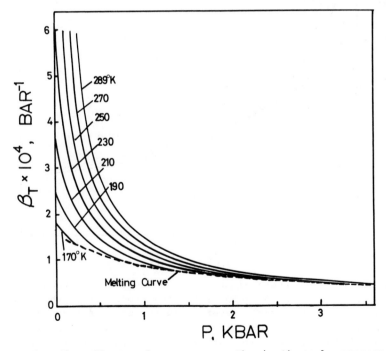

FIG. 2. The effects of pressure on the isothermal compressibility, β_T, of liquid xenon [18].

The qualitative features of the pressure-dependence of the mechanical coefficients are illustrated in Figures 2 - 4, which show
their behavior in liquid xenon [18] at pressures to about 3.5 kbar.
Figures 5 and 6 show the pressure-dependence of the velocity of sound
[19] and the adiabatic compressibility [20] of argon. The effects of
pressure on these properties are determined mainly by the form of the
intermolecular potential function, and since this function is similar
for most ordinary liquids, the behavior shown in Figures 2 - 6 is
characteristic of liquids in general, including those with complex
molecules.

The mechanical coefficients and the adiabatic compressibility
can be used to derive other equilibrium thermodynamic properties
through equations (47) - (50), (55) and (56). The accuracy of the results clearly depends upon the accuracy and precision of the original
PVT and sound velocity data. In PVT measurements it is the volume

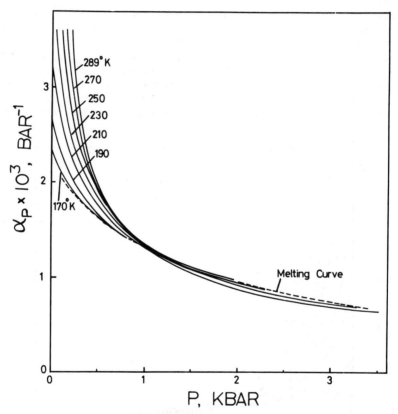

FIG. 3. The effects of pressure on the thermal expansion coefficient, α_p, of liquid xenon [18]. The crossover of the isotherms
at about 1 kbar indicates a reversal in the sign of the second derivative $(\partial^2 V/\partial T^2)_p$, which coincides with a minimum in C_p as a
function of pressure at constant temperature (see Figure 11).

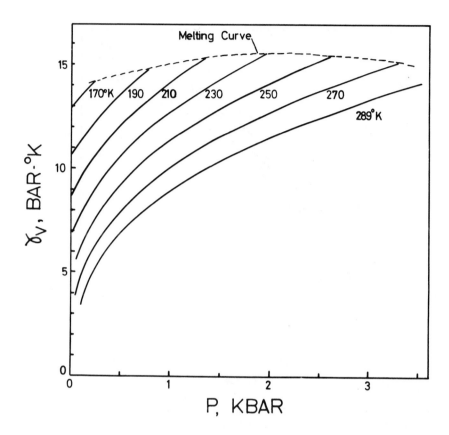

FIG. 4. The effects of pressure on the thermal pressure coefficient, γ_V, of liquid xenon [18].

which almost always contains the largest errors. With reasonable care, pressures and temperatures can be measured with a precision of 0.01 per cent; however, errors in volume are usually of the order of 0.1 to 0.5 per cent, and rarely less than about 0.03 per cent. A rule of thumb for dense fluids is that each successive differentiation of the PVT equation of state results in at least an order of magnitude increase in uncertainty, so that if the uncertainty in the original measurements of volume is, say, 0.1 per cent, the uncertainty in the first derivatives will be about 1-2 per cent and in the second derivatives about 10-20 per cent. This explains why equations (38) and (39) can only be used when the experimental data are of high accuracy and precision. Precision or internal consistency is perhaps the principal factor which determines the accuracy of the derivatives of the PVT equation. The spacing of PVT measurements is also important. The experimental methods most often used in high pressure research are designed to measure volume as a function of pressure at fixed temperature. In many cases, measurements have been carried out at closely spaced intervals of pressure, for a relatively small number of iso-

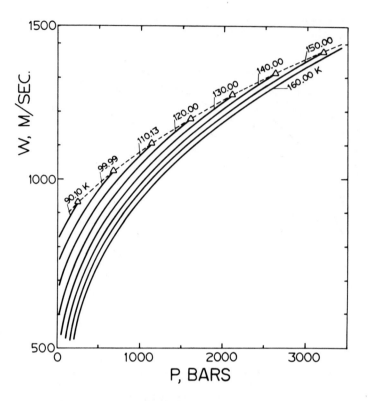

FIG. 5. The effects of pressure on the velocity of sound
in liquid argon [19]. The dashed line is the melting curve.
The steep slopes of the three highest isotherms at low pressures
are associated with the zero in sound velocity at the critical
point. ($T_c \simeq 151°K$ and $P_c \simeq 48$ bars.)

therms. (This is the case, for example, with most of Bridgman's
measurements.) The resulting data inherently contain more information
about the pressure derivatives $(\partial V/\partial P)_T$ than about temperature derivatives $(\partial P/\partial T)_V$ and $(\partial V/\partial T)_P$; that is the calculated values of α_p and
γ_V are less accurate than those of β_T. However, as equations (34)-
(39) and (47) - (50) show, it is the temperature derivatives which appear in equations for calculating the pressure and volume dependence
of entropy, internal energy, enthalpy, and heat capacity. The problem
of uncertainties in the temperature derivatives is aggravated when
isothermal PVT data from several independent experiments are fitted to
a single equation of state, because the inevitable systematic differences between the results of different experiments introduce "wrin-
kles" into the PVT surface. If the original data have been measured
along isotherms, the temperature-derivatives are most strongly affected.

With modern ultrasonic techniques, the velocity of sound in
fluids under pressures up to several tens of kilobars can be measured

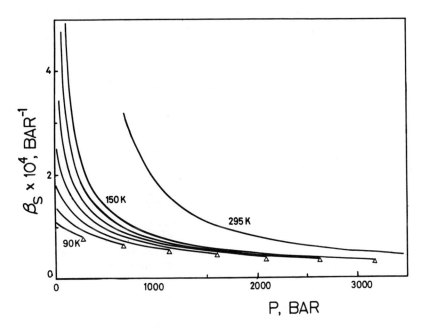

FIG. 6.   The effects of pressure on the adiabatic compressibility, β_S, of liquid argon [20].   Values for 295°K are from the measurements of Leibenberg, et al. [54].

with an accuracy and precision of 0.3 per cent or better. If such measurements are combined with PVT measurements having an uncertainty in volume of, say, 0.1 per cent, then the uncertainty in β_S, calculated from equation (52) will be no more than 0.3-0.5 per cent. In other words, if the PVT equation of state is known with reasonable accuracy, measurements of the velocity of sound provide a direct, accurate measurement of a derivative $(\partial V/\partial P)_S$, of the PVT surface. The usefulness of combining PVT and sound velocity measurements can be seen by comparing equations (38) and (39) to equations (55) and (56). If only the PVT equation is known, the pressure and volume dependence of the heat capacities is accessible only through equations (38) and (39), but these give only the change in heat capacity with pressure and volume, and the error bounds on the results are usually quite large because of uncertainties in the second derivatives $(\partial^2 P/\partial T^2)_V$ and $(\partial^2 V/\partial T^2)_P$. On the other hand, if β_S is known, absolute values of C_P and C_V can be calculated from equations (55) and (56), using only first derivatives of the PVT equation. This method of obtaining heat capacities is not without its own pitfalls, however. In applying equation (56), uncertainties   in α_P are effectively doubled in α_P^2 and uncertainties in β_T and β_S are magnified in taking the difference, $\beta_T - \beta_S$, between two numbers which are usually of the same order of magnitude. Nevertheless, heat capacities derived in this way from PVT and sound velocity data of high accuracy (0.05 per cent in volume and 0.1 per cent in W) can be expected to be accurate to within about 5 per cent.

The advantage of deriving heat capacities by this method is that
it avoids the need for calorimetric measurements (i.e., measurements
of quantities of heat) for fluids at high pressures. Calorimetric
measurements are performed by measuring the temperature rise in a
sample of known mass upon the addition of a known quantity of heat.
The sample must be enclosed in a suitable container, so that in practice the combined heat capacity of sample and container is measured
and the heat capacity of the container, measured separately, is subtracted. The obvious difficulty with measurements at high pressures
is that the heat capacity of the container will, in general, be large
compared to that of the sample, so that the final numerical result is
the difference between two numbers of nearly equal magnitude.

B.   Derived Properties

Changes in the internal energy and entropy of argon [21], with
pressure, are shown in Figures 7 and 8. Increasing pressure leads to
increasing order at the molecular level, hence entropy decreases
monotonically with increasing pressure. Changes in internal energy
with pressure, at constant temperature, are due to changes in the
configurational internal energy -- that part of the internal energy
arising from the forces of interaction between the molecules. Figure
9 shows the form of the force-distance relation for the molecules of
most ordinary liquids. At moderate to large separations the forces
are weakly attractive, but at short separations they become strongly

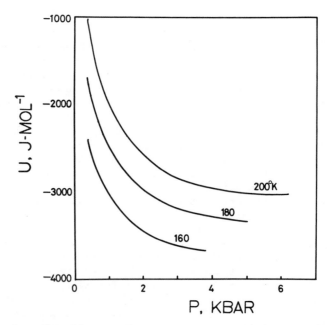

FIG. 7. The effects of pressure on the internal energy, U,
of liquid argon [21]. The 200 K isotherm appears to reach a
minimum in U at about 6 kbar.

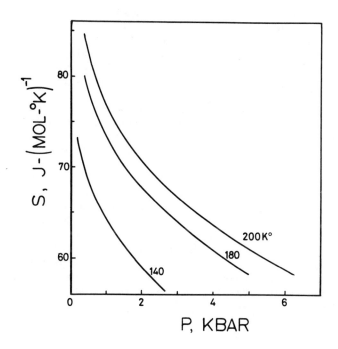

FIG. 8. The effects of pressure on the entropy, S, of liquid argon [21].

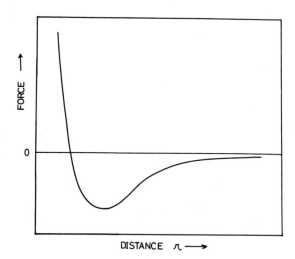

FIG. 9. The force between simple, nonpolar molecules as a function of distance, r. This curve has the same form as a plot of potential energy vs. distance.

repulsive. As a first approximation it can be assumed that the total configurational internal energy is the sum of pairwise interactions between the molecules, hence comparisons of Figures 7 and 9 indicate that at 200 K the interactions between argon molecules are predominantly attractive at pressures up to about 6 kbar. The minimum in the U-P curve at this point occurs because of the increasing importance of the repulsive forces.

At higher pressures the internal energy increases monotonically with pressure. This effect can be seen in Figure 10 which shows the internal energy of a Lennard-Jones fluid as a function of pressure at a reduced temperature of 1.17 (which corresponds to a temperature of about 140 K in argon) from the Monte Carlo computer calculations of Streett, Raveche and Mountain [22]. Since the intermolecular potential functions of many fluids are similar in form to the Lennard-Jones potential, it is reasonable to expect that, at constant temperature, internal energy will initially decrease with increasing pressure but eventually pass through a minimum and thereafter increase. Similar conclusions follow from consideration of macroscopic behavior and equation (30). At low pressures, $(\partial U/\partial V)_T$ is positive and it follows that $(\partial U/\partial P)_T$ is negative. However, as pressure increases, the rate of increase of $(\partial P/\partial T)_V$ is less than that of the pressure, and the reversal in the sign of $(\partial U/\partial P)_T$ -- the minimum in the U-P curve -- occurs when $P = T(\partial P/\partial T)_V = T\gamma_V$.

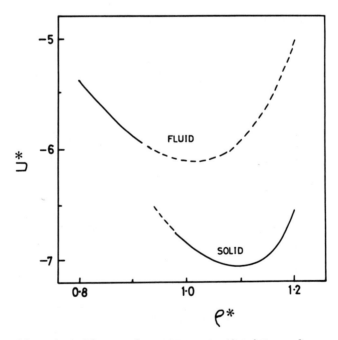

FIG. 10. The effects of pressure on the internal energy of a Lennard-Jones fluid at a reduced temperature T* = 1.17, from Monte Carlo computer simulations [22]. The dashed portion of the fluid branch is a region of metastable or sub-cooled liquid.

   Bridgman [3] calculated the heat capacities of twelve liquids, at pressures up to 12 kbar, from PVT data, using equations (38) and (39). He reported that "examples of nearly every conceivable type of behavior may be found; ...the specific heats may increase or decrease with increasing temperature or pressure, and the curves for different temperatures may cross and recross in the most bewildering way." He attributed this mainly to complexity of molecular structure; however, in retrospect it seems likely that at least part of this unusual behavior is not real, but is the result of small errors in his PVT data which resulted in large errors in the derivatives $(\partial^2 V/\partial T^2)_P$ and $(\partial^2 P/\partial T^2)_V$ used to calculate the heat capacities.

   The heat capacity of n-octane at 45°C and pressures up to 4.7 kbar, calculated by Benson, et al. [23] from their PVT measurements is shown in Figure 11. The behavior shown is characteristic of liquids in general, and plots of C_V vs. P at constant T have a similar form. For liquids, both C_p and C_V at first decrease with increasing pressure, passing through a minimum at a pressure between about 0.5 and 3.0 kbar, and thereafter increase. The significance of the C_V minima in relation to the shape of the PVT surface has been discussed by Verbeke and van Itterbeek [24] and more recently in much greater detail by Stephenson [25]. From equations (38) and (39) it follows that the minima in C_V and C_p correspond, respectively, to points at which $(\partial^2 P/\partial T^2)_V = 0$ and $(\partial^2 V/\partial T^2)_P = 0$; that is, to inflection points in isochores and isobars on the PVT surface. An analytical equation of state must be capable of reproducing these subtle features if it is to provide an accurate representation of real PVT behavior.

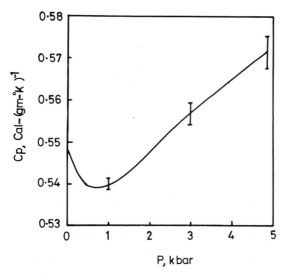

FIG. 11. The effects of pressure on the constant pressure heat capacity, C_p, of n-octane at 45°C, calculated by Benson, et al [23] from PVT data, using equation (39). Vertical lines indicate estimated errors.

C.   Corresponding States

The classical or macroscopic theory of corresponding states was derived by van der Waals from his well-known equation of state. According to this theory, the PVT surfaces of all fluids are identical if described in terms of reduced variables $P_r = P/P_c$, $V_r = V/V_c$, and $T_r = T/T_c$, where subscript c refers to the critical point. Since there is a fixed relation between P_c, T_c and V_c, only two of the three reducing parameters are independent. A more general form of the theory is obtained from an approach based on statistical mechanics [26]. The four assumptions upon which this theory is based are [27]:

1.   The potential energy between two molecules depends only on the distance between them and not on their orientations.

2.   The reduced potential energy $\phi(r)/\varepsilon$, where ε is a characteristic energy for each substance, is a universal function of the reduced length, r/σ, where σ is a characteristic length for each substance.

3.   The particles obey classical statistical mechanics (quantum effects are negligible).

4.   The total potential energy of an assembly of molecules is the sum of the potential energies of all pairs in the assembly.

It follows that the intermolecular potential functions of the substances which conform to this theory have the form

$$\phi_1(r) = \varepsilon_{11} \, F \, (r_1/\sigma_{11}) \text{ and} \qquad (82)$$

$$\phi_2(r) = \varepsilon_{22} \, F \, (r/\sigma_{22}), \qquad (83)$$

where ε and σ are energy and length parameters and F is a function of the reduced distance r/σ.

An example of a potential of the form of equations (82) and (83) is the well known Lennard-Jones 6-12 potential

$$\phi(r) = 4\varepsilon \quad [\left(\frac{\sigma}{r}\right)^{12} - \left(\frac{\sigma}{r}\right)^6]. \qquad (84)$$

It is to be emphasized, however, that the statistical derivation of the theory of corresponding states is independent of the specific form of the potential, so long as it has the general form of an energy parameter multiplied by a universal function of reduced distance. The reducing parameters for P, V, and T are, respectively, ε/σ^3, σ^3 and ε. It follows that the classical derivation of van der Waals is a special case of the general theory. There is no compelling theoretical reason to prefer the properties of the critical point, as reducing parameters, to those of any other unique point, such as the triple point. There is an important practical reason, however, in that there exists a large body of engineering data based on critical constants as reducing parameters.

The statistical formulation of the theory of corresponding states leads to a general method for examining corresponding-states behavior, known as the method of scale factors [40]. In this method the potential energy of substance i is written in the form

$$\phi_i(r) = f_{i0} \, \phi_0 \, (r/g_{i0}) \tag{85}$$

where

$$f_{i0} = \frac{\varepsilon_{ii}}{\varepsilon_{00}} \, , \, g_{i0} = \frac{\sigma_{ii}}{\sigma_{00}} \tag{86}$$

and ϕ_0 is the potential function of a reference substance, which can be any of the substances to be compared. The reducing parameters for P, V, and T now become f_{i0}/g_{i0}^3, and f_{i0}. It follows that in comparing the reduced properties of substance i to the reference substance 0, the scale factors f_{i0} and g_{i0} are dimensionless ratios of the energy and length parameters of substance i to those of the reference substance. Values of these ratios which result in the "best" corresponding-states agreement (in some well defined sense) can then be determined by a suitable trial and error method, without knowing the values of any of the individual parameters. The advantage of this method is that it avoids uncertainties which result from uncertainties in the reducing parameters themselves, and it thus provides a more meaningful test of how closely two or more substances approach true corresponding-states behavior. Using this method, Streett and Staveley [28] have shown that with suitably chosen scale factors the reduced volumes of argon, krypton, and xenon agree to within one or two parts per thousand, which is comparable with the accuracy of the experimental data, at pressures up to 3.5 kbar. Similar agreement has been found for reduced sound velocities of argon and krypton [19], although the comparison in this case is limited to pressures below about 0.7 kbar.

The theory of corresponding states expressed by equation (82) and the four assumptions listed above is limited to those substances whose molecular interactions can be adequately described by a function containing only two parameters. In addition to the noble gases, small nonpolar (or slightly polar) molecules, such as nitrogen, methane (CH_4), oxygen, and carbon monoxide (CO) fall into this category. For more complex molecules it is necessary to introduce an additional parameter into the potential function; that is, to construct a three-parameter theory of corresponding states. This is most conveniently done by dividing fluids into classes according to the extent of their deviation from simple-molecule behavior [29].

D.   Generalized Properties of Fluids

The classical theory of corresponding states has been used as a basis for developing generalized property charts and tables, for use

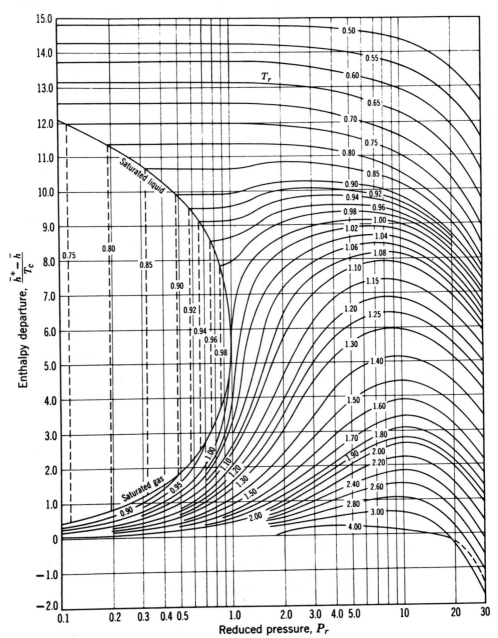

FIG. 12. Generalized enthalpy departure from ideal gas
behavior (per mole of gas or liquid, $z_C = 0.27$). Reproduced by
permission from G. J. Van Wylen and R. E. Sontag, Introduction
to Classical Thermodynamics, John Wiley & Sons, New York, 1965.

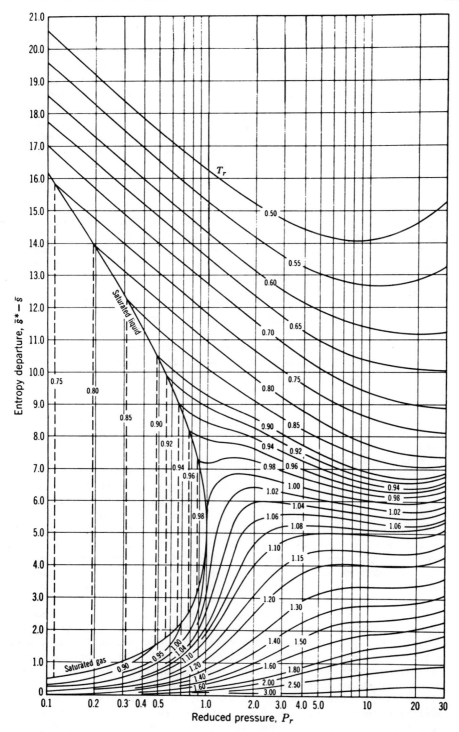

FIG. 13. Generalized entropy departure from ideal gas behavior (per mole of gas or liquid, Z_C = 0.27). (Van Wylen & Sonntag see Fig. 12.)

in estimating the properties of real fluids in regions where there
are no experimental data. According to this theory, all substances
exhibit the same departures from ideal gas behavior when examined at
the same reduced conditions of pressure and temperature. Reduced
property correlations based on the two-parameter corresponding-states
theory were found to be useful for only a small class of fluids, and
later correlations have been based on a three-parameter theory.
Several different third parameters have been proposed. One of the
most successful is the parameter Z_c, the compressibility factor at
the critical point. With this additional parameter the theory of
corresponding states becomes

$$X_r = f(P_r, T_r, Z_c),$$ (87)

where X is any property of state. Extensive tables and charts of reduced properties have been developed by Lyderson, et al. [60] using
the relation in equation (87). A discussion of the methods used to
develop these tables and charts has been given by Hougen and Watson
[61]. Examples of generalized enthalpy and entropy charts are shown
in Figures 12 and 13.

IV.   TRANSPORT PROPERTIES

Transport properties are properties which depend upon the motions of atoms or molecules. These particles possess mass, velocity,
momentum and energy. When a non-equilibrium state results from the
presence of a gradient in the mean velocity of the particles, a
transport of momentum occurs, governed by the coefficient of viscosity, which is a kind of coefficient of internal friction. Temperature gradients produce a transport of kinetic energy, governed by the
coefficient of thermal conductivity, and concentration gradients produce a transport of mass, governed by the coefficient of diffusion.
Mass transport in a mixture can also result from a temperature gradient, and is governed by the coefficient of thermal diffusion.

It was shown in the previous section that the first and second
laws of thermodynamics provide a framework for the study of the effects of pressure on equilibrium properties. These effects depend
primarily upon the forces between molecules and the manner in which
these forces change with distance. The subject of irreversible thermodynamics [30] provides the corresponding framework for the study of
the effects of pressure on transport properties. It is more complex
than equilibrium thermodynamics, mainly because of the additional
variables needed to account for molecular motions and collisions. In
systems of polyatomic molecules momentum and energy are transferred
in part through vibrational and rotational degrees of freedom, hence
the transport properties are sensitive to molecular characteristics
which are unimportant in equilibrium calculations. Moreover, statistical theory shows that transport properties are more sensitive than
equilibrium properties to the details of the intermolecular potential
function; consequently, advances in the theory of transport properties have been closely linked with the development of methods of obtaining precise information about these functions. Theoretical meth-

ods for predicting and correlating transport coefficients for dense fluids are considerably less advanced than those used for equilibrium properties [56].

A theorem of corresponding states has been derived for transport properties [31]. The expressions for the reduced coefficients of viscosity, thermal conductivity, and diffusion are, respectively,

$$\eta^* = \eta\sigma^2/(M\varepsilon)^{\frac{1}{2}} \qquad\qquad (88)$$

$$\lambda^* = (\lambda\sigma^2/k)\ (M/\varepsilon)^{\frac{1}{2}},\ \text{and} \qquad\qquad (89)$$

$$D^* = DM^{\frac{1}{2}}/\varepsilon^{\frac{1}{2}}\sigma, \qquad\qquad (90)$$

where M is the molecular mass, k is Boltzmann's constant, and ε and σ are the energy and length parameters of the intermolecular potential function. As in the case of equilibrium properties, the theorem states that the reduced transport coefficients of fluids at the same reduced values of any two independent properties (e.g., temperature and pressure) should be equal if the molecular interaction potentials have the same functional form. Because of the greater sensitivity of transport properties to the details of the intermolecular potential function, corresponding states correlations of these properties, for fluids at high pressures, have been less successful than correlations of the equilibrium properties. Some success has been achieved, however, in corresponding-states treatment of limited classes of substances of similar molecular structure. Thodos and co-workers [32] have undertaken an extensive program to develop reduced-state correlations of the transport properties of a number of pure substances of industrial importance, including hydrogen, methane, ethylene, ammonia, water, and others. McLaughlin [33] and Gubbins [35] have reviewed the theory of transport properties in dense fluids.

A.   Viscosity

The most extensive systematic studies of the effects of pressure on viscosity are those of Bridgman, who studied more than forty substances at pressures to 12 kbar and several others to 30 kbar [4]. More recent experiments have been carried out at pressures as high as 60 kbar [34]. In some cases Bridgman's measurements were carried out for sub-cooled liquids -- that is, liquids in a metastable state at pressures and temperatures for which the solid is the thermodynamically stable phase. With the notable exception of water at low pressures, and at temperatures near 0°C, the viscosities of all liquids increase with increasing pressure, although there are very great differences between the rates of increase for different liquids. At pressures above about 10 kbar, plots of the logarithm of viscosity vs. pressure are concave-upward for some liquids, indicating that an arithmetic increase of pressure is accompanied by a geometric (or even more rapid) increase of viscosity. This behavior is quite different from that observed in most of the equilibrium properties, for

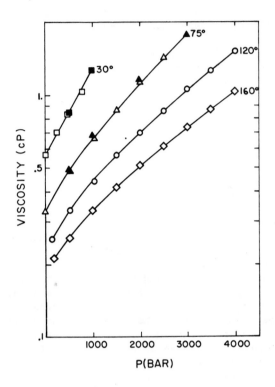

FIG. 14. The viscosity of benzene as a function of pressure
and temperature. The full symbols are Bridgman's measurements.
Reproduced by permission from H. J. Parkhurst and J. Jonas,
J. Chem. Phys., 63: 2705 (1975).

which there are qualitative and quantitative similarities for a variety of liquids, and for which the numerical values appear to approach
a limiting value at high pressures. Over the range 0 to 12 kbar, the
change in viscosity ranges from a factor of less than 2 for mercury
to a factor of more than 10^7 for eugenol. In contrast, changes in
compressibility of ordinary liquids over the same range are less than
a factor of about 15, and changes in thermal expansion are a factor
of 2 or 3. Experimental measurements also show that viscosity decreases with increasing temperature, but that the relative change
with temperature becomes markedly greater at high pressures, a behavior which again is abnormal in comparison with equilibrium properties. Recent measurements of the effects of pressure and temperature
on the viscosity of benzene [58], together with some of the measurements of Bridgman, are shown in Figure 14. A list of references to
experimental data on pressure dependence of viscosity in gases is contained in Table 2.
     Although some success has been achieved in explaining these phe-
nomena in terms of molecular theory [35], the simple heuristic ex-

Table 2.   References to experimental data on pressure dependence of
           viscosity in gases (ref. 63)

Gas	Reference	Gas	Reference	Gas	References
Ar	96-101	C_2H_4	111	H_2/He	119
Ar/H_2	98	C_2H_6	112,113	H_2/N_2	120
Ar/He	102,103	C_3H_8	112	H_2O	121,122
Ar/N_2	103	$\underline{n}-C_4H_{10}$	112	He	100,119,123
Ar/NH_3	104	$\underline{n}-C_5H_{12}$	114	Kr	124
Air	99,105,106	CO	115	N_2	99,100,115,120,125, 128
CH_4	107	CO_2	96,116	N_2/O_2	129
CH_4/C_3H_8	108	D_2	117	Ne	130
CH_4/C_4H_{10}	109	Freons	118	NH_3	96,131
CH_4/H_2	110	H_2	100,117	O_2	96,129,132
C_2H_2	96				

Experimental data for the variation viscosities of many liquids are
given by Bridgman, refs. 3, 4.

Table 3.   References to experimental data on pressure dependence of
           thermal conductivity in liquids and gases (ref. 63)

Substance	References	Substance	References
(a) Liquids		Toluene	78
Ar	89,90	1-Undecene	85
CH_4	90	Water	92,95
Cs	91	Xenon	90
CO_2	88		
Ethyl acetate	79	(b) Gases	
$\underline{n}$-Heptadecane	87	Ar	65
$\underline{n}$-Heptane	83	Air	64
$\underline{n}$-Hexane	82	C_3H_6	66
Iso-Propyl Alcohol	81	C_3H_8	67,68
Kr	90	C_2H_6/CO_2	69
$\underline{n}$-Octane	82	C_2H_6/N_2	69
1-Octene	84	CO_2	70-73
$\underline{n}$-Pentadecane	86	CO_2/H_2	69
Propane	68	He	74
Propyl alcohol	81	N_2	75,76
Propyl formate	80	Ne	77

planation proposed by Bridgman [3] is still very useful. He observed
that pressure effects are largest for substances with the most complex molecules, and suggested that the increase in viscosity with
pressure is a result of the interlocking of complex structures, which
impedes the slipping of molecules past one another. According to
this idea a rise in temperature, which increases the kinetic energy
of the molecules, tends to unlock the structure and decrease the viscosity. The extent of interlocking, and hence the rate of increase
of viscosity with pressure, should increase with the complexity of
the molecules. All of these effects are observed, and provide an in-
sight into viscosity differences for different classes of fluids.
The concept of the "complexity of a molecule" is a hazy one, however,
as Bridgman himself admitted, and it has been of limited usefulness
in predicting and correlating viscosities at high pressures.

B.   Thermal Conductivity

The effect of pressure on the thermal conductivities of several
fluids is shown in Figure 15, based on measurements of Le Neindre
[62]. The general effect of pressure is the same for most fluids,
with thermal conductivities increasing by a factor of about 2 or 3
over a range of 10 kbar. The rate of change decreases with increasing pressure, as in the case of the equilibrium properties. The
change between 6 and 10 kbar is only about half that between 0 and 6
kbar. A list of references to experimental data on pressure dependence of thermal conductivity in liquids and gases is contained in
Table 3.

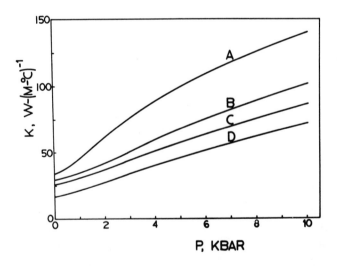

FIG. 15. The thermal conductivity, K, as a function of
pressure for several gases, from measurements by Le Neindre
[62]. Legend: A, methane at 25°C;   B, carbon dioxide at 200°C;
C, nitrogen at 25°C;   D, Argon at 25°C.

C. Diffusion

Measurements of self-diffusion in dense fluids are mostly of recent origin, having been carried out by the use of radio-active tracers or by the decay properties of induced nuclear magnetism. Measurements at high pressures are relatively scarce.

Diffusion in dense fluids can be related to viscosity through the Stokes-Einstein theory [35], which gives the coefficient of self-diffusion for molecules of diameter σ in a medium of viscosity η as

$$D = kT/C\pi\sigma\eta, \tag{91}$$

where k is Boltzmann's constant and C is a numerical factor with a value between about 3 and 6. For monatomic and small polyatomic molecules σ can be taken to be a constant (the hard sphere approximation) and it follows that diffusion varies inversely as viscosity. Figure 16 shows recent measurements of self-diffusion and viscosity, and calculated values of the factor $kT/\pi D\eta\sigma$ for tetramethylsilane, reported by Parkhurst and Jonas [58]. The Stokes-Einstein relation holds very well in this case.

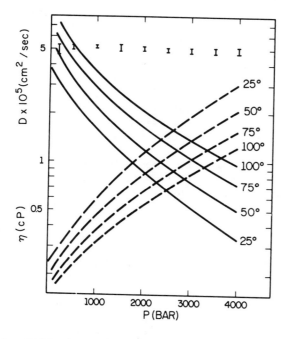

FIG. 16. Diffusion, viscosity and Stokes-Einstein relation for tetramethylsilane. The solid lines show the experimental results for self-diffusion and the dashed lines the experimental results for viscosity. The vertical bars indicate the range of the calculated values of $kT/\pi D\eta\sigma$ from 25 to 100°C. Reproduced by permission from H. J. Parkhurst and J. Jonas, J. Chem. Phys., 63: 2705 (1975).

V.   PHASE EQUILIBRIA IN FLUID MIXTURES

Compared to pure fluids, relatively little is known about the effects of pressure on mixtures. An exception to this rule is phase equilibria, for which extensive systematic experiments have been carried out on fluid mixtures in recent years. Phase behavior is one of the fundamental properties of a mixture. In addition to its practical importance in separation processes, knowledge of the phase behavior of a mixture is a prerequisite for other types of high-pressure experiments, inasmuch as it establishes the pressure-temperature-composition boundaries which define the limits of existence and coexistence of separate phases.

Excellent descriptions of equipment and techniques used in high pressure phase equilibria experiments have been presented by Tsiklis [12] and Schneider [36].

The phase rule of Gibbs provides the framework for the description of phase diagrams. This rule is expressed by the equation $f = c + 2 - p$, where f is the number of independent variables in an equilibrium system of c components and p phases. If we consider the variables P, V, and T, it follows that a single phase, which is characterized by two independent variables, is represented in PVT space by a surface, having an equation of the form $V = V(P,T)$, or $T = T(P, V)$. The coexistence of two phases is described by two lines, and that of three phases by three points. In other words the vapor pressure, sublimation, and melting curves, which appear as single lines in two-dimensional P-T diagrams, each have two branches in three dimensional PVT space, because the coexisting phases have different specific volumes. Similarly, the triple point becomes three points in PVT space, because the coexisting solid, liquid, and gas phases have different volumes. Since these phases coexist at the same pressure and temperature, each pair of lines and each triplet of points has a common projection on the P-T plane. The addition of a second component adds another degree of freedom to the system, and another dimension to the geometry of the figures required to describe the equilibrium between a fixed number of phases.

In describing the phase behavior of two-component systems, the most convenient variables are pressure, temperature, and composition (P,T,X). A two-component system can be completely described by a three-dimensional PTX diagram. These diagrams are often highly complex and difficult to visualize in three dimensions. Characteristics which can aid in interpreting these diagrams can be summarized as follows:

1. In the PTX phase diagram of a two-component system, a single phase is represented by a volume, equilibrium between two phases by two surfaces, equilibrium between three phases by three lines, and equilibrium between four phases by four points.

2. Among the key features of these diagrams are the boundary lines at which the system possesses a single degree of freedom. These are: (a) the pure component two-phase boundary lines (vapor pressure, melting and sublimation curves) in the two P-T faces of the diagram;

             (b) critical lines, at which two phases become iden-
             tical; and (c) three-phase lines, which describe the
             coexistence of three distinct phases.
     3.   A three-phase, two-component system has a single degree
             of freedom -- that is, only one independent intensive
             property -- hence the pressure at which three phases
             coexist is a unique function of temperature for a
             particular set of three phases. The consequential
             geometry requires that the three lines representing
             these phases lie in a ruled surface perpendicular to
             the P-T faces of the diagram. They therefore have
             common projections upon those faces.
     In single-component systems, equilibrium between two fluid phases
is normally limited to liquid-gas equilibrium, but in two-component
systems there are two additional types, liquid-liquid and gas-gas.

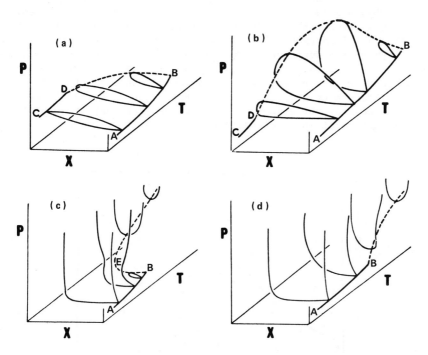

     FIG. 17. P-T-X sketches showing several types of critical
lines in binary systems. In (a) and (b) the critical line is a
continuous line in PTX space joining the two pure-component
critical points B and D. Figs. (c) and (d) show critical lines
for systems which exhibit gas-gas equilibria. In these systems
the critical point of the second component lies at lower temperatures, out of the range of the diagram. (See sec. V.C. below.)

A.   Liquid-Gas Equilibria

Figures 17a and b show gas-liquid critical lines for two systems with continuous critical lines. AB and CD are the vapor pressure curves of the pure components, with critical points B and D. Two surfaces, representing the two coexisting phases, extend across the diagram between these two lines. The remaining solid lines within each diagram are lines cut in these surfaces by isotherms. The dashed line BD is the mixture critical line, a boundary at which the liquid phase, represented by the upper surface, becomes identical to the gas phase, represented by the lower surface. Critical points on the individual isotherms always occur at pressure maxima (or minima). The various types of continuous critical lines are summarized in Figure 18, where they are shown projected onto the P-T plane. These lines may pass monotonously from one critical point to the other, or they may exhibit maxima or minima in temperature or maxima in pressure. A large number of continuous critical lines for specific systems are shown in Figure 6.14 of the book by Rowlinson [40] (first edition only). In systems with continuous critical lines, the maximum critical pressure seldom exceeds a few hundred bars. The critical line is not always continuous, but is sometimes interrupted by the appearance of other liquid or solid phases. The intersection of a critical line and a three-phase region results in the termination of both, and is usually called a critical end point. Rowlinson [40] has described some of the many variations of critical and three-phase combinations in fluid systems at low and moderate pressures.

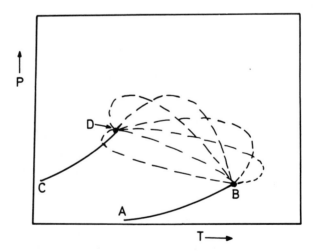

FIG. 18.   P-T projection showing different types of continuous gas-liquid critical lines (the dashed lines) in binary systems. AB and CD are the vapor pressure curves of the two components, with critical points B and D.

B.   Liquid-Liquid Equilibria

The effects of high pressures on liquid-liquid equilibria have
been extensively investigated by Schneider and co-workers [37-39], beginning in 1963, for pressures up to about 7 kbar. Several types of
hitherto unknown pressure dependence of immiscibility phenomena have
been observed. The different types of pressure- and temperature-de-
pendence of liquid-liquid immiscibility are summarized in Figure 19.
In the left hand column, Figures a to d show the four different types
of temperature-composition diagrams for constant pressure. Type a exhibits an upper critical solution temperature, T_C^U, type b a lower
critical solution temperature, T_C^L, and types c and d exhibit both T_C^U
and T_C^L. The uppermost horizontal row, e to h, shows that the same
types of diagrams occur in pressure-composition space at constant
temperature. The remaining rows show the three-dimensional characteristics of different types of liquid-liquid equilibria found or presumed to exist by Schneider. In these figures the dark lines extending from left to right across the figures are liquid-liquid critical

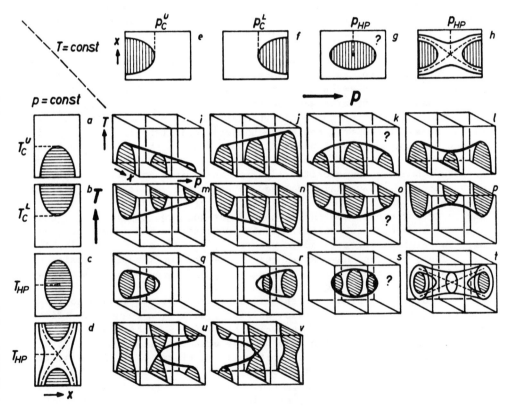

FIG. 19. The effects of pressure on various types of liquid-
liquid equilibria. Reproduced by permission from G. M. Schneider,
Ber. Bunsenges. Phys. Chem., 76: 325, (1972). See text for discussion.

lines -- that is, lines at which the two coexisting phases become identical. The two coexisting phases are represented by two surfaces which meet at the critical lines. The dark boundaries of the shaded areas are T-X slices through these surfaces. The second row, i to ℓ, shows that with increasing pressure upper critical solution temperatures (UCST's) can either decrease (i), increase (j), or pass through a minimum (ℓ). The question mark in k indicates that systems with maxima in the UCST curve have not yet been observed. The third horizontal row, q to t, shows that closed loops in T-X space can either expand (r) or contract (q) with increasing pressure, and sometimes disappear completely. Figure 19t shows the behavior of the system 2-methylpyridine-water. This system exhibits a closed T-X loop at low pressures which disappears at approximately 200 bars and reappears again at approximately 2000 bars. Further details of liquid-liquid phase behavior at high pressures are described in reviews by Schneider [37-39].

C.   Gas-Gas Equilibria

Near the turn of the century, van der Waals [41], applying his equation of state to the study of the AVX surface of mixtures (A is Helmholtz free energy) predicted the existence of phase separations in binary fluid mixtures at temperatures above the critical temperatures of both components. Kamerlingh-Onnes and Keesom [42] discussed these phase separations in more detail, describing them as limited miscibility in the gas phase. The first experimental confirmation of van der Waals' prediction was reported in 1940 by Krichevskii [43], who observed that mixtures of nitrogen and ammonia separate into two fluid phases at temperatures above the critical temperatures of nitrogen and ammonia, at pressures above about 1 kbar. In subsequent experiments, at pressures up to 15 kbar, Krichevskii, Tsiklis and co-workers observed supercritical phase separations in more than 20 binary mixtures. They have called these phase separations gas-gas equilibria, and this rather misleading term has gained wide acceptance. More recently the phenomenon of gas-gas equilibrium has been studied in many more systems, mainly by Schneider and co-workers [37-39], and also be Streett and co-workers [44 - 46] and others [5, 47, 48]. These studies have shown that there are continuous transitions between phase diagrams of the gas-gas type and those of the more familiar gas-liquid and liquid-liquid types.

Two different kinds of transitions from liquid-liquid and gas-liquid equilibria to gas-gas equilibria have been found. The first of these, which has been extensively studied by Schneider and co-workers in mixtures of carbon dioxide or water with families of hydrocarbons, is illustrated schematically in Figure 20. In Fig. 20a the system n-octane-carbon dioxide exhibits a liquid-liquid phase separation of the type shown in Fig. 19j, and a liquid-gas critical line similar to Fig. 17b. With increasing complexity of the hydrocarbon component, the liquid-liquid critical line moves to higher temperatures and eventually merges with the gas-liquid critical line from the critical point of the hydrocarbon component as in Fig. 20b (n-hexadecane-carbon dioxide). The critical line from the critical

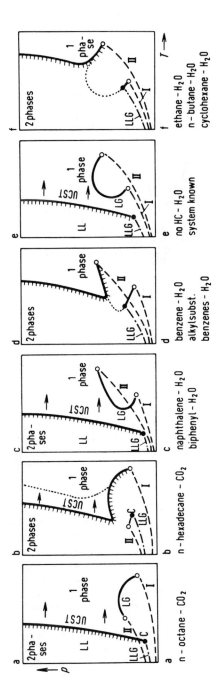

FIG. 20. P-T diagrams showing transitions from liquid-liquid and gas liquid equilibria to gas-gas equilibria reported by Schneider. Legend: ———— and шшшш, critical lines; c, pure component critical points; o, critical end points (the intersection of a critical line and a three-phase region); ---, pure component vapor pressure curves; - - —, three phase regions. These figures are meant to be interpreted in pairs: a + b, c + d, and e + f. Each pair shows how a phase diagram with separate gas-liquid and liquid-liquid critical lines (a, c, and e) is transformed into a system with a gas-liquid critical end point and a gas-gas critical line (b, d, and f). The difference between a, c, and e is the type of gas-liquid critical line. Reproduced by permission from G. M. Schneider, Ber. Bunsenges. Phys. Chem., 76: 325, (1972).

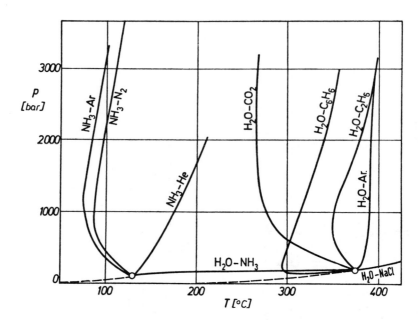

FIG. 21.   Critical lines in ammonia and water systems.
The critical line for ammonia-water is a continuous line between the two critical points of the pure components. The remaining lines (excluding H_2O-NaCl) are gas-gas type critical
lines for binary mixtures in which either water or ammonia is
the less volatile component. Reproduced by permission from
G. M. Schneider, Ber. Bunsenges. Phys. Chem., 76: 325 (1972).

point of carbon dioxide ends in an intersection with the three-phase
region liquid-liquid-gas, forming an upper critical end point C.
Figures 20c, d, e, and f illustrate similar transitions for slightly
different kinds of gas-liquid critical lines. Figure 21 shows critical P-T curves for water and ammonia systems at pressures up to 3
kbar. In each system the region to the left of the critical line is
a region of two fluid phases and the region to the right is a homogeneous single phase. Systems in which the critical lines always have
a positive slope in the P-T projection have been called gas-gas equilibria of the first type (e.g., NH_3 -He, H_2O-Ar, H_2O-NaCl), while
those in which the critical line passes through a temperature minimum
(NH_3-Ar, NH_3-N_2, H_2O-C_6H_6, etc.) have been called gas-gas equilibria
of the second type. The three-dimensional features of these two
types are shown in Figs. 17(d) and 17(c), respectively. The distinction between them is of no fundamental significance.
     The second type of transition to gas-gas equilibria has been ob-
served mainly by Streett and co-workers [44 - 46] in mixtures of light
gases at high pressures and low temperatures. These systems develop
in the sequence shown in Figure 22. Figure 22a is a P-T projection of
the important boundary lines of a system in completely miscible liquid
phases solidify to form partially miscible solid phases. AB and CD

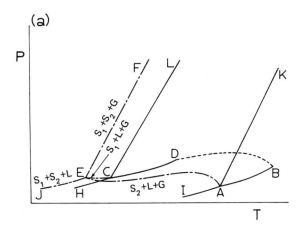

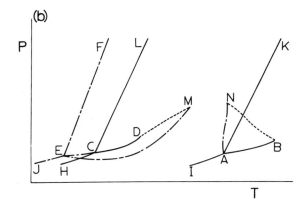

FIG. 22.   P-T diagrams showing how the transition from
gas-liquid to gas-gas equilibria occurs in mixtures of light
gases at low temperatures and high pressures [44-46].   See
text for discussion.

are the vapor pressure curves, AK and CL the melting curves, and IA
and HC the sublimation curves.   DB is the critical line, and the remaining lines JE, EF, EC, and EA are three-phase lines.   (Each of the
latter consists of three separate lines in PTX space, but each triplet has a common P-T projection.)   E is a quadruple point.   There is
no liquid-liquid phase separation in these systems.   If the temperature of D, the critical point of the first component, lies far below
that of A, the triple point of the second component, both the critical line and the three-phase region solid-liquid-gas (EA) curve upward, and eventually intersect to form two critical end points, M and
N, as in Fig. 22b.   It is the right hand portion of this figure which
is of interest here.   The triangular region ABN is the region of coexistence of two fluid phases, bounded by the vapor pressure curve

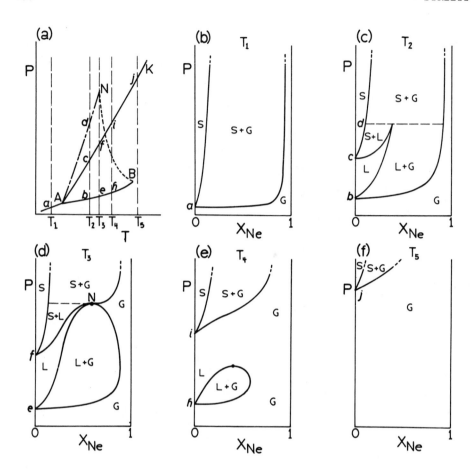

FIG. 23.   P-T and P-X diagrams for the system neon-argon
[49] at pressures to about 1 kbar (not to scale). See text for
discussion.

AB, the three-phase region AN, and the critical line BN. For systems
with progressively greater differences between the critical temperatures of the pure components, the critical line becomes steeper and
its intersection with AN rises to higher temperatures and pressures,
until finally the critical line and the three-phase region do not intersect, and the gas-gas region extends to very high pressures at
high temperatures. This sequence of transitions is illustrated in
Figures 23 to 26 which show, respectively, phase diagrams for neon-
argon [49], helium-argon [44], helium hydrogen [46], and helium
methane [45].
     Figure 23(a) is a projection of the principal phase boundary
lines of the neon-argon system on the P-T plane. (These are the
lines at which the system has a single degree of freedom.) The
three-phase region (AN) and the critical line (BN) end at their intersection, the critical end point N. AB and AK are the vapor pres-

sure and melting curves of argon, which lie in one P-T face of the PTX
diagram.   The remaining diagrams, Figs. 23(b) to 23 (f), are P-X isotherms corresponding to the vertical dashed lines T_1-T_5 in (a).   In
(b)-(f), single-phase regions are indicated by a single letter (L, S,
or G),   and two-phase regions by two letters connected by a plus sign
(S + G, etc).   A point within one of the two-phase regions represents
a mixture of fixed total composition at fixed temperature and pressure.   The compositions of the two coexisting phases are found at the
points where a horizontal line through this point intersects the
boundaries of the two-phase region.   (Detailed explanations of the
interpretation of binary phase diagrams for fluid systems can be found
in most standard textbooks on chemical engineering thermodynamics.
See also Ch. 4, Vol. II.)   The horizontal dashed lines in (c) and (d)
mark the three-phase pressures for isotherms T_2 and T_3.   Figure (d)
shows that at T_3,   the temperature of the critical end-point, the
liquid-gas region (L + G) forms a dome, on which the pressure maximum

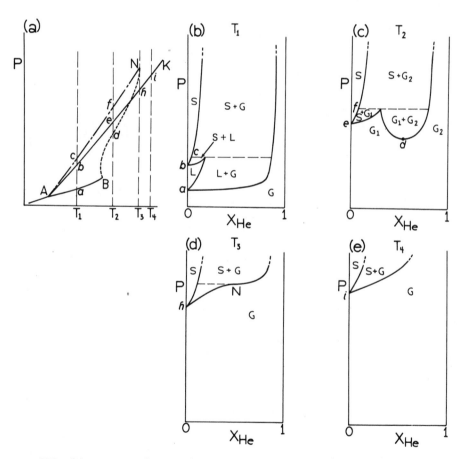

FIG. 24.   P-T and P-X diagrams for the system helium-argon
[44] at pressures to about 11 kbar (not to scale).   The explanation follows that for Fig. 23 (see text).

is a gas-liquid critical point. The dome is tangent to the L and G branches of the remaining two phase regions, S + L and S + G, which become indistinguishable at higher temperatures. At temperatures above T_3, the L + G and S + G regions are separated by a region of complete miscibility. The L + G phase separation dies out as the temperature approaches the critical temperature of argon. This system does not exhibit gas-gas equilibrium, since the region of two fluid phases, ABN in Figure 23(a), is confined to temperatures below the critical temperature of argon. The critical end-point N marks the highest pressure at which two fluid phases coexist. This point lies at a pressure of about 1 kbar and at a temperature of 93.3 K -- about 58° below the argon critical temperature. The critical line rises steeply as it approaches N, and it might well have taken on a positive slope at high pressures (leading to gas-gas equilibrium) had it not been terminated by the appearance of a solid phase -- that is, by the solidification of the argon-rich liquid phase. Further details of this system can be found in the original paper by Streett and Hill [49].

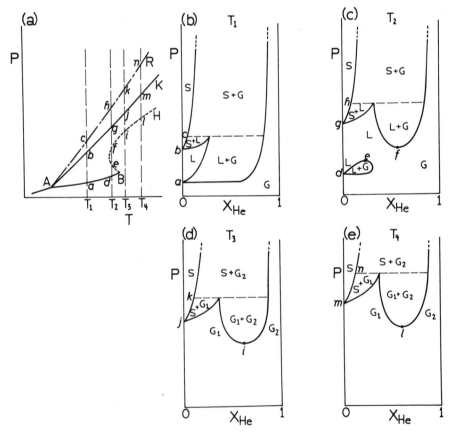

FIG. 25.   P-T and P-X diagrams for the system helium-hydrogen at pressures to about 10 kbar (not to scale).   The explanation follows that of Figs. 23 and 24 (see text).

     The helium-argon system, Figure 24, is similar to neon-argon
(Figure 21), but the critical end-point N now lies at a temperature of
about 205 K (54° above the critical temperature of argon) and at a
pressure of about 11 kbar. The reversal in the slope of the critical
line with increasing pressure opens the region of coexistence of two
fluid phases to higher pressures and temperatures, but it is eventually terminated at the critical end-point N. Further details of this
system can be found in the original paper by Streett and Erickson
[44].

     In the hydrogen-helium system, Figure 25, the critical line BH
turns away from the three phase region AR, and the region of coexistence of two fluid phases is unbounded at the highest pressures (~10
kbar) and temperatures (~100 K) at which the system has been studied.
Comparison of Figs. (d) and (e) shows that the pressure depth of the
gas-gas region increases with increasing temperature; that is, the
gas-gas phase separation is expanding at the limits of the experimental measurements and shows no signs of terminating in a critical end-
point. Further details of this system can be found in the original
paper by Streett [46]. The helium methane system [45], Figure 26, is
similar to hydrogen-helium, except that the slope of the critical line
is everywhere positive in the P-T diagram. The critical line for this
system has been followed experimentally to 290 K (about 100° above the
critical temperature of methane) at which the critical pressure is approximately 8 kbar. In these systems the gas-gas phase separation apparently persists to very high pressures and temperatures, perhaps
even to the limits of stability of the molecular phases.

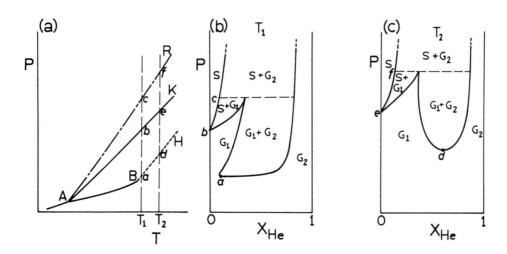

FIG. 26. P-T and P-X diagrams for the system helium-
methane [45] at pressures to about 10 kbar (not to scale).
See text for discussion.

VI.   FUTURE PROGRESS

The rapid growth of high pressure research during the last 25 years has made available a wide variety of experimental equipment and techniques for the study of fluids under pressure. Valves, fittings, compressors, and other devices, suitable for use with fluids at pressures up to 10 kbar or higher, are now available from commercial sources. The accumulation of experimental data from high pressure research on fluids can be expected to continue, and to accelerate as high pressure processes come into more widespread use in industry.

In the immediate future, the most important advances in the field of fluids at high pressures are likely to come from advances in the statistical mechanical theory of dense fluids. One of the principal goals of the science of materials is the development of theories and methods for predicting the properties of matter at the macroscopic level from knowledge of properties and behavior at the microscopic or molecular level. In the cases of dilute gases and crystalline solids, significant progress toward this goal has been achieved. The statistical mechanical problems are simplified in the former case because the effects of intermolecular forces can be ignored, and in the latter case because of the high degree of order in the crystalline lattice. Progress in the molecular theory of dense fluids has been slow, both because of a lack of precise knowledge of the forces between individual molecules and because the complex integral equations which arise in the statistical mechanical theory have defied general solution. The primary emphasis in theoretical work on dense fluids has been on classical systems of simple spherical molecules, which are most closely approximated in real fluids by the heavy noble gases (argon, krypton and xenon). It is necessary for the equilibrium theory of these simple fluids to be satisfactorily developed before proceeding to real systems of complex molecules. During the past decade this groundwork has been reasonably well established [14, 50].

A key factor in recent progress in the molecular theory of dense fluids has been the development of computer simulations of fluids using Monte Carlo [6] and molecular dynamics [7] methods. These techniques, which first came into use in 1957, have brought important insights into structure and motions at the molecular level, and have made possible computer "experiments" in which the properties of hypothetical substances, whose molecules interact according to well-defined potential functions, can be calculated. Accurate data on the properties of these model systems has contributed to the success of the so-called perturbation theories [51], which have been reasonably successful in predicting properties of real fluids. Computer simulations have illuminated the central role of the steep repulsive force between molecules in determining the properties of dense fluids. This point has been well made by Alder and Hoover [7]:

> "...the high-temperature behavior of a fluid can
> be used as the 'ideal' state, which brings with
> it the great simplification that only the repul-
> sive part of the potential need be considered in
> determining the structure and many other proper-
> ties of the fluid. ...For example, even the

structure of metallic and ionic systems is nearly
the same as that of insulating materials at corresponding densities. Thus the crucial idealiza-
tion of all liquids is that their primary behavior
is determined by the steep repulsive potential
which is best idealized by an infinitely steep one,
namely a hard sphere potential. ...The overall impact of computer studies is then to revive two very
old models, van der Waals' and Enskog's, for calculating the equilibrium and transport properties
of fluids."

These comments have important implications for the study of
fluids at high pressures. They suggest that if the details of the repulsive parts of intermolecular potential functions can be measured or
calculated, then computer simulations and theoretical methods can be
used to predict the bulk properties of fluids at any desired pressure
and temperature. The prospect of using these methods to calculate
properties of fluids at extreme pressures and temperatures, beyond the
range of laboratory experiments, is particularly attractive, and some
attempts along these lines have already been made. Ross and Alder
[52] have used the Monte-Carlo method to study melting at high pressures and temperatures, and Hubbard and Slattery [53] have used it to
simulate molecular hydrogen at temperatures as high as 3500 K and
pressures in excess of 1 megabar -- conditions which are presumed to
exist in the interior of the planet Jupiter.

REFERENCES

1. T. Andrews, Phil. Trans., 159: 575 (1869).
2. E. H. Amagat, Ann. Chim. Phys., 29: 68 (1893).
3. P. W. Bridgman, Physics of High Pressure, Bell & Sons,
    London (1949).
4. P. W. Bridgman, Collected Experimental Papers, Vols. I-VII,
    Harvard University Press, Cambridge, Mass. (1964).
5. S. Takenouchi and G. C. Kennedy, Amer. J. Sci., 262: 1055
    (1964).
6. W. W. Wood in Physics of Simple Liquids (H.N.V. Temperley,
    G. S. Rushbrooke and J. S. Rowlinson, eds.), Wiley,
    New York (1968).
7. B. J. Alder in Physics of Simple Liquids (H.N.V. Temperley,
    G. S. Rushbrooke and J. S. Rowlinson, eds.), Wiley,
    New York (1968).
8. L. Merrill (ed.), High Pressure Bibliography, 1900-1968,
    Vols. I & II, High Pressure Data Center, Brigham Young
    University, Provo, Utah (1970).
9. J. F. Cannon and L. Merrill (eds.), A Compilation of Volumes
    I-IV of Bibliography of High Pressure Research, Vols. I and
    II, HIgh Pressure Data Center, Brigham Young University,
    Provo, Utah (1972).

10. Y. S. Touloukian (ed.), Thermophysical Properties of Matter, IFI/Plenum, New York, 1970; and Y. S. Touloukian, J. K. Gerritsen and W. H. Shafer (eds.), Thermophysical Properties Research Retrieval Guide, IFI/Plenum, New York (1973).

11. D. R. Lide and S. A. Rossmassler in Annual Review of Physical Chemistry, Vol. 24 (H. Eyring, ed.), Annual Reviews, Inc., Palo Alto, Calif. (1973).

12. D. S. Tsiklis, Handbook of Techniques in High Pressure Research and Engineering, Plenum, New York (1968).

13. J.M.H. Levelt-Sengers in Physics of High Pressures and the Condensed Phase, (A. van Itterbeek, ed.), North Holland, Amsterdam,(1965).

14. P. A. Eglestaff in Annual Review of Physical Chemistry, Vol. 24, (H. Eyring, ed.), Annual Reviews, Inc., Palo Alto, Calif. (1973).

15. J. J. Martin, Ind. Eng. Chem., 59: 34 (1967).

16. J. B. Ott, J. R. Goates and H. T. Hall, J. Chem. Ed., 48: 515 (1971).

17. J. G. Hust and R. D. McCarty, Cryogenics, 7: 200 (1970).

18. W. B. Streett, L. S. Sagan, and L.A.K. Staveley, J. Chem. Thermo., 5: 633 (1973).

19. W. B. Streett and M. S. Costantino, Physica, 75: 283 (1974).

20. W. B. Streett, Physica, 76: 59 (1974).

21. R. K. Crawford and W. B. Daniels, J. Chem. Phys., 50: 3171 (1971).

22. W. B. Streett, H. J. Raveche and R. D. Mountain, J. Chem. Phys., 61: 1960 (1974).

23. M. S. Benson, P. S. Snyder and J. Winnick, J. Chem. Thermo., 3: 891 (1971).

24. O. Verbeke and A. van Itterbeek in Physics of High Pressure and the Condensed Phase, (A. van Itterbeek, ed.), North Holland, Amsterdam (1965).

25. J. Stephenson, Can. J. Phys. 53: 1367 (1975).

26. K. S. Pitzer, J. Chem. Phys., 7: 583 (1939).

27. J. M. Prausnitz, Molecular Thermodynamics of Fluid-Phase Equilibria, Prentice-Hall, Inc., Englewood Cliffs, N.J. (1969).

28. W. B. Streett and L.A.K. Staveley, Physica, 71: 51 (1974).

29. G. N. Lewis, M. Randall, K. S. Pitzer and L. Brewer, Thermodynamics (2nd ed.) McGraw-Hill, New York (1961).

30. I. Prigogine, Thermodynamics of Irreversible Processes, Thomas, Springfield (1955).

31. E. Helfand and S. A. Rice, J. Chem. Phys., 32: 1642 (1960).

32. L. I. Stiel and G. Thodos in Progress in International Research on Thermodynamic and Transport Properties, (J. F. Masi and D. H. Tsai, eds.), Academic Press, New York (1962).

33. E. McLaughlin in Progress in International Research on Thermodynamic and Transport Properties (J. F. Masi and D. H. Tsai, eds.), Academic Press, New York (1962).

34. J. D. Barnett and C. D. Bosco, J. Appl. Phys., 40: 3144 (1969).

35. K. E. Gubbins in Statistical Mechanics, Vol. 1, (K. Singer, ed.), The Chemical Society, London (1973).

36. G. M. Schneider in Experimental Thermodynamics, International
    Union of Pure and Applied Chemistry (to be published).
37. G. M. Schneider, Advances in Chemical Physics, 17: 1 (1970).
38. G. M. Schneider, Fortschr. Chem. Fortschr., 13: 559 (1970).
39. G. M. Schneider, Ber. Bunsenges. Phys. Chem., 76: 325 (1972).
40. J. S. Rowlinson, Liquids and Liquid Mixtures (1st ed.)
    Butterworths, London (1959).
41. J. D. van der Waals, Zittinsvers. K. Akad. Wet. Amst., p. 133
    (1894).
42. H. Kamerlingh-Onnes and W. H. Keesom, Commun. Phys. Lab.
    Univ. Leiden, Suppl. No. 15 (1907).
43. I. R. Krichevskii, Acta. Phys. Chem. USSR, 12: 480 (1940).
44. W. B. Streett and A. L. Erickson, Phys. Earth and Planet.
    Interiors, 5: 357 (1972).
45. W. B. Streett, A. L. Erickson and J.L.E. Hill, Phys. Earth
    and Planet. Interiors, 6: 69 (1972).
46. W. B. Streett, Astrophys. J., 186: 1107 (1973).
47. J. de Swaan-Arons and G.A.M. Diepen, J. Chem. Phys., 44:
    2322 (1966).
48. N. J. Trappeniers and J. A. Schouten, Phys. Lett. 27A: 340
    (1968).
49. W. B. Streett and J.L.E. Hill, J. Chem. Phys., 54: 5088 (1971).
50. G. S. Rushbrooke in Physics of Simple Liquids, (H.N.V.
    Temperley, G. S. Rushbrooke and J. S. Rowlinson, eds.),
    North Holland, Amsterdam (1968).
51. W. R. Smith in Statistical Mechanics, Vol. 1 (K. Singer, ed.),
    The Chemical Society, London (1973).
52. M. Ross and B. J. Alder, Phys. Rev. Lett., 16: 1077 (1966).
53. W. L. Slattery and W. B. Hubbard, Astrophys. J., 181: 1032
    (1973).
54. D. H. Liebenberg, R. L. Mills and J. C. Bronson, J. Appl.
    Phys. 45: 741 (1974).
55. S. D. Hammann, Physico-Chemical Effects of Pressure, London,
    Butterworths (1957).
56. T. M. Reed and K. E. Gubbins, Applied Statistical Mechanics,
    McGraw Hill, New York (1973).
57. J. R. McDonald, Revs. Mod. Phys., 41: 316 (1969).
58. H. J. Parkhurst and J. Jonas, J. Chem. Phys., 63: 2705 (1975).
59. G. J. Van Wylen and R. E. Sonntag, Introduction to Classical
    Thermodynamics, John Wiley and Sons, New York (1965).
60. A. L. Lyderson, R. A. Greenkorn and O. A. Hougen, "Generalized
    Thermodynamic Properties of Pure Fluids", Univ. Wisconsin Eng.
    Exp. Sta. Rept. 4 (Oct. 1955).
61. O. A. Hougen, K. M. Watson and R. A. Ragatz, Chemical Process
    Principles, Part II, Chap. 14, John Wiley and Sons, New York
    (1959).
62. B. Le Neindre, Int. J. Heat Mass Transfer, 15: 1 (1972).
63. P. Bolsaitis (private communication).
64. D. L. Carrol, H. L. Lo, and L. E. Stiel, J. Chem. Eng. Data,
    13: 53 (1968).
65. A. Michels, et al., Physica, 22: 121 (1955), ibid. 29: 149 (1963).
66. Y. M. Naziev and A. A. Abasov, Gazov Prom., 15: 37 (1970).

67. D. E. Leng and E. W. Comings, Ind. Eng. Chem. (High Press. Symp. Am. Chem. Soc.), 49: 2042 (1957).

68. L. T. Carmichael, J. Jacobs and B. H. Sage, J. Chem. Eng. Data, 13: 40 (1968).

69. T. F. Gilmore and E. W. Comings, AICHE J., 12: 1172 (1966).

70. A. Michels and J. V. Sengers, Physica, 28: 1238 (1962).

71. A. Michels, J. V. Sengers and P. S. Van der Gulik, Physica, 28: 1201, 1216 (1962).

72. V. V. Altunin and M. A. Sakhabetdinov, Teploenergetica, 5: 85 (1973).

73. L. A. Guildner, J. Res. Nat. Bur. St. (U.S.A.), 66A: 341 (1962).

74. N. B. Vaigaftic and N. H. Zimina, At. Energ. (USSR), 19: 300 (1965).

75. A. Michels and A. Botzen, Physica, 19: 585 (1953).

76. P. Johannin, Compt. Rend., 244: 2700 (1957).

77. J. V. Sengers, W. T. Wolk and C. J. Stigter, Physica, 30: 1018 (1964).

78. R. Kaudiyoti, E. McLaughlin and J.F.T. Pittman, J. Chem. Soc. Faraday Trans., 1: 1953 (1973).

79. K. D. Guseinov and K. Nadzhidov, Teplofiz. Vys. Temp., 11: 215 (1973).

80. K. D. Guseinov and K. Magerramov, Izv. Vyssh. Ucheb. Zared. Neft. Gaz., 16: 75 (1973).

81. T. N. Vasilkovskaya and I. L. Galubev, Teploenergetica, 16: 84 (1969).

82. Y. M. Naziev and A. A. Nurbedyev, Chem. Tech (Leipzig), 24: 546 (1972).

83. Y. M. Naziev, A. A. Nurbedyev and A. A. Abasov, Chem. Tech. (Leipzig), 23: 738 (1971).

84. Y. M. Naziev and A. A. Abasov, Izv. Vyssh. Ucheb. Zaved. Neft. Gaz., 12: 65 (1969).

85. R. A. Mustafaev, Zh. Fiz. Khim., 47: 2716 (1973).

86. R. A. Mustafaev, Zh. Fiz. Khim., 47: 1043 (1973).

87. R. A. Mustafaev, Izv. Vys. Ucheb. Zaved. Fiz., 11: 159 (1973).

88. V. V. Altunin and M. A. Sakhabetdinov, Teploenergetica, 5: 85 (1973).

89. H. Ziebland and J.T.A. Burton, Brit. J. Appl. Phys., 9: 52 (1958).

90. L. D. Ikenberry and S. A. Rice, J. Chem. Phys., 39: 1561 (1963).

91. J. M. Hochman and C. R. Bonilla, Nucl. Sci. Eng., 22: 434 (1965).

92. U. L. Rastorguev and V. V. Pugach, Teploenergetica, 17: 77 (1970).

93. A. W. Lawson, R. C. Lowell and A. L. Jain, J. Chem. Phys., 30: 643 (1959).

94. A. M. Mamedov, T. S. Akhundov and D. S. Ismailov, Teplofiz. Vys. Temp., 10: , 1329 (1972).

95. B. Leneindre, P. Johannin and B. Vodar, Compt. Rend, 258: 3277 (1964).

96. R. Kiyama and T. Makita, Rev. Phys. Chem. Japan, 22: 49 (1952).

97. A. Michels, A. Botzen and W. Schurrman, Physica, 20: 1141 (1954).

98. H. Iwasaki, Bull. Chem. Res. Inst. Non Aqueous Sol., Tohoku Univ., 6: 61 (1956).

99. T. Makita, Rev. Phys. Chem. Japan, 27: 16 (1957).

100. J. A. Gracki, G. P. Flynn and J. Ross, J. Chem. Phys., 51: 3856 (1969).

101. J. Vermesse and D. Vidal, C. R. Acad. Sci. B, 277: 191 (1973).

102. H. Iwasaki and J. Kestin, Physica, 29: 1345 (1963).

103. N. E. Gnezdilov and I. F. Golubev, Teploenergetica, 14: 89 (1966).

104. H. Iwasaki, J. Kestin and A. Nagashima, J. Chem. Phys., 40: 2988 (1964).

105. H. Iwasaki, Sci. Rept. Res. Inst., Tohoko Univ., 3: 2 (1951).

106. H. Y. Lo, D. L. Carroll and L. I. Stiel, J. Chem. Eng. Data, 11: 540 (1966).

107. H. Iwasaki and H. Takahasi, Kogyo Kagaku Zasshi, 62: 918 (1959).

108. E.T.S. Huang, S. W. Swift and F. Kunata, AICHE J., 13: 846 (1967).

109. J. P. Dolan, R. T. Ellington and A. L. Lee, J. Chem. Eng. Data, 9: 484 (1964).

110. H. Iwasaki and H. Takahasi, Bull. Chem. Res. Inst. Non Aqueous Solutions, Tohoku Univ., 10: 1 (1961).

111. M. G. Gonikberg and L. F. Vereshchagin, Dokl. Acad. Nauk. USSR, 55: 813 (1947).

112. K. E. Starling, et al., in "Progress in International Research on Thermodynamic and Transport Properties" ASME, New York (1961).

113. B. E. Eakin, et al., J. Chem. Eng. Data, 7: 33 (1962).

114. N. A. Agaev and I. F. Golubev, Gaz. Prom., 8: 45 (1963).

115. C. Chierici and A. Paretella, AICHE J., 15: 786 (1969).

116. A. Michels, A. Botzen and W. Schurrman, Physica, 23: 95 (1957).

117. A. Michels, A.C.J. Schipper and W. H. Rintoul, Physica, 19: 1011 (1953).

118. T. Makita, Rev. Phys. Chem. Japan, 24: 74 (1954).

119. I. F. Golubev and N. E. Genzoilov, Gaz. Prom., 10: 38 (1965).

120. H. Iwasaki, Sci. Rept. Res. Inst., Tohoku Univ., 6: 3 (1954).

121. A. Nagashima and I. Tanishita, Bull. Japan Soc. Mech. Eng., 12: 1467 (1969).

122. S. L. Rivkin, A. Y. Levin and L. B. Izrailevski, Teploenergetica, 15: 74 (1963).

123. N. S. Runenko and A. I. Kaznus, Ukr. Fiz. Zh., 13: 1332 (1968).

124. N. J. Trappeniers, et al., Physica, 31: 945 (1965).

125. A. Michels and R. O. Gibson, Proc. Roy. Soc. (London), A134: 288 (1932).

126. F. Lazarre and B. Vodar, Compt. Rend., 243: 487 (1956).

127. J. Vermesse, P. Joliannin and B. Vodar, Compt. Rend., 3014 (1963).

128. J. Vermesse, Ann. Phys. (Paris), 4: 245 (1969).

129. H. Iwasaki and H. Takashashi, Bull. Chem. Res. Inst. of Non Aqueous Sol., 7: 77 (1958).

130. N. J. Trappeniers, et al., Physica, 30: 985 (1964).

131. H. Iwasaki and H. Takahaski, Rev. Phys. Chem. Japan, 38: 18 (1968).

132. T. Makita, Rev. Phys. Chem. Japan, 26: 74 (1956).

Chapter 13

PROPERTIES OF MATERIALS AT HIGH PRESSURE

P. Bolsaitis

Instituto Venezolano de Investigaciones Cientificas (IVIC)
Center of Engineering and Computation
Caracas, Venezuela

Ian L. Spain

Laboratory for High Pressure Science and
Engineering Materials Program
Department of Chemical Engineering
University of Maryland
College Park, Maryland 20742

I.   INTRODUCTION

The purpose of this chapter is to present a broad overview of the known effects of pressure on some of the more commonly useful properties of solid materials. The treatment is necessarily of a condensed nature since a complete discussion would be paramount to presenting an encyclopedia of properties of matter, since none is immune to the effects of changing pressure.

From a fundamental point of view, the effects of pressure are a result of changes in the band structure and of the energy content of a material produced by the reduction of interatomic spacing under applied pressure. In this regard condensed phases differ from gases in that applied pressure leads to complex changes in the electronic structure and energies by virtue of strong interaction between atoms, while in gases the effects of pressure and decreasing interatomic distances can generally be accounted for by relatively simple equations of state. It must be pointed out, however, that fairly substantial pressures are required (i.e. of the order of hundreds or thousands of bars) to produce appreciable effects on the properties of condensed phases. In recent years commercial equipment capable of producing hydrostatic pressures of up to about 70,000 bars has become available, pressures of 600,000 bars have been attained in specially designed equipment, and many researchers are setting their sights on the range up to 1 Mbar.

This chapter will discuss the compressibility and equation of state of materials and the influence of pressure on some physical properties including transport properties. The effects of pressure on mechanical properties and phase equilibria are topics of Chapters 14 in this volume and Chapter 4 in Vol. II respectively.

II.   EQUATIONS OF STATE: COMPRESSIBILITY AND ELASTIC CONSTANTS

A.   Some General Considerations

The equation of state of a material is a relationship which expresses the variation of the volume of a material with pressure and temperature, V(P,T). In practice, the form of the equation of state most frequently measured is the relationship between the volume of a material and the applied hydrostatic pressure, at a specified temperature, $P_T(V)$, called the isotherm, whose derivative is the isothermal compressibility, defined in the same way as for fluid materials, i.e.,

$$K_T = - \frac{1}{V} \left(\frac{\partial V}{\partial P} \right)_T \equiv \frac{1}{B_T} \tag{1}$$

where B_T is the isothermal bulk modulus.

Solids differ from fluids in that they do not assume the shape of their container and that their shape may change under an applied stress. This results from the fact that the binding energy, hence the force-displacement relationships, may be different along different directions in a crystalline solid (see for example refs. 1, 2). For an "isotropic" material, in the deformation range where Hooke's law is applicable, only one constant (the "shear modulus"), in addition to the compressibility, is necessary to describe the volume and shape of a material for any possible state of stress. The number of such constants increases with the degree of anisotropy of the crystalline structure. Most materials used in engineering applications are, however, polycrystalline in nature, i.e. they consist of a very large number of individual crystals oriented at random. Since in such materials any applied stress samples all possible crystalline directions, they behave as isotropic materials even though the individual crystals may not be isotropic.

Experimentally, the equation of an isotherm (T = constant) is most usually measured, and equations then fitted to the data. Examples of such data [3] for a highly compressible solid, argon, are shown in Figure 1(a) for several temperatures. From such data the compressibility, or its reciprocal, the bulk modulus (see eqn. 1) can be derived, and this is plotted in Figure 1(b) for argon, krypton, neon. It is to be noted that by plotting data for substances near their triple points (see Chap. 4, Vol. II) the possibility of a reduced equation of state is suggested for argon, krypton and xenon, while data for the lighter element, neon, do not fall on the same curve.

These rare gas solids are obviously not "engineering materials". However, they have been widely studied because experiments can be made to relatively high values of the compression $V_o - V(P) / V_o$. Also, the description of the forces between atoms should be particularly simple in these solids, so that the equation of state should be capable of theoretical interpretation.

By analogy with the reduced equation of state of fluids, it is instructive to express isotherm data for solids in reduced form. One approach is to use reduced pressures, P/B_o, and reduced volumes, V/V_o, where B_o and V_o are the bulk modulus and volume at $P = 0$ respectively. In section II-C, specific forms for such reduced equations will be discussed.

Once the isotherm for a given temperature is known, data is often required at another temperature. The correction can best be understood as a thermal pressure, with reference to Figures 1(c) and 1(d). In 1(c) the increase of pressure that occurs when a sample of argon is held at constant volume is plotted for several molar volumes. From the thermodynamic equation (see eqn. (11) in previous chapter)

$$P = - \left(\frac{\partial A}{\partial V}\right)_T = - \frac{\partial}{\partial V}(U-TS)_T \qquad (2a)$$

which may be rearranged as follows

$$P = - \frac{dU_o}{dV} - \frac{\partial}{\partial V}[U(T) - U_o - TS]_T \qquad (2b)$$

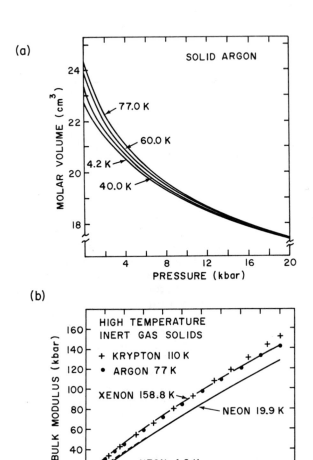

FIG. 1. (a) Isotherms for solid argon at several temperatures.
(b) The isothermal bulk modulus of argon, krypton and xenon near
their triple points showing the similarity of results, and indicating the possibility of a reduced equation of state for these
solids. Data for the lighter element, neon, is shown for comparison.

then the pressure may be written as

$$P \equiv P(T=0) + P^*(T) \tag{2c}$$

where U_o is the internal energy at T=0. This implies that the pressure can be divided into a part which is dependent on the interatomic

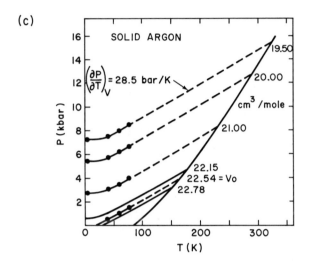

(c)

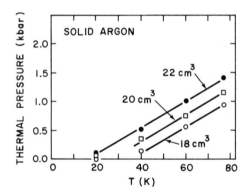

(d)

FIG. 1. (c) Isochores for argon. The melting line is also
indicated. (d) The thermal pressure derived from (c). (Figures
from Anderson and Swenson, ref. 3.)

forces, P(O), and a thermal pressure, P*(T), which arises from the
lattice vibrations or other thermally-activated excitations. If the
material were held at constant pressure (_isobaric_) conditions, then
the volume would increase, whereas at constant volume (isochoric) conditions the pressure increases. The thermal contribution to the pressure is given in Figure 1(d) for argon, for several values of the
molar volume.

        In the following section some experimental methods for obtaining
equation of state data in solids are discussed; models for the isotherm are developed in Section C and simple considerations of the

thermal pressure in D.   Section E discusses elastic constant data and
its relationship to the equation of state.

B.   Experimental Methods for Determining Compressibilities
or Isotherm Data

Ideally, from eqn. 2b it is advantageous to measure experimentally the isotherm at T=O. In practice, however, a finite temperature
(usually T≈295K) is chosen.

A number of different techniques can be used to measure the isotherm relationship $P_T(V)$ or compressibilities. Major types can be
classified in the following way:

1. Dilatometers. The linear compressibility in one direction can be
directly related to the volume compressibility for an isotropic material, while for anisotropic solids such measurements must be made according to the symmetry of the crystal structure. Bridgman [4] developed a mechanical lever system which could be contained within a pressure vessel, giving strain sensitivities $\sim 2.10^{-5}$ mm. Since the method
gives dilations relative to that of iron used in the lever system, the
compressibility of iron needed to be obtained accurately [5, 6].
Bridgman's earlier value [5] differed from his 1940 determination [6],
and Rotter and Smith [7] have shown that his absolute values for iron
are incorrect.   Results for many substances are listed in Bridgman's
book, "The Physics of High Pressure" [4], and corrected values tabulated by Birch, Schairer and Spicer [8].

Length changes can be measured directly using optical sighting
onto markers on a rod-like specimen [9]. X-ray shadowing techniques
have also been used [10]. Displacement can also be sensitively measured using magnetic markers at the ends of rod-like specimens [11]. A
recent version of this technique [12] gives random errors in strain of
$\pm 6 \times 10^{-8}$ and possible systematic error of less than 4×10^{-7}.

2. Piston-Cylinder Methods [4, 13-15] have been used up to about 50
kb. The volume change is estimated from the piston displacement, correcting for dilation of the piston and expansion of the cylinder.
These corrections can be sizeable. Since the material is subjected to
an uniaxially applied force, some corrections are necessary for stress
gradients within the sample. Also, friction forces cause hysteresis
effects which produce uncertainty in the pressure measurement (see
Chapter 11). The method is particularly suited to "soft" solids such
as the rare-gas solids, alkali metals, etc. Typical data obtained by
this method are shown in Figure 2.

3. Lattice Constant Measurement Using X-ray Diffraction Techniques.
Many different arrangements have been proposed. For relatively low
pressures, the sample can be contained in a conventional pressure vessel, constructed entirely of material transparent to x-rays (e.g. Be,
B_4C) or utilizing "windows" of these materials. At higher pressure,
the sample may be compressed between anvils using either a transparent
gasket (e.g. B) or transparent anvils (diamond cell, see Chapter 11).
Such techniques are reviewed by Banus [16]. Similar information can
be obtained from neutron-diffraction techniques [17-20].

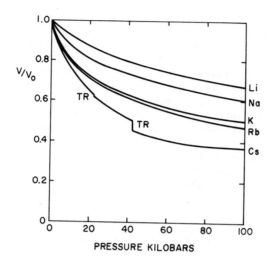

FIG. 2. The volume compression of Li, Na, K, Rb, Cs, up
to 100 kb, from Bridgman [31]. A phase change occurs for Cs at
~23kb and again at ~42kb.

4. Elastic Constant Measurements directly yield the compressibility
of a sample. Such elastic constants can most readily be determined
from measurements of the velocities of elastic waves. For this,
quartz transducers are usually bonded to the sample, enabling ultrasonic waves to be propagated in specified directions. Using this
technique, the variation of compressibility with pressure can also be
readily measured up to ~20 kb (see ref. 21 for example) while some
work has been reported up to 50 kb [22-24]. The technique essentially measures the velocity of elastic waves in the sample, and can give
very precise values of the compressibility and its derivative with
pressure. Since most useful data can be obtained from this source,
it will be dealt with in some detail in Section F.

5. Shock Techniques. For volumetric data at the highest pressures,
data may be obtained from experiments in which the material is shocked
to a high density. Since the temperature of the material is also
raised during the shock compression, equation of state data must subsequently be reduced to isothermal conditions. Several reviews of
these techniques have been given [25-29] and a discussion is given in
Chapter 14, Vol. II, while methods used to reduce data are discussed
in Section E.

6. Comparison of Results from Different Methods. Finally some comment is necessary of the differences obtained in the experimental values of compressibilities using different techniques. X-ray diffraction techniques measure directly the interatomic spacings. Thus, the
derived value of the compressibility may be smaller than the value obtained from a bulk sample because of the presence of defects, which

may be removed at high pressure. These differences can be appreciable if the sample is polycrystalline or, even worse, prepared by compacting a powder. For such compacts, the measured compressibility may be significantly higher than the intrinsic value. (See for example ref. 30 for discussion of MoS_2.) In addition, volume changes measured in polycrystalline, or compacted materials may not be reversible as a result of permanent mechanical damage. This is most likely to occur at contact points between crystallites or neighboring particles, where stresses resulting from applied external pressure may be very high. Finally it is to be noted that ultrasonic velocity measurements yield adiabatic (isentropic) values of the compressibility. Care should be taken to determine that tabulated values have been corrected to isothermal conditions (see further discussion in Section E).

C.   Empirical Relationships for the Equation of the Isotherm

As seen from Section A, the main problem in the equation of state of solids is to obtain the equation for the isotherm at a convenient temperature (usually at ~25°C). At low pressures the volume-pressure relationship is expected to be roughly linear, so that for most engineering applications, knowledge of the compressibility (K_o) or bulk modulus (B_o) at P=0 is all that is required to estimate volume changes at pressure. At higher pressures the rate of change of volume becomes less rapid as illustrated in Figure 3 so that higher-order terms are required. The data drawn in Figure 3 are for equations given in Table 1 with typical values for $B_o' = 5$ and $B_o'' = -B_o'/B_o$.

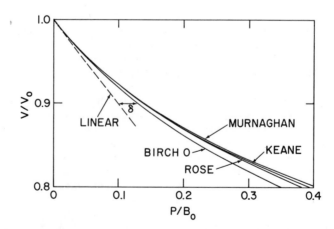

FIG. 3. Graph of the relative volume versus reduced pressure for several of the theoretical equations given in Table 1. The value of B_o' is taken to be 5, which is a typical value. In the Rose, Keane and Birch 2nd order equations, B_o'' is taken equal to $-B_o'/B_o$.

Table 1.   Some Empirical Equations Useful for Describing Isotherm Data $P_T(V)$:

Polynomial	$\dfrac{\Delta V}{V_o} = aP + bP^2 + CP^3 +$
Murnaghan	$B_T(P) = B_o + B_o' \, P'$ $P/B_o = \dfrac{1}{B_o'} \left[\left(\dfrac{V}{V_o}\right)^{-B_o'} - 1 \right]$
Rose	$B_T(P) = B_o + B_o'P + 1/2\, B_o''P^2$ $P/B_o = \dfrac{2}{B_o'} \left[\left(\dfrac{V}{V_o}\right)^{-qB_o'} - 1 \right] \left[1 + q - (1-q)\left(\dfrac{V}{V_o}\right)^{-qB_o'} \right]^{-1}$ $q \equiv \left[1 - 2B_o B_o'' / (B_o')^2 \right]^{1/2}$
Birch 1st Order (Birch-Murnaghan)	$2\varepsilon \equiv x = \left[\left(\dfrac{V}{V_o}\right)^{-2/3} - 1 \right] \quad (\varepsilon = \text{strain})$ $\dfrac{P}{B_o} = \dfrac{3}{2} x (1+x)^{5/2}(1-\xi x)$ $\xi \equiv -\dfrac{3}{4}(B_o'-4)$
Birch 2nd Order	$\dfrac{P}{B_o} = \dfrac{3}{2} x (1+x)^{5/2}(1-\xi x + \eta x^2)$ $\eta = \dfrac{3}{8}\left[B_o B_o'' + (B_o')^2 - 7 B_o' + 143/9 \right]$
Keane	$\dfrac{P}{B_o} = \dfrac{B_o'}{B_\infty'^2}\left[\left(\dfrac{V}{V_o}\right)^{-B_\infty'} - 1 \right] + \left[\dfrac{B_o'}{B_\infty'} - 1 \right] \ln \dfrac{V}{V_o}$ $B_\infty' = B_o' + B_o B_o'' / B_o' \equiv B(P \to \infty)$
Tait	$\dfrac{P}{B_o} = (e^{\alpha \Delta V / V_o} - 1)/\alpha \qquad (\Delta V \equiv V_o - V)$
Modified Tait	$\dfrac{P}{B_o} = \dfrac{1}{\alpha+1}\left(\dfrac{V_o}{V} e^{\alpha \Delta V / V_o} - 1 \right)$
Generalized Lennard-Jones	$\dfrac{P}{B_o} = \dfrac{3}{n-m}\left[\left(\dfrac{V}{V_o}\right)^{\frac{n+3}{3}} - \left(\dfrac{V}{V_o}\right)^{\frac{m+3}{3}} \right]$

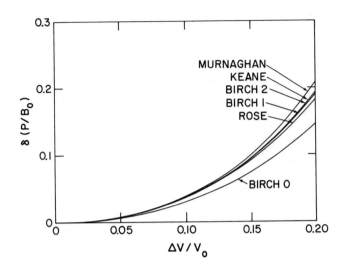

FIG. 4. A sketch of the reduced pressure differences
between data shown in Fig. 3 and a straight line extrapolation
($V/V_o = 1-B_o P$), plotted as a function of $\Delta V/V_o \equiv (V_o -V)/V_o$.
Note that the linear extrapolation can be used to $\Delta V/V_o \sim 0.02$
($P/B_o \sim 0.02$) and that all equations agree well up to
$V/V_o \sim 0.05$ and most up to $\Delta V/V_o \sim 0.1$.

What is meant by high or low pressure? This may be readily an-
swered by comparing pressure, P, with the bulk modulus, B_o. If
$P \lesssim 0.02\ B_o$, then the isotherm equations are all linear to $\sim 0.5\%$:
if $P/B_o \lesssim 0.1$ then a linear approximation is accurate to about 3%.
These data are summarized in Figure 4. It will be seen that many
different isotherm equations have been proposed, but provided P/B_o
$\lesssim 0.1$, then all equations are virtually indistinguishable. Above
this value differences can become significant. Thus, the question
of whether a particular pressure is "high" or "low" can only be an-
swered with respect to the particular material.

Table 1 gives several equations which have been used to describe
the isotherm equation for solids. For engineering applications at
low pressure ($P \lesssim 0.05 B_o$) a simple expression such as the polynomial
with only the first two terms (Table 1) can be used. Bridgman obtained such data for a wide range of materials. Conveniently, Bridgman's papers have been collected [31] so that this data may be readily obtained. Alternatively, elastic-constant data can be used to obtain values for K_o. For work of the highest precision it should be
recalled (Chapter 11) that the pressure scale has been modified since
Bridgman collected his data, so that appropriate corrections should
be made.

For anisotropic substances, a similar equation to the polynomial
expression can be used for linear changes in dimension

$$- \frac{\Delta L'}{L_o} = a_L P + b_L P^2 + \text{---} \tag{3}$$

where the coefficients a_2 and b_2 will depend on crystalographic direction and again be related to elastic constants and their derivatives with pressure.

The Birch equations [32] (zero, first and second-order) are based on the theory of finite strain developed by Murnaghan [33, 34]. The strain is defined by

$$\varepsilon = \frac{-1}{2} \left\{ \left(\frac{V_o}{V} \right)^{2/3} - 1 \right\} \tag{4}$$

and the strain energy-density is then written as a power series in the strain

$$\delta \tilde{U} = \sum_{n=1}^{\infty} a_n \varepsilon^n \tag{5}$$

By terminating the expression at the cubic term the 2nd-order Birch equation, called often the Birch-Murnaghan equation, is obtained, using (2a). If the adjustable parameter ($\xi = 3/4(B_o' - 4)$) is set equal to zero, the equation is called the Birch zero-order equation. Knopoff [35] has reviewed these equations and questioned the uniqueness of the definition used for strain. However, the Birch-Murnagham equation is perhaps the most widely used equation for fitting isotherm data to moderately high values of compression (e.g. $|\Delta v / v_o| \leq 0.2$).

The Murnaghan [33] and Rose [37] equations are based on power-expansions of the bulk modulus to first- and second-order respectively. The Murnaghan equation tends to be too "hard" (i.e. gives too small a value for $|\Delta v/v_o|$ for a given pressure); whilst the Birch zero-order equation is too "soft" (Figure 3).

A general difficulty with the above equations is that the bulk modulus becomes negative (B_o'' is negative) at sufficiently high values of pressure. Keane [38] derived an equation for which the bulk modulus is a monotonously increasing function of pressure, and its pressure-coefficient a monotonously decreasing function of pressure. Its application has been discussed by Anderson [39].

Grover, Getting and Kennedy [40] showed that the isothermal compression curve of metallic solids can be represented in a "strikingly simple way" for specific volume changes up to 40%, corresponding to pressures of nearly twice the bulk modulus. They found a linear relationship between the bulk modulus and the relative volume change of the form:

$$\ln B_T = \ln B_{T,0} + \alpha \left(\frac{\Delta V}{V} \right) \tag{6a}$$

where α is a constant. Since this equation could not be integrated to give an expression for $P_T(V)$ in terms of simple functions, comparison was made with other isotherm equations. The best fit could be obtained with a modified Tait equation of the form:

$$P = P_o + \frac{B_o}{\alpha+1} \left[\frac{V_o}{V} e^{\alpha \Delta V / V_o} - 1 \right] \tag{6b}$$

Equations for the isotherm can also be derived from simple expressions for the inter-atomic potential V(r) in solids. For instance, if a generalized Lennard-Jones type of potential is used:

$$V(r) = \frac{-A}{r^m} + \frac{B}{r^n} \tag{7}$$

where r is the interatomic distance, A, B, m and n are constants, then an equation of the form

$$\frac{P}{B_O^{'}} = \frac{3}{(n-m)}\left[\left(\frac{V_O}{V}\right)^{\frac{n+3}{3}} - \left(\frac{V_O}{V}\right)^{\frac{m+3}{3}}\right] \tag{8}$$

is obtained. If n = 4, m = 2, then the first term in Birch's equation is obtained. Towle [41] has also shown that an equation of this form can be obtained if a generalized expression for the strain is used in place of eqn. (4). Gilvarry [42] has also discussed this equation.

Over a limited range of pressure, extremely precise data are required to choose the "best" isotherm with which to fit the data. A critical examination of several equations of state has been given by MacDonald [43, 44], who also discusses statistical methods for analyzing data. A good test of the appropriateness of an equation to the data can be obtained by fitting to data over the lower half of the pressure range used, then comparing the extrapolated curve against the experimental data in the upper half. A good example of this test and of careful experimentation and analysis of data is given by Anderson and Swenson [3] for rare-gas solids, whose data have been illustrated in Figure 1.

D.   Correction of Isotherm Data to Different Temperatures

The problem of correcting isotherm data obtained at one temperature to a different temperature essentially involves the calculation of the difference of the thermal pressure between the two temperatures (eqn. 2(c)).   The Mie-Gruneisen approximation for P*(T) [45] will be described briefly, although another approximation suggested by Hildebrand [46] is sometimes used, and is similar in concept.

Using simple thermodynamic concepts, the variation of molar entropy (s) with temperature can be written:

$$s(v,T) = \int_O^T \frac{C_v(T')dT'}{T'} \tag{9}$$

where $C_v(T')$ is the specific heat at constant volume. In order to proceed further, a functional form for $C_v(T')$ is required. It is assumed that its volume dependence can be expressed in the reduced-coordinate form

$$C_v (v,T) \equiv C_v(T/\phi) \tag{10a}$$

where $\phi = \phi(v \text{ only})$ \hspace{1cm} (10b)

where ϕ is a characteristic temperature of the material, related to vibrational properties. Then P*(T) can be obtained by integrating the second term in (2b)

$$P(v_1, T) = P(v_1,0) + \frac{\Gamma}{v_1} \int_0^T C_{v_1}(T')dT' \qquad (11)$$

$$= P_o + \frac{\Gamma}{v_1} u^*(T) \qquad (12)$$

where $u_1^*(T)$ is the thermal contribution to the molar internal energy, v_1 the molar volume at which the correction is applied, and Γ is Gruneisen's parameter, defined as:

$$\Gamma \equiv - \frac{d\ln\phi}{d\ln V} \equiv \frac{d\phi}{dV} \equiv \Gamma(v \text{ only}) \qquad (13)$$

For most engineering applications it is usual to assume a specific form for the variation of the specific heat with temperature known as the Debye form, in which case $\phi \equiv \Theta_D$, where Θ_D is the Debye temperature. A discussion of Debye's theory [47, 48] is given by Kittel [2], for example, and values of related functions are given in Landolt-Bornstein [49]. Values for Θ_D can be obtained from a number of measurements [50, 51] and are listed in several compilations. If data are not avilable, a relationship between Θ_D and B_o may be used for isotropic solids [52, 53].

$$\Theta_D = \left(\frac{h}{k} \frac{9N}{4\Pi}\right)^{1/3} \frac{B_o^{1/2}}{M^{1/3}\rho^{1/6}f^{1/3}(\sigma)} \qquad (14)$$

where

$$f(\sigma) = \left\{\frac{1+\sigma}{3(1-\sigma)}\right\}^{3/2} + 2\left\{\frac{2(1+\sigma)}{3(1-2\sigma)}\right\}^{3/2} \qquad (15)$$

In these equations, h is Planck's constant, k is Boltzmann's constant, N is Avagadro's number, M, the atomic or molecular weight, ρ, the density, and σ, Poisson's ratio (see Section E for further discussion of σ).

There are several similar equations for Θ_D which give similar numerical values, but which are appropriate for different measurements of θ_D (e.g., specific heat, electrical resistivity, influence of temperature or x-ray diffraction, etc.). Further discussion is given by Blackman [50], Herbstein [51], James [52] and Anderson [53] for example.

Once Θ_D is known, u*(T) can be obtained from tables for the required value of T/Θ_D. Gruneisen's parameter can most readily be obtained from the thermodynamic identity

$$\Gamma = \frac{\beta v}{C_v K_T} \qquad (16)$$

where β is the volume coefficient of the thermal expansivity, $\tilde{v}$ the molar volume, and K_T the isothermal compressibility. Values of Γ for a few materials can be obtained from the literature. A compilation for elements is given by Gschneider [54] and selected values for elements and compounds by Collins and White [55], for example. An alternative formula for Γ given by Dugdale and McDonald [56] is

$$\Gamma = 1/2 (B'_{o,s} - 1) \qquad (17)$$

where $B'_{o,s}$ is the isothermal pressure derivative of the adiabatic bulk modulus $(\partial B_s/\partial P)_{T,P=0}$, where $B_s = -V(P)(\partial P/\partial V)_s$, which can be obtained from elastic constant data, discussed in Section F. An alternative source of information from shock-data is discussed in Chapter 14, Vol. II. There are several alternative ways of obtaining Γ from the same data, giving slightly different numerical results, discussed in ref. 1, 56, 57. The formulae [16, 17] are the most useful for engineering applications, and, although Γ is strictly speaking a function of volume, it may be considered a constant, for small values of the compression, so that the functional form of Θ_D is, by integrating (13):

$$\Theta_D = \Theta_D(V_o) \left(\frac{V_o}{V}\right)^{-\Gamma} \qquad (18)$$

The Debye temperature and its dependence on volume are important for several other properties, to be discussed later in the chapter.

For the Debye model discussed, the specific form predicted for $P^*(T)$ is:

$$P^*(T) = \begin{cases} \dfrac{3\Gamma R}{v} \cdot T & (T >> \Theta_D) \qquad (19) \\[2ex] \dfrac{3\Gamma}{5v} \pi^4 R\theta \left(\dfrac{T}{\theta}\right)^4 & (T << \Theta_D) \qquad (20) \end{cases}$$

The results depicted in Figure 1(d) show sharp curvature at low temperature, then a linear relationship at higher temperature, in accord with eqn. 19 and 20. Note that $(\partial P/\partial T)_v$ and $P^*(T)$ must tend to zero as T tends to absolute zero -- a prediction of the Third Law of Thermodynamics. Also, the thermal pressure at a given temperature is greatest for "soft" solids, which have the lowest Debye temperature. For "hard" solids, such as diamond, corrections for thermal pressure can often be neglected at ambient temperature.

Although the Debye temperature should be only a function of volume, values obtained from specific heat measurements show a characteristic variation with temperature, of the order of 10% [50, 51]. If Θ_D can be fitted to a reduced plot of the form

$$\Theta_D(V,T) = \Theta_o(V) f(T/\Theta_o) \qquad (21)$$

it can be shown [58] that the Mie-Gruneisen equation (12) still holds, with ϕ in (13) replaced by Θ_o.

Thus, the equation of an isotherm together with isochoric equations of the form (19, 20) can be used to determine the equation of state, P(V,T), over the range of pressure and temperature for which data are known, or can be usefully extrapolated.

At finite temperature, simplified statistical-mechanical models can be used to justify the Mie-Gruneisen equation. For a solid composed of N atoms, there are 3N degrees of vibrational freedom. For harmonic interatomic potentials, the energy levels of the j'th oscillation are

$$\varepsilon_{j,n} = (n+\tfrac{1}{2})h\nu_j \tag{22}$$

where n is an integer (0,1----) and ν_j the characteristic vibrational frequency of the oscillation. Then, the thermal vibrational contribution to the internal energy can be written (for a more detailed account, see ref. 1, 2, 35, 50) for example

$$u_{th.}(T) = \frac{1}{2} \sum_{j=1}^{3N} h\nu_j + \sum_{j=1}^{3N} \frac{h\nu_i}{(\exp h\nu_j/kT-1)} \tag{23}$$

where the first term represents the zero-point (i.e. T=0K) contribution, and the second term the temperature-dependent part, previously denoted U*(T). From the expression for the Helmholtz free energy, derived from (23), the thermal pressure P*(T) can be written

$$P*(T) = -\frac{1}{V} \sum_{j=1}^{3N} \frac{h\nu_i}{(\exp h\nu_i/kT-1)} \frac{d\ln\nu_i}{d\ln V} \tag{24}$$

If the frequencies are all assumed to depend on volume in the same way, then, by comparing (23) and (24), expression (13) is obtained immediately.

In many solids the vibrational frequencies do not all behave in the same way. In fact certain modes can decrease in frequency (soften) as pressure is increased (negative value of the mode Gruneisen constant Γ). This behavior is often associated with a phase transition, and the softening of the mode is said to "drive" the transition. This subject is being intensively investigated at the present time, particularly in materials with high superconducting temperatures which undergo martensitic transformations (see Chapter 4, Vol. II).

A discussion of the suitability of the approximations leading to the Mie-Gruneisen equation of state is beyond the scope of this chapter where the aim is to illustrate the concepts behind the more commonly used representations of P-V-T relations in solids. A number of classic articles [1, 35, 39, 45, 59-63] on the subject may serve the interested reader to further explore this subject.

E. Elastic Constants and Their Variation with Pressure

It is apparent from comments made earlier in the chapter that elastic constants and their variation with pressure afford a valuable source of information for the equation of state of materials. This relationship is summarized by Anderson [64] who shows that such data evaluated at modest pressures (e.g., 10,000 bars) can be used to predict the equation of state to relatively high values of the compression. In particular these data allow us to obtain values for B_o and B'_o, and even higher derivatives, which are of prime interest for isotherm equations (see Table 1). One must distinguish between isotropic (amorphous or equiaxial polycrystalline) and single crystal materials. The elastic properties of isotropic materials can be represented in terms of a bulk modulus B, a modulus of elasticity E (also called the Young's modulus), a shear modulus G, and the Poisson ratio σ. These four elastic parameters are related by the equations:

$$G = \frac{E}{2(1+\sigma)} \tag{25}$$

and

$$B = \frac{E}{3(1-2\sigma)} \tag{26}$$

hence, only two of the four parameters are independent.

For such an isotropic material, there are two distinguishable elastic waves -- one compressional, or longitudinal, and one shear or transverse. The velocity of the longitudinal wave, w_ℓ is (see for example ref. 2):

$$w_\ell = (E/\rho)^{1/2} \tag{27}$$

and that of transverse waves, w_t, by

$$w_t = (G/\rho)^{1/2} \tag{28}$$

where ρ is the density of the material. Thus, the pressure-derivatives of the sound velocities together with the compressibility can be used to determine the pressure derivatives of the modulus of elasticity and of the shear modulus.

In the case of single crystals the elastic properties are represented by a tensor (see ref. 65, for example) relating three normal (X_x, Y_y, and Z_z) and three shear stresses (X_y, Y_z, and Z_x) to an equal number of normal strains (e_{xx}, e_{yy} and e_{zz}) and shear strains (e_{xy}, e_{yz}, and e_{zx}). As a result of general thermodynamic considerations, the 36 possible values of the stiffness tensor components ($C_{\alpha\beta\gamma\delta}$) expressing Hooke's Law for the solid may be reduced to only 21 independent values. It is conventional to use a shortened matrix notation (Voigt scheme) [65] to describe these coefficients ($C_{\alpha\beta\gamma\delta} \rightarrow C_{ij}$). If the crystal contains symmetry elements, then the number of independent coefficients is reduced further. However, for most engineering applications, either cubic or isotropic solids are of interest. For crystals of other symmetries relationships between elastic constants may be found in Nye [65] while Love [66] gives the general relationship between the elastic constants and the bulk modulus.

The elastic constants can be measured by a number of techniques, but the one most widely used is based on the measurement of velocities of ultrasonic waves travelling in specific crystallographic directions with given polarizations. For cubic crystals it is convenient, for example, to measure the velocities of the three principal waves prop- agating in the [110] direction of the crystal with polarizations in the [110] (compressional wave), [100] (shear wave), and [$\bar{1}$10] (shear wave) directions. Their velocities are [2]:

$$w_{\ell,110} = \left[\frac{C_{11} + C_{12} + 2C_{44}}{2\rho} \right]^{\frac{1}{2}} \tag{29a}$$

$$w_{t,100} = \left(\frac{C_{44}}{\rho} \right)^{\frac{1}{2}} \tag{29b}$$

$$w_{t,\bar{1}10} = \left(\frac{C_{11} - C_{12}}{2\rho} \right)^{\frac{1}{2}} \tag{29c}$$

These three equations may be solved easily to give the three independent stiffness constants, C_{11}, C_{12}, C_{44} in terms of the measured velocities. For a cubic solid the relationship between bulk modulus and stiffness constants is simple:

$$B = 1/3 \ (C_{11} + 2 \ C_{12}) \tag{30}$$

The elastic constants and their pressure derivatives can be ob- tained with high precision using modern ultrasonic techniques (see ref. 21 for review). There are several ways to mathematically formu- late the pressure variation, depending on details of definitions used for strain and other parameters. A thorough review is given by Bar- ron and Klein [67].

However, the main problem with obtaining B_o and B_o' from ultra- sonic data is that the ultrasonic wave compresses the medium under approximately adiabatic (no heat flow) conditions, while static meas- urements are made under isothermal conditions (T = constant). The adiabatic (K_S) and isothermal (K_t) compressibilities are related by the thermodynamic identity:

$$K_T = K_S[1 + \frac{v\beta^2 T}{K_S C_p}] \equiv K_S[1 + \beta\Gamma T] \tag{31}$$

where β is the (isobaric) thermal expansion coefficient, and C_p the specific heat at constant pressure and Γ is Gruneisen's parameter. Usually this is a fairly small correction and we will assume that isothermal compressibilities or bulk moduli, or their pressure de- rivatives, can be obtained satisfactorily from the data. A more de- tailed discussion is given in Appendix A to this chapter.

Copious data from elastic constants of materials can be found in ref. 68. However, following the pioneering studies of Lazarus [69] many measurements of pressure derivatives have been made, which have not been previously summarized. A compilation of such data is given in Tables 1 - 3 of the Appendix.

F.   Concluding Remarks

In the previous sections, a largely empirical approach has been taken to the equation of state. From a theoretical point of view it is possible, in principle, to calculate the equation of state from first principles, using quantum mechanics. However, in practice this is impossible. Simple interatomic potential models can be used to derive the OK isotherm, particularly for simple crystals such as those in which bonding is predominantly ionic in character [1, 2, 70] (e.g., alkali halides) or can be described using fluctuating dipolar interactions [71] (van der Waals dispersive forces applicable to the rare gas solids) (see ref. 2, for example). Calculations can also be made for simple metals using simplified schemes for computing the electronic energies [72], such as pseudopotential theory (see for instance ref. 73 for aluminum).

The consistency of equations of state useful for use to high temperature has been considered by Thomsen and Anderson [74]. They show that neither the Birch nor Birch-Murnaghan equations (Table 1) are consistent with the Mie-Gruneisen equation (11) and (12). Consistency is only realized under special conditions, for which experimental verification is not available. This may pose problems for model equations of state developed for NaCl and CsCl by Decker [75, 76] for high pressure calibration (see Chapter 11).

In summary, much information is now available to describe the compression of solids to even high values of $\Delta V/V_o$. In the future, progress will probably be made in obtaining more accurate data, and in theoretical description of solids under pressure.

III.   THE EFFECT OF PRESSURE ON THE ELECTRICAL CONDUCTIVITY OF MATERIALS

The conductivity is defined by Ohm's Law relating the current density to the applied field. It may be expressed in terms of the density of the charge carriers (n), their mobility (μ), and the charge carried by them (q), and is related to the resistivity (ρ) by:

$$\sigma = n|q|\mu \tag{32}$$

If more than one type of charge carrier is responsible for the conduction, then the total conductivity is obtained by directly summing all contributions.

It is convenient to divide types of material conduction into (a) metallic (b) semi-metallic (c) semiconductor (d) ionic. Since ionic conduction is directly related to atomic diffusion, it will be considered in the next section. All other types are electronic conductors ($|q| \equiv e =|$ electronic charge$|$).

The essential, distinguishing, feature between metals, semi-metals and semiconductors concerns the manner in which the electronic energy levels broaden as the atoms are brought close together into the condensed phase. Electrons from neighboring atoms overlap, and their interaction broadens the original atomic levels into energy bands. The distinct possibilities are indicated in Figure 5.

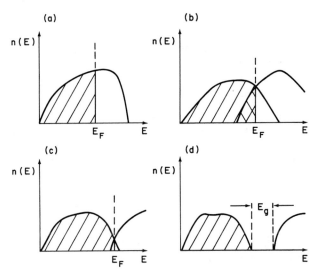

FIG. 5. Sketch of the density of electronic states (no.
of electrons with energy lying per unit energy increment) for
(a) monovalent metal; (b) divalent metal; (c) semimetal; (d)
semiconductor with band-gap E_g. Shaded areas show occupied
energy levels. The Fermi energy in a, b, c represents the
highest filled level at T=0K.

- (a) Monovalent metal - the energy band is half-filled
      (maximum filled energy is called the Fermi energy,
      E_F).

- (b) Divalent metal - Valence and conduction bands overlap
      appreciably, so that the valence band is not com-
      pletely filled (maximum energy is E_F).

- (c) Semimetal - Overlap between conduction and valence
      bands of divalent material is very small (e.g. $\lesssim 0.1$ev).

- (d) Semiconductor or insulator - The valence of an elemental
      semiconductor is even so that the valence band is
      filled and separated from the conduction band levels
      by an energy gap (E_g).

The density of carriers in (32) does not consist of all electrons
in the energy band except for the monovalent metals. For a semiconductor or semi-metal, for example, it is effectively equal to the number of electrons in the conduction band plus the number of vacant
levels in the valence band (holes). This convention is dictated by
the dynamic relationships which exist between applied fields and the
motion of the carriers (see for example ref. 2). In the case of the
semiconductor, empty valence states, or filled conduction band states,
can only be created by thermal excitation across the band gap or from
impurities, or other lattice imperfections.

A.   Elemental Metals

At ambient temperature the conductivity of the metal is given by the Bloch-Gruneisen formula (see for example ref. 77):

$$\sigma = \text{constant} \ \frac{\theta_D^2}{T} \qquad (T > \theta_D) \tag{33}$$

where θ_D is the Debye temperature and the constant includes terms that express the coupling between the electrons and lattice vibrations that produce the resistance to the passage of electrons (in a perfect lattice at T=0K, the conductivity is infinite). Then

$$\frac{d\ln\sigma}{dP} = \frac{d\ln\theta^2}{dP} = 2\Gamma K_T \tag{34}$$

where Γ is Gruneisen's parameter and K_T the isothermal compressibility, as before. This is the "normal" effect of pressure on resistivity. As pressure increases, it stiffens the lattice, so that the amplitude of thermal vibration decreases. This in turn leads to an increase in the mobility, and thus in the conductivity.

For divalent metals (s- and p-shells) another important contribution to the pressure derivative $d\ln\sigma/dP$ can arise from the relative shifting of the bands with respect to each other. For metals with unfilled d- or f-bands, the electrons in this band do not usually contribute appreciably to the conductivity, since the band is narrow (high effective mass, and low mobility). However, the high density of states increases the scattering probability for the s- or p-band electrons, so that the conductivity is relatively low and the pressure coefficient complicated.

In spite of the complicated nature of the electronic states and scattering mechanisms in metals, an extension of equation (31) can be used to fit data for a wide variety of metals, with one adjustable parameter [78, 79]:

$$\frac{d\ln\sigma}{dP} = K_T \left[\left(2 + \frac{\theta^2}{9T^2} \right) \Gamma + C \right] \tag{35}$$

where the parameter C may take a value between -4/3 and +2/3 depending on the model assumed for the binding of electrons in the material [79].

Equation (35) predicts an increase in conductivity with increasing pressure, hence a decrease in resistance. This type of behavior is in fact observed for a large number of metals (Be, Na, Mg, Al, K, Mn, Cr, Fe, Co, Ni, Cu, Zn, Ga, Rb, Zr, Cb, Mo, Rh, Pd, Ag, Cd, In, Sm, Te, Cs, Ba, La, Ce, Pr, Nd, Ta, W, Hf, Ir, Pt, Au, Hg, Th, Pb, Th, U) at moderate pressures, the observed values falling generally between those calculated by equation (35) within the extremal values of the constant C. Such "normal" behavior is illustrated by the relative resistivity of the noble metals, Al, In, Tl and shown in Figure 6.

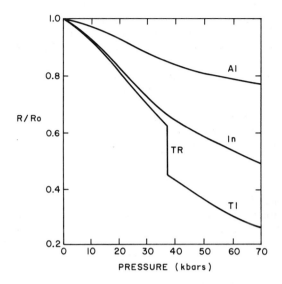

FIG. 6. Bridgman's data [31] for the change of resistance with pressure for Al, In, Tl. Note that the pressure scale has been adjusted to give the Thallium transition at ~37.5 kb, in contrast to Bridgman's original estimate of 44 kb.

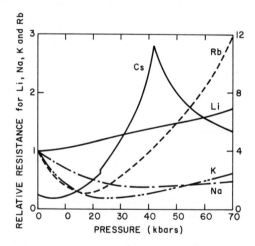

FIG. 7. Bridgman's data [31] for the change of resistance with pressure for Li, Na, K, Rb, Cs. Note the phase changes in cesium at ~23 and 42 kb, seen in volumetric data in Figure 2.

There are, however, notable exceptions to this behavior. Lithium, for instance, exhibits a decrease in conductivity with pressure, and the other alkali metals, after reaching a maximum conductivity, re-

verse the sign of their pressure dependences (e.g. see Figure 7).
Similarly the alkaline earths calcium, barium and strontium also exhibit a marked decrease in conductivity with pressure. This abnormal
behavior may be attributed to band structure changes on compression
[80-82].

Note also the abrupt changes in resistance that occur in Tl (Figure 6) and Cs (Figure 7) as a result of phase changes (see Chapter 4,
Vol. II).

References to data on the electrical resistivity of metals and
semimetals under pressure are given in Table 4 of the Appendix. General reviews are given in refs. 78, 83-86.

B.   Semimetals

Typical semimetals are Bi, Sb, As, Graphite. For such conductors
a large contribution to $d\ln\sigma/dP$ can arise from both $d\ln n/dP$ and $d\ln\mu/dP$ (see eqn. 32). Graphite is an interesting case in which these two
terms approximately cancel one another out at ambient temperature for
basal conduction in crystalline material [87, 88] ($d\ln\sigma_a/dP \lesssim -10^{-3}kb^{-1}$,
$d\ln n/dP \approx -d\ln\mu/dP \sim 0.04kb^{-1}$) However, the decrease in the mobility with
pressure arises from the change in the density of states with pres-
sure. This effect is much larger than the increase in the mobility
resulting from lattice stiffening.

Experimental data for graphite also illustrate two other points.
Firstly, that conductivity is a tensor property, (directionally dependent) so that pressure effects are also described by a tensor. For
graphite, $d\ln\sigma_c/dP \approx 0.028kb^{-1}$ [89]. Secondly, the results for polycrystalline material (e.g., nuclear-pile graphite) are sample-depend-
ent and normally the pressure coefficient of the conductivity is very
different for this type of material than for crystalline. The resistance can largely be attributed to inter-crystalline effects which are
strongly dependent on pressure ($d\ln\sigma/dP$ can be higher than $0.1 \ kb^{-1}$).
Also, the conductivity-pressure curves can show large hysteresis effects. For such polycrystalline material, values for $d\ln\sigma/dP$ must be
determined for each sample.

C.   Alloys

Effects similar to those encountered in pure metals are also
found in alloys, with the added feature of more complex scattering
mechanisms, a larger number of high pressure phases, and ordering or
clustering effects that may occur in alloys. In Fe-Ni alloys, for example, Bridgman [90] found a decrease in resistivity with pressure in
iron-rich alloys but the opposite effect in nickel-rich alloys. Table
5 in the Appendix summarizes the references to experimental measurements of conductivities (or resistivities) of alloys and intermetallic
compounds at high pressures.

Several alloys with low temperature-coefficients of resistance
have been proposed for secondary pressure gauges (manganin, xeranin,
gold-chromium, see Chapter 8). However our theoretical understanding
of either the temperature or pressure-dependence of their resistance
is scant. Another low temperature-coefficient of resistance alloy,
Evanohm, has a remarkably low pressure coefficient [91].

D.   Semiconductors

The main effect of pressure on the conductivity of pure semiconductors arises from the change of the energy band gap (E_g) with pressure, which therefore influences the number of electrons (n) in the
conduction band and empty states (holes, p) in the otherwise filled
valence band. For an "intrinsic" semiconductor, n=p, given by the
mass-action formula

$$n = p = A_c A_v \exp{-E_g/2kT} \qquad (36)$$

where A_c and A_v are parameters related to the density of   electronic
states in the conduction and valence bands. Thus:

$$\frac{d\ln\sigma}{dp} = \frac{d\ln A_c}{dp} + \frac{d\ln A_v}{dp} - \frac{1}{2kT}\frac{dE_g}{dp} + \frac{d\ln\mu}{dp} \qquad (37)$$

The energy band gap variation with pressure usually dominates in this
equation [92]. In some semiconductors the band gap is reduced by the
application of pressure, in others (e.g., Ge) it increases.

The study of the variation of band gaps with pressure has been
of technological importance in furthering our understanding of the
electronic energy states in semiconductors. For the group IV elemental and III-V compound semiconductors it is found that there are
three conduction band minima of importance (Figure 8). (A minimum
occurs whenever the electronic energy reaches a minimum value with
respect to its wave vector $k = \frac{2\pi}{\lambda}$ where λ is the electron wavelength).
The band gap E_g is defined as the energy difference between the high-
est valence band energy level and the energy of the lowest conduction
band minimum, labelled Γ, X or L in Figure 8. It is found that for
all of these technologically important semiconductors, the pressure
dependence of each gap (either $d(E_\Gamma - E_v)/dP$, $d(E_L - E_v)/dP$ or $d(E_x - E_v)dP$)
is similar [93] (see Table 2).

For Ge, the minimum band gap is $Eg = (E_L - E_v)$ so that this gap in-
creases with pressure ($dE_g/dP \approx 4\times10^{-3}$ eV/kb). Thus, at 350K, $d\ln\sigma/dP$ $-0.06kb^{-1}$.

Semiconductors may also contain defects (e.g. impurities, point
defects, dislocations etc.) which either donate electrons to the conduction band (donors) or accept electrons from the valence band (ac-
ceptors). In this case the number of electrons and holes are not
equal (n≠p) and (36) may not be used to estimate n or p. (It may
still be used to determine $(np)^{1/2}$ (see for example ref. 2)). In this
case the conductivity and its pressure dependence is similar to that
of a semimetal. Thus, the main effect of pressure on the conductivity
will be through the term $d\ln\mu/dP$, since n or p will usually be insensitive to pressure. Consequently, the pressure coefficient of resistance will be much smaller. More detailed reviews of the effect of
pressure on semiconductor behavior have been given by Paul [83], Paul
and Warschauer [92], Landwehr [84], and Paul [93].

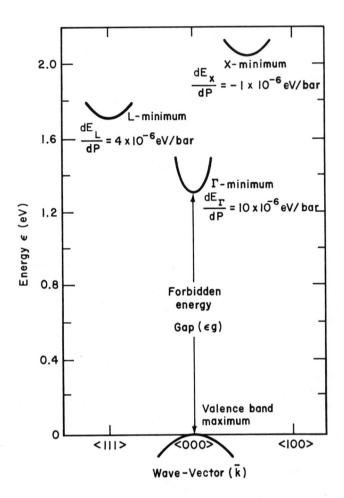

FIG. 8. A diagram of the dispersion relationship for a typical III-V compound semiconductor. The variation of energy (ε) with wave vector (k) is shown for two principal crystallographic directions for which minima occur in the $\varepsilon(k)$ relationship. The top of the valence band at k = 0 is separated from the lowest lying conduction band minima (also at k = 0) by the band gap (E_g). Other higher lying minima are shown at L and X. The differential change of the energy of these minima with pressure are indicated. The case drawn is for InP. Similar figures are obtained for other group III-V compounds (see ref. 93).

The change of resistance of Ge and Si with pressure is shown in Figure 9. The abrupt decrease is associated with a phase change to a metallic state (see Chapter 4, Vol. II). Table 6 in the Appendix summarizes references to measurements of the effect of pressure on the conductivity or resistivity of various semiconductors.

Table 2

The Effect of Pressure on Electronic Energy Band Minima of
Some Group V, Elemental, and Groups III-V, II-VI Compound
Semiconductors

Semiconductor	Compressibility $(\times 10^{-6}\ bar^{-1})$	$\dfrac{dE_\Gamma}{dP}$ 10^{-6} ev/bar	$\dfrac{dE_\Gamma}{dP}$ 10^{-6} ev/bar	$\dfrac{dE_{\Gamma L}}{dP}$ 10^{-6} ev/bar
Si	1.02	–	-1.5	–
Ge	1.33	14	-1.5	5.0
GaP	1.12	11	-1	–
GaAs	1.25	11.7	–	–
GaSb	1.77	14.5	–	5
InSb	2.14	15	–	5
InAs	1.72	10	–	–
ZnTe	2.36	6	–	–
CdTe	2.36	8	–	–

Table Adapted from W. Paul[93]

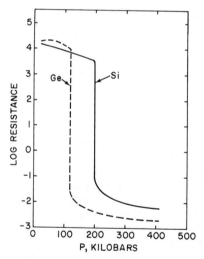

FIG. 9. The variation with pressure of the resistance of
germanium and silicon, which both undergo a semi-conductor to
metal transition. The current estimates for these transition
pressures are 92 and 148 kb respectively. The initial increase
in the resistance of Ge is due to the increase of the band-gap
E_g with pressure. (Figure adapted from Minomura and Drickamer,
ref. 94.)

IV.   EFFECT OF PRESSURE ON ATOMIC TRANSPORT PROCESSES

In this section a brief outline of the effect of pressure on diffusion and ionic conductivity is given. In both cases the transport mechanism is atomic.

A.   Diffusion

The effect of pressure on diffusion, as well as on other rate processes, is ordinarily interpreted in terms of the classical absolute rate theory first articulated by Eyring [96]. It is assumed that the diffusing atoms move by discrete jumps and that the diffusion coefficient may be expressed in terms of the length (ΔX) of these jumps and the jump frequency (Ξ), i.e. [97]:

$$D' = \frac{1}{2} \Xi (\Delta X)^2 \tag{38}$$

The jump frequency Ξ, interpreted in terms of the absolute rate theory, is determined by the frequency of attempted jumps, ν, and the activation energy barrier, ΔG_m^*, that the diffusing atom must surmount in order to move from one site to another. The use of this formulation leads to expressions for the diffusion coefficient of the form [98, 99]:

$$D' = f \, a^2 \nu \, \exp \, (-\Delta G_m^*/RT) \tag{39}$$

where 'f' is a geometrical factor characteristic of the lattice structure, 'a' the lattice parameter and ν a vibrational frequency frequently associated with the Debye or Einstein frequency ($k \theta = h\nu$).

Several objections may be raised to the representation given by equation (39), chief among which is the fact that the absolute rate theory presupposes that the transition state ("activated complex") has a sufficiently long lifetime to be treated as a thermodynamic state. Nevertheless, experimental results have shown the adequacy of this representation for most phenomena of interest, e.g. the temperature and pressure dependence of diffusivities.

In solid materials it is found that the main mechanisms for diffusion inside crystalline lattices are interstitial diffusion and vacancy diffusion. The former of these mechanisms is prevalent for impurity atoms located in interstitial positions of the host lattice (e.g., C,O,N in Fe) while in many metals and alloys the movement occurs most frequently by jumps of atoms into vacant neighboring lattice sites [98]. It follows, then, that for this latter mechanism the diffusion coefficient should also be proportional to the number of vacant sites present in the lattice. It can be shown, however, by simple thermodynamic arguments that the concentration of vacancies is also proportional to an exponential function of the free energy of formation of vacancies, i.e.

$$n_v = n_o \, \exp \, (-\Delta G_v/RT) \tag{40}$$

Thus, combining equations (39) and (40)

$$D = \frac{D'n_v}{n_o} = a^2 \nu \exp\left(-\frac{\Delta G_m^* + \Delta G_v}{RT}\right) = fa^2\nu \exp\left(-\frac{\Delta G^*}{RT}\right) \qquad (41)$$

Under some conditions, e.g. polycrystalline materials at low temperatures, the diffusion of atoms along grain boundaries may become the
predominant mechanism (which implies a change in the magnitude of the
activation energy), and in some particular cases the diffusion along
dislocations and subgrain boundaries may contribute to the diffusion
process.

In terms of the formulation derived from the absolute rate theory the effect of pressure on the diffusion coefficient (as well as
of any other thermally-activated process) is defined in terms of the
pressure dependence of the activation energy, called the "activation
volume", derived from the thermodynamic relation

$$\Delta V = \left(\frac{\partial \Delta G}{\partial P}\right)_T \qquad (42)$$

Applying this relation to ΔG^* of equation (41), the activation
volume is given by (see, for example, ref. 99):

$$\Delta V_{act} = -RT\left(\frac{\partial \ln D}{\partial P}\right)_T + RTf\left(\frac{\partial \ln a^2 \nu}{\partial P}\right)_T \qquad (43a)$$

$$= \left(\frac{\partial \Delta G_m^*}{\partial P}\right)_T + \left(\frac{\partial \Delta G_v}{\partial P}\right)_T \qquad (43b)$$

The activation volume thus consists of two contributions: one
due to the volume change associated with the activated, intermediate
state along the displacement path of the diffusing atom, and the
other, due to the volume change resulting from the presence of a vacancy (or interstitial) in the crystal lattice. The second term on
the righthand side of equation (43a) can be shown to be generally
much smaller than the first term, thus the activation volume gives,
approximately, the pressure dependence of the diffusion coefficient.

The activation volume may also be estimated directly from the
activation energy ΔG^* by means of a relationship derived by Keyes
[100] on the basis of a continuum model:

$$\Delta V_{act} = 2[\Gamma - 1/3]K_T \Delta G^* \qquad (44)$$

where Γ is the Gruneisen constant (see Section II-D) and K_T the isothermal compressibility. Since the parameters on the righthand side
of equation (44) are known for many materials, it provides a way of
estimating the pressure dependence of diffusion coefficients.

By the same model it can also be shown that

$$\Delta V_{act} = \left[1 + 2\left(\Gamma - \frac{1}{3}\right)\beta T\right]^{-1}\left(\frac{\partial \Delta H^*}{\partial P}\right)_T \qquad (45)$$

where β is the thermal expansion coefficient and ΔH^* the enthalpy of
activation. Thus a positive change in the activation enthalpy with
pressure implies that the activation volume is also positive.

The activation volume is the most succint parameter for expressing the pressure dependence of diffusion coefficients. For most materials the activation volume is positive, reflecting a decrease in the mobility of atoms with pressures. However, it has been found that in some materials the activation volume is negative. This has been explained on the basis that in such materials the increase in number of defects (e.g. interstitialcies) with pressure outweighs the decrease in mobility. Such conditions might occur in materials that have a negative volume of melting.

The relationship between diffusion and melting can be expressed more precisely. It has been found experimentally for many solids that (see for example refs. 99 and 100):

$$D = D_o \exp \frac{-aT_m}{T} \qquad (46)$$

where a is a constant and T_m the melting temperature, which may be considered a function of pressure. Thus, the effect of pressure on the diffusion rate is summarized compactly in this equation. Furthermore, comparing (46) and (41) it may be shown that,

$$\Delta H^* = aT_m \qquad (47)$$

in which $\Delta G^* = \Delta H^* - T\Delta S^*$, and a is approximately constant for a number of materials in which diffusion takes place by vacancy motion [101]. The above relationships have been considered theoretically by Rice and Nachtrieb [102] and Lawson [103].

Several detailed reviews of the effect of pressure on diffusion rates have been given by Lawson [99], Lazarus and Nachtrieb [104], Shewmon [101], and Girifalco [105].

B.   Ionic Conductivity

Ionic conductors are materials in which the flow of electricity occurs by the migration of ions rather than by flow of electrons. Such materials typically have an electron band structure characteristic of insulating materials, i.e. a wide forbidden energy gap between a filled valence band and an empty conduction band, thus allowing no appreciable electronic contribution to the conductivity. Ionic conductors do, however, conduct electricity by the movement of ions under an applied electric field, through a mechanism equivalent to diffusion. The movement of ions may take place through interstices of the crystal lattice or by jumps into adjacent vacant sites in the lattice. The number of vacant sites or of ions in interstitial sites can be increased by properly doping the particular ionic compound with ions of a different valence (i.e. charge). In principle both cations and anions can conduct electricity under an applied electric field, but in practice it is found that generally one or the other species has a much higher mobility and thus dominates the overall conduction process.

Formally, the ionic conductivity and diffusion rate are related by Einstein's equation:

$$\sigma = \frac{q^2 D}{kT} \qquad (48)$$

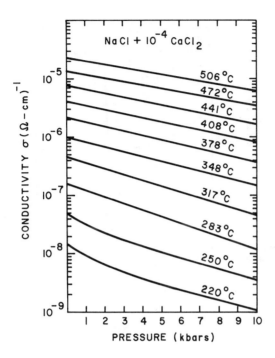

FIG. 10. The effect of pressure on the ionic conductivity
in NaCl doped with 10^{-4} mole faction $CaCl_2$. As a result of this
doping, conduction arises predominantly from anions. (Figure
from Pierce, ref. 17, Appendix Table 8.)

so that the conductivity is of the form [99]

$$\sigma = q^2 \ (Cn_f \nu/T) \exp \ (-\Delta G_m^*/kT) \tag{49}$$

where C is a proportionality constant related to the type of lattice
structure, n_f the concentration of ions of charge 'q' available for
conduction, ν a characteristic vibrational frequency of the ions and
ΔG_m^* a free energy of activation for the jump process.
Although applied pressure may also affect the conductivity
through changes in n_f and ν, its main effect is to increase the free
energy of activation, reducing thereby the ionic conductivity of the
material. This effect is illustrated in Figure 10 for NaCl. As in
the case of other materials, changes in crystal structure may occur
at high pressures, accompanied by an abrupt change in resistivity, as
is observed, for instance in RbCl (Figure 11). Another pressure effect may be a sufficient narrowing of the band gap at high pressures
which eventually increases the electronic contribution to the conductivity, thereby increasing the total conductivity of the material.
This effect is illustrated for the case of AgI in Figure 12. References to experimental measurements of pressure effects on the conduc-

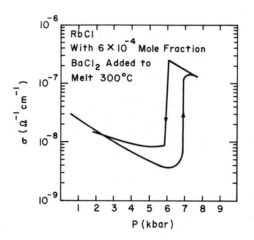

FIG. 11.   Conductivity vs. pressure through the transition
pressure for RbCl. (Figure from Pierce, ref. 17, Appendix
Table 8.)

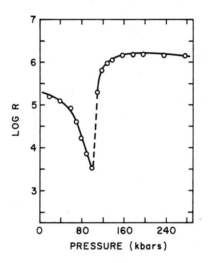

FIG. 12.   Resistance versus pressure for AgI, showing the
behavior near a phase transformation. (Figure from Riggelman
and Drickamer, ref. 10 in Appendix Table 8.)

tivities or resistivities of materials of predominantly ionic conductivity are summarized in Table 8 of the Appendix.

    Many compounds of partly covalent and partly ionic bonding also
exhibit a combined ionic and electronic conductivity. The effects of
pressure on the conductivity of such materials exhibit a combination

of the features derived from band structure and ionic mobility changes under pressure. For instance, in some glasses of mixed conductivity such as AsSe and Fe_2O_3-P_2O_5 glasses, the conductivity increases with pressure, while for other glasses such as Na_2O-B_2O_3, the conductivity decreases with pressure [107]. Other materials of mixed conductivity in which the pressure effect on conductivity has been studied are Sapphire [108], AgCN [109], $CeAl_3$ and $LaAl_3$ [110], $CeAl_2$ and $LeAl_2$ [111], and CuCl [112].

V.   EFFECT OF PRESSURE ON THERMAL CONDUCTIVITY OF MATERIALS

The change in thermal conductivity with pressure is of interest both from the theoretical point of view, since it relates to energy transport mechanisms in materials, as well as for practical consider- ations in the design of high pressure systems. The measurement of thermal conductivities of solids at high hydrostatic pressures pre- sents considerable experimental difficulties because the corrections for heat loss by conduction and convection to the pressure transmit- ting medium are generally much larger than the effect being measured. Therefore, much of the early work on the pressure dependence of ther- mal conductivity has been found to be in error [113].

For metals, the electrons are mainly responsible for transporting heat, and the thermal conductivity (κ) is related to the electrical conductivity (σ) by the Wiedemann-Franz law (see ref. 2, for example):

$$\frac{\kappa}{\sigma} = LT \tag{50}$$

where L is the Lorenz number which has a theoretical value (for "free electrons") of $2.45 \cdot 10^{-8}$ Watt-ohms/K^2. Typical experimental values of the Lorenz number range from $2.23 \cdot 10^{-8}$ W-Ω/K^2 for copper to $3.04 \cdot 10^{-8}$ W-Ω/K^2 for tungsten [31], the deviations from the free electron value being generally larger for transition metals and at temperatures below the Debye temperature, θ_D. Experimental values for the change of thermal conductivity of metals can therefore be interpreted on the basis of a constant value for the Lorenz number, with a small correc- tion which results from an increase in the lattice contribution, to be discussed next.

The thermal conductivity of dielectric or insulating materials is determined primarily by the amount of heat transmitted through lattice vibrations (quanta of vibrations are called phonons) [114]. The ther- mal energy thus transferred depends primarily on scattering processes among phonons and, at low temperatures, on the scattering by lattice defects such as grain boundaries, vacancies, and impurity atoms.

By analogy with the kinetic theory of gases [115], the lattice thermal conductivity κ_L may be written as

$$\kappa_L = \frac{(C_v\rho)wl_t}{3} \tag{51}$$

where $C_v\rho$ is the specific heat per unit volume, w the sound velocity and l_t the mean free path of phonons (inversely proportional to exp $(-\theta_D/T)$, the number of scattering phonons present) [114]. Specific forms for the variation of C_v and ℓ_t can be used for different temperature ranges.   A general form for κ_L is:

$$\kappa_L = C_o \; f \; (\Theta_D/T) \tag{52}$$

where C_o is a proportionality constant, and $f(\Theta_D/T) \approx (T/\Theta_D)^3 \; \exp(\Theta_D/kT)$ when $T \ll \Theta_D$ and $f(\Theta_D/T) \approx \Theta_D/T$ when $T > \Theta_D$.

Dugdale and McDonald [116] have shown that the proportionality constant in equation (52) can be approximated by

$$C_o = C_v \rho v \; a/3K_T \Gamma \Theta_D \sim 8(K_T/h)^3 \left(\frac{a^4 \rho \Theta_D^2}{\Gamma^2} \right) \tag{53}$$

where a is the lattice parameter, K_T the isothermal compressibility, Γ the Gruneisen constant, and Θ_D the Debye temperature. Combining equations (52) and (53) it is thus possible to estimate the thermal conductivity and its pressure-dependence from other, often more read- ily available data. Equations similar to (53) have also been derived by Liebfried and Schlomann [117] and Lawson [118].

The application of pressure causes a stiffening of the lattice and an increase in the Debye temperature resulting, in accordance with the above equations, in an increase in thermal conductivity. This ef- fect is confirmed by experimental measurements on non-metallic solids.

It should be pointed out that the above equations for κ_L are based on a monoatomic lattice model and do not include lattice scat- tering by optical modes. Thus although thermal conductivities and their pressure derivatives calculated from these equations may give good approximations for many materials, caution should be exercised in applying them to polyatomic materials, especially where the atomic weights of the constituent atoms are widely different. It should also be noted that pressure-induced phase transformations may be accompa- nied by abrupt changes in thermal conductivity.

Table 10 in the Appendix summarizes results of experimental measurements of the pressure-dependence of the thermal conductivity of several dielectric materials. It is found that the increase in thermal conductivity with pressure is approximately linear, with a slight decrease in slope at high pressures. Some early results which showed an abnormally large rise in conductivity at low pressures have been questioned on the basis of later results, the abnormal effect being attributed to improper experimental procedures [119].

Appendix Table 10 also includes data on a number of amorphous materials such as glasses and polymers. The previous equations for κ_L are of doubtful applicability to these materials since parameters such as the Debye temperature and the Gruneisen constant are ill-de- fined for amorphous materials.   Based on some simplifying assump- tions, Bohlin and Anderson [120] have derived an equation for estima- ting the pressure-dependence of the thermal conductivity of these ma- terials:

$$\left(\frac{\partial\,(\kappa/\kappa_{_O})}{\partial p}\right)_{T} = (\gamma_{_O} + 1/6)/K_{_T} \qquad (54)$$

where $K_{_T}$ is the isothermal compressibility and $\gamma_{_O} = 0.5(\partial k_{_T}/\partial P)$. The results obtained by Andersson and Backstrom [121, 122] on the pressure-dependence of the thermal conductivity of low and high density polyethylene also indicate that it is a sensitive function of the degree of crystallinity of the polymer.

The most marked effect of pressure on thermal conductivity results from the uptake of pressurizing fluid into a porous insulator, such as sintered alumina. The insulating properties depend on thermal resistance between crystallites. When fluid separates them, this inter-crystalline resistance drops markedly and increases of several times the initial conductivity can result. This has particularly important consequences in the design of internal furnaces in pressure vessels. Unfortunately the size of the effect cannot be predicted, and depends both on the microstructure of the insulator and the penetrating fluid.

VI.   MISCELLANEOUS PROPERTIES

In the preceeding sections the effects of pressure on some of the most technologically important properties have been considered. Any physical property of a material is dependent on interatomic spacing, and thus on the pressure. Thus, any chapter discussing such effects must be highly selective.

A classic example of a property dependent on interatomic spacing is that of magnetism. On an atomistic scale, the exchange-interaction, J, is responsible for the parallel (ferromagnetic) or anti-parallel (anti-ferromagnetic) alignment of magnetic moments on neighboring ions. The Bethe-Slater curve [123, 124] for the dependence of J on normalized interatomic spacing is shown in Figure 13. The circles in this figure correspond to data obtained on elements and alloys (e.g. Mn-Mn spacing extrapolated from data on Mn-rich Mn-Cu alloys in the face-centered-tetragonal structure) in their normal structures at room temperature.

Since a positive exchange interaction corresponds to ferromagnetic ordering, a negative value to anti-ferromagnetic, then increased pressure ultimately reduces ferromagnetic materials to an anti-ferromagnetic state. The particular behavior observed for a given material at modest pressure, (which can only change the exchange interaction by a small amount) will depend on its relative position on the Bethe-Slater curve at P=0. The values of J for these materials relate directly to the Curie (T_{c}) or Neel (T_{N}) temperatures (temperatures at which ferro- or anti-ferromagnetic materials become respectively paramagnetic). Thus pressure will also affect T_{c} and T_{N}, thus altering the magnetic phase diagram [125]. An example is given in Figure 14 for a ternary alloy 50% Fe, 46% Rh, 4% Ir [126], in which paramagnetic, ferromagnetic, and anti-ferromagnetic phases are present at different pressure and temperature conditions.

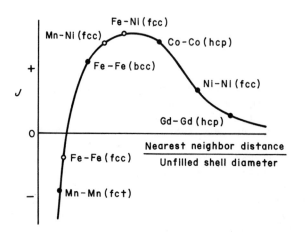

FIG. 13. The exchange energy J schematically drawn as a function of the ratio of nearest-neighbor distance to the diameter of the unfilled electron shell. (Figure redrawn from Kouvel, ref. 127.)

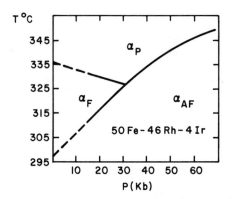

FIG. 14. Magnetic phase diagram for the alloy 50Fe-46Rh-4Ir. Ferromagnetic, anti-ferromagnetic and paramagnetic phases are shown. (Figure from Leger, Susse, and Vodar, ref. 126.)

For many technologically important materials, another factor which controls their magnetic properties is the magnetic domain structure, which depends on both atomistic (exchange interaction, magnetic anisotropy) and macroscopic properties (grain size, dislocations etc.). The effects of macroscopic defects cannot be treated analytically, so that most high pressure studies have concentrated on studying its effect on magnetic moments, Curie temperatures, magnetocrystalline anisotropy constants, etc. For general reviews, ref. 125, 127-129 may be consulted.

Neutron-diffraction techniques can be particularly useful for these studies, allowing information about crystal structure and volume, and magnetic properties to be explored through the interaction of neutrons with magnetic moments of the electrons [17]. Such techniques have been used to study fundamental problems relating to magnetic phase transitions (see also Chapters 4 and 5, Vol. II).

By analogy with magnetic properties, ferro-electric and dielectric properties have been studied in detail at high pressure so that properties related to atomistic interactions are fairly well understood [130]. Problems relating to phase-transitions in these materials have been well studied [131, 132]. However, it is difficult to model the effect of pressure on the domain behavior, related to macroscopic properties. Thus, such technologically important effects as the ageing of piezo electric ceramics when subjected to pressure cycling (of importance in underwater technology) are only partly understood.

The importance of the energy density of electronic states, $n(E)$, in a material has already been stressed in Section III, discussing conductivity. This parameter is also of importance in determining the thermo-electric power (or Seebeck coefficient) (see for instance, ref. 77). The technologically important subject of the effect of pressure on thermocouple output is dealt with in Chapter 8.

The density of electronic states is also of importance in the phenomenon of superconductivity, as evidenced by the Bardeen-Cooper-Schrieffer equation [134] for the superconducting transition temperature, T_c (see for example ref. 2).

$$T_c = 1.14\Theta_D \exp\left[-1/\bar{u}n(E_F)\right] \tag{55}$$

where Θ, is the Debye Temperature, $\bar{u}$ the electron-lattice interaction energy and $n(E_F)$ the density of states at the Fermi energy E_F. All of the quantities in (55) may be changed by pressure.

This equation has not been useful for the computation of superconducting transition temperatures absolutely, mainly because of the difficulty of calculating $\bar{u}$. However, it is useful for the understanding of trends; for instance, the variation of T_c with pressure. Pressure can both increase and decrease T_c. Also, materials not normally superconducting may become so at high pressure (e.g. Bi, Ba, Sb, Te) usually accompanied by a phase change (Figure 15). In some cases it has been possible to retain the high pressure, superconducting phase as a metastable material. An example of this is InSb [135, 136].

Another important parameter associated with superconductivity, is the initial magnetic field (B_c) required to destroy superconductivity when the temperature is lower than T_c. B_c is related to T_c for Type I superconductors so that the dependence of B_c on pressure can be related to $T_c(P)$. The shift of the $B_c(T)$ curve with pressure for cadmium is shown in Figure 16. However, in these "clean" superconductors, B_c is usually less than 0.1T (1 kg), and it is only in Type II (or "dirty") superconductors that technologically useful values of B_c can be obtained (e.g. $B_c > 50T$ (500 kg)) [137]. In these superconductors, quanta of flux penetrate the sample below B_c,

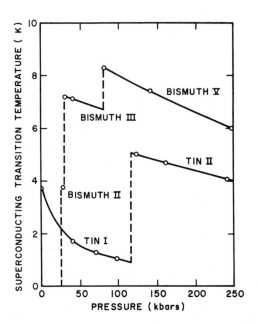

FIG. 15.   The effect of pressure on the superconducting
transition temperature, T_c, for tin and bismuth. Note that
bismuth is only superconducting in high pressure polymorphs.
(Figure redrawn from Brandt and Ginzburg, ref. 142.)

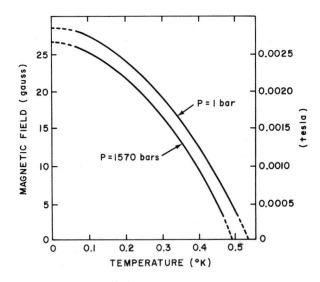

FIG. 16.   Critical field-temperature curve for cadmium at
P = 1 bar and 1570 bars. (Figure redrawn from Alekseevski and
Gaidukov, ref. 143.)

so that it is not perfectly diamagnetic, as it is for Type I super-conductors up to B_c. The behavior of these technologically inter-esting materials (e.g., Zr-Hf alloys, Nb_3Sn, V_3Ga, etc.) is thus again controlled by macroscopic material properties (dislocations, grain-boundaries). Again, by analogy with magnetic or electric ma-terials, these macroscopic factors do not allow simple theories to be applied or quantitative models to be used. Further reading on this topic is contained in refs. 138-141.

As mentioned previously, this chapter can only serve to intro-duce the reader to a vast subject. References to other properties and their dependence on pressure are given in other chapters, or may be obtained from the Bibliography of High Pressure, already re-ferred to in earlier chapters.

APPENDIX A

Thermodynamic Boundary Conditions for Characterization of Pressure Derivatives of Elastic Constants

The pressure derivatives of elastic constants are dependent on the conditions under which they are measured. Principal derivatives are:

(a) Isothermal derivatives of isothermal elastic constants, i.e., $(\partial C_{ij}^T/\partial P)_T$

(b) Isothermal pressure derivatives of adiabatic elastic constants, i.e., $(\partial C_{ij}^S/\partial P)_T$

(c) Adiabatic pressure derivatives of adiabatic elastic constants, i.e., $(\partial C_{ij}^S/\partial P)_S$

Thus, the ultrasonically-determined pressure-derivatives of elas-tic constants correspond to $(\partial C_{ij}^S/\partial P)_T$. On the other hand, the elastic-wave propagation in statically-stressed solids corresponds to adiabatic derivatives of isothermal constants and such is the case in the propagation of seismic waves in the interior of the earth, for ex-ample. Meanwhile statically determined, or theoretically calculated elastic constants and pressure derivatives are, generally, of the isothermal type.

The following relationships can be derived between the different types of elastic constants and their pressure derivatives:

The isothermal and adiabatic elastic constants are related by [144]:

$$C_{11}^T - C_{11}^S = C_{12}^T - C_{12}^S = B^T - B^S \tag{A-1a}$$

$$C_{44}^T = C_{44}^S \tag{A-1b}$$

Where B^T and B^S are bulk moduli related through equation (31), (B=1/K).

The isothermal pressure derivatives of the isothermal and adiabatic elastic constants are related as follows [145]:

$$\left(\frac{\partial C_{11}^T}{\partial P}\right)_T - \left(\frac{\partial C_{11}^S}{\partial P}\right)_T = \left(\frac{\partial C_{12}^T}{\partial P}\right)_T - \left(\frac{\partial C_{12}^S}{\partial P}\right)_T = \left(\frac{\partial (B^T - B^S)}{\partial P}\right)_T \qquad (A-2a)$$

$$\left(\frac{\partial C_{44}^T}{\partial P}\right)_T = \left(\frac{\partial C_{44}^S}{\partial P}\right)_T \qquad (A-2b)$$

Thus the conversion from one type of derivative to the other can be made if information on the pressure-dependence of the thermal expansion coefficient, compressibility and heat capacity is known or can be calculated.

Similarly the adiabatic and isothermal derivatives of the isothermal elastic constants are related by the equation [146]:

$$\left(\frac{\partial C_{ij}^S}{\partial P}\right)_S - \left(\frac{\partial C_{ij}^S}{\partial P}\right)_T = \frac{v\beta T}{C_p}\left(\frac{\partial C_{ij}^S}{\partial T}\right)_P \qquad (A-3)$$

The differences between the various pressure derivatives of a large number of materials (*) have been examined by Barsch and Chang [147]. These calculations indicate that the difference $(\partial (C_{11}^S - C_{11}^T)/\partial P)_T$ is less than 5% of the pressure derivatives for all the materials considered except CuZn(β), being less than 1% for Ag, Fe, Si, Ge, KBr, KI, RbBr, MgO and Si, while CuZn has an exceptionally high value at 12.5%. The difference $(\partial C_{11}^S/\partial P)_S - (\partial C_{11}^S/\partial P)_T$ is somewhat larger, being less than 8% for the materials considered except for NaCl (at 523°K) where it amounts to 14% of the isothermal pressure derivative. It is also found that the magnitudes of the derivatives fall in the order $(\partial C_{11}^S/\partial P)_S < (\partial C_{11}^S/\partial P)_T < (\partial C_{11}^T/\partial P)_T$, except for K, NaCl (at 523K), NaBr and NaI, where the last inequality is reversed, and for SiO$_2$ for which both inequalities are reversed.

For the elastic constant C_{12} the difference $(\partial (C_{12}^S - C_{12}^T)/\partial P)_T$ is less than 5% for most materials and between 5% and 9% for Al, β-CuZn, KBr, KI, CsCl, CsBr and CsI. The difference $(\partial C_{12}^S/\partial P)_S - (\partial C_{12}^S/\partial P)_T$ is less than 6% of the isothermal derivative except Al, KCl, CsBr, CsI and β-CuZn, being 13% for this latter material.

For the constant C_{44} the inequality $(\partial C_{44}/\partial P)_S < (\partial C_{44}/\partial P)_T$ holds for all of the materials considered, except SiO$_2$.

The difference between the two derivatives is less than 5% of the isothermal derivative for Si, SiO$_2$, Al, Cu, Ag, Au, Fe, Ge, RbI, and MgO; between 5% and 10% for LiF, KCl, RbBr, CsCl, CsBr, CsI and β-CuZn, between 10% and 15% for NaF and NaCl and between 20% and 40% for the other materials studied.

(*) Al, Cu, Ag, Au, Na, K, Fe, Si, Ge, LiF, NaF, NaCl (at 298 and 523K), NaBr, KCl, DBr, KI, RbBr, RbI, MgO, CsCl, CsBr, CsI, β-CuZn and SiO$_2$.

APPENDIX   B

References for Tables 1, 2 and 3

1.   A. L. Jain, Phys. Rev., 123, 1234 (1961).

2.   J. P. Day & A. L. Ruoff, Phys. Stat. Solidi, A25, 205 (1974).

3.   W. B. Daniels, Phys. Rev., 119, 1246 (1960).

4.   R. H. Martinson, Phys. Rev., 178, 902 (1969).

5.   P. A. Smith & C. S. Smith, J. Phys. Chem. Solids, 26, 279 (1965).

6.   C. A. Rotter & C. S. Smith, J. Phys. Chem. Solids, 27, 267 (1966).

7.   M. W. Guinan & D. N. Beshers, J. Phys. Chem. Solids, 29, 541 (1968).

8.   W. B. Daniels & C. S. Smith, Phys. Rev., 111, 713 (1958).

9.   P. S. Ho, J. P. Poirier & A. L. Ruoff, Phys. Stat. Solidi, 35, 1017 (1969).

10. R. E. Schmunk & C. S. Smith, J. Phys. Chem. Solids, 9, 100 (1959).

11. P. S. Ho & A. L. Ruoff, J. Appl. Phys., 40, 3151 (1969).

12. R. E. Schmunk and C. S. Smith, J. Phys. Chem. Solids, 9, 100 (1959).

13. H. J. McSkimin & P. Andreatch, J. Appl. Phys., 43, 2944 (1972).

14. H. J. McSkimin & P. Andreatch, J. Appl. Phys., 35, 2161 (1964).

15. A. G. Beattie & J. E. Shirber, Phys. Rev., B1, 1548 (1970).

16. H. J. McSkimin & P. Andreatch, J. Appl. Phys., 34, 651 (1963).

17. R. A. Miller & D. E. Schuele, J. Phys. Chem. Solids, 30, 589 (1969).

18. B. Golding, S. C. Moss & B. L. Averbach, Phys. Rev., 158, 637 (1967).

19. R. Chiarodo, J. Green, I. L. Spain & P. Bolsaitis, J. Phys. Chem. Solids, 33, 1905 (1972).

20. R. Chiarodo, I. L. Spain & P. Bolsaitis, J. Phys. Chem. Solids, 35, 762 (1974).

21. D. Lazarus, Phys. Rev., 76, 545 (1949).

22. R. A. Miller & C. S. Smith, J. Phys. Chem. Solids, 25, 1279 (1964).

23. L. S. Ching, J. P. Day & A. L. Ruoff, J. Appl. Phys., 44, 1017 (1973).

24. R. A. Bartels & D. E. Schuele, J. Phys. Chem. Solids, 26, 537 (1965).

25. M. Ghafelehbashi & K. M. Koliwad, J. Appl. Phys., 41, 4010 (1970).

26. K. M. Koliwad, P. B. Ghate & A. L. Ruoff, Phys. Stat. Solidi, 21, 507 (1967).

27. G. R. Barsch & H. E. Shull, Phys. Stat. Solidi, B43, 637 (1971).

28. P. J. Reddy & A. L. Ruoff in "Physics of Solids at High Pressures", (C. T. Tomizuka and R. M. Emrick, eds.) (Academic Press, New York, 1965).

29. Z. P. Chang & G. R. Barsch, J. Phys. Chem. Solids, 32, 27 (1971).

30. M. Ghafelehbashi, D. P. Dandekar & A. L. Ruoff, J. Appl. Phys., 41, 652 (1970).

31.   Z. P. Chang, G. R. Barsch & D. L. Miller, Phys. Stat.
      Solidi, 23, 577 (1967).

32.   K. F. Loje & D. E. Schuele, J. Phys. Chem. Solids, 31, 2051
      (1970).

33.   P. S. Ho & A. L. Ruoff, Phys. Rev., 161, 864 (1967).

34.   C. Wong & D. E. Schuele, J. Phys. Chem. Solids, 29, 1309
      (1968).

35.   P. F. Carcia & G. R. Barsch, Phys. Stat. Solidi, B59, 595
      (1973).

36.   R. E. Larsen & A. L. Ruoff, J. Appl. Phys., 44, 1021 (1973).

37.   P. F. Carcia, G. R. Barsch & L. R. Testardi, Phys. Rev. Lett.,
      27, 944 (1971).

38.   P. F. Carcia & G. R. Barsch, Phys. Rev., B8, 2505 (1973).

39.   B. H. Lee, J. Appl. Phys., 41, 2984, 2988 (1970).

40.   M. Kodamo, S. Saito & S. Minomura, J. Phys. Soc. Japan,
      33, 1361 (1972).

41.   G. E. Morse & A. W. Lawson, J. Phys. Chem. Solids, 28, 939
      (1967).

42.   E. H. Bogardus, J. Appl. Phys., 36, 2504 (1965).

43.   O. L. Anderson & P. Andreatch, J. Am. Ceram. Soc., 49, 404
      (1966).

44.   P. R. Son & R. A. Bartels, J. Phys. Chem. Solids 38, 819 (1972).

45.   A. L. Dragoo & I. L. Spain (to be published in J. Phys. Chem.
      Solids (1977)).

46.   E. Schreiber, J. Appl. Phys., 38, 2508 (1967).

47.   A. G. Beattie & G. A. Samara, J. Appl. Phys., 42, 2376
      (1971).

48.   D. Gerlich & C. S. Smith, J. Phys. Chem. Solids, 35, 1587
      (1971).

49.   H. J. McSkimin, A. Jayaraman & P. Andreatch, J. Appl. Phys.,
      38, 2362 (1967).

50.   D. A. Swyt & D. E. Schuele, Case Western Reserve Univ. (1971),
      quoted by M. R. Vukcevich, Phys. Stat. Solidi, B54, 219 (1972).

51.   G. I. Peresana, Sov. Phys. Solid State, 14, 1546 (1972).

52.   J. A. Corll, Office of Naval Research Tech. Rep. No. 6
      Contr. NONR 1141(05) (1962).

53.   E. S. Fisher, M. H. Manghnani & T. J. Sokolowski, J. Phys.,
      41, 2991 (1970).

54.   J. F. Green, P. Bolsaitis & I. L. Spain, J. Phys. Chem. Solids,
      34, 1927 (1973).

55.   D. J. Silversmith & B. L. Averbach, Phys. Rev., B1, 567 (1970).

56.   E. S. Fisher, M. H. Manghnani & R. Kikuta, J. Phys. Chem.
      Solids, 34, 687 (1973).

57.   H. S. Yoon & R. E. Newham, Acta. Cryst., A29, 507 (1973).

58.   M. H. Manghnani, E. S. Fisher & W. S. Bower, Jr., J. Phys.
      Chem. Solids, 33, 2149 (1972).

59.   I. J. Fritz, J. Phys. Chem. Solids, 35, 817 (1974).

60.   J. H. Gieske & G. R. Barsch, Phys. Stat. Solidi, 29, 121,
      (1968).

61.   H. J. McSkimin, P. Andreatch & R. N. Thurston, J. Appl. Phys.,
      36, 1624 (1965).

62.   D. P. Dandekar, Phys. Rev., 172, 873 (1968).

63. J. A. Corll, Phys. Rev., $\underline{157}$, 623 (1967).

64. E. Chang & G. R. Barsch, J. Phys. Chem Solids, $\underline{34}$, 1543 (1973).

65. F. F. Voronov & D. V. Stalgarova, Zh. Exp. Teor. Fiz., $\underline{49}$, 755 (1965).

66. F. F. Voronov, L. F. Vereshchagin & V. A. Gonchorova, Dokl. Akad. Nauk SSSR, $\underline{135}$, 1104 (1960).

67. F. F. Voronov & D. V. Stalgorova, Fiz. Metal. Metalloved, $\underline{34}$, 296 (1972).

68. F. F. Voronov & V. A. Goncharova, Sov. Phys. Sol. State, $\underline{13}$, 3146 (1972).

69. N. Soga, J. Amer. Ceram. Soc., $\underline{52}$, 246 (1969).

70. N. Soga, J. Geophys. Res., $\underline{73}$, 5385 (1968).

71. D. H. Chung & G. Simmons, J. Appl. Phys., $\underline{39}$, 5316 (1968).

72. N. Soga, J. Appl. Phys., $\underline{40}$, 3382 (1969).

73. R. C. Liebermann, Phys. Earth, Planet Int., $\underline{7}$, 461 (1973).

74. R. C. Liebermann, Phys. Earth Planet Int., $\underline{6}$, 360 (1972).

75. L. Peselnick, R. Meister & W. H. Wilson, J. Phys. Chem. Solids, $\underline{28}$, 635 (1967).

76. M. H. Manghnani & W. M. Benzing, J. Phys. Chem. Solids, $\underline{30}$, 2241 (1969).

77. M. H. Manghnani, Bull. Am. Ceram. Soc., $\underline{44}$, 434 (1969).

78. M. H. Manghnani, J. Amer. Ceram. Soc., $\underline{55}$, 360 (1972).

79. J. P. Day & A. L. Ruoff, J. Appl. Phys., $\underline{44}$, 2447 (1973).

80. D. S. Hughes, E. B. Blankenship & R. L. Mims, J. Appl. Phys., $\underline{21}$, 294 (1950).

81. M. H. Manghnani & C. S. Rai, Bull. Amer. Ceram. Soc., $\underline{52}$, 390 (1973).

82. N. Soga, M. Kanugi & R. Ota, J. Phys. Chem. Solids, $\underline{34}$, 2143 (1973).

83. S. N. Vaidya, S. Bailey, T. Pasternok & G. C. Kennedy, J. Geophys. Res., $\underline{78}$, 6893 (1973).

84. O. L. Anderson, E. Schreiber, R. C. Liebermann & N. Soga, Rev. Geophys., $\underline{6}$, 491 (1968).

85. K. Fielitz, Zeit. Geophys., $\underline{37}$, 943 (1971).

Appendix   Table 1 Single Crystal Elastic Constants (in units of 10^{10} Pa, or 10^{11} dynes/cm^2) and Their Pressure Derivatives (dimensionless) of Materials of Cubic Structure. (Isothermal derivatives of adiabatic elastic constants)

MATERIAL	CRYSTAL STRUCTURE	TEMP (°K)	PRESS RANGE (kBAR)	C_{11}^S	$(dC_{11}^S/dP)_T$	C_{12}^S	$(dC_{12}^S/dP)_T$	C_{44}^S	$(dC_{44}^S/dP)_T$	REF.
Li	A2	298	0-2	1.31		1.10		0.88	1.02	[1,2]
Na	A2	298	0.9	0.737-0.769	3.91-4.12	0.621-0.647	3.46-3.64	0.419-0.431	1.64-1.69	[3,4]
		78	0.9	0.857	4.10	0.711	3.57	0.587	1.17	[4]
K	A2	298	0-1.3	0.370	4.20	0.314	3.79	0.188	1.60	[5]
α-Fe	A2	298	0-10	23.00-23.14	6.72-7.51	13.46	4.57-5.19	11.64-11.66	2.59-2.63	[6,7]
Cu	A1	298	0-10	16.81	6.37	12.14	5.21	7.51	2.36	[8]
Ag	A1	298	0-10	12.40	6.79-7.18	9.34	5.48-5.84	4.61	2.32-2.37	[8,9]
		77	0-4	13.05	6.81	9.70	5.53	5.02	2.27	[9]
Au	A1	298	0-10	19.22	7.00	16.28	6.13	4.20	1.78	[8]
Al	A1	298	0-4	10.688-10.734	7.22-7.35	6.073-6.096	3.93-4.11	2.832	2.31-2.39	[10,11,12]
C(diamond)	A4	298	0-20	107.9	5.98	12.4	3.06	57.8	2.98	[13]
Si	A4	298	0-2.1	16.578	4.32	6.394	4.08	7.963	0.801	[14]
		4	0-10	16.754	-4.19	6.492	4.02	8.024	0.80	[15]
Ge	A4	298	0-2.1	12.85	5.06	4.83	4.33	6.68	1.41	[16]
Pb	A1	296	0-3.5	4.966	5.94	4.231	5.33	1.4977	2.06	[17]

Material	Structure	Temp	Range							Ref
Au-0.0972Ni	Al	298	0-10	19.23	6.80	16.09	5.92	4.497	1.865	[18]
Au-0.4242Ni	Al			19.95	6.58	15.60	5.67	6.186	2.106	
.8Au-.2Cu	Al	298	0-7	19.13	6.66	15.63	5.80	4.75	2.15	[19]
.5Au-.5Cu	Al			18.83	7.20	15.03	6.26	4.15	1.54	
.25Au-.75Cu	Al			17.67	6.53	13.37	5.64	6.55	1.92	
.81Cu-.19Zn	Al	298	0-7	15.91	6.37	12.13	5.41	7.37	2.37	[20]
.71Cu-.29Zn	Al		0-7	15.21	8.34	11.39	7.30	7.19	2.49	
Cu-Zn	A2	298	0-10	12.9	3.08	10.97	3.33	8.24	1.89	[21]
LiF	B1	298	0-7.5	11.37	9.92	4.76	2.72	6.37	1.38	[22]
LiCl	B1	298	0-3.5	4.927	10.42	2.281	2.86	2.495	1.70	[23]
LiBr	B1	298	0-3.5	3.929	10.42	1.907	2.78	1.902	1.67	[23]
NaF	B1	298	0-7.5	9.70	11.59	2.38	1.99	2.82	0.205	[22]
NaCl	B1	298	0-4	4.899-4.936	11.66-11.83	1.257-1.288	1.197-2.08	1.272-1.278	0.37	[24,25]
NaBr	B1	298	0-6	4.037	11.50	1.013	1.68	1.015	0.423	[26]
NaI	B1	298	0-10	3.035	11.8	0.915	2.30	0.742	0.59	[27]
KF	B1	298	0-3	6.485	11.74	1.427	1.66	1.281	-0.452	[26]
KCl	B1	298	0-4	4.05	12.77	0.698	1.61	0.630	-0.392	[24]
KBr	B1	298	0-10	3.419	13.47	0.520	1.61	0.508	-0.296	[28]
KI	B1	298	0-10	2.677-2.771	12.91-14.56	0.405-0.436	1.19-2.45	0.369-0.373	-0.204-0.227	[27,28]
RbCl	B1	298	0-4.5	3.659	13.16-13.72	0.615	1.34-1.45	0.475	-0.605-0.615	[29,30]
RbBr	B1	298	0-4.5	3.155-3.163	13.38-13.62	0.467-0.493	1.26-3.05	0.380-0.384	-0.361-0.587	[28,29,30]
RbI	B1	298	0-4.5	2.573	13.39-13.56	0.378	1.31-1.33	0.279	-0.994-0.522	[29,30]

(continued)

Appendix Table 1. cont.

MATERIAL	CRYSTAL STRUCTURE	TEMP (°K)	PRESS RANGE (kBAR)	C_{11}^S	$(dC_{11}^S/dP)_T$	C_{12}^S	$(dC_{12}^S/dP)_T$	C_{44}^S	$(dC_{44}^S/dP)_T$	REF.
CsCl	B2	298	0-4	3.68	6.82	0.893	5.05	0.817	3.56	[31]
CsBr	B2	298	0-10	2.97-3.08	5.56-6.30	0.719-0.827	4.16-4.93	0.747-0.760	3.20-3.68	[28,31]
CsI	B2	298	0-6	2.46	6.46	0.659	4.78	0.644	3.74	[31]
AgCl	B1	300	0-10	5.985	10.9	3.611	4.38	0.624-0.622	-0.507	[32]
AgBr	B1	300	0-10	5.610	11.9	3.27	4.60	0.724	-0.261	[32]
CaF$_2$	C1	300	0-4	16.42-16.51	6.05-6.62	4.33-4.40	4.35-6.08	3.37-3.38	1.31-1.33	[33,34]
BaF$_2$	C1	300	0-2	9.20	4.84	4.16	5.18	2.57	0.777	[34]
V$_3$Si	A15	298 / 77	0-10 / 0-2	28.73 / 23.36	5.38 / 4.51	12.09 / 15.10	3.76 / 4.60	8.099 / 7.71	1.03 / 0.770	[35,36] / [37]
V$_3$Ge	A15	298 / 77	0-10 / 0-2	29.14 / 28.66	5.73 / 4.45	10.82 / 11.33	2.84 / 5.48	6.955 / 7.066	1.02 / 0.82	[37,38]
ZnTe	B3	295	0-6	7.11	4.81	4.07	5.12	3.13	0.45	[39]
ZnSe	B3	295	0-6	8.59	4.44	5.06	4.93	4.06	0.43	[39]

Material	Structure	Temp	Range								Ref
TlCl	Cubic P43m	298	0–4	4.015	7.14±.05	1.537	6.47±.10	0.784	2.54 ±.05		[40]
TlBr	Cubic	298	0–5	3.760	8.84	1.458	6.05	0.757	3.33		[41]
MgO	B1	296	0–2	29.65–29.71	8.93–9.48	9.51–9.54	1.76–1.99	15.59–15.61	1.16–1.20		[42,43]
		77	0–2	30.62	9.98	9.38	2.17	15.76	1.41		[43]
CaO	B1	298	0–6	22.3	10.5	5.9	3.7	8.1	0.6±0.1		[44]
		295	0–6.6	22.62±.09	11±2	6.24±.9	1±4	8.06±.03	1.0±0.5		[45]
MgO/2.6Al$_2$O$_3$	Spinel Fd3m	298	0–2	29.86	4.90	15.37	3.90	15.76	0.85		[46]
SrTiO$_3$	Cub. perov, Pm3m	295	0–22	31.76	10.21±.15	10.25	3.51±.15	12.35	1.24±.05		[47]
LiH	B1	298	0–1.6	6.720	9.05±.18	1.493	1.17±.18	4.637	2.06±.07		[48]
GaAs	B3	298	0–2	11.88	4.63	5.372	4.42	5.944	1.10		[49]
InAs	B3	298			4.52		4.92	—	0.41		[50]
InSb*	B3	298	0–5.5	30.04	4.43	4.65	4.65	4.17	0.467		[51]

*The values quoted in this table are based on a positive value for $(C_{11}-C_{12})$, rather than the negative value quoted in ref.51

Appendix Table 2

Single Crystal Elastic Constants (in units of 10^{10}Pa, or 10" dynes/cm^2) and their Pressure Derivatives (dimensionless) of Materials of non-cubic Structure (Isothermal derivatives of adiabatic elastic constants, all data at room temperature)

Material and structure	M^S / $(\partial M^S/\partial p)_T$	Pressure range (Kbar)	C_{11}	C_{12}	C_{13}	C_{14}	C_{33}	C_{44}	C_{66}	Ref.
Cd(hcp) P6$_3$/mmc	M^S	0-5	11.38	3.96	4.01	9.58	5.11	2.00	3.74	[52]
	$(\partial M^S/\partial p)_T$		9.29	4.10	5.66	5.27	7.26	2.38	2.59	
Mg(hcp) P6$_3$/mmc	M^S	0-6.5	5.974	2.61	2.17		6.17	1.639		[12]
	$(\partial M^S/\partial p)_T$		6.13	3.41	2.57		7.22	1.58		
Zr(hcp) P6$_3$/mmc	M^S	0-4.7	14.37	7.304	6.588		16.52	3.214		[53]
	$(\partial M^S/\partial p)_T$		3.93	3.42	4.25		5.49	-0.22		
C-graphite (hex)	M^S	0-7	-	-	-		34.7-35.6	3.19-4.31		[54]
	$(\partial M^S/\partial p)_T$						11.3-16.6	-		
Be(hcp) P6$_3$/mmc	M^S	0-6	29.54	2.59	-0.1		35.61	17.06		[55]
	$(\partial M^S/\partial p)_T$		6.92	2.76	3.3		8.98	2.55		
Be-2.4%Cu (hcp)	M^S	0-6	29.31	3.01	0.3		35.52	17.31		[56]
	$(\partial M^S/\partial p)_T$		12.01	7.73	-2.8		13.42	2.58		
Gd(hcp) P6$_3$/mmc	M^S	0-5	6.67	2.50	2.13		7.19	2.07		[56]
	$(\partial M^S/\partial p)_T$		3.018	2.26	3.53		5.726	0.185		
Dy(hcp) P6$_3$/mmc	M^S	0-5	7.47	2.62	2.23		7.87	2.43		[56]
	$(\partial M^S/\partial p)_T$		3.092	2.277	3.32		5.331	0.434		
Er(hcp) P6$_3$/mmc	M^S	0-5	8.63	3.05	2.27		8.55	2.81		[56]
	$(\partial M^S/\partial p)_T$		4.785	3.062	2.16		5.448	0.949		

Material		Pressure range							Ref.
$3BeO \cdot Al_2O_3 \cdot 6SiO_2$ (hex.) $P6_3/mmc$	M^S	0–10	30.85	12.89	11.85		28.34	6.61	[57]
	$(\partial M^S/\partial p)_T$		4.47	3.96	3.78		3.43	−0.18	
TiO_2 (tetr) $P4/mnm$	M^S	0–20	27.01–27.14	17.66–17.80	14.80–14.96	48.19–48.40	12.39–12.44	19.30–19.48	[58,59]
	$(\partial M^S/\partial p)_T$		6.29–6.47	9.02–9.10	5.02–5.57	8.13–8.34	1.08–1.10	5.91–6.43	
Al_2O_3 (corund) hexagonal	M^S	0–10	49.760	16.26	11.718	−2.290	50.185	14.724	[60]
	$(\partial M^S/\partial p)_T$		6.174	3.282	3.653	0.130	4.998	2.243	
SiO_2 (α-quartz) hexagonal	M^S	0–2	8.680	0.704	1.191	−1.804	10.575	5.820	[61]
	$(\partial M^S/\partial p)_T$		3.28	8.66	5.97	1.93	10.84	2.66	
$CaCO_3$ (ortho) Pcmn	M^S	0–6	14.63	5.97	5.08	−2.08	8.53	3.40	[62]
	$(\partial M^S/\partial p)_T$		1.18	0.48	11.35	−1.02	−0.22	0.73	
CdS (hex) $P6_3mc$	M^S	0–4	8.565	5.321	4.616		9.361	1.487	[63]
	$(\partial M^S/\partial p)_T$		3.05	1.72	4.62		3.22	−0.628	
ZnS (hex) $P6_3mc$	M^S	0–10	12.34	5.85	4.55		13.96	2.885	[64]
	$(\partial M^S/\partial p)_T$		4.25	4.55	4.15		5.11	−0.085	

Appendix Table 3   Elastic Constants (in units of 10^{10} Pa, or 10^{11} dynes/cm^2) and Their Pressure Derivatives (dimensionless) for Polycrystalline and Amorphous Materials (Isothermal Derivatives of Adiabatic Elastic Constants.)

MATERIAL	TEMP. (°K)	PRESSURE RANGE (Kbar)	E or B_S (at p=1atm)	$(\partial E$ or $B_s/\partial P)_T$ (at P=1atm)	(at p=1atm)	$(\partial G/\partial p)_T$ (at P=1atm)	(at p=1atm)	$(\partial \nu/\partial p)_T$	REF.
Ba	298	0-6	1.588(E)	3.71(ave)	0.646	1.5 (ave)	0.229	$3 \cdot 10^{-4}$	[65]
Ce	298	0-6	1.45(E)	-8.0(ave)	1.2	1.3 (ave)	0.25	$-1.33 \cdot 10^{-2}$	[66]
Bi	298	0-25.4	3.25(E)	5.33	1.22	1.99	0.3336	$1.14 \cdot 10^{-4}$	[67]
In	298	0-20	1.51(E)	2.97	0.523	1.04	0.441	$-3.28 \cdot 10^{-4}$	[68]
BeO	298	0-28	38.99(E)	2.12	16.18	0.88	0.205	$4.84 \cdot 10^{-4}$	[69]
CaO	298	0-2	10.59(B_S)	5.23	7.61	1.75	0.210	6.50×10^{-4}	[70]
Al_2O_3	298	0-10	25.51(B_S)	4.41	16.362	1.79	0.233	—	[71]
GeO_2 (vitr)	298	0-30	4.33(E)	0.69	1.81	-1.17	0.197	—	[72]
	77	0-30	4.50(E)	1.31	1.89	-0.50	0.190	—	
(poly cr.)	293	0-7.5	26.5(B_S)	(see ref.73)	14.6	(see ref.73)	0.266	—	[73]
SnO_2	293	0-7.5	20.3(B_S)	(see ref.73)	9.8	(see ref.73)	0.291	—	[73]
$NiFe_2O_4$	298	0-3	18.77(B_S)	4.53	7.34	0.39	0.327	$2.9 \cdot 10^{-4}$	[74]
SiO_2 (100%)	298	0-8	3.65(B_S)	-6.15	3.12	-3.25	—	$-1.67 \cdot 10^{-3}$	[75,76]

Material									Ref
SiO$_2$ (96.5%) Vycor	298	0–8	2.62(B$_s$)	-4.72	2.24	-2.67	0.167	$-1.63 \cdot 10^{-3}$	[76]
SiO$_2$ (80.7%) Pyrex	298	0–8	—	-4.05	—	-2.39	—	$-0.50 \cdot 10^{-3}$	[76,77]
NaO$_2$-TiO$_2$-SiO$_2$* (glasses)	300	0–7	4.939-5.295(B$_s$)	2.59 to 4.85	3.021-3.495	-0.06 to +0.53	0.224-0.250	$1.19-1.57 \cdot 10^{-3}$	[78]
WC/3%Co	298	0–2	6.54(E)	5.20	2.75	1.85	—	—	[79]
Polystyrene ρ=1.046g/cm^3	304	0–15	0.3627	0.30**	0.1358	0.092**	0.336	$1.6 \cdot 10^{-2}$**	[80]‡
	365	0–15	0.3237	0.56**	0.1208	0.10**	0.340	$1.6 \cdot 10^{-2}$**	
Lucite ρ=1.171g/cm^3	306	0–15	0.5226	0.91**	0.1968	0.28**	0.338	~0	[80]
Alkali-Silicate† glasses	~300-520	0–5	—	—	—	—	—	—	[81]
Se glass	298	0-0.5	94.6(B$_s$)	8.5	—	2.7	0.331	—	[82]
As$_2$Se$_3$ glass	298	0-0.5	143.7(B$_s$)	8.1	—	2.1	0.294	—	[82]
Minerals of geophysical interest.	These references contain a useful compilation of data on several mineral types								[83-85]

(*) 20-25% TiO$_2$, SiO$_2$/Na$_2$O (molar ratio) 1.67 to 3.54 Range of values reported.

(**) Estimated over 0-2.5 Kbar range.

(†) Elastic properties studied as a function of glass composition

(‡) This reference also includes limited data for polyethylene.

References for Table 4

1. P. W. Bridgman, Proc. Am. Acad. Arts Sci. 81, 169 (1952).
2. G. M. Gandelman et al, Zh. Eksp. Teor. Fiz. 65, 226 (1973).
3. B. Vasvari & V. Heine, Phil. Mag. 15, 731 (1967).
4. A. R. Moodenbaugh & Z. Fisk, Phys. Lett. A43, 479 (1973).
5. P. W. Bridgman, Proc. Am. Acad. Arts Sci., 83, 1 (1954).
6. M. B. Brodsky & R. J. Friddle, Phys. Rev., B7, 3255 (1973).
7. M. Nicolas-Francillon & D. Jerome, Solid State Comm. 12, 523 (1973).
8. P.J.A. Fuller & J. H. Price, Nature 193, 262 (1962).
9. R. A. Beyerlein & D. Lazarus, Phys. Rev. B7, 511 (1973).
10. I. K. Kikoin & A. P. Senchenkov, Fiz. Metal Metalloved 24, 743 (1967).
11. S. H. Groves & W. Paul, Bull. Am. Phys. Soc., 7, 184 (1962).
12. G. A. Samara & H. G. Drickamer, J. Chem. Phys. 37, 471 (1962).
13. R. G. Arkhipov, V. V. Kechin, A. I. Likhter, Yu A. Pospelov, Soviet Phys., JETP 17, 1321 (1963).
14. M. L. Yeoman & D. A. Young, J. Phys. (London), C2, 1742 (1969).
15. I. L. Spain, N.B.S. Spec. Publ. No. 323, 717 (1971).
16. K. Noto & T. Tsuzuku, J. Phys. Soc., Japan 35, 1649 (1973).
17. E. M. Bobrova, A. I. Ivanov, V. V. Kechin and A. I. Likhter, Fiz. Tverd. Tela 16, 1526 (1974).

References for Table 5

1. P. W. Bridgman, Proc. Am. Acad. Arts Sci., 81, 165 (1952).
2. P. W. Bridgman, Proc. Am. Acad. Arts Sci., 84, 179 (1957).
3. A. Michels & T. Wassenaar, Physica, 14, 61 (1948).
4. L. R. Edwards, Phys. Stat. Sol. B51, 537 (1972).
5. P. W. Bridgman, Proc. Am. Acad. Arts Sci., 84, 43 (1956).
6. P. W. Bridgman, Proc. Am. Acad. Arts Sci., 84, 131 (1957).
7. H. Ebert & J Gielessen, Phys. Ann., 6, 229 (1947).
8. E. E. Pippig, Ann. Phys., 8, 318 (1961).
9. J. Cisowski & W. Zdanowicz, Phys. Stat. Sol., A19, 741 (1973).
10. T. C. Wilson, Phys. Rev., 56, 598 (1939).
11. M. H. Lenssen & A. Michels, Proc. Roy. Soc. Amsterdam, 32, 1379 (1929).
12. P. W. Bridgman, Proc. Am. Acad. Arts Sci., 82, 71 (1953).
13. C. W. Ufford, Proc. Am. Acad. Arts Sci., 63, 309 (1928).
14. P. W. Bridgman, Proc. Am. Acad. Arts Sci., 84, 1 (1956).
15. P. W. Bridgman, Proc. Am. Acad. Arts Sci., 83, 149 (1954).
16. E. King & I. R. Harris, J. Less Comm. Met., 27, 51 (1972).
17. C. Villain & C. Lorriers-Susse, J. Phys. (Paris), 34, 441 (1973).
18. E. King & I. R. Harris, J. Less Comm. Met., 20, 237 (1970).
19. N. Mori, J. Phys. Soc. Japan, 37, 1285 (1974).
20. A. Michels & J. W. Van Sonte, Physica, 9, 737 (1942).
21. Y. I. Kondorsky & V. L. Sedov, Zh. Exper. Theor. Fiz., 35, 53 (1958).
22. E. King & I. R. Harris, J. Less Comm. Met., 21, 275 (1970).
23. T. C. Wilson, J. Chem. Phys., 8, 13 (1940).
24. R. Harris & M. J. Zuckermann, Phys. Rev., B8, 2360 (1973).

25. B. Viswanathan, K. Govindarajan & A. Jayaraman, Proc. Nucl. Phys. Sol. St. Phys. Symp., C14, 149 (1972).

26. P. W. Bridgman, Proc. Am. Acad. Arts Sci., 47, 321 (1911).

27. L. H. Adams, R. W. Goranson & R. E. Gibson, Rev. Sci. Instr., 8, 230 (1937).

28. S. E. Babb, Jr. in "High Pressure Measurement" (A. A. Giardini and E. C. Lloyd, eds.) (Butterworths, London, 1963) p. 115.

29. A. W. Birks & C. A. Gall, Strain, p. 1, April 1973.

30. P. Andersson & G. Bäckström, Rev. Sci. Instr., 46, 1292 (1975).

References to Table 6

1. F. C. Champion & J. R. Prior, Nature, 182, 1079 (1958).

2. A.W.S. Williams, E. C. Lightowlers & A. T. Collins, J. Phys., C3, 1727 (1970).

3. W. Paul & G. L. Pearson, Phys. Rev., 98, 1755 (1955).

4. W. Paul & D. M. Warschauer, J. Phys. Chem. Solids, 5, 102 (1958).

5. T. E. Slykhouse & H. G. Drickamer, J. Phys. Chem. Solids, 7, 210 (1958).

6. L. J. Neuringer, Phys. Rev., 1113, 1495 (1959).

7. M. G. Holland & W. Paul, Phys. Rev., 128, 30 (1962).

8. M. I. Nathan & W. Paul, Phys. Rev., 128, 38 (1962).

9. O. Shimomura, et al., Phil. Mag., 29, 547 (1974).

10. P. H. Miller, Jr. & J. H. Taylor, Phys. Rev., 76, 179 (1949).

11. J. H. Taylor, Phys. Rev. 80, 919 (1950).

12. P. W. Bridgman, Proc. Am. Acad. Arts Sci., 79, 129 (1951).

13. H. H. Hall, J. Bardeen & G. L. Pearson, Phys. Rev., 84, 129 (1951).

14. W. Paul, Phys. Rev., 90, 336 (1953).

15. W. Paul & H. Brooks, Phys. Rev., 94, 1128 (1954).

16. D. M. Warschauer, W. Paul & H. Brooks, Phys. Rev., 98, 1193 (1955).

17. D. Long, Phys. Rev., 101, 1256 (1956).

18. W. Paul & D. M. Warschauer, J. Phys. Chem. Solids, 5, 89 (1958).

19. A. Michels, et al., J. Phys. Chem. Solids, 10, 12 (1959).

20. M. G. Holland & W. Paul, Phys. Rev., 128, 43 (1962).

21. S. Groves & W. Paul, Phys. Rev. Lett., 11, 194 (1963).

22. A. L. Edwards & H. G. Drickamer, Phys. Rev., 122, 1149 (1961).

23. A. L. Edwards, T. E. Slykhouse & H. G. Drickmaer, J. Phys. Chem. Solids, 11, 140 (1959).

24. A. Onodera, N. Kawai, K. Ishizaki & I. L. Spain, Solid State Comm., 14, 803 (1974).

25. R. J. Sladek, Bull. Am. Phys. Soc., 9, 258 (1964).

26. G. A. Babonas, R. A. Bendoryus & A. Yu. Shileika, Sov. Phys. Semicond., 5, 392 (1971).

27. I. F. Sviridov, et al., Izv. Vyssh. Ucheb. Zaved. Fiz., 15, 124 (1972).

28. A. Sagar, Phys. Rev., 117, 93 (1960).

29. R. W. Keyes & M. Pollak, Phys. Rev., 118, 1001 (1960).

30. A. I. Likhter & E. G. Pel, Sov. Phys. Semicond., 5, 1508 (1972).

31. J. H. Taylor, Phys. Rev., 100, 1593 (1958).

32. G. D. Pitt & M.K.R. Vyas, J. Phys., $\underline{C6}$, 274 (1973).

33. R. W. Keyes, Phys. Rev., $\underline{99}$, 490 (1955).

34. H. A. Gebbie, et al., Nature, $\underline{188}$, 1095 (1960).

35. G. C. Vezzoli, Phys. Stat. Solidi, $\underline{B60}$, K31 (1973).

36. G. A. Samara & H. G. Drickamer, J. Phys. Chem. Solids, $\underline{23}$, 457 (1962).

37. G. A. Samara & A. A. Giardini, Phys. Rev., $\underline{A140}$, 388 (1965).

38. M. I. Daunov & E. L. Broyda, Phys. Stat. Solidi, $\underline{B55}$, 155K (1973).

39. C. T. Elliot, et al., Solid State Res., MIT Lincoln Lab., $\underline{1}$, 29 (1971).

40. J. Stankiewicz & W. Giriat, Phys. Stat. Solidi, $\underline{B48}$, 467 (1971); ibid $\underline{49}$, 387 (1972).

41. R. W. Keyes, Phys. Rev., $\underline{92}$, 580 (1953).

42. D. B. McWhan, T. M. Rice & P. H. Schmidt, Phys. Rev., $\underline{177}$, 1063 (1969).

43. P. W. Bridgman, Proc. Am. Acad. Arts. Sci., $\underline{72}$, 200 (1938).

44. J. Robin, J. Phys. Radium, $\underline{20}$, 506 (1959).

45. W. Becker, W. Fuhs & J. Stuke, Phys. Stat. Solidi, $\underline{B44}$, 147 (1971).

46. A. Koma, T. Tani & S. Tanaka, Phys. Stat. Solidi, $\underline{B66}$, 669 (1974).

47. B. T. Kolcmiets & E. M. Raspopova, High Temp.-High Press., $\underline{6}$, 111, (1974).

48. G. Pfister, Phys. Rev. Lett., $\underline{33}$, 1474 (1974).

49. M.K.R. Vyas & G. D. Pitt, J. Phys., $\underline{C7}$, L423 (1974).

50. N. A. Goryunova, S. V. Popova & L. G. Khvostantsev, Dokl. Akad. Nauk. SSSR $\underline{186}$, 592 (1969).

51. M. I. Daunov & A. B. Magonedov, Fiz. Tekh. Poluprov, $\underline{8}$, 45 (1974).

52. R. Keller, J. Fenner & W. B. Holzapfel, Mat. Res. Bull., $\underline{9}$, 1363 (1974).

53. G. D. Pitt & M.K.R. Vyas, Solid State Comm., $\underline{15}$, 899 (1974).

54. A. Stella, et al., Phys. Stat. Solidi, $\underline{23}$, 697 (1967).

55. P. H. Thrasher & R. J. Kearney, Phys. Stat. Solidi, $\underline{B53}$, 623 (1973).

56. N. Mori, T. Mitsui & S. Yomo, Solid State Comm., $\underline{13}$, 1083 (1973).

57. G. A. Samara & H. G. Drickamer, J. Chem. Phys., $\underline{37}$, 1159 (1962).

58. A. I. Likhter, E. G. Pel & S. I. Prysyazhnyuk, Phys. Stat. Solidi, $\underline{A14}$, 265 (1972).

59. A. A. Averkin, B. B. Ya. Moizhes & I. A. Smirnov, Sov. Phys. Sol. State, $\underline{3}$, 1354 (1961).

60. J. Melngailis, et al., Solid State Res. MIT Lincoln Lab., $\underline{4}$, 57 (1971); J. Nonmetals $\underline{1}$, 329 (1973).

61. B. Viswanathan, S. Usha-Devi & G.N.R. Rao, Pramana, $\underline{1}$, 48 (1973).

62. C. N. Berglund & A. Jayaraman, Phys. Rev., $\underline{185}$, 1034 (1969).

63. D. B. McWhan & T. M. Rice, Phys. Rev. Lett., $\underline{22}$, 887 (1969).

64. A. R. Hutson, et al., Z. Physik, $\underline{158}$, 151 (1960).

65. Y. Harada, et al., Bull, Chem. Soc. Japan, $\underline{37}$, 1378 (1964).

66. P. T. Kozyrev & D. N. Nazledov, Dokl. Akad. Nauk. SSSR, $\underline{110}$, 207 (1956).

Appendix   TABLE 4

References to Data on the Resistance of Metals
and Semimetals at High Pressure

Material	Li	Na	K	Rb	Cs	Be	Mg
Reference	[1]	[1]	[1,2]	[1,2]	[1]	[1]	[1]

Material	Ca	Sr	Ba	Y	La	Ce	Pr
Reference	[1,3]	[1,3]	[1,3,4]	[4]	[5]	[5,6,7]	[5]

Material	Nd	Sm	Gd	Dy	Ho	Er	Tm
Reference	[5]	[5]	[5]	[5]	[5]	[5]	[5]

Material	Yb	Lu	Ti	Zr	Hf	Th	V
Reference	[5]	[5]	[1]	[1]	[1]	[1]	[1]

Material	Nb	Ta	Cr	Mo	W	U	Mn
Reference	[1]	[1]	[1]	[1]	[1]	[1]	[1]

Material	Re	Fe	Co	Rh	Ir	Ni	Pd
Reference	[1]	[1,8]	[1]	[1]	[1]	[1]	[1,9]

Material	Pt	Cu	Ag	Au	Zn	Cd	Hg
Reference	[1]	[1]	[1]	[1]	[1]	[1]	[1,10]

Material	Al	In	Tl	Sn	Pb	As	Sb
Reference	[1]	[1]	[1]	[1,11]	[1]	[1]	[17]

Material	Bi	Graphite
Reference	[1]	[12-16]

References for this Table are given on page 526.

TABLE 5

References to Experimental Data on Pressure Dependence of
Electrical Conductivity or Resistivity of Alloys and
Intermetallic Compounds

Alloy	Ref.	Alloy	Ref.	Alloy	Ref.
Ag_2Al	[1]	Bi/Sb	[5]	Fe/Ni	[2,17,20,21]
Ag/Au	[2,3,4]	Bi/Sn	[12,13]	Fe/Si	[6]
Ag/Bi	[5]	Bi/Te	[5]	Gd/Pr	[22]
Ag/Cd	[2]	Bi/Tl	[14]	Ho/Pr	[22]
Ag_5Cd_8	[1]	Bi/Zn	[5]	In/Pb	[15]
Ag/Cu	[6]	Ca/Cd	[15]	In/Tl	[14]
Ag/In	[2]	Ca/Mg	[15]	Li/Mg	[15]
Ag/Mn	[2,7]	Ca/Pb	[13,15]	Li/Pb	[23]
Ag/Mn/Sn	[7]	Cd_5Cu_8	[1]	Li/Sn	[13,15]
Ag/Pd	[6]	Cd/Pb	[15]	Mn/Ni	[6,17]
Ag/Pt	[2]	Cd/Sn	[15]	Mn/Pt	[7]
Ag/Re	[8]	Cd/Tl	[14]	Ni/Pd	[24]
Ag/Sn	[7]	Cd/Zn	[15]	Ni/Si	[6]
Ag/Zn	[2]	Ce/La	[16]	Ni/Ti	[17]
Ag_5Zn_8	[1]	Co/Fe	[6]	Ni/V	[17]
Al/Cu	[6]	Co/Ni	[17]	Pb/Sb	[15]
Al/Mg	[6]	Cr/Cu	[2,7]	Pb/Sn	[15]
Al/Zn	[6]	Cr/Cu/Ni	[7]	Pb/Tl	[14]
As/Cd/Zn	[9]	Cr/Ni	[17]	Pb/Zn	[15]
Au/Cu	[6]	Cu/Ga	[6]	Pr/Tb	[22]
Au/Cu	[10]	Cu/Ge	[6]	Pr/Th	[22]
Au/Cu_3	[10]	Cu/Ni	[6]	Pr/Yb	[22]
Au-0.55%Co	[7]	Cu/Mn	[2]	Sn/Tl	[14]
Au-2.1%Cr[†]	[11]	Cu/Pd	[2]	Sn/Zn	[15]
Au/Mn	[2,7]	Cu/Pt	[2]	V/Ru	[25]
Au/Pd	[2]	Cu/Re	[8]	Constantan	[7]
Au/Pt	[2]	Cu/Si	[6]	Invar	[2,7]
Au/Re	[8]	Cu/Zn	[6,10]	Manganin[†]	[7,26-28]
Au/Zn	[1]	CuZn	[1]	Steels (Fe-C)	[2]
Bi/Cd	[12]	Cu_5Zn_8	[1]	Zeranin[†]	[29]
Bi/In	[5]	Eu/Yb	[18]	Evanohm	[30]
Bi/Pb	[5]	Fe/Mn	[19]		

[†]For further references and discussion of these materials used for
secondary pressure gauges, see Chapter 8.

References for this table are given on pages 526 and 527.

Appendix-TABLE 6

References to Experimental Measurements Related to
Conductivity of Semiconductors at High Pressures

Material	References	Material	References
Group IV Elements:		Other Semiconductors:	
C (diam.)	[1,2]	P	[41]
Si	[3-8]	Se	[66]
Si(Amorph.)	[9]	Sr $^{(*)}$	[42]
Ge	[10-20]	Te	[17,43-46]
Ge(Amorph.)	[9]	Yb $^{(*)}$	[42]
Sn	[21]	As_2Se_3(Amorph)	[47,48]
III-V Compounds		$CdGeAs_2$	[49]
AℓSb	[22]	$CdGeP_2$	[50]
GaP	[23,24]	$CdSiP_2$	[49,50]
GaAs	[22,23,25-27]	$CdSnAs_2$	[51]
GaSb	[17,22,28,29]	$Cs_2Au_2Cl_6$	[52]
$GaAs_{0.6}P_{0.4}$	[30]	$CuFeS_2$	[53]
InP	[22]	Mg_2Ge	[54]
InAs	[17,22,31,32]	Mg_2Si	[54]
InSb	[17,33-35]	Mg_2Sn	[17,54,55]
II-VI Compounds		NiS_2	[56]
ZnS	[36]	PbS	[57]
ZnSe	[36]	PbSe	[59]
ZnTe	[36]	$Pb_{(1-x)}Sn_{(x)}Se$	[60]
CdS	[37]	PbTe	[57]
CdTe	[26,36]	$SnSe_2$	[58]
HgTe	[38]	Ti_2O_3	[61]
$Hg_{(1-x)}Cd_{(x)}Te$	[39,40]	V_2O_3	[62,63]
		ZnO	[64]
		$ZnGeP_2$	[50]
		$ZnSiP_2$	[49,50]
		$ZnSnP_2$	[50]
		Organic Semiconductors	[65]

$^{(*)}$Metal-semiconductor transition at high pressures.

References to Table 7

1. C. T. Tomizuka, R. C. Lowell & A. W. Lawson, Bull. Am. Phys. Soc., $\underline{5}$, 181 (1960).
2. F. R. Bonanno & C. T. Tomizuka, Phys. Rev., $\underline{137}$, A1264 (1965).
3. M. Beyeler & Y. Adda, J. Phys. (Paris), $\underline{29}$, 345 (1968).
4. K. L. DeVries & P. Gibbs, J. Appl. Phys., $\underline{34}$, 3119 (1963).
5. A. L. Ruoff in "Physics of Solids at High Pressures", (C. T. Tomizuka & R. M. Emrick, eds.) (Academic Press, New York, 1965).
6. C. R. Kohler & A. L. Ruoff, J. Appl. Phys., $\underline{36}$, 2444 (1965).
7. B. M. Butcher, H. Hutto & A. L. Ruoff, Appl. Phys. Lett., $\underline{7}$, 34 (1965).
8. R. H. Dickerson, R. C. Lowell & C. T. Tomizuka, Phys. Rev., $\underline{137}$, A613 (1965).
9. A. Ott & A. Norden-Ott, J. App. Phys., $\underline{42}$, 3745 (1971).
10. R. A. Hultsch & R. G. Barnes, Phys. Rev., $\underline{125}$, 1832 (1962).
11. N. H. Nachtrieb, J. A. Weil, E. Catalano & A. W. Lawson, J. Chem. Phys., $\underline{20}$, 1189 (1952).
12. N. H. Nachtrieb & A. W. Lawson, J. Chem. Phys., $\underline{23}$, 1193 (1955).
13. B. M. Butcher & A. L. Ruoff, J. Appl. Phys., $\underline{32}$, 2036 (1961).
14. K. L. DeVries, G. S. Baker & P. Gibbs, J. Appl. Phys., $\underline{34}$, 2254 (1963).
15. N. H. Nachtrieb, H. A. Resing & S. A. Rice, J. Chem. Phys., $\underline{31}$, 135 (1959).
16. J. A. Cornet, J. Phys. Chem. Solids, $\underline{32}$, 1489 (1971).
17. K. L. DeVries, G. S. Baker & P. Gibbs, J. Appl. Phys., $\underline{34}$, 2258 (1963).
18. C. Coston & N. H. Nachtrieb, J. Phys. Chem., $\underline{68}$, 2219 (1964).
19. R. N. Jeffery, Phys. Rev., $\underline{B3}$, 4044 (1971).
20. T. Liu & H. G. Drickamer, J. Chem. Phys., $\underline{22}$, 312 (1954).
21. L. C. Chhabildas & H. M. Gilder, Phys. Rev. $\underline{B5}$, 2135 (1972).
22. A. J. Bosman, et al., Physica, $\underline{23}$, 1001 (1957); $\underline{26}$, 533 (1960).
23. C. G. Homan & J. F. Cox in "Physics of Solids at High Pressures: (C. T. Tomizuka & R. M. Emrick, eds.) (Academic Press, New York, 1965); also Phys. Rev. $\underline{B5}$, 4755 (1972).
24. J. Keiser & M. Wuttig, Phys. Rev. $\underline{B5}$, 985 (1972).
25. R. G. Hickman, J. Less Comm. Met., $\underline{19}$, 369 (1969).
26. M. Kuballa & B. Baranowski, Ber. Bunsenges. Phys. Chem., $\underline{78}$, 335 (1974).
27. G. W. Tichelaar, R. V. Coleman & D. Lazarus, Phys. Rev., $\underline{121}$, 748 (1961).
28. E. D. Albrecht & C. T. Tomizuka, J. App. Phys., $\underline{35}$, 3560 (1964).
29. G. W. Tichelaar & D. Lazarus, Phys. Rev., $\underline{113}$, 438 (1959).
30. A. Ott, Phys. Stat. Solidi, $\underline{B43}$, 213 (1971).
31. H. R. Curtin, D. L. Decker & H. B. Vanfleet, Phys. Rev., $\underline{139}$, A1552 (1965).
32. J. A. Weyland, D. L. Decker & H. B. Vanfleet, Phys. Rev., $\underline{B4}$, 4225 (1971); ibid. $\underline{B5}$, 3370 (1972).
33. R. N. Jeffery & D. Gupta, Phys. Rev. $\underline{B6}$, 4432 (1972).
34. J. Combronde, Scripta Met., $\underline{6}$, 801 (1972).

35. E. Schmidtmann & K. H. Doerner, Arch. Eisenhuettenw., 39, 469 (1968).

36. N. M. Gumen & L. P. Podus, Fiz. Metal. Metalloved., 34, 776 (1972).

37. C. T. Candland, D. L. Decker & H. B. Vanfleet, Phys. Rev., B5, 2085 (1972).

38. S. G. Fishman & R. N. Jeffery, Phys. Rev., B3, 4424 (1971).

39. R. F. Peart, Phys. Stat. Solidi, 20, 545 (1967).

40. R. E. Hanneman, R. E. Ogilvie & H. C. Gatos, Trans. AIME, 233, 691 (1965).

41. N. M. Gumen & M. A. Rarog, Fiz. Metal. Metalloved, 23, 786 (1971).

42. W. Jost & K. Wagner, Z. Physik Chem. (N.F.), 19, 121 (1959).

43. C. T. Candland & H. B. Vanfleet, Phys. Rev. B7, 575 (1973).

44. K. Roy, Ber. Bunsenges, Physik Chem., 70, 1151 (1966).

45. S. W. Kurnick, J. Chem. Phys., 20, 218 (1952).

46. C. B. Pierce, Phys. Rev., 123, 744 (1961).

47. G. Martin, D. Lazarus & J. L. Mitchell, Phys. Rev. B8, 1726 (1973).

Appendix   TABLE 7

Activation Volumes For Diffusion

System	Structure	Pressure Range	Temperature	Activation Volume (cm^3/mol)	Exp. Method	Ref.
Self Diffusion:						
Ag	fcc	0-8k.b.		9.3	tracer	[1]
		0-9k.b.	910°C	9.6	tracer	[2]
Aℓ	fcc	0-10k.b.	400-610°C	12.9	tracer	[3,4]
			270°C	13.6±0.6	creep	[5,7]
Au	fcc	2-9k.b.	860-910°C	7.2±0.4	tracer	[8]
		0-10k.b.	700-990°C	7.3	tracer	[3]
Cd	hcp	0-8k.b.	0-57°C	8.2 at 1bar 4.6 at 8bar	creep	[4]
In	fct	1.2-9.6k.b.	118-148°C	8.1(1)	tracer	[9]
K	bcc	0-1k.b.	15-45°C	25.2±0.5	creep	[6]
Li	bcc	0-7k.b.	35-80°C	3.6±0.3	NMR	[10]
Na	bcc	0-12k.b.	35-130°C	7.3(10k.b.)	tracer	[11]
		0-3.5k.b.	-45°C	12.4(0k.b.)		
				9.6±0.5	NMR	[10]
P	Orthorh.	0-4k.b.	30-41.3°C	30(at 30°C) ~210(at 41°C)	tracer	[12]
Pb	fcc	0-10k.b.	70-80°C	13.9	creep	[13]
		0-10k.b.	0-57°C	12.6	creep	[14]
		0- 8k.b.	253°C	15.4±1.7	tracer	[15]
		0- 8k.b.	301°C	13.0±0.8	tracer	[15]
ε-Pu	bcc	0-20k.b.	1000°C	-4.9	tracer	[16]
Sn	bct	0- 8k.b.	0-57°C	5.1	creep	[17]
		0-10k.b.	160-228°C	5.3±0.3	tracer	[18]
β-Ti	bcc	0- 6k.b.	1000°C	3.6±1.0	tracer	[19]
Zn(2)	hcp[(∥)]	0-10k.b.	307°C	16.9 at 1k.b.		
	hcp[(⊥)]			3.6 at 10k.b. 5.0 at 1 bar	tracer	[20,21]
				2.7 at 10k.b.		
	hcp	0- 8k.b.	27-57°C	6.0 at 1 bar 2.8 at 8k.b.	creep	[4]
Cu	fcc	0-10k.b.	700-900°C	6.4	tracer	[3]

Table 7 cont.

System	Structure	Pressure Range	Temperature	Activation Volume (cm^3/mol)	Exp. Method	Ref.
Fe Ni[3]	fcc	0-20.8k.b.	1100°C	4.8(50%Ni)	microprobe	[35]
			1400°C	6.0(50%Ni)		
		20.8-33.8k.b.	1100-1400°C	4.5(50%Ni)		
Fe in β-Ti	bcc	0-4k.b.	808°C	6.0(0-10%Fe)	tracer	[39]
Fe V (x_V = 0.007-0.30)	bcc	0-40k.b.	>1100°C	5.5 for x_V=0.007 3.3 for x_V=0.30		[40]
FeZn	hex./ monoclinic	0-14k.b.	300-400°C	1.4	Kirkendall effect	[41]
In in Ag	fcc	0-9k.b.	910°C	9.1±0.1	tracer	[2]
In-Tℓ		0-5.4k.b.	150°C	────	tracer	[42]
Ni in Pb	fcc	0-50k.b.	208-591°C	1.3(25°,50k.b.) 2.0(25°,0k.b.) 2.4(327°,0k.b.)	tracer	[43]
Ni-Sb[3,5]		2-14k.b.	400-550°C	-11.3 at 1bar ~0 at 12k.b.	X-ray anal.	[36]
			590°C	-11.3 at 1bar ~0 at 4k.b.		
NiZn	Cubic tetragonal	0-14k.b.	300-400°C	~0	Kirkendall effect	[41]
Pb-Tℓ[3]	fcc	0-5k.b.	330°C	12.0(Pb),12.0(Tℓ) (20 and 40%Tℓ)	tracer	[44]
Diffusion in Ionic Crystals:						
AgVr(Cd^{++})	NaCl	0-8k.b.	202-289°C	2.7(Ag^+int.) 7.4(Ag^+vac.)	Ion. cond.	[45]
NaCl(Ca^{++})	NaCl	0-9k.b.	200-500°C	7.7	Ion. cond.	[46]
KCl(Sr^{++})	NaCl	0-9k.b.	200-500°C	7.0	Ion. cond.	[46]
Na in NaCl	NaCl	0-6k.b.	663-731°C	37 at ~0k.b. 25 at ~6k.b.	tracer	[47]

(1) Independent of temperature and crystal orientation.
(2) ‖ and ⊥ denote parallel and normal, respectively, to the basal plane.
(3) Activation volumes for interdiffusion coefficient.
(4) Reactive diffusion, with Co_2Si, CoSi, $CoSi_2$ in diffusion zone.
(5) Reactive diffusion, with NiSb, $NiSb_{2+x}$ in diffusion zone.

Table 7 cont.

System	Structure	Pressure Range	Temperature	Activation Volume (cm^3/mol)	Exp. Method	Ref.
Diffusion of Interstitials:						
C in Fe	bcc	0-3k.b.	239-253K	0.0	Mag. relax	[22]
		0-4.8k.b.	727°C	14.3	tracer	[23]
C in Fe/Ni slloys	fcc	0-6k.b.	68-90°C	2.59(63%Ni) 3.90(36%Ni) 1.35(31%Ni)	Mag. relax	[24]
D in Ag/Pd alloys	fcc	0-~70bar	300-500°C	Not Stated	permeation through cylinder	[25]
H in β-Pd		0-22.7k.b.	25°C	——	Resist. relax	[26]
N in Fe	bcc	0-3k.b.	-39°C	0.0	Mag. relax.	[22]
N in V	bcc	0-9k.b.	157-163°C	1.1	Anel. relax.	[27]
O in V	bcc	0-9k.b.	83-98°C	1.7	Anel. relax.	[27]
Diffusion in Substitutional Alloys:						
Ag-Au[3]	fcc	2.1-8.3k.b.	860-960°C	7.2(Ag),7.5(Au)	tracer	[28]
Ag-Zn[3]	fcc	0-9k.b.	110-150°C	5.36 at 27.7%Zn	Anel.relax.	[29]
Ag in In	fct	0.9-7.7k.b.	96-112°C	6.4±0.4	tracer	[30]
Ag in Pb	fcc	0-11.9k.b. 11.9-39k.b. 0-46k.b.	556-769K 25-327°C	8.7-9.9 6.9-7.1 6.2-7.2	tracer tracer tracer	[31] [32]
Au-Zn[3]	CsCℓ	0-5.2k.b.	521°C	8.5 at 50% Au 3.5 at 49% Au and 51% Au	tracer	[33]
Au in Pb	fcc	0-46k.b.	25°C 327°C	5.9 5.1	tracer	[32]
Cd in Mg	hcp	0-8 k.b.	498°C	$\Delta V_{11} = 14.4\pm1.4$ $\Delta V_{1} = 15.3\pm1.2$	electronic	[34]
Co-Ni[3]	fcc	0-20.8k.b. 20.8-33.8k.b.	1100°C 1400°C 1100°C 1400°C	5.3(50%Ni) 5.8(50%Ni) 4.9(50%Ni) 5.0(50%Ni)	microprobe	[35]
Co-Si[3,4]	fcc	2-14k.b.	900-1100°C	~14 at 1 bar ~ 0 at 10 k.b. ~17 at 14k.b.	X-ray anal.	[36]
Cu in Pb	fcc	0-56k.b.	327°C	0.7±0.5	tracer	[37]
Fe in Co	bcc	0-4.55k.b.	945°C	4.4±0.9(50%Fe)	tracer	[38]

References for Table 8

1. S. W. Kurnick, J. Chem. Phys., 20, 218 (1952).
2. A. N. Murin, B. G. Lur'e & N. A. Lebedev, Fiz. Tverd. Tela, 2, 2606 (1960).
3. K. Wagner, Z. Phys. Chem. (N.F.), 23, 305 (1960).
4. W. Jost & G. Nehlep, Z. Phys. Chem., B34, 348 (1936).
5. K. Shimizu, Rev. Phys. Chem. Japan, 30, 73 (1960).
6. A. E. Abey & C. T. Tomizuka, J. Phys. Chem. Solids, 27, 1149 (1966).
7. A. N. Murin, B. G. Lur'e & I. V. Murin, Fiz. Tverd. Tela, 9, 2350 (1967).
8. A. N. Murin, I. V. Murin & V. P. Sivkov, Fiz. Tverd. Tela, 15, 142 (1973).
9. A. N. Murin, B. G. Lur'e & Yu. P. Tarlakov, Fiz. Tverd. Tela, 3, 3299 (1961).
10. B. M. Riggleman & H. G. Drickamer, J. Chem. Phys., 37, 446 (1962); ibid. 38, 2721 (1963).
11. H. Hoshino & M. Shimoji, J. Phys. Chem. Solids, 33, 2303 (1972).
12. M. Hara, T. Mori & M. Ishiguro, Japan J. Appl. Phys., 12, 343 (1973).
13. H. Hoshino, S. Makino & M. Shimoji, J. Phys. Chem. Solids, 35, 667 (1974).
14. R. H. Radzilowski & J. T. Kummer, J. Electrochem. Soc., 118, 714 (1971).
15. R. S. Bradley, D. C. Munro & P. N. Spencer, Trans. Faraday Soc., 65, 1920 (1969).
16. I. V. Murin & B. F. Kornev, Zh. Fiz. Khim., 48, 2517 (1974).
17. C. B. Pierce, Phys. Rev., 123, 744 (1961).
18. W. H. Taylor, W. B. Daniels, B.S.H. Royce & R. Smoluchowski, J. Phys. Chem. Solids, 27, 39 (1966).
19. B. Cleaver, Z. Phys. Chem., 45, 359 (1965).
20. H. Lallemand, C. R. Acad. Sci. Paris Ser AB., 267B, 715 (1968).
21. A. Kirfel & A. Neuhaus, Z. Phys. Chem., 91, 121 (1974).
22. M. Beyeler & D. Lazarus, Solid State Comm., 7, 1487 (1969).
23. K. Shimizu, Rev. Phys. Chem. Japan, 31, 67 (1961).
24. I. V. Murin & B. T. Kornev, Zh. Fiz. Khim., 47, 2405 (1974).

Appendix Table 8

References to Measurements of Pressure Effects On

The Conductivity of Ionic Conductors

MATERIAL	REFERENCES	MATERIAL	REFERENCES
AgBr	(1,2,3)	KCL	[17,18]
AgCl	(4-8)	KNO_3	[19]
$AgCl/MnCl_2$	(8)	LiF	[20]
AgI	(3,6,9-13)	$MgGeO_3$	[21]
$\beta-Al_2O_3$(doped)	(14)	NaCl	[17,22,23]
AuCN	(15)	$NaNO_3$	[16]
$CsNO_3$	(16)	RbCl	[17]
CuCN	(15)	$RbNO_3$	[25]

Appendix

TABLE 9

Effect of Pressure On Thermal Conductivity of Metals At

Ambient, or Room, Temperature

METAL	PRESSURE RANGE	PRESS. COEFF. OF THERMAL COND. $(\Delta K/K_o)/\Delta p$ bar^{-1}	PRESS. COEFF. OF WIEDEMANN-FRANZ RATIO (bar^{-1})	REFERENCE
Ag	0-12Kbar	$(4.5\pm0.3)10^{-6}$	$(1.1\pm0.3)10^{-6}$	[1]
Au	0-12Kbar	$(4.0\pm0.3)10^{-6}$	$(1.0\pm0.4)10^{-6}$	[1]
Cu	0-12Kbar	$(3.0\pm0.3)10^{-6}$	$(1.1\pm0.3)10^{-6}$	[1]
Bi	0-25Kbar	$-5.7.10^{-6}$	$4.4.10^{-6}$	[2]
Pb	0-12Kbar	$17.3.10^{-6}$	6.10^{-6}	[3]
Sn	0-12Kbar	$12.2.10^{-6}$	3.10^{-6}	[3]
Te*	0-3.5Kbar			[4]

*Temperature and pressure dependence of the conductivity described by the empirical relation.

$$\lambda = \frac{C_1}{T+T_o} + \frac{C_2}{P+P_o};$$

$C_1 = 2414 - 3437$ cal cm^{-1}sec^{-1} (12-150°C).

$C_2 = 3.651$ cal bar cm^{-1}sec^{-1}deg^{-1}

$T_o = 273°C$

$P_o = 2$Kbar

References to Table 9

1.   C. Starr, Phys. Rev. 54, 210 (1938).
2.   T. Rosander & G. Backström, Phys. Lett. A29, 517 (1969).
3.   P. W. Bridgman, "The Physics of High Pressures" (G. Bell & Sons, London, 1949).
4.   K. I. Amirkhanov, Y. B. Magomedov & S. N. Emir, Fiz. Tverd Tela. 15, 1512 (1973).

Appendix - Table 10

Effect of Pressure on the Thermal Conductivity of

Some Non-Metallic Solids

MATERIAL	PRESSURE RANGE	PRESS. COEFF. OF THERMAL COND. $(\Delta K/K_o)/\Delta p \ bar^{-1}$	REFERENCE
NaCl	0-12k.b.	$3.6.10^{-5}$	[1]
KCl	0-19k.b.	$(3.3\pm0.3).10^{-5}$	[2]
AgCl	0-40k.b.	$1.1.10^{-5}$	[3]
Pyrex	0-12k.b.	$3.8.10^{-6}$	[1]
Nylon-6	0-35k.b.	$1.3.10^{-4}$	[4]
PMMA	0-350bar	$1.5.10^{-4}$	[4]
Teflon		$1.2.10^{-4}$	[5]
Epoxy resin	0-25k.b.	$1.8x10^{-4}$	[6]
Polyethylene	0-25k.b.	$\sim 11x10^{-5} \ p\rightarrow0$ (high density) $7.5x10^{-5}p\rightarrow0$ (low density)	[6]

References to Table 10

1. P. W. Bridgman, "The Physics of High Pressures (G. Bell & Sons, London, 1949).
2. O. Alm & G. Backström, J. Phys. Chem. Solids, 35, 421 (1974).
3. T. Rosander & G. Backström, Phys. Scripta, 1, 269 (1970).
4. P. Lohe, Kolloid-Z. Polymere, 203, 115 (1965); ibid 204, 7 (1965).
5. R. E. Barker & R.Y.S. Chen, J. Chem. Phys., 53, 2616 (1970).
6. P. Andersson & G. Backström, J. Appl. Phys., 44, 705 (1973); ibid 44, 2601 (1973).

REFERENCES

1. J. C. Slater, "Introduction to Chemical Physics", (McGraw-Hill, New York, 1939).
2. C. Kittel, "Introduction to Solid State Physics" (John Wiley and Sons, 3rd edition, New York & London, 1966).
3. M. S. Anderson & C. A. Swenson, J. Phys. Chem. Solids, 36, 145 (1975).
4. P. W. Bridgman, "The Physics of High Pressure" (G. Bell & Sons, London, 1949).
5. P. W. Bridgman, Proc. Am. Acad. Arts Sci., 58, 165 (1923).
6. P. W. Bridgman, Proc. Am. Acad. Arts Sci., 74, 11 (1940).
7. C. A. Rotter & C. S. Smith, J. Phys. Chem. Solids, 27, 267 (1966).
8. F. Birch, J. Schairer and H. C. Spicer, "Handbook of Physical Constants", Special Papers #36, (Geological Society of America, 1942).
9. H. Ebert, Physik Z., 36, 388 (1935).
10. A. H. Smith, N. A. Riley, A. W. Lawson, Rev. Sci. Instr., 22, 138 (1951).
11. J. Reitzel, I. Simon, J. A. Walker, Rev. Sci. Instr., 28, 828 (1957).
12. R. C. Lincoln, A. L. Ruoff, Rev. Sci. Instr., 44, 1239 (1973).
13. J. W. Stewart & C. A. Swenson, Phys. Rev., 94, 1069 (1954).
14. J. W. Stewart, J. Phys. Chem. Solids, 1, 146 (1956).
15. S. N. Vaidya & G. C. Kennedy, J. Phys. Chem. Solids, 31, 2329 (1970).
16. M. D. Banus, High Temp.-High Pressure, 1, 483 (1969).
17. G. E. Bacon, "Neutron Diffraction" (Oxford Univ. Press, 2nd ed., 1962).
18. Many designs for neutron diffraction cells have been published of which Refs. 19, 20 are very recent, giving refs. to earlier work.
19. J. Paureau & C. Vettier, Rev. Sci. Instr., 46, 1484 (1975).
20. D. Bloch, J. Paureau, J. Voiron, G. Parisot, Rev. Sci. Instr., 47, 296 (1976).
21. E. R. Fuller, A. V. Granato, J. Holder & E. R. Naimon, "Methods of Exp. Physics", Vol. 11, Chapter 7 (R. V. Coleman, ed.) (Academic Press, New York, 1974).
22. T. J. Ahrens & S. Katz, J. Geophys. Res., 67, 2935 (1962).
23. P.L.M. Heydemann & J. C. Houck, J. Appl. Phys., 40, 1609 (1969).
24. P.L.M. Heydemann & J. C. Houck, NBS Specl. Publ. 326, p. 11 (E. C. Lloyd, ed.) (1971).
25. M. H. Rice, R. G. McQueen & J. M. Walsh, Solid State Phys., 6, pp. 1-63 (Academic Press, New York, 1958).
26. R. G. McQueen, "Laboratory Techniques for Very High Pressures", (K. A. Gschneidner, Jr., M. T. Hepworth & N.A.D. Parlee, eds.) Metall. Soc. Conf., Vol. 22 (Gordon & Breach, New York, 1964).
27. Ya. B. Zel'dovich & Yu. P. Raizer, "Physics of Shock Waves and High-Temperature Hydrodynamic Phenomena", Vol. 2, pp. 685-784, (W. D. Hayes & R. F. Probstein, eds.) (Academic Press, New York, 1967).

28. R. G. McQueen, S. P. Marsh, J. W. Taylor, J. N. Fritz & W. J. Carter, "The Equation of State of Solids from Shock-Wave Studies" in High-Velocity Impact Phenomena, (R. Kinslow, ed.) (Academic Press, New York, 1970).

29. G. E. Duvall and successive articles by R. N. Keeler and E. B. Royce in "Shock Waves in Condensed Media", (P. Caldirola & H. Knoepfel, eds.), Proceedings of the International School of Physics "Enrico Fermi", Course XLVIII (Academic Press, New York, 1971).

30. A. W. Webb, J. L. Feldman, E. F. Skelton, L. C. Towle, C. Y. Liu & I. L. Spain, J. Phys. Chem. Solids, 37, 329 (1976).

31. P. W. Bridgman, "Collected Experimental Papers" (Harvard University Press, Cambridge, 1964).

32. F. Birch, J. Geophys. Res., 57, 227 (1952).

33. F. D. Murnaghan, Amer. J. Math., 59, 235 (1937).

34. F. D. Murnaghan, "Finite Deformation of an Elastic Solid", (John Wiley & Sons, New York, 1951).

35. L. Knopoff in "High Pressure Physics and Chemistry", Chapter 5(i), p. 227 (R. S. Bradley, ed.) (Academic Press, 1963).

36. F. D. Murnaghan, Proc. Natl. Acad. Sci., 30, 244 (1944).

37. M. F. Rose, Phys. Stat. Sol., 21, 235 (1967).

38. A. Keane, Australian J. Physics, 7, 322 (1954).

39. O. L. Anderson, Phys. Earth & Planetary Interiors, 1, 169 (1968).

40. R. Grover, I. C. Getting, G. C. Kennedy, Phys. Rev. B7, 567 (1973).

41. L. C. Towle, Appl. Phys. 8, 117 (1975).

42. J. J. Gilvarry, J. Appl. Phys. 28, 1253 (1957).

43. J. R. McDonald, Revs. Mod. Phys., 38, 669 (1966).

44. J. R. McDonald, Revs. Mod. Phys., 41, 316 (1969).

45. E. Grüneisen in "Handbuch der Physik", Vol. 10, p. 1 (Springer, Berlin, 1926).

46. J. H. Hildebrand, Z. Physik, 67, 127 (1931).

47. P. Debye, Ann. Phys., 39, 789 (1912).

48. P. Debye, Ann. Phys., 43, 49 (1914).

49. Landolt-Bornstein, "Tables" 6th ed., Vol. 2, Part 4, (J. Springer, Berlin, 1961).

50. M. Blackman, Proc. Phys. Soc. (London), 74, 17 (1959).

51. F. H. Herbstein, Adv. in Phys., 10, 313 (1961).

52. R. W. James, "The Optical Principles of X-Ray Diffraction", (G. Bell & Sons, Ltd., London, 1954).

53. O. L. Anderson, J. Phys. Chem. Solids, 24, 909 (1963).

54. K. A. Gschneider, Solid State Physics 16, 275 (F. Seitz, D. Turnbull, eds.) (Academic Press, New York, 1964).

55. J. G. Collins & G. K. White, Chapter IX, p. 450 in "Progress in Low Temperature Physics" Vol. IV, (C. J. Gorter, ed.) (John Wiley & Sons, New York, 1964).

56. J. S. Dugdale & D.K.C. MacDonald, Phys. Rev. 89, 832 (1953).

57. D. J. Pastine, Phys. Rev., 138A, 767 (1965).

58. I. L. Spain & S. Segal, Cryogenics, 11, 26 (1971).

59. M. Born, J. Chem. Phys., 7, 591 (1939).

60. L. Knopoff in "High Pressure Physics and Chemistry", Chapter 5ii, p. 247 (R. S. Bradley, ed.) (Academic Press, London and New York, 1963).

61. T.H.K. Barron, Ann. Phys., $\underline{1}$, 77 (1957).

62. F. G. Fumi & M. P. Tosi, J. Phys. Chem. Solids, $\underline{23}$, 395 (1962).

63. N. Bernardes & C. A. Swenson in "Solids Under Pressure", (W. Paul & D. M. Warschauer, eds.) (McGraw-Hill, New York, 1963).

64. O. L. Anderson, J. Phys. Chem. Solids, $\underline{27}$, 547 (1966).

65. J. F. Nye, "Physical Properties of Crystals" (Oxford Univ. Press, London, 1964).

66. A.E.H. Love, "Treatise on the Mathematical Theory of Elasticity" (Dover Publ. Co., 4th ed. 1944).

67. T.H.K. Barron & M. L. Klein, Proc. Phys. Soc. (London), $\underline{85}$, 523 (1965).

68. Landolt-Börnstein, "Numerical Data and Functional Relationships in Science and Technology", Group III, Vols. 1 & 2, (Springer-Verlag, Berlin, 1966 (Vol. 1) and 1969 (Vol. 2)).

69. D. Lazarus, Phys. Rev., $\underline{76}$, 545 (1949).

70. M. P. Tosi, Solid State Physics, $\underline{16}$, 1 (F. Seitz & D. Turnbull, eds.) (Academic Press, 1964).

71. J. E. Jones & A. E. Ingham, Proc. Roy. Soc. (London) $\underline{A107}$, 636 (1925).

72. V. Heine & D. Weaire, Solid State Physics, $\underline{24}$, 249 (F. Seitz & D. Turnbull, eds.) (Academic Press, 1970).

73. C. Friedli & N. W. Ashcroft, Phys. Rev., $\underline{B12}$, 5552 (1975).

74. L. Thomsen & O. L. Anderson, NBS Specl. Publ., 326, p. 209, (E. C. Lloyd, ed.) (1971).

75. D. L. Decker, J. Appl. Phys., $\underline{37}$, 5012 (1966).

76. D. L. Decker, J. Appl. Phys., $\underline{42}$, 3239 (1971).

77. N. F. Mott & H. Jones, "The Theory and Properties of Metals and Alloys", (Oxford Univ. Press, London, 1936).

78. A. W. Lawson in "Metal Physics", p. 56, (B. Chalmers & R. King, eds.) (Pergamon Press, New York, 1956).

79. M. H. Lennsen & A. Michels, Physica $\underline{2}$, 1091 (1935).

80. M. H. Frank, Phys. Rev., $\underline{47}$, 282 (1935).

81. N. F. Mott, Proc. Phys. Soc. (London), $\underline{46}$, 680 (1934).

82. M. F. Manning & H. M. Krutter, Phys. Rev., $\underline{51}$, 761 (1937).

83. W. Paul in "High Pressure Physics and Chemistry", Chapter 5iv, p. 299 (R. S. Bradley, ed.) (Academic Press, London and New York, 1963).

84. G. Landwehr in "Physics of High Pressures and the Condensed Phase", Chapter 14, p. 556 (A. Van Itterbeek, ed.) (North Holland Publ. Co., Amsterdam and Interscience Publ., New York, 1965).

85. F. P. Bundy & H. M. Strong, Solid State Phys., $\underline{13}$, 81, (F. Seitz & D. Turnbull, eds.) (Academic Press, New York, 1962).

86. N. H. March in "Advances in High Pressure Research", Vol. 13, Chapter 4, p. 241, (R. S. Bradley, ed.) (Academic Press, New York, 1969).

87. V. V. Kechin, A. I. Likhter, G. N. Stepanov, Sov. Phys. Solid State, $\underline{10}$, 987 (1968).

88. I. L. Spain, Nat. Bur. Stand. (U.S.) Spec. Publ. 323, p. 717
    (L. H. Bennett, eds.) (1971).
89. M. Yoeman & D. A. Young, J. Phys. $C2$, 1742 (1969).
90. P. W. Bridgman, Proc. Am. Acad. Arts Sci., 84, 179 (1957).
91. P. Andersson & G. Backström, Rev. Sci. Instr., 46, 1292 (1975).
92. W. Paul & D. M. Warschauer in "Solids Under Pressure", Chapter 8,
    p. 179, (W. Paul & D. M. Warschauer, eds.) (McGraw-Hill, 1963).
93. W. Paul, "Pressure Effects on Band Structure" unpublished re-
    port from Summer School, (Delft, 1970) and
    D. L. Camphausen, G.A.N. Connell, W. Paul, Phys. Rev. Lett.,
    26, 184 (1971).
94. S. Minomura & H. G. Drickamer, J. Phys. Chem. Solids, 23,
    451 (1962).
95. G. D. Pitt, J. Phys. (Gt. Britain), $E1$, 915 (1968).
96. S. Glasstone, K. J. Laidler & H. Eyring, "The Theory of Rate
    Processes" (McGraw-Hill, New York, 1941).
97. S. Chandrasekhar, Revs. Mod. Phys., 15, 1 (1943).
98. C. Zener in "Imperfections in Nearly Perfect Crystals",
    (John Wiley & Sons, New York, 1952).
99. A. W. Lawson in "Solids Under Pressure", p. 15, (W. Paul &
    D. M. Warschauer, eds.) (McGraw-Hill, New York, 1963).
100. R. W. Keyes in "Solids Under Pressure", p. 71, (W. Paul &
    D. M. Warschauer, eds.) (McGraw-Hill, New York, 1963).
101. P. G. Shewmon, "Diffusion in Solids" (McGraw-Hill, New York,
    1968).
102. S. A. Rice & N. H. Nachtrieb, J. Chem. Phys., 31, 139 (1959).
103. A. W. Lawson, J. Chem. Phys., 30, 1114 (1959).
104. D. Lazarus & N. H. Nachtrieb in "Solids Under Pressure", p.43,
    (W. Paul & D. M. Warschauer, eds.) (McGraw-Hill, New York,
    1963).
105. L. H. Girifalco in "Metallurgy at High Pressures and High
    Temperatures", Chapter 10, p. 280, (K. A. Gschneider, Jr.,
    M. T. Hepworth & N.A.D. Parlee, eds.) (Gordon & Breach,
    New York, 1963).
106. C. W. McCombie & A. B. Lidiard, Phys. Rev., 101, 1210 (1956).
107. K. Arai, et al., J. Non-Cryst. Solids, 13, 131 (1973).
108. E. Piche & G. A. Rubin, Can. J. Phys., 51, 9, (1973).
109. R. S. Bradley, D. C. Munro & P. N. Spencer, Trans. Faraday
    Soc., 65, 1920 (1969).
110. A. Percheron, et al., Solid State Comm., 12, 1289 (1973).
111. M. Nicolas-Francillon, et al., Solid State Comm., 11, 845
    (1972).
112. R. S. Bradley, D. C. Munro & A. N. Spencer, Trans. Faraday
    Soc., 65, 1912 (1969).
113. C. Starr, Phys. Rev., 54, 210 (1938).
114. R. Peierls, Ann. Physik, [5], 3, 1055 (1929).
115. J. O. Hirschfelder, C. F. Curtiss, R. B. Bird, "Molecular
    Theory of Gases & Liquids", (John Wiley, New York, 1954).
116. J. S. Dugdale & D.K.C. MacDonald, Phys. Rev., 96, 57 (1954).
117. G. Liebfried & E. Schlomann, Nachr. Akad. Wiss. Gottingen
    Math. Phys., $2a$, 71 (1954).
118. A. W. Lawson, J. Phys. Chem. Solids, 3, 155 (1957).

119. O. Alm & G. Bäckström, J. Phys. Chem. Solids, $\underline{35}$, 421 (1974).
120. L. Bohlin & P. Andersson, Solid State Comm., $\underline{14}$, 711 (1974).
121. P. Andersson & G. Bäckström, J. Appl. Phys., $\underline{44}$, 705 (1973).
122. P. Andersson & G. Bäckström, J. Appl. Phys., $\underline{44}$, 2601 (1973).
123. J. C. Slater, Phys. Rev., $\underline{36}$, 57 (1930).
124. H. Bethe, "Handbuch der Physik", Vol. 24, p. 273, (A. Smekal, ed.) (Springer-Verlag, Berlin, 1933).
125. J. S. Kouvel in "Solids Under Pressure", Chapter 10, p. 277, (W. Paul & D. M. Warschauer, eds.) (McGraw-Hill, New York, 1963).
126. J. M. Leger, C. Susse, B. Vodar, Nat. Bur, of Stand. Specl. Publ. 326, p. 251 (E. C. Lloyd, ed.) (1971).
127. J. S. Kouvel in "Metallurgy at High Pressures and High Temperatures", Chapter 8, p. 229, (K. A. Gschneider, M. T. Hepworth & N.A.D. Parlee, eds.) (Gordon & Breach, New York, 1963).
128. D. Bloch, High Temp.-High Pressure, $\underline{1}$, 1 (1969).
129. D. Bloch & A. S. Pavlovic, "Advances in High Pressure Research", Vol. 3, p. 41 (R. S. Bradley, ed.) (Academic Press, New York, 1969).
130. G. A. Samara, "Advances in High Pressure Research", Vol. 3, p. 155 (R. S. Bradley, ed.) (Academic Press, New York, 1969).
131. G. A. Samara, Proc. IV Int. Conf. High Pressure, Kyoto, Japan, p. 247, (J. Osugi, editor-in-chief) (Kawakita Printing Co., Kyoto, 1975).
132. B. Jaffe, W. R. Cook & H. Jaffe, "Piezoelectric Ceramics", (Academic Press, London, New York, 1971).
133. O. E. Mattiat (ed.), "Ultrasonic Transducer Materials", (Plenum Press, New York, 1971).
134. J. Bardeen, L. N. Cooper & J. R. Schrieffer, Phys. Rev. $\underline{106}$, 162 (1957) and $\underline{108}$, 1175 (1957).
135. A. J. Darnell & W. F. Libby, Science, $\underline{139}$, 1301 (1963).
136. H. E. Bömmel, A. J. Darnell, W. F. Libby, B. R. Tittman, Science, $\underline{139}$, 1301 (1963).
137. S. Foner & B. B. Schwartz (eds.), "Superconducting Machines and Devices", (Plenum Press, New York, 1973).
138. M. Levy & J. R. Olsen in "Physics of High Pressure and the Condensed Phase", Chapter 13, p. 525, (A. Van Itterbeek, ed.) (North Holland Publ. Co., Amsterdam, 1965).
139. D. H. Bowen in "High Pressure Physics and Chemistry", Chapter 5, p. 355, (R. S. Bradley, ed.) (Academic Press, 1963).
140. T. F. Smith, J. Low Temp. Phys., $\underline{6}$, 171 (1972).
141. R. I. Boughton, J. L. Olsen, C. Palmy, "Progress in Low Temperature Physics", Vol. VI, Chapter IV, p. 163, (C. J. Gorter, ed.) (American Elsevier Publ. Co., 1970).
142. N. B. Brandt & N. I. Ginzburg, Sci. Am., p. 83 (April 1971).
143. N. E. Alekseevski & Y. P. Gaidukov, J. Exp. Theor. Phys. (USSR) $\underline{29}$, 898 (1955); Sov. Phys. JETP $\underline{2}$, 762 (1956).
144. G. Leibfried & W. Ludwig, Sol. State Phys., $\underline{12}$, 275 (1961).
145. R. A. Bartels & D. E. Schuele, J. Phys. Chem. Solids, $\underline{26}$, 537 (1965).
146. G. R. Barsch, Phys. Stat. Solidi, $\underline{19}$, 129 (1967).
147. G. R. Barsch & Z. P. Chang, Phys. Stat. Solidi, $\underline{19}$, 139 (1967).

Chapter 14

MECHANICAL PROPERTIES OF SOLIDS UNDER PRESSURE

H. Ll. D. Pugh and E. F. Chandler

Hydrostatic Extrusion and High Pressure
Engineering Division, National Engineering Laboratory
East Kilbride, Scotland

I.   INTRODUCTION

The mechanical properties of materials have been the subject of
many studies over a considerable period of time. This interest stems
from the fact that a knowledge of the stress and strain characteristics of engineering materials is required for safe and economic design. Most of this information has been obtained from uniaxial tension, compression and torsion tests.

In practice, however, stresses in structures and materials undergoing working processes are more complex and the particular stress
system can influence the mechanical behaviour of the material. It is
therefore of major   importance to study the mechanical properties of
materials under controlled conditions of combined stress.   Now any
combined stress system can be split into two components. One is a deviatoric component, which is a function of the differences of principal stresses and, in many wrought metals, governs the onset of plastic deformation. The other is a hydrostatic component, which is the
average of the principal stresses and has a considerable influence on
ductility. Thus combined stress systems can be simulated by simple
mechanical tests in the presence of a hydrostatic pressure so that the
two components of stress can be varied independently.

Amongst the earliest published works on the effect of hydrostatic
pressure on the mechanical properties of materials are those of Voigt
[1] who, in 1893, studied sodium chloride crystals and of Von Karman
[2] who, in 1911, carried out classical investigations on the fracture
of marble and sandstone cylinders under hydrostatic pressure. The
earliest published report on metals tested under pressure was that of
Böker [3] who, in 1914, carried out work on zinc. This work was followed by compression tests on cast iron carried out under hydrostatic
pressure by Ros and Eichinger [4] in 1929, while in 1934 Cook [5] reported his findings on torsion tests on copper and mild steel under a
superimposed hydrostatic pressure. However Bridgman [6, 7] is generally regarded as the pioneer of investigations into the properties of
materials under pressure and the experimental techniques and apparatus
which he designed and developed are still used by some workers.

This chapter provides a brief review of apparatus and methods
that are being used for carrying out tension, torsion and compression
tests under a superimposed hydrostatic pressure and the results from
tests on metals and polymers.

II.   EXPERIMENTAL EQUIPMENT AND METHODS

Practically every research team investigating the effect of a
superimposed pressure on the mechanical properties of materials has
equipment with some distinctive feature of its own. It is possible,
however, to divide into distinct groups most of the equipments and operating techniques.

A.   Tension and Compression Apparatus

The most commonly used tensile loading system is based on the
'yoke' apparatus originated by Bridgman [6] in the early 1940's.

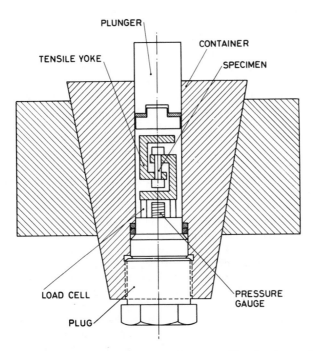

FIG. 1. Bridgman Yoke Apparatus for tensile tests under
hydrostatic pressure [6].

Bridgman's particular high pressure container was capable of operating
up to test pressures of 30 kbar and his complete apparatus is illustrated diagrammatically in Figure 1. It consists basically of the
high pressure container sealed at the lower end by a plug and at the
upper end by a pressurizing plunger. The dumb-bell type tensile specimen is held in an opposed yoke arrangement the lower end of which
rests on a load cell located on the sealed end. The top part of the
yoke is contacted by the pressurizing plunger which forces it downwards thus applying a tensile force to the specimen. Specimen extension is obtained by measuring the movement of the pressurizing/loading
plunger and making suitable allowances for the deflection of the
plunger, yoke and load cell. The pressurizing fluid is normally a
liquid, with isopentane being used for the highest test pressures, although nitrogen has been used successfully up to 14 kbar by Carpentier
et al [8]. The hydrostatic pressure is usually determined by measuring the change in electrical resistance of a manganin wire gauge [9]
or by a standard pressure gauge for tests in the lower pressure range.
     A major shortcoming of the equipment as described above is that
pressure will increase throughout the tensile test. Thus every part
of the stress-strain curve obtained will be correlated to a different
superimposed hydrostatic pressure. Several methods have been developed to ensure a constant pressure during the pulling of the specimen.
Of these the best are (a) coupling the top plunger mechanically outside the container to an identical lower plunger [10], (b) piping a

constant pressure from an adjacent intensifier into the test chamber
[11], and (c) sealing the lower end of the container by a billet in a
die which is hydrostatically extruded at an approximately constant
pressure [12, 13].

Another tensile testing design widely used is one in which one
end of the tensile test specimen is held securely with respect to the
container whilst the other end is connected to the bottom plunger.
As pressure is built up inside the container the outward movement of
the plunger is resisted by a press ram or jack. When the test pressure is reached the ram is withdrawn at the required speed, the
plunger follows and a tensile load is applied to the specimen. Constant test pressure can be maintained by using either an external
pumping system or by controlling the relative movement of the top and
bottom plungers. By using the appropriate grips and connectors most
forms of tensile specimen can be tested in this type of equipment.
There are various high hydrostatic pressure testing equipments [14-16]
which use the above method of loading and one of the more sophisti-
cated types [14] is shown in Figure 2.

This particular design, developed by Pugh and his co-workers at
the National Engineering Laboratory, Scotland, enables the true
stress-true strain relations of the test piece to be determined. It
has an internal load gauge which consists of two parts. The load-
measuring part (active gauge) is a flanged, thin tube upon which are
mounted longitudinally a number of equispaced electrical resistance
foil strain gauges. To compensate for pressure and temperature effects a second tube carrying a similar arrangement of strain gauges
fits inside the first tube but is not subjected to the applied load.
Load calibrations of this type of load gauge within the pressure
range 0 - 10 kbar were found to agree within $\pm$1 1/2 per cent [17].

The apparatus also has facilities for recording the changing diameter of the specimen under test. This is accomplished by using a
cross-bored container fitted with windows and an optical system which
enables a silhouette of the deforming specimen to be projected on a
screen. By placing a galvanometer displaying load adjacent to this
screen a continuous photographic record of both load and specimen profile is obtainable (Figure 3). Toughened plate glass windows are
suitable for pressures up to 13 kbar whilst sapphire windows can be
used for even higher pressures.

Where photographic facilities do not exist there are alternative
methods of obtaining measurements of the change in diameter of the
specimen. One such method is that of the 'interrupted' test first
used by Bridgman [6]. A certain load is applied to the specimen at
the appropriate test pressure, the load and pressure are removed, the
specimen removed from the container and the new diameter of the specimen is measured. This procedure is repeated for a number of load in-
crements and quite obviously is both slow and cumbersome. An alternative method suggested by MacGregor [18] is to use a tapered specimen
having a series of circumferential marks along its length. After a
known load has been applied to the specimen, the new diameters at each
circular mark are measured and the true stress and strain at each position evaluated. These, when plotted, make up the stress/strain
curve.

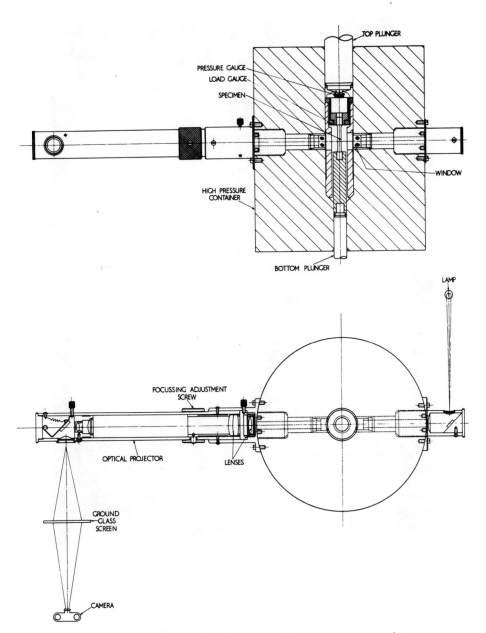

Fig. 2.   Tensile apparatus for obtaining true stress/true strain curves under hydrostatic pressure [14].

The simplest form of compression apparatus is one in which a static lower plunger or plug supports the specimen with the upper plunger pressurizing the fluid as well as deforming the specimen [6]. The pressure will, of course, continue to increase throughout the

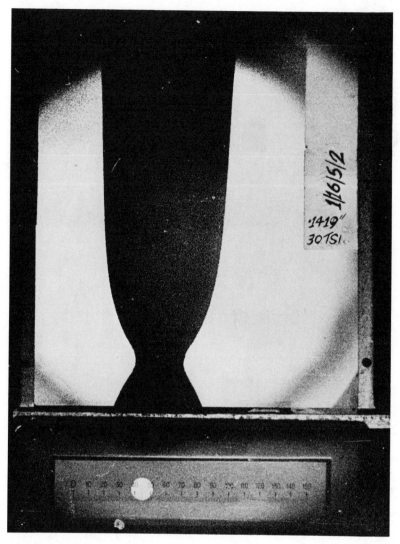

FIG. 3. Tensile specimen being deformed under a hydrostatic pressure.

test. An improvement on this equipment is to use a stepped-bore container with a retractable lower plunger to maintain a constant test
pressure [19].

     An alternative arrangement to the above is that originated by
Griggs [20] and shown schematically in Figure 4. A stepped bore container is again used but the upper and lower plungers have the same
diameter. The high pressure container is pressurized by the upward
movement of the lower plunger, the upper plunger being fixed. Once
the test pressure has been reached the plungers can be assumed to be
mechanically coupled so that the distance between them will not alter
during the test. The container is then raised so that the specimen

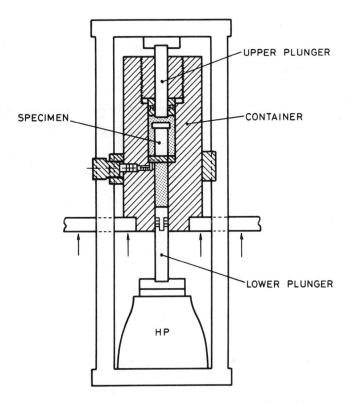

FIG. 4. Griggs' apparatus for compression tests under
hydrostatic pressure [20].

contacts the upper plunger and is deformed with the fluid pressure remaining constant. Variations on this method are to have the plungers
yoked together, pressurized fluid piped into the container, and the
specimen loaded by movement of either the yoke [21] or the container
[22]. A further variation of Griggs' basic concept is one in which
the plunger for applying the compressive load is located inside the
high pressure containers [23].

Compressive strain, which is a function of reduction of height of
the specimen is obtained by recording externally the movement of the
loading plunger or by a displacement transducer located within the
chamber [24]. Although the use of internal load cells with strain-
gauges in both tension and compression rigs is increasing, many workers still use an external load cell. This, although easier to operate, has a number of shortcomings. It measures not only the load being applied to the specimen but also the friction at the plunger seal
and the load on the plunger due to the pressure in the container.
Consequently, at high pressures the applied load may be only a small
portion of the load actually measured by the external load cell. Variation of pressure during the actual test will, of course, make accurate determination of the applied load extremely difficult. To reduce
friction and thus improve the accuracy of the load measurement a number of equipments use the Johnson and Newhall [25] controlled clear-

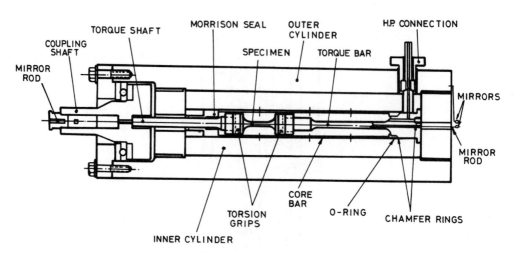

FIG. 5. Apparataus for torsion tests under hydrostatic
pressure - external drive [27].

ance seal (see Chapter 8) or the Morrison [26] seal for the loading
plunger.

B.   Torsion Apparatus

Apparatus for carrying out torsion tests under a superimposed
hydrostatic pressure can be divided into two basic designs, the power
unit applying the torque to the test piece being located either outside or inside the high pressure container.

There are a number of different versions of the first type of
apparatus [26-30]. In all of them, however, one end of the torsion
specimen is held secure with respect to the high pressure container
whilst the other end is connected to a shaft or piston which passes
through a low friction, high pressure seal (Figure 5). The shaft is
then either (a) connected to a motor gearbox arrangement such as a
conventional torsion testing machine [26-28] or (b) secured to a fixed
platen and the container rotated by some driving arrangement [29, 30].
The torque being applied is measured by a torque tube placed in series
with the specimen and is located either in or outside the pressure
chamber.   Most torque tubes consist of chevron-shaped foil strain-
gauges mounted on to a thin tube or rod. In an alternative design
[26] the torque tube is positioned so that its outside surface is subjected to the test pressure and its bore to the atmosphere (Figure 5).
This requires the tube not only to remain elastic and have adequate
sensitivity to torque but also to withstand the external pressure.
Torque in the tube, and thus the specimen, is determined by measuring
optically the relative twist of the calibrated tube. Torsional strain
is derived from recording the rotation of the driving head or container.

Pressurized fluid is normally piped in the container from a secondary intensifier or pump. The maximum test pressure is essentially
governed by the strength of the container and difficulties with friction at the moving seal.

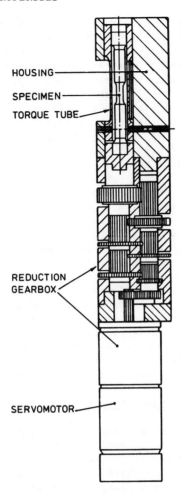

HOUSING

SPECIMEN

TORQUE TUBE

REDUCTION
GEARBOX

SERVOMOTOR

FIG. 6. Apparatus for torsion tests under hydrostatic
pressure - internal drive [31].

    With the second design [31, 32], the internal motor drive, the
complete torsion apparatus is accommodated inside the container and
subjected to the test pressure. This type of rig, shown in Figure 6,
consists of a small synchronous motor, a gearbox to reduce the speed
of rotation, a specimen housing and a strain-gauged torque tube.   To
reduce the overall length of the torsion unit it is customary to fit
the specimen inside the torque tube. The angle of twist is determined
from a make and break contact arrangement on one of the gear wheels.
    Control and operation of this type of apparatus is easier than
that of the previously mentioned system. It has, however, some limitations. Practical limits on the size of the pressure container control the size of the motor and thus the maximum load capacity. Because of the viscous drag on the motor and gearing due to the pressure
transmitting fluid, fluids with low viscosity at high pressures, such
as isopentane or kerosene, have to be used which have poor lubrication
properties. These limitations restrict the practical maximum pressure
for the operation of these equipments to 10 kbar.

III.   EFFECT OF PRESSURIZATION AND PRESSURE MEDIA
ON THE PROPERTIES OF MATERIALS

Considerable care must be exercised in the interpretation of experimental results obtained from mechanical tests carried out under
hydrostatic pressure particularly when such factors as criteria of
fracture are being investigated since it is essential that the material remains the same during the investigations. However, it is known
that under sufficiently high pressures many metals exhibit phase
changes, which may be reversible or irreversible [33]. A good example
is bismuth, which undergoes several reversible phase changes between
about 24 kbar and about 100 kbar (see Chap. 4, Vol. II). These phase
changes may be accompanied by changes in the microstructure of the
metal on removal of the pressure.

Since many of the investigations of the mechanical properties
under pressure are directed towards the effect of different stress
systems on the behaviour of materials it is essential that the material remains the same both during and at the end of tests. Insufficient
attention has been given by many workers in the past to this point and
it is necessary in this work to differentiate between those reversible
effects which occur during the application of the hydrostatic pressure
and those permanent microstructural changes which may occur as a result of the application of a pure, hydrostatic pressure. Another
often neglected factor is the possible interaction between the pressure transmitting medium and the test piece which may have a significant effect on the material's mechanical performance under pressure.

A.   Structural Changes Under Pressure

For single crystals, whether they be isotropic in elastic constants as in the cubic metals or anisotropic as in the hexagonal metals such as zinc or magnesium, it is generally accepted that under
true hydrostatic compression there will be no plastic flow unless
there is a permanent change in density or some other form of irreversible phase transformation. However, for polycrystalline materials,
high hydrostatic pressures can cause local plastic deformation if the
individual crystals are anisotropic. Thus, under the influence of hydrostatic pressure, two single anisotropic crystals of different orientations would, in isolation, change dimensions to different degrees
in any given direction. When adjoining each other in polycrystalline
form, they will exert a restraint on each other and shear stresses
will be set up along the grain boundaries. Local plastic deformation
will occur if the anisotropy in compressibility or the boundary angle
is sufficiently large that the shear stresses exceed the critical
shear stress for slip. Davidson [34] and Johannin and Vu [35, 36] determined the degree of plastic deformation developed as a result of
pressure soaking by a microscopic examination of the electro-polished
surfaces of various metal specimens before and after the application
of pressure. Davidson found a definite correlation existed between
the degree of anisotropy in linear compressibility and the occurrence
of deformation in polycrystalline metals induced by hydrostatic pressure. Unfortunately, assessing the actual effect of these particular

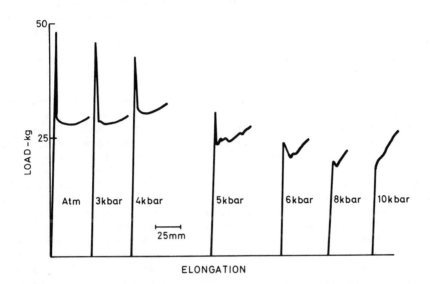

FIG. 7. Effect of different pressurizations on the
load/elongation curve of Armco iron [38].

induced changes on the stress/strain characteristics of metals being
tested under pressure is extremely difficult. Measurement of mechanical properties after a pressurization, however, suggest these effects
are probably quite small. After being subjected to pressures up to
25 kbar the stress/strain relations at atmospheric pressure of zinc,
a metal having a high linear compressibility ratio, is unchanged [37].
      Second-phase particles, which can cause local plastic deformation
and generate new dislocation centres at the boundary between the hard
inclusion and even an isotropic matrix, have been found to have a far
greater effect on the tensile properties, particularly of body centred
cubic metals following pressure soaking. Bullen, Henderson and Wain
[38] used the Hutchinson centre-anneal technique to make accurate
measurements of the change in magnitude of the load drop on discontinuous yielding in annealed wire specimens of a commercially pure iron
(Armco 0.03 per cent C) and a high purity iron (0.007 per cent C)
after pressure soaking up to 10 kbar. They showed how pressure generated more dislocations at second phase particles, thus lowering the
yield and flow stresses (Figure 7).   The effect was particularly
marked on the load drop on discontinuous yielding. Radcliffe [33] observed changes in the magnitude of the lower yield or flow stress for
a series of annealed plain carbon steels containing 0.2, 0.4, 0.6 and
0.95 per cent C after pressure soakings up to 40 kbar. Both sets of
investigators found that, after application of a sufficiently high hydrostatic pressure, the discontinuous yield point in iron/carbon alloys was eliminated, whilst, in contrast, the load drop in a high purity iron was reduced only slightly, thereby confirming the role of
the second phase particles. Bullen [38] explained his results by the
presence of second phase particles whose compressibility was markedly

different from the Armco iron matrix. Consequently the results may be
expected to depend on purity. The deformation resulted in the localized creation of new dislocations, which were not locked by the atmospheres of the solute atoms. Hence yield could take place by the
movement of these free dislocations at a lower stress. The effect
could be detected because of the extreme sensitivity of the discontinuous yield to the presence of free dislocations.

Bullen et al [39] also found that at room temperature and pressure recrystallized chromium fractured without detectable plastic deformation whereas similar specimens, after being subjected to a pressure of 10 kbar, attained an elongation of over 60 per cent before
fracture. (This phenomenon was limited to chromium containing a relatively high percentage of nitrogen impurities and tested at low
straining rates.) These authors concluded that the application of hydrostatic pressure to chromium had resulted in the generation of free
dislocations so that yield at a lower stress was possible.   As
Cottrell [40] and Petch [41] have pointed out, there is a close association between yield point and brittleness. Brittle fracture occurs
when the yield stress is higher than that required to initiate and to
propagate cracks. Hence the superimposition of pressure induced ductility by lowering the yield stress below that for fracture. The
effect of pressure soaking of chromium in Bullen's experiment [59]
therefore was to reduce the brittle-ductile transition temperature for
that particular batch of material to below room temperature.

B.   Pressure Media Effects

One of the problems of pressure testing is the choice of the
pressure transmitting media. Liquids are usually preferred because of
problems due to the high compressibility of gases and resulting potential energy as well as the difficulties of sealing, whilst even soft
solids can only given an approximation to hydrostatic pressure conditions.   The number and range of liquids, however, are limited. Many
become extremely viscous and freeze below 10 kbar so true hydrostatic
pressures may not be obtained at these pressures and consequently non
hydrostatic stresses can be imposed on the test specimen.   In some
cases this can cause a reduction in strength and in others raise the
yield stress by work hardening the material.

It has been known for some time that sheathing of the specimens
is necessary if consistent results are to be obtained with certain materials, in tests under pressure. However, care has to be exercised
over the choice and thickness of the material used for sheathing since
a relatively thick or strong sheath will support an appreciable proportion of the applied load and may even apply an additional compressive stress to the specimen. Many types of rubber exhibit a change
from a rubbery to glassy state at about 5 kbar with resulting increases in strength and Young's modulus but a marked decrease in ductility [42], changes which may be reflected in the behaviour of the
test specimen.   A sheathing now being used with some success is a
proprietary rubber solution which can be applied thinly to test specimens of all shapes and sizes [43]. It serves the purpose of preventing contact between specimen and liquid thereby inhibiting the pene-

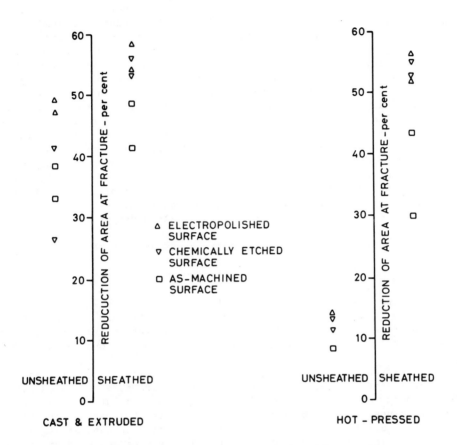

FIG. 8. Effect of surface finish and sheathing on the
ductility of beryllium tested under a hydrostatic pressure of
7.7 kbar [45, 46].

tration of the liquid into cracks and pores but, at the same time,
allowing the hydrostatic pressure to be applied to the specimen.

In general, it is found that cast materials or materials with a
porous structure show marked differences in tensile tests under pressure when the specimens are in the sheathed and in the unsheathed condition.

It has been suggested [44] that sheathing is not necessary if the
surfaces of such metals are electropolished to reduce surface damage
to a minimum. Figure 8 summarizes the effect on ductility of different surface preparations of hot-pressed powder and extruded ingot beryllium specimens, tensile tested under a pressure of 7.7 kbar [45,
46]. This graph demonstrates quite clearly that sheathing is the dominant factor in preventing premature fracture of this type of material. It is interesting to note that in hydrostatic pressure fatigue
tests on steel vessels [47] their fatigue life was increased if the
vessel bore was given an initial rubber coating.

In studies of the effect of hydrostatic pressure on the mechanical properties of polymers few investigators have used sheathed test
specimens and, in investigations where the specimens have been jacketed, the principal reason for jacketing has been to prevent possible
bulk plasticization of the test material. One investigation [48],
however, involving infra-red spectroscopy studies of polystyrene specimens that had been pressurized to 7.7 kbar disclosed no signs of bulk
plasticization by the oil. There is evidence [37, 49] however, that
the strain-to-fracture under pressure of polymers is substantially increased if the specimen gauge length is sheathed. The surface finish
of polymer specimens also plays a major role in determining their
strain-to-fracture since an increase in pressure changes many polymers
from a plastic to a 'glassy' state so that machine marks and other defects on the surface now become critical stress-raisers. A great deal
of work still remains to be carried out on the relationship between
pressure media and the properties of materials under pressure.

IV.   EFFECT OF HYDROSTATIC PRESSURE ON METALS

A.   Ductility

The increase in ductility of metals under triaxial compressive
stress is a well established fact and it is utilized in many metal-
working processes. For this reason, the quantitative relationship between ductility and hydrostatic pressure for various metals is of
value.

Tests on a wide range of metals have resulted in general agreement that, for metals, ductility or strain-to-fracture increases with
hydrostatic pressure.   There is, however, some variation in the ductility values obtained by various investigators at a particular pressure for similar metals. This must be expected as the metals tested
could vary in grain size, fabrication and heat treatment and the tests
could be carried out at different strain rates. This was clearly demonstrated by Tanaka [50] in investigations on zinc of different grain
sizes, d, tested at different strain rates, $\dot{\varepsilon}$. From Figure 9 it can
be seen, however, that the general form of the ductility/pressure
curves for that particular metal is fairly consistent although they
are displaced relative to one another.

The effect of pressure on the ductility of wrought metals has
been studied by a number of investigators and some of the experimental
results are shown in Figure 10. Bridgman [6] found that the ductility, or strain-to-fracture, was a linear function of pressure for all
but two of the steels he tested. Ryabinin, Livshitz and Vereschagin
[51] confirmed this finding for a range of steels of different carbon
content, for high carbon steel heat-treated to different hardnesses,
for duralumin, and a heat-resisting steel. Other workers, for example
Pugh [14], Yajima and Ishii [52], Brandes [53], Nolan and Davidson
[54] and French and Weinrich [55] have reported similar findings for
other steels.   However, as may be seen from Figure 10, a deviation
from linearity does occur for some wrought metals, particularly copper
and aluminium.   In some tests this deviation is in the direction of
larger strains Pugh [14], Brandes [53],but Ryabinin [56] found for six

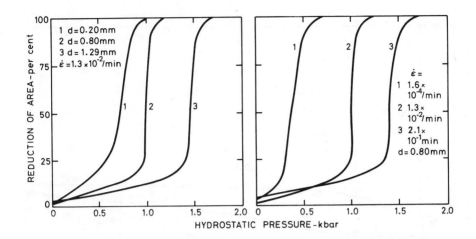

FIG. 9. Effect of grain size, d, and strain rate, $\dot{\varepsilon}$, on the ductility of zinc under pressure [50].

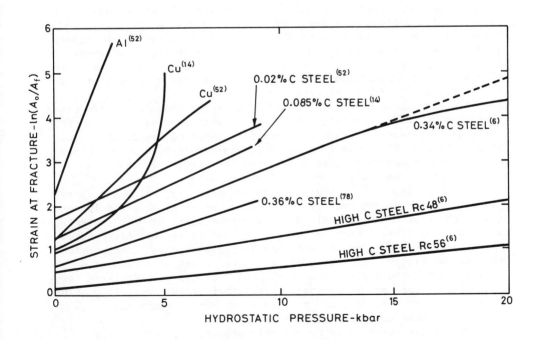

FIG. 10. Effect of pressure on the ductility of wrought metals.

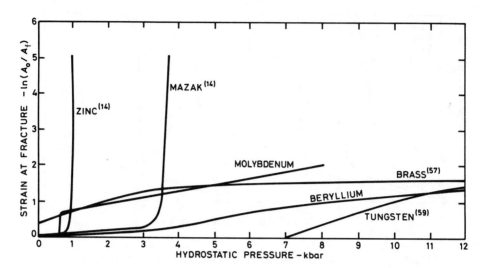

FIG. 11. Ductility transitions in metals tensile tested
under pressure.

of Bridgman's softer steels a decrease in the rate of increased ductility at the highest pressure. He obtained a similar behaviour in a
0.46 per cent carbon steel as did Yajima [52] for copper and aluminium. For these reasonably ductile materials, accurate measurements of
the area of fracture of the highly deformed fracture surfaces of the
test specimens are difficult to obtain and this could account at least
partly for the different experimental findings.

Beresnev and his colleagues [57] also found that a 60/40 brass
exhibited an initial increase in ductility with pressure, but at pressures greater than 3.5 kbar the ductility remained practically constant (Figure 11). This form of ductility/pressure curve for brasses
has been confirmed by Hu [58] and Yajima and Ishii [52] who also obtained similar results for copper/germanium alloys.

The behaviour of the hexagonal metals zinc, magnesium and beryllium as well as the body-centered cubic metals tungsten, molybdenum
[59], and chromium is markedly different from that of the materials
described earlier. These materials are all fairly brittle at atmospheric pressure and initial increases in pressure have little effect
on their ductility. At a critical pressure, which is dependent on the
actual metal being tested, there is rapid increase in ductility (Figure 11). Those materials that exhibit this brittle-ductile pressure
transition also undergo similar transitions when temperature rather
than pressure is the variable parameter.

In torsion tests under hydrostatic pressure, Crossland [26] obtained a linear relation between increase in ductility and pressure
for mild steel, copper and a silicon/aluminium alloy (Figure 12) but
no apparent increase in ductility for zinc and Mazak under pressure.
Erbel [29] investigated the effect of hydrostatic pressure on the
strain-to-fracture of steel, copper and aluminium tubular specimens

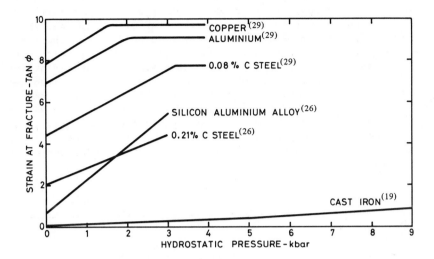

FIG. 12. Effect of pressure on the torsional ductility of
some metals.

and found an initial linear correlation, but at higher pressures the
fracture strain remained constant. Ohmori and Yoshinaga [28] also
found that the ductility of 70-30 brass and mild steel increased linearly with pressure. Their results for aluminium agreed with Erbel's.

Pugh [76] suggested that the apparent levelling-off of the torsional fracture strain-pressure relationship was probably due to an
effect analagous to necking in tension, that is, at some point uniform
twisting ceased and twisting was concentrated into a very short length
of specimen. Under these conditions the measurement of overall twist
would be of as little value as elongation in a tensile test; there
would be an initial increase in ductility with pressure, but once a
critical strain was reached, increasing pressure would have little or
no apparent effect on the material's ductility. In work on cast iron
[19], it was shown that up to maximum torque (strain 0.428), strain
was uniform along the gauge length, but when the apparent strain had
reached 0.49, the major part of the deformation was restricted to 14
per cent of the original gauge length.

There are few published ductility results on compression tests
under hydrostatic pressure. The major difficulty is the determination of the precise point at which fracture occurs, particularly in
materials displaying large amounts of deformation. Consequently tests
have been generally confined to materials which are extremely brittle
at atmospheric pressure. It has been found [19] that the fracture
strain of cast iron and magnesium alloy AZ91 increases linearly with
pressure, although that of AZ91 levels off at the higher pressures as
in the case of similar alloys in tension tests. For both these materials sheathing appeared to have no effect on their ductility. For
aluminium bronze, pressure appeared to have no effect on the fracture
strain and a similar result was obtained with tension tests under
pressure on this same material.

Identifying a general relationship between the strain-to-fracture and pressure for a whole range of materials is difficult. However it does appear that most of these materials have one common
characteristic, namely a transition pressure above which the rate of
increase of strain-to-fracture increases fairly markedly with pressure. In those metals, such as ductile steels, which exhibit a linear
relation between strain-to-fracture and pressure, this transition
pressure may be assumed to be less than atmospheric pressure. The
transition in the slope of the strain-to-fracture/pressure curve from
a low value to a higher value may take place over a relatively wide
pressure range, as in copper and aluminium, or over a very narrow
pressure range, as in zinc and Mazak. This point is discussed in a
later section (IV.G.) of the chapter.

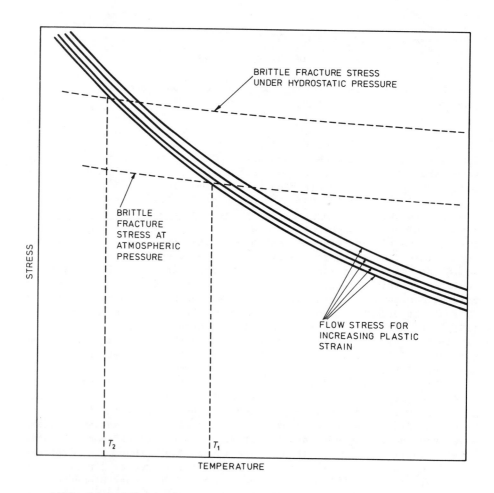

FIG. 13. Effect of pressure on the brittle-ductile
transition temperature of metals.

B.   Brittle-Ductile Transition Temperature

The sharp transition in zinc, mazak etc. is analagous to the sharp transition obtained in some materials when the temperature is varied. The effect of pressure on the brittle-ductile transition temperature can be explained by an expansion of the concepts suggested by Davidenkov and Wittman [60] for the existence of a transition temperature. The yield or flow stress and the brittle-fracture stress change at different rate with temperature. For a tensile test at atmospheric pressure these stresses are shown diagrammatically in Figure 13. The transition temperature is T_1 since, above that temperature, the flow stress is lower than the brittle fracture stress whereas below T_1 stress, the material achieves the value for brittle fracture before that for yielding or flow. Consider the same tensile test carried out in the presence of a high hydrostatic pressure. The criterion of yield or flow, which is dependent on the value of the differences of principal stresses, is unaffected by hydrostatic pressure and hence the flow stress curve is unchanged. However, the criterion of brittle fracture is usually found to be dependent on the absolute value of the maximum tensile stress, therefore the material under hy-

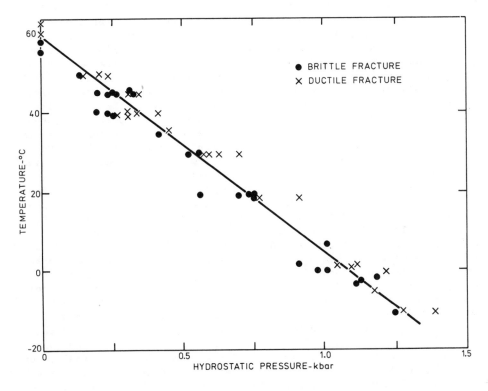

FIG. 14. Effect of temperature and pressure on the brittle-ductile transition of zinc.

drostatic pressure will require a larger superimposed tensile stress
to initiate brittle fracture.   Consequently, the brittle fracture
curve will be raised giving a lower transition temperature T_2.   It
follows that there will be a relation between the brittle-ductile
transition temperatures and pressures, which will depend on the effect
of pressure in raising the brittle fracture curve and on the shape of
the flow stress-temperature curve.   Further, the higher the hydrostatic pressure, the greater the reduction in the transition temperature for a given material.   Such a relation has been determined for
zinc, as shown in Figure 14, and over the range investigated it appears to be linear. Galli and Gibbs [61] found a similar lowering of
the brittle-ductile transition temperature for molybdenum from 50°C
to 2°C when a hydrostatic pressure of 1.4 kbar was applied.

C.   Residual Ductility After Pre-strain Under Pressure

The increase in ductility of metals with hydrostatic pressure is
clearly of importance in relation to forming processes.   Another of
the results of the study of properties under pressure which is of
significance in the properties of cold-formed products is the residual ductility of metals after pre-strain under pressure.
Using a variety of steels, Bridgman [6] investigated the effect
of a given amount of pre-strain applied under hydrostatic pressure
on the residual ductility obtained in tests at atmospheric pressure.
He tried pre-straining under pressure both in tension and compression,
and the subsequent tests at atmospheric pressure were either in
tension, compression or torsion.   The tests showed that the total
strain-to-fracture increased with the amount of pre-strain at a given
pressure, and further, that for a given pre-strain, the total strain
increased with pressure.   His results for a tempered pearlitic steel
pre-strained in tension tests over a range of pressures and pulled to
fracture at atmospheric pressure are presented in Figure 15.   These
results have been confirmed by Pugh [14] who carried out compression
tests on a 11 per cent silicon aluminium alloy.
These experiments suggest that for a given strain which is identical both in magnitude and in change of geometry, the internal damage is less if that pre-strain has been carried out in the presence of
a hydrostatic pressure and hence the specimen whould then be able to
sustain a larger additional strain at atmospheric pressure.   The truth
of this general conclusion has been confirmed in a number of metal-
working operations; thus in the hydrostatic extrusion of magnesium
alloy ZW3 it was found that the loss of ductility due to normal cold
working could be virtually eliminated if the product emerged into a
pressurized liquid [62].   Similar improved properties have been obtained in flanges formed hydrostatically on tubes and on hydrostatically "deep drawn" cups [63]. Confirmation that the improved properties are due to the inhibition of the formation of internal voids
and other forms of damage has been obtained by Rogers and Coffin [64],
who carried out sheet-drawing tests with and without a superimposed
hydrostatic pressure. Using the superimposed hydrostatic pressure
they showed conclusively that the density of the drawn sheet did not
decrease with increasing reduction, as it did when no hydrostatic

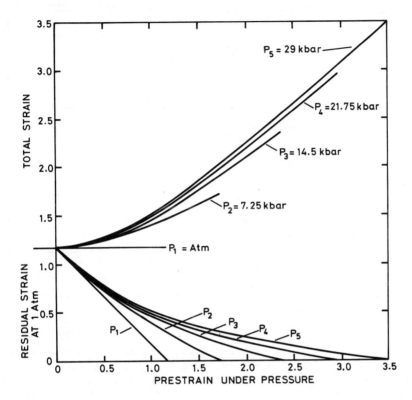

FIG. 15. Residual ductility of a pearlitic steel prestrained under different hydrostatic pressures [6].

pressure was applied, and further that the properties of the sheet drawn under hydrostatic pressure were considerably superior.

D. Tensile Strength and Necking

Bridgman's initial studies on the stress-strain properties of steels under pressure led him to conclude that hydrostatic pressure caused a linear increase in tensile strength of about 6.5 per cent of the increase in pressure. Similar results were obtained for a range of other materials such as nickel, niobium, molybdenum and tungsten although the increase was limited to about 4 per cent. He also reported that the strain at the maximum load appeared to be independent of pressure. Ratner [11] has also reported an increase in tensile strength of 41 per cent and 21 per cent for, respectively, a magnesium alloy and beryllium bronze with increasing pressure, whilst Pelczynski [65] reported an increase in the strain at the maximum load point for steel, brass, copper and aluminium with increase in pressure, a result contrary to that of Bridgman's.

Pugh [14] however, using the equipment described earlier for obtaining a continuous photographic record of stress and strain found

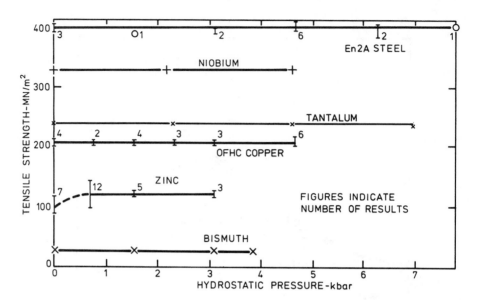

FIG. 16. Effect of hydrostatic pressure on the tensile
strength of wrought metals.

that both the instability strain and the tensile strength of wrought
metals were independent of pressure. The suggestion in Figure 16 of
an effect for zinc (shown by a dotted line) in the pressure range 0 to
0.7 kbar is due to the fact that, in this pressure range, zinc had
fractured before reaching the maximum load, (or instability) point.
This part of the curve thus refers to its nominal fracture stress.
    As a consequence Pugh re-examined the widely accepted theoretical
solution [66] that a transverse pressure superimposed on a specimen in
tension delayed the onset of necking and resulted in an increased uniform elongation. He found the original solution to be erroneous and
when corrected showed that hydrostatic pressure has no effect on the
instability point.
    This result is of prime importance because it clearly indicated
that the application of a hydrostatic pressure or an increased hydrostatic component of stress is unable to result in practical advantages in those metal-working processes that are limited by necking,
for example most drawing processes, since the increased ductility
under pressure will only be apparent in the local reduction of the
neck and not in uniform deformation.
    Some cast and powder compacted materials do however show an in-
crease in tensile strength (Figure 17), but, as the test pressure increases, the tensile strength tends to a constant value. This initial
increase is probably due to further compaction of these materials.
    The stress system in a tensile specimen becomes complicated· with
the onset of localized necking. The stress distribution in this region has been investigated by Siebel [67], Bridgman [6] and by Davidenkov and Spiridonova [68]. In each analysis the relation between
maximum and average stresses has been evaluated by correction factors

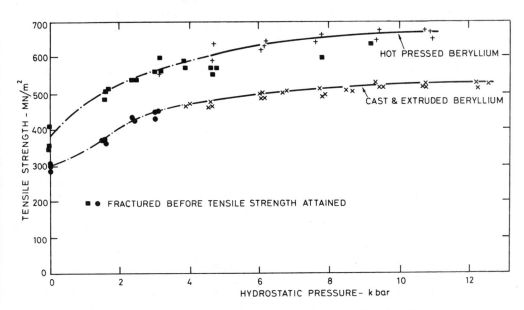

FIG. 17. Effect of hydrostatic pressure on the tensile strength of some cast and powder compacted metals.

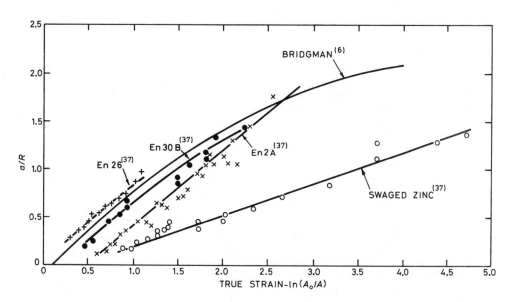

FIG. 18. Relationship between neck profile and true strain for several metals.

based on the radius of curvature at the neck, R, and the radius of cross section of the neck, a. Experimental investigations show that a/R for a particular material is a function of the strain in the test material and is independent of ambient pressure. The relationship between a/R and strain, however, is different for different materials (Figure 18) although the relationships for different steels are approximately the same.

E.   Stress-Strain Relationship

Once necking starts, the true stress-true strain relations of a tensile specimen can be obtained only if the pertinent cross-sectional area is known. As described earlier this is particularly difficult when the tests are carried out inside a high pressure container. Most workers therefore have restricted their investigations to the effect of pressure on the nominal stress-elongation curve of metals.

The stress/strain results of earlier workers differ. Thus Ratner [11] reported that yield stresses of beryllium bronze and two magnesium alloys were raised by 20-40 per cent at pressures of 2.2 kbar whilst there was hardly any change for copper. Gladkovski and Oleinik [15] found that the flow stress of beryllium bronze was considerably raised by pressures of 3 kbar but that the rate of strain hardening was reduced. Haasen and Lawson [69] found an increase in yield stress of alpha brass single crystals at 5 kbar, but they also discovered that the application of high pressure alone was sufficient to alter the stress-strain curve by the same amount. Results for the tensile stress-strain curve of single crystals of aluminium, copper, nickel and polycrystalline nickel indicated no significant change in the flow stress at low strains but an increased flow stress at higher strains. Hu and his colleagues [70] reported no change in the yield stress of an aluminium alloy or of a mild steel at 3.4 kbar and an examination of their data on pre-stressed mild steel specimens also indicates no definite effect of pressure. In work on Nittany No. 2 brass (61.5 per cent copper, 35 per cent zinc, 3.25 per cent lead) Hu [58] found that yield stress increased with pressure but, in later tests on pre-stressed specimens, discovered that the yield stress decreased with increasing pressure up to 7 kbar.

Investigations by other workers however suggest that the stress-strain relations for most wrought metals are unaffected by hydrostatic pressure. The earlier work by Ros and Eichinger [4], in 1929, on cast iron and by Cook [5], in 1934, on mild steel and copper indicated no significant effect whilst those of Bridgman [6] confirmed that the flow-stress at low strains for a number of materials was independent of pressure. More recently, Pugh [14] and his colleagues, using the method involving the continuous photographic recording of the deforming specimen, showed that hydrostatic pressure had no significant effect on the stress-strain relationship for wrought metal including a number of steels, copper, zinc, and tungsten (Figure 19). Similar conclusions were obtained by Brandes [53] for Armco iron, copper and a number of steels, Carpentier and Francois [71] for zinc and cobalt, French and Weinrich [55], Ohmori et al [72] and Nobuki and Oguchi [73] for different steels. Torsion tests on a number of metals in-

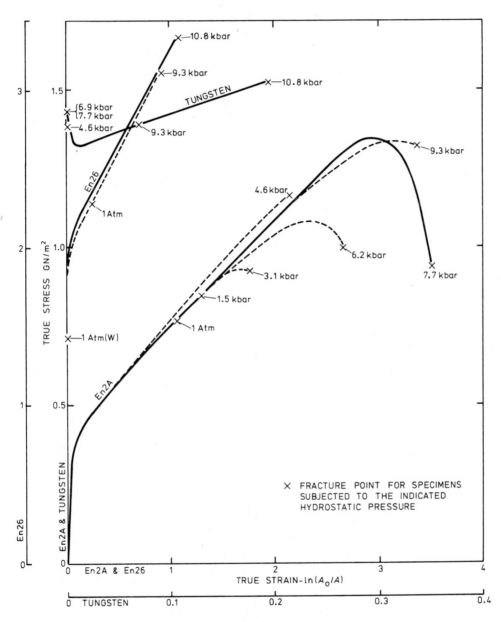

FIG. 19. True stress/true strain curves of a number of metals at various hydrostatic pressures.

cluding copper, zinc, Mazak, silicon aluminium [26], magnesium alloy [31] and steels [43, 72] have produced similar findings, namely that the stress-strain curve is generally unaltered by hydrostatic pressure other than an increase in strain-to-fracture. Since the yield represents one part of a stress-strain curve this confirms the universally

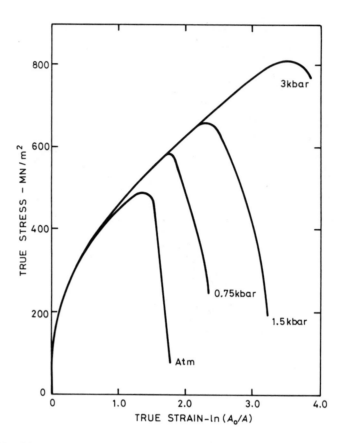

FIG. 20 True stress/true strain curve of OFHC copper under various pressures.

accepted criterion of yielding of wrought metals, namely that yielding is independent of the hydrostatic component of stress and is dependent only on the deviatoric components of stress, i.e. the shear stresses.

It has been found for metals such as copper, which are reasonably ductile at atmospheric pressure, that true stress increases with increasing strain, but apparently just before fracture starts to decrease. The effect of increasing pressure is to extend the stress-strain curve and to delay this 'turnover' (Figure 20). The deviation of the curves for different pressures near fracture is due to the fact that, after necking, the non-uniform stress distribution across the neck results in the largest hydrostatic tension occurring longitudinally on the axis and this causes the formation of cavities, which enlarge and coalesce and form cracks inside the neck as the extension increases [74]. The rapid decrease of the true stress after reaching the maximum point is attributed to the formation and coalescence of these internal cracks, which make the actual area over which the load is transmitted smaller than it appears from the cross-sectional area of the neck. The effect of hydrostatic pressure is, of course, to in-

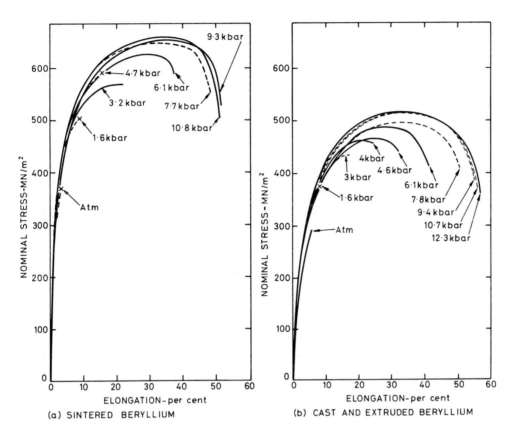

FIG. 21. Effect of different pressures on the nominal
stress-strain curves of: (a) hot-pressed beryllium, (b) extruded
ingot beryllium.

hibit and defer the formation of cavities and therefore, for a given
strain, the number of cavities or voids inside the specimen is smaller the higher the superimposed hydrostatic pressure.

It has been shown that hydrostatic pressure does have a definite
and significant effect on the stress-strain relation of some cast and
pressed-powder materials [37] (Figure 21). The stress-strain curves
were raised by increasing pressure but these increases became less
until a point was reached at which further increases in pressure had
negligible effect. This would appear to be caused by hydrostatic
pressure compacting the specimens further and reducing the volume of
internal voids, giving a denser and, generally speaking, a stronger
material.

F.   Fracture Stress

Bridgman [6] who carried out tensile tests on a range of steels
and tried to find a fracture-stress criterion based on various com-

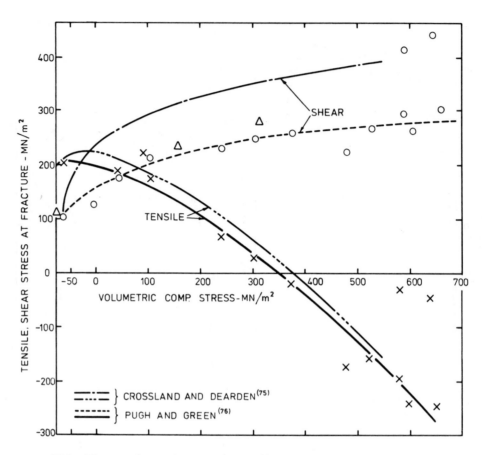

FIG. 22. Maximum shear and tensile stresses at fracture
for cast iron tested under hydrostatic pressure [76].

binations of stresses concluded that 'there is no universally valid
criterion of fracture in terms of stresses only'. The two criteria
which gave the closest agreement were that fracture occurred when (a)
the maximum tensile stress at the centre of the neck reached a critical value and (b) the mean total hydrostatic component of stress
reached a critical value.

Many investigators have studied the mechanical properties of materials under pressure subsequent to Bridgman, but it is still true to
say that no universally valid criterion of fracture has been found.
Indeed this may not be surprising if the criterion of fracture is
meant to apply to all materials. Even the well-established criterion
of yield is applicable only to wrought metals and not to cast materials or polymers.

An extensive analysis of the fracture of cast iron under hydrostatic pressure has been carried out by Crossland and Dearden [75]
using torsion tests and by Pugh and Green [76] using tension tests.
The cast iron used by both sets of investigators was identical. The

results of an analysis of this experimental work is shown in Figure 22. The results are plotted for both the maximum shear stress and the maximum tensile stress acting on the test specimens at fracture as a function of the volumetric stress. The volumetric stress is taken as the average of the principal stresses at fracture. It may be seen from this figure that a maximum shear-stress criterion is not valid for these results, but the maximum tensile-stress criterion closely represents the results for fracture stress both for torsion tests and for tension tests under a range of hydrostatic pressures. It was subsequently pointed out by Pugh [77] that, for the stress systems involved in tension tests under pressure, the maximum deviatoric-stress criterion was identical with the maximum tensile-stress criterion and therefore was equally satisfactory as a criterion of fracture for these results. However these criteria gave different predictions for compression tests under pressure. From compression and tension tests under hydrostatic pressure for a different cast iron it was concluded that the maximum tensile-stress criterion did not represent the data, which were in fact in good agreement with the maximum deviatoric-stress criterion as may be seen from Figure 23. As this criterion is also a good representation of the results of torsion tests and tension tests under pressure, it would appear to be a satisfactory criterion for cast iron.

Investigations on ductile metals [14, 72, 78] have suggested that the average tensile stress at fracture (σ_f) increased linearly with increasing pressure (ρ) but the net axial stress ($\sigma_f - \rho$) remained contant (as shown in Figure 24). These results of course represent a maximum tensile-stress criterion of fracture, i.e., when the net tensile stress achieves a critical value fracture takes place.

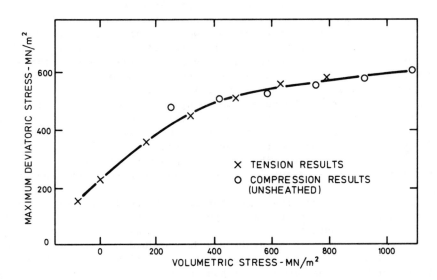

FIG. 23. Deviatoric stresses at fracture for cast iron tested under hydrostatic pressure [77].

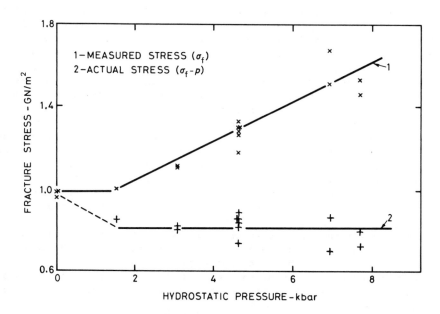

FIG. 24. Effect of pressure on the fracture stress of a
ductile steel En2A [14].

In tensile tests on tool steels [79] which were still brittle at
very high pressure the net axial stress at fracture decreased linearly
with increasing pressure (Figure 25). Similar results have been obtained for AZ91 and intermetallic compounds TiNi and NiAl.

It is possible that there may be two criteria of fracture which
apply to the results of tensile tests under increasing pressure. The
first criterion is for materials failing in a brittle manner, that is
with negligible deformation under pressure. The second criterion, for
ductile failure, applies when the material in question behaves in a
ductile manner under the superimposed hydrostatic pressure. If this
suggestion is correct, one would in general expect for a material
which is brittle at atmospheric pressure and whose strain to fracture
increases under pressure that the average net axial stress at fracture
would initially decrease with increasing pressure and after a certain
critical hydrostatic pressure would be unchanged with increasing pressure.

It would seem from Figure 26 that for hot-pressed powder materials neither of the above criteria apply, the net axial tensile stress
being no longer constant but increasing with hydrostatic pressure.
It was suggested that this result was due to voids in the hot-pressed
material. In a test to verify this, material was taken from the original hot-pressed and swaged molybdenum and extruded. The resulting
product was then tested under a range of hydrostatic pressures and the
results for the net axial stress at fracture is also shown in Figure
26. It may be seen that the results now conform to the results obtained for the wrought materials, namely that the net axial, tensile

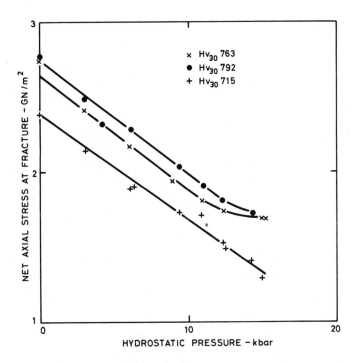

FIG. 25. Effect of pressure on the net axial fracture stress of M2 tool steel.

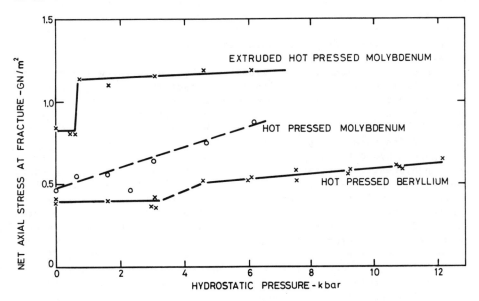

FIG. 26. Effect of pressure on the fracture stress of hot-pressed powder metals.

stress at fracture is almost independent of pressure. These results
and similar results for hot-pressed beryllium suggest that working the
original material under a compressive-stress system eliminates the
voids, thereby increasing the densities, with the result that the criterion of fracture approximates to that of wrought metals. These results are in agreement with the experimental work of Rogers and Coffin
[64] on sheet-drawing under pressure which demonstrated conclusively
that voids were eliminated by working under pressure and that the density was increased accordingly.

This suppression of defects has been confirmed by Bulychev and
his colleagues [80]. In their tests defects in the form of voids were
introduced into wrought copper by carrying out creep tests at elevated
temperature, the voids being 0.005 mm to 0.01 mm diameter. Tensile
specimens were then prepared from this defective material and the control specimens from the original defect-free material. Tests carried
out on these specimens at atmospheric temperature showed that the defective material had a much smaller strain-to-fracture and consequently fracture stress than the defect-free material. However for tests
carried out at increasing pressures the difference between the two became smaller and eventually disappeared. The results were confirmed
by metallurgical observations showing that the defects gradually disappeared in the tension tests at an increasing hydrostatic pressure.

G.   Transitions in Ductility

As indicated earlier it has been observed by several workers that
for many materials there seems to be a transition in the ductility
when plotted as a function of pressure, that is the rate of change of
ductility with pressure changes fairly abruptly from a very low value
to quite a high value giving a knee in the curve (Figure 9). It was
suggested above that there is a direct relation between the transition
in the ductility-pressure curve and the stress-strain relation and the
criterion of fracture for that material.

This suggestion can be illustrated by a material which conforms
to a maximum tensile stress criterion of fracture as in the case of
wrought, ductile metals. At atmospheric pressure, it may well have a
stress-strain curve like that illustrated diagrammatically in Figure
27 with fracture at the point marked one atmosphere. With this criterion of fracture, the point at which it would fracture under a superimposed hydrostatic pressure P would be at the point marked P and
under a hydrostatic pressure 2P at the point marked 2P and so on up to
the point marked 4P where the intervals on the stress axis are all e-
qual. When these fracture strains are projected on to a strain-to-
fracture axis on the lower diagram and plotted against the pressures
P, 2P and so on, the curve shown in the lower diagram is obtained.
This is fairly characteristic of a ductile material and its curvature
would reflect the curvature in the stress-strain curve. Indeed if the
stress-strain curve is fairly linear then, with this criterion of
fracture, the relation between ductility and pressure would also be
fairly linear. In general however, the true stress-strain curve for
materials deviates from linearity in the direction of giving a decreasing slope with increasing strain and, consequently, it may be

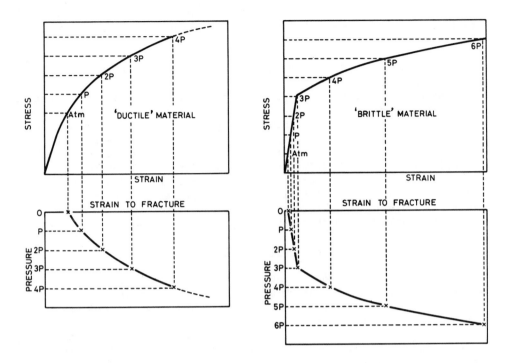

FIG. 27. Relationship between the stress–strain and
ductility-pressure curves of nominally ductile and brittle
materials.

seen that the curve of the strain-to-fracture against pressure would
be one which deviates from linearity in the direction of an increasing
strain-to-fracture. It is therefore difficult to understand reported
results which give a deviation from linearity in the ductility-pressure curve, which is in the opposite direction. As already mentioned,
this may be due to the inaccuracies involved in measuring small areas
of fracture of ductile materials tested under pressure.

For a so-called brittle material, a stress–strain curve similar
to that shown in Figure 27 may be assumed as the property of that material, but at atmospheric pressure it would fracture in the elastic
range at the point marked one atmosphere, as indeed has been observed
in tests on tungsten (see Figure 19). From a similar construction
based on an identical argument, a strain-to-fracture versus pressure
relationship shown in the lower figure is obtained and, in this case,
there is clearly a very marked transition in slope of the curve at a
particular pressure. Such curves are typical of so-called brittle materials and with some materials, such as zinc and Mazak, very sharp
changes in ductility are obtained at a particular critical or transition pressure indicating a flat-topped stress–strain curve.

Although the above argument is based on a particular criterion of
fracture, it is suggested that with a more general criterion of fracture a similar argument will apply, although the relationship between
the ductility-pressure relation and that of the stress–strain curve

would be rather more complicated. Nevertheless there would be a direct connection between the two curves through the criterion of fracture.

V.   EFFECT OF HYDROSTATIC PRESSURE ON POLYMERS

With the development of polymer science and the polymer industry there has been a rapid increase in the use of a wide range of different polymers for structural purposes. In many instances they are subjected to triaxial stress systems but the design of the structures are based on the mechanical properties obtained from uniaxial tests. There is also technological interest in the development of new forming processes in which the polymers will be deformed by high compressive stresses. A study of their mechanical properties under controlled triaxial stresses is therefore of paramount importance in assessing their performance in service or in forming processes.

A.   Stress-Strain Relationship

It has been found that subjecting polymers to a short duration pressurization has no apparent effect on their mechanical properties when they are subsequently tested at atmospheric pressure. Any changes in the stress-strain curve of a material under pressure therefore are due not intrinsically to the hydrostatic pressure but to the deformation being carried out in the presence of a high hydrostatic compressive stress system.

Tension, torsion and compression tests under hydrostatic pressure have been carried out on a wide range of polymers [48, 81-84]. The polymers examined include polystyrene, low and high density polyethylene, polymethyl methacrylate, polypropylene, polyoxymethylene, polytetrafluoroethylene, polycarbonate and polyurethane. In all of these investigations the effect of increasing hydrostatic pressure was to raise the stress-strain curve. It has been found, however, that some polymers increased in ductility whilst others exhibited more brittle behaviour with increasing pressure. Figures 28 and 29 are typical examples of the stress-strain curves of a polymer nominally (a) brittle at atmospheric pressure, and (b) ductile at atmospheric pressure.

B.   Ductility and Instability

For metals the only meaningful expression of ductility is based on the change in cross-sectional area, that is, $\ln A_o/A$, where A_o is the original, and A the actual area of the specimen. With polymers, ductility is far better represented in terms of the change of length. Polymer tensile specimens either extend with the diameter reducing uniformly along the gauge length or, after necking down to a particular diameter, the neck then runs along the length of the specimen.

In the section on Pressure Media Effects it was mentioned how the strain-to-fracture of polymers tested under pressure was substantially increased if the specimen gauge length was sheathed to prevent fluid penetration into surface flaws and cracks [37, 49]. It was also pointed out that surface defects, arising from specimen preparation,

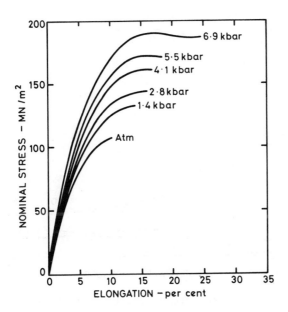

FIG. 28.   Nominal tensile stress-strain curves for polyimide at various pressures [94].

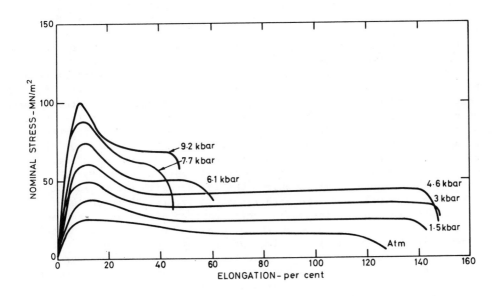

FIG. 29.   Nominal tensile stress-strain curves for HD polyethylene at various pressures.

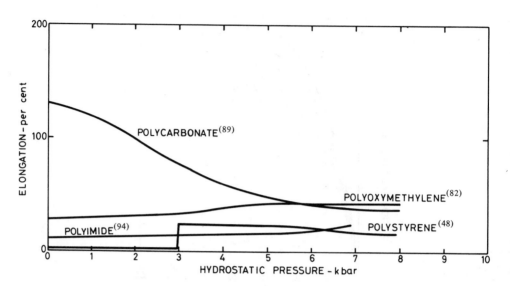

FIG. 30. Effect of pressure on the elongation of
several polymers.

which had little effect on ductility at atmospheric pressure became
critical stress raisers under pressure. These effects are without
doubt responsible for some of the conflicting ductility results obtained from different investigations. These factors must always be
taken into consideration when examining ductility and fracture stress
results of polymers tested under pressure.

The effects of hydrostatic pressure on the elongation-to-fracture
of a number of polymers are shown in Figure 30. It was found that
most polymers which are normally brittle at atmospheric pressure, such
as polystyrene, exhibit little initial ductility increase as the test
pressure increases. At some critical pressure, however, they behave
rather like zinc by showing a sharp increase in elongation and signs
of localized necking. At higher pressures they show no further increase and sometimes even a decrease in elongation-to-fracture. In
the case of those polymers which exhibit extensive elongation at atmospheric pressure, e.g. polyethylenes, polytetrafluoroethylene, it
was found that under increasing test pressure elongation was drastically reduced. With some polymers the drastic reduction in elongation-to-fracture with increasing pressure is due to a difference in
the mode of behaviour in the tensile test at atmospheric pressure and
under pressure. At atmospheric pressure it was found that when the
specimen started to neck in tension the neck proceeded to run along
the length of the specimen, that is the polymer was 'cold drawn'.
When tested under hydrostatic pressure the specimen behaved rather
like a metal, that is after necking was initiated nearly all the subsequent deformation was concentrated in the neck. The actual reduction in area was unaffected by pressure.

There   was   a   marked change in appearance of some of the polymer
specimens. At atmospheric pressure the necked, or cold drawn portion
of   the   specimen,   became   opaque   due   to the creation and growth of
voids. Under a superimposed pressure the deformed section   was   semi-
transparent   because   the   pressure   inhibited the formation of voids.

The strain at which yield occurs and necking commences is dependent   not   only on the type of polymeric material but also on the pressure under which it is tested. For polyethylenes of various densities,   nylon   and polypropylene, the strain at which necking commenced
was reduced by increasing pressure.   For   most   polymers,   including
polycarbonate, polymethylmethacrylate, polyvinyl chloride, polyurethane and polyethylene terephthalate, the effect of increasing pressure was to increase the instability strain.

C.   Yield Stress and Fracture Stress

All   investigations   on   polymers   have   shown   that   their yield
stresses increased with increasing pressure and that in most cases a
linear relationship existed between yield stress and superimposed hydrostatic pressure, Figure 31. As mentioned above the upper yield
stress for polymers is the stress at which necking is initiated.   For

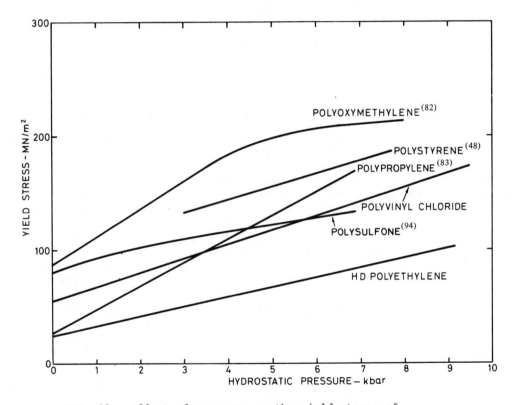

FIG. 31. Effect of pressure on the yield stress of
several polymers.

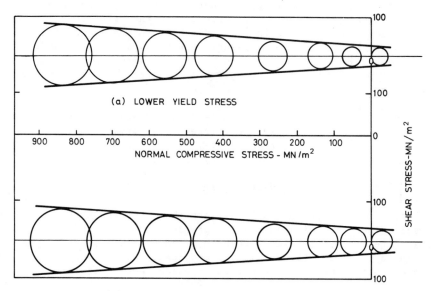

FIG. 32.   Effect of hydrostatic pressure on the Mohr
circles for the upper and lower yield stresses of polyvinyl
chloride [37].

metals, of course, the onset of necking occurs when the tensile
strength is attained.

It is therefore evident that polymers do not obey the pressure-
independent Tresca or Von Mises criteria for yielding which is applicable to ductile, wrought metals. Any yield criterion for polymers
must take into account the superimposed pressure, or mean normal
stress on the yield plane. This point is illustrated more clearly in
Figure 32, in which the upper and lower yield stresses of polyvinyl-
chloride for various superimposed pressures have been plotted in the
form of Mohr circles. The envelope of these Mohr circles is linear,
thus giving the yield criterion as a linear function between the shear
yield stress and the normal stress on the yield plane. This criterion
is of the Coulomb-Navier type and is similar to that used to represent
yielding in soils.

It was shown earlier that for wrought, ductile metals the net
longitudinal tensile stress at fracture, that is the algebraic sum of
the applied tensile stress and the hydrostatic pressure, was constant.
All the polymer investigations so far, however, have indicated that
this particular fracture stress decreased with increasing pressure.
As was mentioned in the previous section the ductility of a polymer
tested under pressure can be significantly affected by contact with
the pressure medium and the specimen preparation. There will, of
course, be a corresponding effect on the evaluated fracture stress.

D.   Young's Modulus

With the existing experimental equipments it has not been possible to obtain absolute values of the Young's modulus for the materials
tested since the strain is measured by the overall extension. Provided, however, the specimen length is constant for a series of tests,
it is possible to obtain relative values of Young's modulus and to obtain information on the general effect of pressure on modulus.

All investigations have shown that Young's modulus is increased
by pressure. Paterson [42] found that for vulcanized, natural rubber
the modulus increased by about 30 per cent up to a pressure of 4 kbar
after which there was a rapid increase by a factor of 1000.   Several
synthetic rubbers exhibited the same order of modulus increase but
with the transition regions occurring at different pressure levels for
the different rubbers. These large changes in modulus were attributed
to the raising of the glass transition temperature by the application
of pressure. For polymers the change in modulus has been found to be
much smaller, see Figure 33. The relative change in elastic modulus
is very small for the amorphous polymers such as polycarbonate and
polystyrene with the modulus increasing by less than 5 per cent/kbar
increase in pressure. Generally the more crystalline polymers exhibit
a greater pressure effect with the elastic modulus of low density
polyethylene increasing by about 70 per cent/kbar. Over the range of
pressure investigated there appeared to be a linear relation between
Young's modulus and pressure for most of the materials. Polytetrafluoroethylene showed a deviation from this trend at about 4 kbar and
at 5 kbar its Young's modulus actually started to decrease. This
change is attributed [89] to a solid-solid phase transition in this
pressure region.   It has been found that there is no correlation between the rates of change of Young's modulus and yield stress of

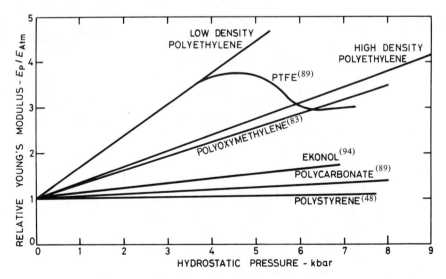

FIG. 33. Relative Young's modulus as a function of pressure
for several polymers.

polymers tested under pressure, the ratio of the two varying from material to material. This means that the factors controlling the modulus of polymers are not entirely the same as those governing plastic
flow.

VI.   CONCLUSIONS

Studies of the effect of a superimposed hydrostatic pressure on
the mechanical properties of materials have shown the following:
1.   Hydrostatic pressure increases the strain-to-fracture of
metals.   The relationship between fracture strain and pressure varies
with different metals. It is suggested that this relationship is governed by the shape of the true stress-true strain curve and the criterion of fracture of the metal. An increase in hydrostatic pressure
also reduces the brittle-ductile transition temperature of metals.
The reason for the increase in ductility is that the imposed compressive stress system inhibits the propagation of cracks and other defects.
2.   Apart from the increased strain-to-fracture the general
shape of the stress-strain curves of wrought metals is unaffected by
hydrostatic pressure.   Some cast and sintered metals exhibit an initial raising of the stress-strain curve with increasing pressure.
3.   The fracture of a number of wrought metals appears to occur
when the net axial tensile stress reaches a critical value. The criterion of fracture for cast iron seems to be the maximum deviation
stress. There is a suggestion, however, that two criteria of failure
operate, one for brittle failure and the other for ductile failure.
The fact that the criterion of fracture for hot-pressed powder materials has not conformed to that for wrought materials is thought to be
due to the presence of voids.
4.   Pressure soaking can alter the subsequent atmospheric properties of certain metals, particularly polycrystalline anisotropic
metals and metals with hard, second-phase particles. Hydrostatic
pressure generates dislocation sources at the interfaces between anisotropic grains and second phase particles.
5.   Pressure has a marked effect on the stress-strain curve of
polymers with both yield stress and Young's modulus being increased
by increasing pressure. The ductility of normally brittle polymers is
initially increased by pressure whilst that of ductile polymers show
an apparent decrease under pressure.

REFERENCES

1.   W. Voigt, "Observations on the Strain of Homogeneous
     Deformation", Göttinger Nachr., p. 521 (1893).
2.   Th. Von Karman, "Strength Investigations Under Hydro-
     static Pressure", (in German), Z. Ver. dtsch. Ing,
     pp. 55, 1749 (1911).
3.   R. Böker, "The Mechanism of Plastic Deformation in
     Crystalline Bodies", (in German), Dissn tech. Hochsch. Z.,
     Aachen Rheinisch-Westfälische Hochschule (1914).

4. M. Ros and A. Eichinger, "Experiments for Clarifying the
    Question of Fracture-risk Metals", (in German),
    Versuchanstalt Ind. Bauw. Gewerbe (34), Zurich (1929).
    English Translation: NEL Translation No. 449, East
    Kilbride, Glasgow, National Engineering Laboratory.

5. G. Cook, "The Effect of Fluid Pressure on the Permanent
    Deformation of Metals by Shear", Selected Engineering
    Papers - No. 170, Institution of Civil Engineers, London
    (1934).

6. P. W. Bridgman, "Studies in Large Plastic Flow and
    Fracture", McGraw-Hill, New York (1952).

7. P. W. Bridgman, "The Effect of Pressure on the Tensile
    Properties of Several Metals and other Materials",
    J. Appl. Phys. 24(5), pp. 560 - 570 (1953).

8. D. Carpentier, M. Contre, R. Daumas, J. P. Rech and
    D. Francois, "The Deformation of Beryllium Under Hydro-
    static Pressure of up to 15 kbars", (in French),
    Conf. Int. sur la metallurgie du béryllium, Presses
    Universitaires de France, Paris (1966).

9. P. W. Bridgman, "The Physics of High Pressure", G. Bell
    and Sons, London (1949).

10. M. Nishihara, K. Tanaka and T. Moramatsu, "Effect of
    Hydrostatic Pressure on Mechanical Behaviour of Metals",
    SOC. MAT. SCI. Proc. 7th Japan Congress on Testing Materials,
    Soc. Mat. Sci., Kyoto, p. 154 (1964).

11. S. I. Ratner, "Change of Mechanical Properties of Metals
    Under Hydrostatic Pressure", (in Russian), Zh. Tekhn, Fiz.,
    19(3), p. 408 (1949).

12. T. Pelczynski and I. Pawlak, "Tensile Testing of Some
    Metals Under High Hydrostatic Pressure", Pr. Zakl. Plast.
    Polytech. Warsz., (1959).

13. B. I. Beresnev, D.K. Bulychev and E. D. Martynov, E. D.
    Zawodsk Lab., 8, 1017 (1964).

14. H. Ll. D. Pugh, "The Mechanical Properties and Deformation
    Characteristics of Metals and Alloys Under Pressure",
    American Society for Testing Materials, Special Tech. Pub.
    No. 374, Phila., Pa. (1964).

15. W. A. Gladkovski and M.I. Oleinik, "Apparatus for Investi-
    gating Mechanical Qualities of Metals under High Hydro-
    static Pressure", (in Russian), Fiz. Metal. i Metalloved,
    4, 531 (1957). (English Translation: Physics Metals
    Metallogr., 4, 118-121, New York (1957).

16. S. Yoshida and A. Oguchi, "An Apparatus for Tensile
    Testing under High Hydrostatic Pressure", Trans. NRIM,
    11, 347, (1969).

17. H. Ll. D. Pugh and D. A. Gunn, "A Strain Gauge Load Cell
    for use under High Hydrostatic Pressure", NEL Report No. 143,
    National Engineering Laboratory, East Kilbride, Glasgow (1964).

18. C. W. Macgregor, "Two-load Method of Determining Average
    True Stress-strain Curve in Tension", J. Appl. Mech., 6(4),
    A156-A158 (1939).

19. E. F. Chandler and W. M. Mair, "Behaviour of Metals under
     Triaxial Compressive Stress", Proc. Instn. Mech. Engrs.,
     182(Pt3C), 122-131 (1967-68).

20. D. T. Griggs, "Deformation of Rocks under High Confining
     Pressures", J. Geol., 44, 541 (1936).

21. M. S. Paterson, "Triaxial Testing of Materials at Pressures
     up to 10 000 kg/sq cm (150 000 lbf/sq in)", J. Instn. Engrs.
     Aust., 36, 23-29 (1964).

22. D. T. Griggs and W. B. Miller, "Deformation of Yule Marble -
     Part I, Compression and Extension Experiments on Dry Yule
     Marble at 10 000 atm confining pressure, room temperature",
     Bull. Geol. Soc. Amer., 62, 853-906 (1951).

23. B. Crossland and C. Ludlow, "Design and Development of a
     Machine for Compression testing of rock samples under a
     Constant Pressure of up to 0.7 GPa (7 kbar)", High Temper-
     atures-High Pressures, 5, 509-514 (1973).

24. E. F. Chandler and M. P. Littleson, "An internal stroke
     gauge for Operation under Hydrostatic Pressure", NEL Re-
     port No. 356, National Engineering Laboratory, East Kilbride,
     Glasgow (1968).

25. D. P. Johnson and D. H. Newhall, "The Piston Gauge as a
     Precise Measuring Instrument", Trans. Amer. Soc. Mech.
     Engrs., 73, 301-310 (1953).

26. B. Crossland, "The Effect of Fluid Pressure on the Shear
     Properties of Metals", Proc. Instn. Mech. Engrs., 168(40),
     935-946 (1954).

27. B. Crossland and R. K. Mitra, "Torsion Machine for use under
     Superimposed Pressures of 1 GPa, and Preliminary Results",
     High Temperatures-High Pressures, 6, 165-172 (1974).

28. M. Ohmori, Y. Yoshinaga, T. Kawahata and Y. Sanemasu,
     "Plastic Deformation of Metals under high Hydrostatic
     Pressure", Proc. 13th Jap. Conf. Matl. Res., pp. 139 -
     142 (1970).

29. S. Erbel, "Behaviour of Materials twisted under Hydro-
     static Pressure", presented at conference organized by
     Department of Metallurgy, Polish Academy of Sciences,
     (April 1966).

30. J. Handin, D. V. Higgs and J. K. O'Brien, "Torsion of Yule
     Marble under Confining Pressure", D. Griggs and J. Handin,
     Eds., Rock Formation, Memoir 79, pp. 245 - 74, Geological
     Society of America, Rochester, New York (1960).

31. H. Ll. D. Pugh and D. Gunn, "An Apparatus for Torsion Tests
     under High Hydrostatic Pressures", NEL Report No. 159,
     National Engineering Laboratory, East Kilbride, Glasgow (1964).

32. A. E. Abey, "Effect of Hydrostatic Pressure on the Stress-
     Strain Curves of OFHC Copper", J. App. Phys., 42(10), 4085 -
     4087 (1971).

33. S. V. Radcliffe, "The Effect of High Pressure and Temperature
     on the Mechanical Properties of Metals and Alloys", ASTM STP
     No. 374, pp. 141 - 162, American Society for Testing and
     Materials, Philadelphia, Pa. (1964).

34. T. E. Davidson, "Effects of Pressure on Structure and Deformation of Materials", Report 199, p. 25, Defense Metals Information Centre, Battelle Memorial Institute Columbus, Ohio (1964).

35. H. Vu and P. Johannin, "Permanent Deformations of Poly-crystalline Solids after exposure to high Hydrostatic Pressure", C. R. Acad. Sci., 241(6), 565-6, Paris (1955).

36. P. Johannin and H. Vu, "Permanent Deformation in Solids after the Application of High Hydrostatic Pressure", C. R. Acad. Sci., 242(21), 2579-81, Paris (1956).

37. H. Ll. D. Pugh and E. F. Chandler, "Mechanical Properties of Materials under Pressure", Professor Galerkin's Anniversary Volume 1974 and NEL Report No. 577, National Engineering Laboratory, East Kilbride, Glasgow (1974).

38. F. P. Bullen, F. Henderson, M. M. Hutchison and H. L. Wain, "The Effect of Hydrostatic Pressure on Yielding in Iron", Phil. Mag., 9(101), 285 (1964).

39. F. P. Bullen, F. Henderson, H. L. Wain and M. S. Paterson, "The Effect of Hydrostatic Pressure on Brittleness in Chromium", Phil. Mag., 9(101), 803 - 815 (1964).

40. A. H. Cottrell, "Theory of Brittle Fracture in Steel and Similar Metals", Trans. Amer. Inst. Min. (metall.) Engrs., 212, 192 - 204 (1958).

41. N. G. Petch, "The Ductile Brittle Transition in the Fracture of αiron", Phil. Mag., 3(34), 1089 -1097 (1958).

42. M. S. Paterson, "Effect of Pressure on Young's Modulus and the Glass Transition in Rubbers", J. Appl. Phys., 35(1), 176-9 (1964).

43. E. F. Chandler, "Effect of Sheathing on the Mechanical Properties of a Material under Hydrostatic Pressure", NEL Report No. 255, National Engineering Laboratory, East Kilbride, Glasgow (1966).

44. E. Aladag, H. Ll. D. Pugh and S. V. Radcliffe, "Mechanical Behaviour of Beryllium at Pressures", Acta Metall., 17, 1467 - 1481 (1969).

45. H. Ll. D. Pugh and E. F. Chandler, "The Effect of Surface Conditions on the Tensile Behaviour of Beryllium under Hydrostatic Pressure", NEL Report No. 537, National Engi-neering Laboratory, East Kilbride, Glasgow (1973).

46. E. Chandler, H. M. Lindsay, H. Ll. D. Pugh and J. S. White, "The Effect of Surface Finish on the Pressure-induced Ductility of Beryllium", J. Mat. Sci., 8, 1788 - 1794 (1973).

47. B. Crossland, "Effect of Pressure on the Fatigue of Metals", Mechanical Behaviour of Materials Under Pressure, Ch. 7, H. Ll. D. Pugh, Ed., Elsevier, London (1970).

48. H. Ll. D. Pugh, E. F. Chandler, L. Holliday, and J. Mann, "The Effect of Hydrostatic Pressure on the Tensile Properties of Plastics", Polymer Engng. Sci., 11(6), 463 - 473 (1971).

49. J. S. Harris, I. M. Ward and J. S. C. Parry, "Shear Strength of Polymers under Hydrostatic Pressure - Surface Coatings Prevent Premature Fracture", J. Mat. Sci., 6(2), 110-114 (1971).

50. K. Tanaka, N. Nagao, and M. Nakashima, "Brittle-ductile Transition of Polycrystalline Zinc under High Pressure", Proc. 13th Jap. Congr. Mat. Res., 13, 135-8 (1970).

51. Yu N. Ryabinin, L. D. Livshitz and L. F. Vereschagin, "Regarding the Plasticity of Certain Alloys at High Pressures", (in Russian), Zhur. Tverd. Tela, (1) 476 (1959).

52. M. Yajima, M. Ishii and M. Kobayashi, "The Effects of Hydrostatic Pressure on the Ductility of Metals and Alloys", Int. J. Fracture Mech., 6(2), 139-150 (1970).

53. M. Brandes, Mechanik, 10, 535 (1966).

54. C. J. Nolan and T. E. Davidson, "The Effects of Cold Reduction by Hydrostatic Fluid Extrusion on the Mechanical Properties of 18 per cent nickel maraging steels", Trans. Am. Soc. Metals, 62, 271 - 277 (1969).

55. I. E. French and P. F. Weinrich, "The Influence of Hydrostatic Pressure on the Tensile Deformation of a Spheroidised 0.5 per cent C steel", Scr. Metall., 8, 87 - 90 (1974).

56. Yu N. Ryabinin, "Effect of all-round Hydrostatic Pressure on the Deformation of metals subjected to Stretching", (in Russian), Inzh. Fiz. Zhur, 1, 90 (1958).

57. B. I. Beresnev, L. F. Vereschagin, Yu N. Ryabinin and L. D. Livshitz, "Some Problems of large plastic Deformation in metals under High Pressure", (in Russian), Dokl. Akad. Nauk SSSR, 1960. English Translation: ASTIA Doc. AD-259, 251, Office of Technical Services, U.S. Dept. of Commerce, Washington, D.C. (1961).

58. L. W. Hu, "Determination of the Plastic Stress-strain Relations in tension of Nittany No. 2 brass under hydrostatic pressure", Proc. 3rd US Nat. Congr. Appl. Mech., pp. 557 - 562, American Society of Mechanical Engineers, New York (1958).

59. B. I. Beresnev and K. P. Rodionov, "The Change in the Structure and Mechanical Properties of brittle materials after Hydrostatic Extrusion", Engineering Solids Under Pressure, pp. 153 - 160, H. Ll. D. Pugh, Ed., Institution of Mechanical Engineers, London (1971).

60. N. N. Davidenkov and N. F. Wittman, "Mechanical Analysis of impact Brittleness", (in Russian), Tech. Phys. U.S.S.R., 4, 308 (1937).

61. J. R. Galli and P. Gibbs, "The Effect of Hydrostatic Pressure on the ductile-brittle transition in Molybdenum", Acta Metall., 12, 775 - 783 (1964).

62. H. Ll. D. Pugh and A. H. Low, "The Hydrostatic Extrusion of Difficult Metals", J. Inst. Metals, 93, 201 - 217 (1964-65).

63. F. J. Fuchs, Jr., "Hydrostatic Pressure - Its Role in Metal Forming", Mech. Engng., 88(4) 34 -40 (1966).

64. H. C. Rogers and L. F. Coffin, Jr., "Structural Damage in Metal Working", Report No. 67-C-047, G.E. Research and Development Centre, Schnectady, N.Y. (1967).

65. T. Pelczynski, "The Influence of Hydrostatic Pressure on the Plastic Deformation Properties of Metals", (in Polish), Arch. Hutnictwa, 7(1), 3 (1962).

66. R. Hill, "Mathematical Theory of Plasticity", Oxford University Press, London (1950).

67. E. Siebel, "Report of Committee on Raw Materials No. 71 of V.D.E., Dusseldorf", Stahl Eisen, 1675 (1924).

68. N. N. Davidenkov and N. I. Spiridonova, "Mechanical Methods of Testing - Analysis of the State of Stress in the Neck of a Tensile Specimen", Proc. Am. Soc. Test. Mater., 46, 1147 - 1175 (1946).

69. P. Haasen and A. W. Lawson, "Effect of Hydrostatic Pressure (5000 atm) on Tensile Deformation of Single Crystals", (in German), Z. Metallk., 49, 280 -291 (1958).

70. L. W. Hu, E. E. Haddington, R. G. Hoffman, J. Schreiber, K. D. Pae and H. E. Shull, "Effect of Triaxial Stresses on Mechanical Properties of Metals under High Pressure", AD 266887, Armed Services Technical Information Agency, Arlington, Va. (1961).

71. D. Carpentier and D. Francois, "Mechanical Properties of Zinc and Cobalt under Pressure", Engineering Solids Under Pressure, pp. 61 - 68, H. Ll. D. Pugh, Ed., Institution of Mechanical Engineers, London (1971).

72. M. Ohmori, Y. Yoshinaga, K. Sukegawa and Y. Maruyama, "Deformation Characteristics of mild steel under Hydrostatic Pressure", Proc. 16th Jap. Congr. Mater. Res., pp. 120 - 125 (1973).

73. M. Nobuki and A. Oguchi, "Effect of Hydrostatic Pressure on the Ductility of a Quenched and Tempered Carbon Steel", J. Jap. Inst. Met., 38, 5, 401-7 (1974).

74. K. E. Puttick, "The shear Component of Ductile Fracture", Phil. Mag., 5(55), 759 - 762 (1960).

75. B. Crossland and W. H. Dearden, "Plastic Flow and Fracture of a 'brittle' material (grey cast iron) with particular reference to the Effect of Fluid Pressure", Proc. Instn. Mech. Engrs., 172(26), 805 - 820 (1958).

76. H. Ll. D. Pugh and D. Green, "The Effect of Hydrostatic Pressure on the plastic flow and fracture of Metals", Proc. Instn. Mech. Engrs., 179(1), 415-437 (1964-65).

77. H. Ll. D. Pugh, "Fracture Criterion for Cast Iron", Nature, 218(5145) 985 (1968).

78. E. F. Chandler, "Tensile Properties of a number of Materials under Hydrostatic Pressure", NEL Report No. 306, National Engineering Laboratory, East Kilbride, Glasgow (1967).

79. H. Ll. D. Pugh and E. F. Chandler, "The Deformation and Fracture of a number of metals used in High Pressure Equipments", 2nd Int. Conf. on High Pressure Engineering, University of Sussex, Brighton (July 8-10, 1975).

80. D. K. Bulychev, B. I. Beresnev, M. G. Gaidukov, E. D. Martynov, K. P. Rodionov and Y. N. Ryabinin, "Structural Defects and Plastic Strain of Copper at High Pressures", (in Russian), Soviet Phys-Dokl. 9(5), 385-387 (1964).

81. S. B. Ainbinder, M. G. Laka and I. Yu Maiors, "Effect of Hydrostatic Pressure on Mechanical Properties of Plastics", Polymer Mech., 50 - 55 (1965).

82. D. Sardar, S. V. Radcliffe and E. Baer, "Effects of High
     Hydrostatic Pressure on the Mechanical Behaviour of a
     Crystalline Polymer - Polyoxymethylene", _Polymer Engng._
     Sci., 8(4), 290 - 301 (1968).

83. K. D. Pae and D. R. Mears, "The Effects of High Pressure
     on Mechanical Behaviour and Properties of Polytetra-
     fluoroethylene and Polyethylene", _Polymer Letters_, 6,
     269 - 273 (1968).

84. G. Biglione, E. Baer and S. V. Radcliffe, "Effect of High
     Hydrostatic Pressure on the Mechanical Behaviour of Homo-
     geneous and Rubber Reinforced Amorphous Polymers", _2nd Int._
     Conf. on Fracture, Brighton (April 1969). (Published by
     Chapman and Hall, London).

85. D. R. Mears, K. D. Pae and J. D. Sauer, "Effects of Hydro-
     static Pressure on the Mechanical Behaviour of Poly-
     ethylene and Polypropylene", _J. Appl. Phys._, 40(11),
     4229 - 4237 (1969).

86. W. J. Vroom and R. F. Westover, "Properties of Polymers at
     High Pressure", _SPE J._, 25, 58 - 61 (1969).

87. D. R. Mears and K. D. Pae, "Deformation and Fracture
     Characteristics of Polycarbonate under High Pressures",
     Polymer Letters, 7, 349 - 352 (1969).

88. S. Rabinowitz, I. M. Ward and J. S. C. Parry, "The Effect
     of Hydrostatic Pressure on the Shear Yield Behaviour of
     Polymers", _J. Mater. Sci._, 5(1), 29 - 39 (1970).

89. A. W. Christiansen, E. Baer and S. V. Radcliffe, "The
     Mechanical Behaviour of Polymers under High Pressure",
     Phil. Mag., 24(188) 451 - 68 (1971).

90. B. I. Beresnev, Yu S. Genshaft, Yu N. Ryabinin and E. D.
     Martynov, "The Mechanical Properties of Polytetra-
     fluoroethylene and Polymethylmethacrylate", _Sov. Phys._
     Dokl., 16(3), 246 - 248 (1971).

91. K. D. Pae and J. A.Sauer, "Influence of Pressure on the
     Elastic Modulus, the Yield Strength and the Deformation
     of Polymers", _Engineering Solids Under Pressure_,
     H. Ll. D. Pugh, Ed., Institution of Mechanical Engineers,
     London (1971).

92. L. A. Davis and C. A. Pampillo, "Deformation of Poly-
     ethylene at High Pressure", _J. App. Phys._, 42(12),
     4659 - 4666 (1971).

93. A. A. Silano, S. K. Bhateja and K. D. Pae, "Effects of
     Hydrostatic Pressure on the Mechanical Behaviour of
     Polymers: Polyurethane, Polyoxymethylene, and Branched
     Polyethylene", _Int. J. Polym. Mater._ (GB), 3(2),
     117 - 131, (1974).

94. S. K. Bhateja, "The Influence of Hydrostatic Pressure on
     the Mechanical Behaviour and Cold-forming of High-
     temperature Polymers", _Ph.D. Dissertation_, Rutgers
     University (1973).

A

B